An Introduction to
Graphene Plasmonics

An Introduction to
Graphene Plasmonics

P.A.D Gonçalves
University of Minho, Portugal

N.M.R Peres
Univeristy of Minho, Portugal

World Scientific

NEW JERSEY · LONDON · SINGAPORE · BEIJING · SHANGHAI · HONG KONG · TAIPEI · CHENNAI · TOKYO

Published by

World Scientific Publishing Co. Pte. Ltd.

5 Toh Tuck Link, Singapore 596224

USA office: 27 Warren Street, Suite 401-402, Hackensack, NJ 07601

UK office: 57 Shelton Street, Covent Garden, London WC2H 9HE

Library of Congress Cataloging-in-Publication Data

Names: Gonçalves, P. A. D., author. | Peres, N. M. R., 1967–   author.
Title: An introduction to graphene plasmonics / P.A.D. Gonçalves (University of Minho, Portugal),
    N.M.R. Peres (University of Minho, Portugal).
Description: Singapore ; Hackensack, NJ : World Scientific Publishing Co.
    Pte. Ltd., [2016] | 2016 | Includes bibliographical references and index.
Identifiers: LCCN 2016003760| ISBN 9789814749978 (hardcover ; alk. paper) |
    ISBN 9814749974 (hardcover ; alk. paper) | ISBN 9789814749985 (softcover ; alk. paper) |
    ISBN 9814749982 (softcover ; alk. paper)
Subjects: LCSH: Plasmons (Physics) | Graphene--Optical properties. | Nanophotonics.
Classification: LCC QC176.8.P55 G66 2016 | DDC 530.4/4--dc23
LC record available at http://lccn.loc.gov/2016003760

British Library Cataloguing-in-Publication Data
A catalogue record for this book is available from the British Library.

In-house Editor: Christopher Teo

Typeset by Stallion Press
Email: enquiries@stallionpress.com

To our families:
(Inês and João Nuno, and Augusta, Augusto, and Filipa)

And our parents:
(Clementina and João Paulo, and Cecília and Luís)

Foreword

The consequences of plasmons, collective excitations of the electron plasma in conductive materials, have been appreciated for centuries — if not millenia, just witness the beautiful ancient colored glasses which are found in museums and churches. There, the colors originate from localized surface plasmonic excitations in nanoscale metallic particles. This old example crystallizes the essence of modern plasmonics: on one hand, one needs surfaces, and on the other hand the active objects should be in the nanometer scale. Graphene, the wonder material of the 21st century, is two-dimensional and therefore all surface; it is just one atomic layer thick and by using modern nanofabrication technologies it can be sculpted laterally to nanoribbons or nanoflakes. Graphene has very special optical properties, which stem from its fascinating Dirac-like electronic dispersion law its charge carriers obey. It is no wonder then that graphene plasmonics is an important and growing area of research. Most importantly, it holds the promise of many practical applications, for example in plasmonic circuitry, nanophotonics, and sensing, just to mention a few.

This is a book on graphene plasmonics. It originates from a group that is world-known for its research in this area — who would be better qualified to write such a book? Students entering graphene plasmonics have had a difficult time in navigating between original research articles, which rarely are particularly pedagogical in their approach, and text books addressing plasmonics (many of which are excellent), but usually having conventional metals as the working material. The present book thus fills in an important gap. Now, all the relevant material is collected in a single volume. What is essential is that the material is thoroughly worked through: the approach is extremely systematic, all equations follow from previous equations in a manner that a dedicated student can easily follow. The technical level of the book is accessible to an advanced undergraduate or a beginning graduate student, but experienced researchers will also benefit from the many insights it offers. It starts with an introduction to Maxwell's equations, and thereafter works its way towards graphene, and the nanostructures made of graphene. The very important issue of interfaces and two-dimensionality is given an exhaustive treatment. Underway the quantum mechanical description of graphene's electronic structure is developed, including effects due to applied magnetic fields, and central topics in many-body physics are discussed (and further explained and elaborated in useful appendices). With the basic concepts at hand, the reader is taken to the frontier of today's research: plasmons in nanostructured two-dimensional materials, such as arrays of microribbons, nanorings, and gratings. The excitation mechanisms needed to launch plasmons and plasmon-polaritons are discussed, as well as methods that are required to observe the excited plasmons. The authors should be congratulated for having accomplished an important task: to offer a self-contained and complete, yet accessible, treatment of a modern area under rapid progress. Armed with the knowledge extracted from this book tomorrow's students are ready to tackle the many exciting opportunities and challenges that plasmonics in two-dimensional materials offer.

Antti-Pekka Jauho,
Technical University of Denmark,
2016

Preface

Over the last few years, graphene plasmonics has emerged as a new research topic, positioned at the interface between condensed matter physics and photonics. This still young but rapidly growing field, resulting from the overlap between graphene physics and plasmonics, deals with the investigation and exploration of graphene surface plasmon-polaritons (GSPs) for controlling light-matter interactions and manipulating light at the nanoscale.

In 2011, experimental research in plasmonics in graphene was in its infancy. Some theoretical papers had already been written, but the field was awaiting for some sort of experimental boost. In 2011 a seminal experimental paper (to be discussed in Section 7.1) emerged, demonstrating that plasmonic effects in graphene could be controlled optically, by shining electromagnetic radiation onto a periodic grid of graphene micro-ribbons. This opened a new avenue in graphene physics, launching the new flourishing field of graphene plasmonics. Since then, the field has been witnessing enormous developments at a rapid pace (to be described in Section 1.2).

This is a book on plasmonics in graphene. It grew out of the authors' own interest in the field. Many of the topics covered here refer to problems that were amongst the first to appear in the scientific literature, to which one of the authors has contributed to some extent. Throughout this monograph we have tried to make this book as self-contained as possible, and to provide substantial detail in all derivations. Indeed, with a pencil, some sheets of paper, a dose of enthusiasm, and, at times, a bit of effort, the interested reader should find it easy to reproduce all the results in the book (the Appendices will be most useful for such endeavor). Therefore, we hope that this book will serve as a springboard for any newcomer in graphene plasmonics. Hopefully, the advanced researcher may also find reasons to keep this book in her/his shelf and to recommend it to her/his students. This is also a book about theoretical techniques to deal with plasmonics in graphene. Many of these techniques have been applied to metal plasmonics before and have been adapted here to tackle plasmonic effects in graphene. The main difference is that metals are described by a complex dielectric function, whereas graphene is described by its complex (non-local whenever necessary) optical conductivity and its sheet (two-dimensional) current.

The book is organized as follows: in Chapter 1 we introduce the theme of the book and present a short review of the literature. We have surely missed some important references and can only hope for the indulgence and comprehension of their authors. This might be specially true in what concerns theoretical and computational work (the literature is already too vast to be covered in a comprehensive way). In Chapter 2 we review some elementary concepts about electronic transport in solids, give a short introduction to graphene's electronic properties, and present some tools to be used later in the book, with particular emphasis on a method to compute the non-local optical conductivity of graphene from the electronic susceptibility in the relaxation-time approximation. Chapter 3 discusses plasmonics at metal-dielectric interfaces and in metal thin films. The results obtained there will be later compared with those found in graphene. Both the semi-infinite metal and the thin metal-film are discussed. In Chapter 4, we finally dive into the field of graphene plasmonics and discuss plasmons both in single- and double-layer graphene. The interesting problem of coupled surface plasmon-phonon-polaritons is also discussed in that Chapter. A short introduction of magneto surface plasmon-polaritons is included, considering both the semi-classical and quantum regimes. We close the chapter with the study of guided (bounded) surface phonon-polaritons in a crystal slab of hexagonal boron-nitride (hBN), a problem very recently considered in the literature and one for which many future developments are expected.

As we shall see, surface plasmons cannot be directly excited in graphene by free propagating electromagnetic waves (as is also the case in planar metal-dielectric interfaces). Thus, in Chapter 5 we review some of the most popular methods of inducing surface plasmon-polaritons in this two-dimensional material (which are also used in traditional metal-based plasmonics). We then move to a detailed study of two methods for exciting plasmons in graphene. These are discussed in Chapters 6 and 7, encompassing the excitation of GSPs by a metallic antenna and by coupling light to GSPs in a patterned array of graphene ribbons. The discussion of the periodic graphene ribbons system sets the stage for the discussion of the spectrum of surface plasmons in graphene micro-ribbons, micro-rings, and disks. This is done in Chapter 8, where the case of a micro-deformation of the graphene sheet is also discussed, taking as an example the problem of a Gaussian groove. This latter case can be considered an example of nano-structured graphene. These three problems are considered in the electrostatic limit, which turns out to be a rather good approximation when dealing with plasmons in graphene. As we shall show they exhibit scale invariance, a property of the electrostatic limit. It then follows, in Chapter 9, a discussion of the excitation of plasmons by a corrugated dielectric grating, which is a generalization of the single micro-deformation studied in the preceding chapter to the case of a periodic deformation. In this case, the calculation takes retardation into account, but could very well be treated within the electrostatic approximation. Despite some differences, the four methods discussed in Chapters 6, 7, 8, and 9 have in common the decomposition of the fields in a superposition of Fourier modes, either using a Fourier series, for periodic structures, or a Fourier integral for non-periodic systems. In Chapter 10 we discuss how an emitting dipole (for example, a dye, a quantum dot, or a Rydberg atom) can excite graphene plasmons by non-radiative energy transfer from the emitter to graphene. We will see that this is in principle possible, and we compute the non-radiative decaying rate of an emitter. The role of particle-hole processes and excitation of plasmons is discussed separately. We end this Chapter with the calculation of the Purcell factor which encodes information about the total decaying rate of the emitter in the presence of graphene. Finally, in Chapter 11, we conclude with a brief outlook and with the discussion of some topics that have been left out in this book, and which we believe will become rather important in the near future. Also, we have written a number of appendices with either details of the calculations or extensions of the materials discussed in the bulk of the book. To get a flavor of the main ideas, reading the appendices is unnecessary. However, they are a requisite for a working knowledge on the topics discussed in the main text.

Our ultimate hope is that readers may find this book a valuable tool for their study and research in the exciting field of graphene plasmonics and nanophotonics, whether they are currently working on the topic or are about to enter on it. The book's intended purpose is to bring together some of the main topics that are scattered throughout the already vast literature, and to do justice to this alluring field of graphene plasmonics.

To conclude, we further hope that the readers may read this book with the same enthusiasm that we had writing it. Comments, suggestions, and corrections aiming at improving the book for future usage will be most welcome (these can be sent to peres@fisica.uminho.pt).

PAULO ANDRÉ DIAS GONÇALVES
NUNO MIGUEL MACHADO REIS PERES

DEPARTMENT OF PHYSICS
UNIVERSITY OF MINHO
BRAGA, 2016

Acknowledgments

The authors acknowledge B. Amorim, Y. V. Bludov, E. V. Castro, A. H. Castro Neto, A. Chaves, A. Ferreira, F. Guinea, J. Lopes dos Santos, V. Pereira, R. M. Ribeiro, J. E. Santos, T. Stauber, and M. I. Vasilevskiy for many years of collaboration and discussions about the physics of graphene in general and on plasmonics in graphene and other materials in particular. We thank B. Amorim, Y. V. Bludov, A. Chaves, A. Ferreira, J. Lopes dos Santos, R. M. Ribeiro, J. E. Santos, and M. I. Vasilevskiy for reading parts of the manuscript at different moments of writing. B. Amorim is acknowledged for providing the derivations in Appendices D, O, and P. A. Chaves is acknowledged for providing the derivation in Appendix A J. E. Santos is acknowledged for providing the derivation in Appendix K. Any error and/or inaccuracy that may remain in the text of these five Appendices (and, for that matter, in the rest of the text) is the authors' own responsibility.

Paulo André Gonçalves acknowledges the Calouste Gulbenkian Foundation for financial support through the grant *"Prémio Estímulo à Investigação 2013"*[1] (No. 132394) and the hospitality of the Centre of Physics of the University of Minho, where most of this book was written. Nuno Peres acknowledges financial support from the Graphene Flagship Project (Contract No. CNECT-ICT-604391).

---

[1]"Gulbenkian Prize for Stimulating Young Researchers 2013".

Contents

List of Figures

List of Tables

Chapter 1

Introduction

Plasmonics holds many promises [Atwater (2007)], such as plasmonic integrated circuitry [Sorger *et al.* (2012)], nanophotonics (e.g., plasmonic antennas) [Dionne and Atwater (2012)], all-optical manipulation of small particles [Quidant (2012)], quantum plasmonics [Jacob (2012); Hanson *et al.* (2015)], spin-polarized plasmonics [Toropov and Shubina (2015)], and plasmonic hot-spots [Weber and Willets (2012)], just to name a few. Whether graphene will be able to fulfil some of these promises remains to be seen, but current prospects are encouraging. This book is about plasmonics in graphene and about the methods to excite and control surface plasmons in this material. Below, we give a brief account of some of the recent developments in the field.

1.1 Plasmonics: Generalities

The recent burst of interest in plasmonics may overshadow the fact that plasmon-related effects can be traced back to the turn of the twentieth century with the works of A. Sommerfeld and J. Zenneck on the propagation of electromagnetic waves along the surface of a conductor [Sommerfeld (1899); Zenneck (1907)]. Concurrently, in 1902, R. H. Wood reported the observation of anomalous intensity drops in the spectrum of visible light diffracted by a metallic grating, which became known as Wood's anomalies [Wood (1902)]. However, Wood was unable to develop any suitable interpretation for this phenomenon. The first effort towards a theoretical interpretation of Wood's results was conducted by Lord Rayleigh [Rayleigh (1907)], albeit with limited success, and was followed by the works of Fano a few decades later [Fano (1941)]. Fano's work was subsequently improved by many [Enoch and Bonod (2012)]. Some of these breakthroughs took place during the 1950s and 1960s [Hessel and Oliner (1965); Ritchie *et al.* (1968)], along with the development of the concept of *surface plasmon-polariton* and its excitation both via the use fast electrons [Powell and Swan (1959); Ritchie (1957)] and through prism coupling[1] [Kretschmann and Raether (1968); Otto (1968); Sambles *et al.* (1991)] (see [Gupta *et al.* (2015)] for an historical perspective on plasmonics).

Surface plasmon-polaritons (SPPs) are electromagnetic surface waves coupled to collective charge excitations of the conduction electrons, propagating along the interface between a dielectric and a conductor. When these coupled excitations occur in metallic nanostructures —such as nanoparticles (NPs)—, the corresponding non-propagating plasmon-polaritons are generally coined as localized surface plasmons (LSPs) [Pelton *et al.* (2008); Pelton and Bryant (2013); Stockman (2011)].

Notwithstanding, modern plasmonics as we know it emerged in the turn of the 21st century with the observation by T. W. Ebbesen *et al.* of an "extraordinary optical transmission through subwavelength hole arrays" in a periodically patterned metal film [Ebbesen *et al.* (1998)]. Shortly after this seminal paper, a theoretical account of this experiment was developed invoking the role SPPs [Martín-Moreno *et al.* (2001)]. One notable feature of SPPs modes is that their momentum is larger than light's momentum corresponding to photons with the same frequency [Maier (2007)]. As a consequence, SPPs are bound modes whose fields decay exponentially away from the interface, exhibiting subwavelength confinement [Barnes *et al.* (2003)] and field enhancement (which can be of several orders of magnitude larger than the value of the incident field) beyond the reach of conventional optics [Barnes *et al.* (2003); Barnes (2006); Maier (2007)]. Thus, plasmonics opened up the possibility for manipulating light and control light-matter interactions at scales well below the diffraction limit (and at the nanoscale), being one of the most promising research topics in the field of nanophotonics and also an important ingredient in the design and implementation of metamaterials [Kabashin *et al.* (2009); Liua and Zhang (2011); Billings (2013)].

Nowadays, plasmonics plays a leading role towards the miniaturization

---

[1]Otto and Kretschmann-Raether are often considered as the first to excite SPPs in metal thin films using a prism. The fact, however, is that Turbadar was the first to excite surface plasmon-polaritons in aluminium (and later in silver and gold) thin films using the prism coupling method [Turbadar (1959)]. However, and contrary to Otto and Kretschmann-Raether, he did not correlated the minimum in the reflectance with the excitation of SPPs.

of photonic structures and circuitry [Ebbesen *et al.* (2008)], along with their integration with current semiconductor technologies [Ozbay (2006)]. For telecommunications and/or information processing photonics offers high speeds and large bandwidths, but albeit at a cost, since bulky components —at least in comparison to chip-based nanoelectrics— are the rule; on the other hand, although electronics provides the much-wanted miniaturization and good carrying capacities, its ability to deal with huge amounts of information is limited by the number of connections, resulting in transmission bottlenecks. Therefore, the hybrid nature of SPPs (being "half-light, half-matter" bosonic quasiparticles, i.e. polaritons) makes them an attractive route to close the gap between photonic and electronic circuits, by combining the advantages of both "worlds".

The revival of the field over the past decade was, to a large extent, due to a rapid development of nanoscale fabrication techniques and experimental tools, together with the implementation of powerful numerical modelling based on full-wave electromagnetic simulations. These triggered the emergence of a plethora of exciting applications (which are often strong motivators by themselves), such as spectroscopy and biochemical sensing [Anker *et al.* (2008); Kabashin *et al.* (2009); Brolo (2012); Haes *et al.* (2005)], optical tweezers [Reece (2008); Juan *et al.* (2011)], photovoltaics and energy harvesting [Green and Pillai (2012); Boriskina *et al.* (2013)], diagnostics and nanomedicine [Aćimović *et al.* (2014); Zheng *et al.* (2012); Vo-Dinh *et al.* (2013)], SPP-based amplifiers and lasers [Berini and Leon (2012); Ma *et al.* (2013)], Fourier and transformation optics [Archambault *et al.* (2009); Huidobro *et al.* (2010); Zhang *et al.* (2011)], etc. Yet, the field of plasmonics is far from being exhausted and continues to thrive. Recently, special attention has been given to forefront research themes including ultrafast plasmonics [MacDonald *et al.* (2009)], non-linear phenomena with SPPs [Kauranen and Zayats (2012)] and quantum plasmonics [Tame *et al.* (2013); van Hulst (2012); Jacob (2012)]. This latter field studies the quantum properties of SPPs, setting up novel ways to investigate new fundamental physics and to fuel the prospection for plasmonic devices operating in the quantum regime [Jacob and Shalaev (2011); Chang *et al.* (2006)]. Several quantum optics experiments with SPPs have been reported [Heeres *et al.* (2013); Fakonas *et al.* (2014); Di Martino *et al.* (2014); Cai *et al.* (2014)], demonstrating the maintenance of quantum coherence and unitarity in (quantum) photonic systems, despite the macroscopic, collective nature (encompassing $\sim 10^6$ electrons) of SPPs [Tame *et al.* (2013)]. New frontiers in metal-semiconductor plasmonics are discussed in [Toropov and Shubina (2015)] and a modern introduction to plasmonics is given in [Maradudin *et al.* (2014)]. The interested reader in localized surface plasmons may want to check [Klimov (2014)].

1.2 Plasmonics: Recent Developments

In the present book we are concerned with graphene as a plasmonic material —graphene plasmonics [Bludov *et al.* (2013); Grigorenko *et al.* (2012); Low and Avouris (2014); de Abajo (2014); Luo *et al.* (2013); Koppens *et al.* (2011); Stauber (2014)]. Isolated in 2004 [Novoselov *et al.* (2004)], graphene —a two-dimensional (2D) crystal made up of carbon atoms arranged in a honeycomb lattice— has been, in the last decade, one of the hottest topics in condensed matter physics, owing to its remarkable electronic, mechanical, and optical properties [Geim (2009); Castro Neto *et al.* (2009)]. As in the case of plasmonics, the number of publications on graphene physics has escalated during the past decade. However, the two fields have evolved quite independently up until very recent times. In fact, the bridge between both these subjects was well-established only as recently as in 2011, motivated by the experimental realization of graphene surface plasmon-polaritons (GSPs) in periodic arrays of graphene ribbons, obtained by patterning a pristine graphene sheet [Ju *et al.* (2011)]. This foundational work has sparked a whole interest in graphene nanoplasmonics, setting the stage for the investigation of graphene as a novel plasmonic material.

Doped graphene is capable of sustaining SPPs [Jablan *et al.* (2009); Bludov *et al.* (2013); Jablan *et al.* (2013); Ju *et al.* (2011); Yan *et al.* (2013); Strait *et al.* (2013); de Abajo (2014); Christensen *et al.* (2012)], which can be actively controlled and tuned through chemical doping or electrical gating, almost in real time. This contrasts with conventional metal-based plasmonics [Toropov and Shubina (2015)], where the tunability is usually limited— it can be only achieved, essentially, by controlling the size, the shape, and the dielectric environment of metallic nanoparticles (the polarization of the electric field also plays a role), the geometry of metal-based nanostructures, or by exploring the thickness of thin films of noble-metals. Furthermore, graphene surface plasmon-polaritons enable higher levels of spatial confinement (within volumes of the order of $\sim \alpha^{-3} \approx 10^6$ below the diffraction limit; here α is the fine structure constant of atomic physics) and are predicted to suffer from relatively low losses[2], therefore having larger lifetimes

---

[2]See [Yan *et al.* (2013)] for a discussion of the damping channels of plasmons in graphene

and propagation lengths, when compared with traditional plasmonic materials. This makes graphene an appealing candidate for nanoplasmonic and nanophotonic applications. The spectral region of graphene plasmonics has been restricted to the terahertz (THz) and mid-infrared (mid-IR), a region that traditional metal-based plasmonics cannot cover, since the SPP's in a continuous metallic film are essentially free radiation at such frequencies, therefore lacking a key advantage of SPP's in graphene, the strong spatial confinement.

As is well known, the electrons in graphene are described by a massless Dirac equation, contrary to the 2D electron gas. The spectrum of SPPs' in graphene provides an indirect confirmation of the form of the dispersion of electrons in this material. Although the frequency of GSPs scales with momentum as $\sqrt{q}$ like in the classic 2DEGs (two-dimensional electron gases), the GSP frequency is proportional to the electronic density as $n_e^{1/4}$, which is specific of graphene (for 2DEG, a $n_e^{1/2}$ scaling is observed instead). This is a direct consequence of the unique electronic properties of graphene, whose electrons behave like relativistic Dirac fermions with a linear dispersion [Castro Neto *et al.* (2009)] (this is true at low energies, that is, in the vicinity of the Dirac points). Another peculiarity of graphene is its ability to support transverse electric[3] (TE) modes [Mikhailov and Ziegler (2007)], which do not occur in ordinary metal/dielectric structures [Maier (2007)].

The frequencies of transverse magnetic (TM) GSP typically lie between the THz and mid-IR regions of the electromagnetic spectrum. While the THz spectral region remains relatively unexplored, THz photonics has become an active field of research over the last few years [Mittleman (2013)]. Here too, graphene plasmonics may contribute to the development of this subject [Tassin *et al.* (2013); Low and Avouris (2014)]. For instance, a geometry akin to the one described in [Ju *et al.* (2011)] together with sub-wavelength bimetallic electrodes has recently been engineered to achieve plasmon-enhanced THz photodetection using graphene [Cai *et al.* (2015)].

Having passed its infancy, one may state that graphene plasmonics is

---

nanostructures; see [Khurgin and Boltasseva (2012)] for a discussion of the same topic in metal-based plasmonics and metamaterials as well as their possible replacement by conducting oxides and transition metal nitrides

[3]We bring to the reader's attention that, up to this point, whenever we mentioned SPP/GSP it was implicit that we were referring to transverse magnetic modes. This will also be the case throughout the rest of the book, unless explicitly stated otherwise.

now entering its young adult life (in fact, it is maturing rather fast). Yet, during these early stages of the field, there were already some remarkable achievements that took place which might be worth mentioning. Apart from the above-referenced excitation of GSPs in periodic gratings made up of graphene stripes [Ju *et al.* (2011); Strait *et al.* (2013); Yan *et al.* (2013); Nene *et al.* (2014)], other experimental landscapes for coupling light to GSPs were also explored, ranging from graphene disks [Fang *et al.* (2013, 2014); Yan *et al.* (2012c)], dimers [Christensen *et al.* (2012); Thongrattanasiri and de Abajo (2013)], resonators [Brar *et al.* (2013, 2014)], rings [Fang *et al.* (2013); Yan *et al.* (2012c)], periodic dot/anti-dot lattices [Yan *et al.* (2012a); Fang *et al.* (2013, 2014); Liu *et al.* (2015b); Yan *et al.* (2012c); Zhu *et al.* (2014)], stacks of graphene disks/ribbons [Yan *et al.* (2012a,c)], periodically modulated or corrugated graphene [Peres *et al.* (2012); Slipchenko *et al.* (2013a)], continuous monolayer graphene resting in dielectric gratings [Gao *et al.* (2013); Zhu *et al.* (2013d); Gao *et al.* (2013)], nano-antennas [Liu *et al.* (2014a); Llatser *et al.* (2012); Yao *et al.* (2013); Fang *et al.* (2012a,b); Alonso-González *et al.* (2014)], among others [Koch *et al.* (2010); Fei *et al.* (2011, 2012); Chen *et al.* (2012)]. Moreover, the unique tunability of GSPs was demonstrated in these experiments, through the shifts of plasmon resonances in response to changes in the amount of doping induced via electrical gating. Indeed, this enables high-speed electro-optical modulation by taking advantage of the capability of significantly (and quickly) alter graphene's optical properties using electric gates [Sensale-Rodriguez *et al.* (2012); Lee *et al.* (2012); Li *et al.* (2008)]. Also, the complete absorption of light in periodic arrays of doped graphene nanodisks, which may have strong implications in photovoltaics and light harvesting platforms, has been predicted in a theoretical work [Thongrattanasiri *et al.* (2012a)]. This constitutes a truly astonishing result taking into account that this is a one-atom-thick material.

A particularly tantalizing and very application-oriented approach, involves the use of GSPs to grant us graphene with bio and chemical sensing capabilities [Wu *et al.* (2010); Vasić *et al.* (2013); Zhao *et al.* (2013); Zhang *et al.* (2013); Wu *et al.* (2014a); Rodrigo *et al.* (2015)] (see also [Brolo (2012)]). The general idea behind the concept of a plasmonic sensor is to explore the subwavelength confinement and enhancement of the electromagnetic field to probe small changes in the local dielectric environment, given that the SPP/LSP resonant frequencies are highly sensitive to the optical constants of the contiguous medium [Anker *et al.* (2008)]. For instance, the adsorption of biomolecules on the surface of graphene promotes a subtle modification of the effective refractive index of the sensing medium, which can be detected by measuring the corresponding spectral shifts of the plasmon resonances [Zhao *et al.* (2013)]. Here, the aim is to target graphene's potential to yield surface plasmons with superior spatial confinement and intense fields to pave the way for cutting-edge, high-performing plasmonic sensors that surpass costumary noble-metal plasmonic devices [Rodrigo *et al.* (2015)].

Another highly appraised and auspicious phenomenon is surface-enhanced Raman scattering (SERS) in plasmonic nanostructures [Kneipp (2007)]. This spectroscopic technique has been one of the most compelling applications of plasmonics, since it enables an exceptionally sensitive detection of low concentration of analytes and even label-free single molecule detection [Sharma *et al.* (2012); Crozier *et al.* (2014); Stiles *et al.* (2008); Schlücker (2014)]. In recent years, it has become popular to build hybrid SERS platforms resulting from the combination of graphene with metallic nanostructures, with the purpose of exploiting graphene's outstanding optoelectronic properties to enhance the Raman signal and also to assist the quantitative assessment of the different SERS mechanisms [Xu *et al.* (2012b); Wang *et al.* (2013)]. Graphene-mediated SERS (G-SERS) was reported in a number of experiments [Wang *et al.* (2012a); Ling *et al.* (2010); Liu *et al.* (2011a); Urich *et al.* (2012); Heeg *et al.* (2013); Kravets *et al.* (2012); Zhao *et al.* (2014); Wang *et al.* (2013); Zhu *et al.* (2013c)], along with the observation of enhancement factors spanning from about 100 [Urich *et al.* (2012)] (for monolayer graphene with Ag nanoislands) or 1000 [Heeg *et al.* (2013); Kravets *et al.* (2012)] (for graphene on top of a Au nano-disk dimmer and plasmonic nanoarrays), up to values as high as 10^6 [Zhao *et al.* (2014)] and 10^{10} [Wang *et al.* (2013)] respectively for the dye molecules Rhodamine B (RhB) and Rhodamine 6G (Rh6G) in a Au-NPs/graphene/Cu composite platform, and for Rh6G and lysozyme in a bio-compatible graphene-Au nano-pyramid hybrid system.

Benefiting from the know-how acquired in the flourishing sub-branch of plasmonics dealing with metallic nanoparticles [Pelton and Bryant (2013); Garcia (2011)], the integration of plasmonic NPs with graphene has emerged as a topic worthy of attention [Fang *et al.* (2012a); Santos *et al.* (2014); Liu *et al.* (2011b); Fang *et al.* (2012b); Xu *et al.* (2012a); Yin *et al.* (2013)]. This offers new possibilities for improving the photodetection of such systems [Liu *et al.* (2011b); Fang *et al.* (2012a)], and also for enhancing the interaction between a polarizable nanoparticle and graphene [Santos *et al.* (2014)] by taking advantage of the appropriate resonances in their spectra, together with graphene's tunability. This enhanced interaction, with its strong dependence upon the NP-graphene distance, makes this system potentially interesting for application in deformation or pressure sensors, as it is technologically feasible to control the deposition of semiconductor or metallic nanoparticles on graphene [Sun *et al.* (2011a); Kamat (2010)]. In addition, shifts of surface plasmon resonances (SPRs) associated with the electromagnetic coupling between graphene and the nanoparticle(s) have been observed [Niu *et al.* (2012)], as well as changes in the SPR linewidth and plasmon-generated hot-electron transfer [Hoggard *et al.* (2013); Fang *et al.* (2012b)].

Beyond traditional noble metal NPs, another research line is connected with SPP-related physical mechanisms operating in compound systems of graphene with quasi-zero-dimensional emitters or absorbers, such as molecules or semiconductor quantum dots (QDs) [Gaudreau *et al.* (2013); Biehs and Agarwal (2013); Velizhanin and Efimov (2011); Chen *et al.* (2010)]. These architectures yield new routes to investigate the interaction among localized excitations such as molecular or QD excitons and propagating SPPs in graphene. It has been shown experimentally that graphene allows for probing the de-excitation dynamics of the emitter [Gaudreau *et al.* (2013)] and causes an enhanced Förster resonance energy transfer from a fluorescent QD [Chen *et al.* (2010)]; at the same time, a hybrid graphene–QD phototransistor with ultrahigh gain has been demonstrated [Konstantatos *et al.* (2012)]. Theoretical proposals include probing graphene-mediated control of the coupling between two emitters (Förster energy transfer [Biehs and Agarwal (2013)] and/or GSP-mediated superradiance [Huidobro *et al.* (2012)]), studies of the GSP dispersion [Velizhanin and Efimov (2011)], and enhancement of radiation absorption in graphene using nanoparticles [Liu *et al.* (2011b); Zhu *et al.* (2013a)].

All of the above has placed graphene on the map as a fertile playground for nanoscale plasmonics. Aside from the research avenues outlined in the preceding paragraphs —including the excitation and control of GSPs in a number graphene-based structures, analysis of the NPs- and QD-graphene electrodynamics, potential applications in biochemical sensing, spectroscopy, electro-optical modulation, and enhanced photodetection and optical absorption—, further developments have also been made in a variety of disciplines, encompassing transformation optics using graphene [Vakil and Engheta (2011)], graphene plasmonic metamaterials [Lee *et al.* (2012); Ju *et al.* (2011); Yan *et al.* (2012a); Ishikawa and Tanaka (2013)], atomically thin waveguiding and optical communications [Christensen *et al.* (2012); Vakil and Engheta (2011)], cloaking devices [Chen and Alù (2011)], broadband graphene polarizers [Bao *et al.* (2011)], excitation of GSPs via surface acoustic waves [Schiefele *et al.* (2013); Farhat *et al.* (2011)], realization of magneto-plasmons in graphene [Crassee *et al.* (2012); Yan *et al.* (2012b)] (see also [Temnov (2012)]), and photothermal therapies for biomedicine [Lim *et al.* (2013)], just to name a few.

As we *begin* to unravel the fundamentals of graphene plasmonics and to take a glimpse at the physics behind graphene's strong light-matter interactions, new research frontiers are already emerging on the horizon. These include, but are not limited to, the investigation of many-body effects, namely the interaction of GSPs with other elementary excitations in graphene (such as plasmon-phonon coupling [Yan *et al.* (2013); Brar *et al.* (2014); Hwang *et al.* (2010)] and plasmarons [Carbotte *et al.* (2012)]); a better understanding of the loss mechanisms affecting GSPs' lifetimes [Yan *et al.* (2013)]; the exploration of strong-coupling regimes in graphene with single quantum emmiters [Koppens *et al.* (2011); Manjavacas *et al.* (2012b); Gullans *et al.* (2013); Cox *et al.* (2012); Manjavacas *et al.* (2012a); Tielrooij *et al.* (2015)]; studies of non-linear phenomena [Gullans *et al.* (2013); Cox and de Abajo (2014); Manzoni *et al.* (201)]; assessment of the importance of quantum finite-size effects in graphene nanostructures supporting GSPs [Thongrattanasiri *et al.* (2012b)], etc. Future perspectives also include the extension of the pioneering research conducted in graphene nanophotonics to other 2D crystals and van der Waals heterostructures (layered structures of atomically thin materials like graphene, MoS_2, WeS_2, hBN, other 2D chalcogenides, and oxides [Geim and Grigorieva (2013)]). Additional challenges are comprised with bringing graphene plasmonics towards the NIR-vis region of the electromagnetic spectrum, along with the demand of cleaner fabrication techniques. The latter are expected to deliver GSPs with lower losses while allowing a better control over the morphology and dimensions of the patterned samples. On the other hand, extreme doping, ultra-narrow graphene ribbons or disks, and nanometer-sized forms of graphene (e.g. chemically synthesized polycyclic aromatic hydrocarbons [Manjavacas *et al.* (2013)]) have been proposed [de Abajo (2014)] to carry out graphene plasmons at NIR-vis frequencies.

Although this young and vibrant field still has some demanding challenges to face, it is now beyond dispute that graphene already deserves its own place in the list of most promising plasmonic materials of the future.

Research on graphene plasmonics is intense and likely to remain one of the hottest topics in nanophotonics in the forthcoming years. In addition, the exploration of the unique properties of graphene surface plasmons may lead to a new gold rush with high impact both in fundamental science as well as in technology and industry, ultimately leading to the appearance of the next generation of devices with tailored optoelectronic properties [Yusoff (2014)].

Chapter 2

Electromagnetic Properties of Solids in a Nutshell

In this chapter, we lay the foundations which will constitute the framework of the problems discussed in the chapters ahead. Here, we shall provide a brief overview on elementary concepts in classical electromagnetism and in graphene physics, including the description of some essential tools to discuss plasmons in the forthcoming chapters.

2.1 Classical Electrodynamics Basics

Nowadays, the quest for a unified theory, commonly known as *"Theory of Everything"*, remains the Holy Grail of High Energy Physics. The first step towards this end was taken by James Clerk Maxwell, in 1864, with the publication of his famous article entitled "A Dynamical Theory of the Electromagnetic Field" [Maxwell (1865)]. By unifying electricity and magnetism, Maxwell headed one of the greatest scientific achievements of the 19th century. In fact, classical electrodynamics can be regarded as the first (classical) unified field theory. Today, within the standard model, classical electromagnetism corresponds to a limit of quantum electrodynamics (QED) [Weinberg (1996)], dubbed by Richard Feynmann as the "jewel" of physics owing to the level of agreement between its predictions and experiment. Still, at macroscopic and mesoscopic scales, provided that the number of photons is large and that momentum and energy transfers are small, the classical formalism often suffices [Jackson (1998)]. This regime is satisfied throughout this book.

2.1.1 *Maxwell's equations*

In the scope of Maxwell's classical electrodynamics, light is described as an electromagnetic wave in the form of a three-dimensional electromagnetic field. Within this framework, light-matter interactions are governed by Maxwell's equations. In SI units, the macroscopic Maxwell's equations read [Jackson (1998)]

$$\nabla \cdot \mathbf{D} = \rho \,, \tag{2.1}$$

$$\nabla \times \mathbf{H} = \frac{\partial \mathbf{D}}{\partial t} + \mathbf{J} \,, \tag{2.2}$$

$$\nabla \times \mathbf{E} = -\frac{\partial \mathbf{B}}{\partial t} \,, \tag{2.3}$$

$$\nabla \cdot \mathbf{B} = 0 \,, \tag{2.4}$$

where $\mathbf{E} \equiv \mathbf{E}(\mathbf{r}, t)$ denotes the electric field, $\mathbf{D} \equiv \mathbf{D}(\mathbf{r}, t)$ the electric displacement, $\mathbf{H} \equiv \mathbf{H}(\mathbf{r}, t)$ the magnetic field and $\mathbf{B} \equiv \mathbf{B}(\mathbf{r}, t)$ the magnetic induction[1]. In turn, $\rho \equiv \rho(\mathbf{r}, t)$ and $\mathbf{J} \equiv \mathbf{J}(\mathbf{r}, t)$ refer to the (free) charge density and current density, respectively, satisfying the continuity equation (charge conservation)

$$\frac{\partial \rho}{\partial t} + \nabla \cdot \mathbf{J} = 0 \,. \tag{2.5}$$

It should be noted that in the case of graphene [or a two-dimensional electron-gas (2DEG)], lying in the $z = 0$ plane, we have

$$\rho = \rho_{2D}\delta(z) \qquad \text{and} \qquad \mathbf{J} = \mathbf{J}_{2D}\delta(z) \,, \tag{2.6}$$

at a given time. Limiting ourselves to media that are linear, uniform and isotropic, and non-magnetic, one can prescribe the following constitutive relations

$$\mathbf{D} = \epsilon\epsilon_0 \mathbf{E} \,, \tag{2.7}$$

$$\mathbf{H} = \frac{1}{\mu_0} \mathbf{B} \,, \tag{2.8}$$

where ϵ_0 is the electric permittivity of vacuum and μ_0 its magnetic permeability; these relate to the speed of light in vacuum via $c^{-2} = \epsilon_0\mu_0$. The quantity ϵ represents the relative permittivity or dielectric constant (in general, it is a function of frequency). Another important constitutive relation, for conducting materials, is the generalized Ohm's law,

$$\mathbf{J} = \sigma\mathbf{E} \,, \tag{2.9}$$

with σ being a frequency-dependent conductivity (in general a tensor). We note that, in general, the linear relationships expressed through equations

---

[1] Remark: we will refer to $\mathbf{B}$ as being the "magnetic" field multiple times throughout the book, though this is technically inaccurate.

(2.7) and (2.9) should be understood as

$$\mathbf{D}(\mathbf{r}, t) = \epsilon_0 \int d\mathbf{r}' dt' \epsilon(\mathbf{r} - \mathbf{r}', t - t') \mathbf{E}(\mathbf{r}', t') , \qquad (2.10)$$

$$\mathbf{J}(\mathbf{r}, t) = \int d\mathbf{r}' dt' \sigma(\mathbf{r} - \mathbf{r}', t - t') \mathbf{E}(\mathbf{r}', t') , \qquad (2.11)$$

where ϵ stands now as a function —the dielectric function. Equations (2.10) and (2.11) simply state that the value of $\mathbf{D}$ and $\mathbf{J}$ at a given position $\mathbf{r}$ and time t, depends on the value of $\mathbf{E}$ at all positions $\mathbf{r}'$ and times $t' < t$ (to preserve causality) via the response functions $\epsilon(\mathbf{r} - \mathbf{r}', t - t')$ and $\sigma(\mathbf{r} - \mathbf{r}', t - t')$, respectively. Consequently, this accounts for both temporal dispersion and non-local effects in space (spatial dispersion), meaning that the linear response is frequency- and momentum-dependent. While the former is usually evident, the latter can be safely neglected in most cases (as long as the corresponding wavevector is much smaller than the electron's Fermi momentum). In Fourier space, the above equations translate to

$$\mathbf{D}(\mathbf{q}, \omega) = \epsilon_0 \epsilon(\mathbf{q}, \omega) \mathbf{E}(\mathbf{q}, \omega) , \qquad (2.12)$$
$$\mathbf{J}(\mathbf{q}, \omega) = \sigma(\mathbf{q}, \omega) \mathbf{E}(\mathbf{q}, \omega) , \qquad (2.13)$$

where the Fourier transform of the electric field is given by

$$\mathbf{E}(\mathbf{r}, \omega) = \int \mathbf{E}(\mathbf{q}, t) e^{i(\mathbf{q} \cdot \mathbf{r} - \omega t)} dt d\mathbf{q} , \qquad (2.14)$$

and similar relations apply for the remaining fields.

Furthermore, notice that in the absence of free charges and currents (as in free space or in a dielectric medium), Maxwell's equations reduce to

$$\nabla \cdot \mathbf{E} = 0 , \qquad (2.15)$$

$$\nabla \times \mathbf{B} = \frac{\epsilon}{c^2} \frac{\partial \mathbf{E}}{\partial t} , \qquad (2.16)$$

$$\nabla \times \mathbf{E} = -\frac{\partial \mathbf{B}}{\partial t} , \qquad (2.17)$$

$$\nabla \cdot \mathbf{B} = 0 . \qquad (2.18)$$

In addition, expressions (2.16) and (2.17) can be written as

$$\nabla \times \mathbf{B} = -i \frac{\epsilon \omega}{c^2} \mathbf{E} , \qquad (2.19)$$

$$\nabla \times \mathbf{E} = i\omega \mathbf{B} , \qquad (2.20)$$

$$(2.21)$$

where a harmonic time dependence in the form of $e^{-i\omega t}$ is assumed [or, equivalently, allowing a decomposition of the fields in terms of a plane-wave expansion – cf. equation (2.14)].

To conclude, let us mention that by combining the curl equations (2.16) and (2.17), one arrives at the following Helmholtz equations for the electric and magnetic fields,

$$\left(\nabla^2 - \frac{1}{v^2}\frac{\partial^2}{\partial t^2}\right)\left\{\begin{array}{c} \mathbf{E}(\mathbf{r},t) \\ \mathbf{B}(\mathbf{r},t) \end{array}\right\} = 0 , \qquad (2.22)$$

which are wave equations describing an electromagnetic wave travelling at speed $v = c/\sqrt{\epsilon}$.

2.1.2 *Boundary conditions*

Up to now we have considered electromagnetic fields in a homogeneous medium. Whenever the physical system under consideration involves more than one material, we must introduce the appropriate boundary conditions at each interface. These are obtained by applying the divergence theorem and Stoke's theorem to Maxwell's equations, yielding [Jackson (1998)]

$$\hat{\mathbf{n}} \cdot (\mathbf{D}_1 - \mathbf{D}_2) = \sigma_s , \qquad (2.23)$$

$$\hat{\mathbf{n}} \cdot (\mathbf{B}_1 - \mathbf{B}_2) = 0 , \qquad (2.24)$$

$$\hat{\mathbf{n}} \times (\mathbf{B}_1 - \mathbf{B}_2) = \mu_0 \mathbf{J}_s , \qquad (2.25)$$

$$\hat{\mathbf{n}} \times (\mathbf{E}_1 - \mathbf{E}_2) = 0 , \qquad (2.26)$$

where $\hat{\mathbf{n}}$ is a unit vector normal to the interface (pointing from medium 2 to medium 1), whereas σ_s and $\mathbf{J}_s$ are the (idealized) surface charge and surface current densities on the boundary, respectively. Note that along the text we use different notations for σ_s and $\mathbf{J}_s$. However no confusion should arise, as these quantities are always properly defined.

2.2 Drude Model

Shortly after J. J. Thomson's discovery of the electron, Paul Drude developed a model for the description of the electrical and thermal conduction in metals, borrowing the basic ideas from the successful kinetic theory of gases [Drude (1909)]. The Drude model attempts to explain the elementary properties of metals by regarding a metal as being essentially a *free electron gas*. Within this description, the loosely bound conduction electrons form an electron gas moving against a lattice of widely spaced, small ionic cores (encompassing both the nucleus and the tightly bound core electrons) in response to an applied electromagnetic field. The only interaction is via (instantaneous) collisions of electrons with the fixed ions, which occur with probability per unit time τ^{-1}, where τ is a phenomenological parameter known as the relaxation time. Therefore, the classical equation of motion for an electron in the metal subjected to an external electric field $\mathbf{E}(t) = \mathbf{E}_0 e^{-i\omega t}$, becomes [Ashcroft and Mermin (1976)]

$$\frac{d\mathbf{p}(t)}{dt} = -\frac{\mathbf{p}(t)}{\tau} - e\mathbf{E}(t) , \tag{2.27}$$

where $e > 0$ is the elementary charge. We are now looking for a steady-state solution exhibiting the same harmonic time dependence of the driving field, so that the Fourier transform of equation (2.27) yields

$$-i\omega\mathbf{p}(\omega) = -\frac{\mathbf{p}(\omega)}{\tau} - e\mathbf{E}(\omega) , \tag{2.28}$$

whose solution is

$$\mathbf{p}(\omega) = \frac{e\mathbf{E}(\omega)}{i\omega - \tau^{-1}} . \tag{2.29}$$

Moreover, notice that the current density due to n_e electrons per unit volume, travelling with net velocity $\mathbf{v}$, is $\mathbf{J} = -en_e\mathbf{v}$. Hence, one can write

$$\mathbf{J}(\omega) = -\frac{en_e\mathbf{p}(\omega)}{m_e} = \frac{e^2 n_e}{m_e}\frac{1}{\gamma - i\omega}\mathbf{E}(\omega) , \tag{2.30}$$

where m_e is the mass of the electron and we have introduced the scattering rate, defined as $\gamma = \tau^{-1}$. By comparing the result (2.30) with the relation (2.9), an expression for the optical conductivity can be determined:

$$\sigma(\omega) = \frac{e^2 n_e/m_e}{\gamma - i\omega} . \tag{2.31}$$

It will be useful for what follows to note that the conductivity and the dielectric function are linked through the following relationship[2]

$$\epsilon(\omega) = 1 + i\frac{\sigma(\omega)}{\epsilon_0\omega} , \tag{2.32}$$

allowing us to write

$$\boxed{\epsilon(\omega) = 1 - \frac{\omega_p^2}{\omega^2 + i\gamma\omega}} , \tag{2.33}$$

where $\omega_p^2 = \frac{e^2 n_e}{m_e \epsilon_0}$ denotes the *plasma frequency* of the free electron gas.

Despite its simplicity, the Drude model provides fairly good results for

---

[2]This can be readily shown by working through Maxwell's equation using the constitutive relation (2.30) [Ashcroft and Mermin (1976)].

many metals (specially alkali metals) in the infrared region of the electromagnetic spectrum, where their optical properties are frequently dominated by the role of intraband transitions. For noble metals, the model fails to a great extent due to the onset of interband transitions in the visible and ultraviolet regions, and to other details of the band structure. The interband transitions can be accounted for, while maintaining the classical picture, by adding a resonant contribution (whose resonant frequency corresponds to the transition energy) to the equation of motion, leading to a dielectric function which also contains a Lorentz-oscillator term.

Nevertheless, a fully and accurate description of the physics governing the optical response of metals requires a quantum mechanical formulation. Still, this is often circumvented by employing clever effective/phenomenological parameters and/or extra terms added "by hand".

Let us end this section with a note on the longitudinal and transverse dielectric functions. From Maxwell's equations it is not difficult to show [Li and Chu (2014)] that each component of the electric field obeys the wave equation $[k^2\delta_{ij} - k_i k_j - \epsilon_{ij}\omega^2/c^2]E_j = 0$, where k is the modulus of the wavevector $\mathbf{k}$, ω is the frequency of the radiation, c is the velocity of light and ϵ_{ij} is the dielectric tensor. Expressing, for an isotropic medium, the dielectric tensor[3] as $\epsilon_{ij} = (\delta_{ij} - k_i k_j/k^2)\epsilon_t + k_i k_j/k^2\epsilon_l$, where ϵ_t and ϵ_l are the transverse and longitudinal components of the dielectric tensor, respectively. We can write a vector equation for the electric field in the form

$$\mathbf{k} \times (\mathbf{k} \times \mathbf{E})(k^2 - \omega^2\epsilon_t/c^2) + \mathbf{k}(\mathbf{k} \cdot \mathbf{E})\epsilon_l = 0. \qquad (2.34)$$

Note that for finite $\mathbf{k}$ the dielectric function is a tensor even for an isotropic system [Wooten (1972)]. From the previous equation it is clear that a nontrivial solution for the existence of longitudinal waves requires the condition $\epsilon_l(\omega, \mathbf{k}) = 0$, where we have made explicit the dependence of the dielectric function on frequency and wavevector.[4] For a two-dimensional system, the solution of the previous condition allow us to find the spectrum of the surface plasmons of the system [see, e.g., Eq. (4.61)]. For transverse fields the condition for their existence is $[k^2 - \omega^2\epsilon_t(\omega, \mathbf{k})/c^2] = 0$.

---

[3]This is the most general tensor that can be built out of the vector $\mathbf{k}$ when the system is isotropic. For another way to intuit this result see footnote in Appendix D.

[4]Note that for a bulk (three-dimensional) system, the previous condition gives the spectrum of the bulk plasmons (more on this on Chapter 3).

For a two-dimensional system, such as graphene, the condition $\epsilon_l(\omega, \mathbf{k}) = 0$ grant us with the spectrum of the surface plasmons.

2.3 Preliminaries to Graphene Plasmonics

Graphene is a two-dimensional allotrope of carbon, where the carbon atoms are arranged in a honeycomb lattice. Being one-atom-thick, it can be regarded as a truly 2D material. Since its isolation in 2004 [Novoselov *et al.* (2004)], graphene has been capitalizing a lot of the scientific community's attention due to its remarkable properties: it is stronger than steel and harder than diamond, but also flexible [Lee *et al.* (2008)]; it exhibits an extremely large thermal conductivity [Balandin *et al.* (2008)], and sustains high current densities; its charge carriers are massless Dirac fermions obeying a Dirac-like relativistic equation (more on this in what follows), showing ballistic transport with large electronic mobilities [Bolotin *et al.* (2008)] even at room temperature; and the list of properties goes on [Geim (2009)]. Owing to these unique properties, graphene has been dubbed a "*wonder material*". Whether graphene will fulfill all the promises for new applications remains to be seen. For a prehistoric perspective about graphene see [Geim (2012)].

2.3.1 *Elementary electronic properties*

For the sake of completeness and self-containedness of the text, this section introduces a few basic notions about graphene's electronic and optical properties. However, this section is not meant to be a comprehensive presentation.

The electronic configuration of a carbon atom is

$$[\mathrm{C}] = 1\mathrm{s}^2 2\mathrm{s}^2 2\mathrm{p}^2 \,. \tag{2.35}$$

Now, in what concerns graphene, three electrons from the $2s$ and $2p$ orbitals participate in the σ-bonds resulting from the hybridization of the sp^2 orbitals, which in turn originate from the hybridization of the $2s$, $2p_x$, and $2p_y$ atomic orbitals. The remaining electron populates the $2p_z$ orbital, and the hybridization of these orbitals originates the π-bonds, which lie above and below the graphene lattice. Thus, from the point of view of the π-orbitals, graphene is a half-filled electronic system, since each carbon atom has an associated electron. The hopping of the electrons along the π-orbitals allow us to define a tight-binding Hamiltonian for the system, from which the electronic and optical properties of the material follow.

The minimal model Hamiltonian considers electron-hopping between nearest neighbors carbon atoms in the graphene lattice (see Fig. 2.1). In

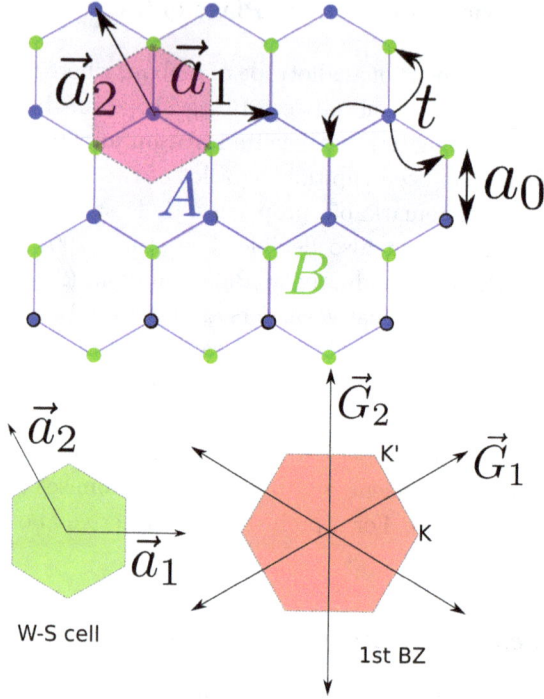

Figure 2.1: Real space and Brillouin zone of graphene. Top: Real space lattice of graphene together with the Wigner-Seitz cell (in pink color). The parameters $t \simeq -2.7$ eV and $a_0 \simeq 1.42$ Å are the hopping integral and the carbon-carbon distance, respectively. The vectors $\boldsymbol{a_1} = a(1,0)$ and $\boldsymbol{a_2} = a(-1/2, \sqrt{3}/2)$ are the primitive lattice-vectors, where $a = a_0\sqrt{3}$. Bottom: Wigner-Seitz cell (in green) and Brillouin zone (in pink) of graphene. The two special points $\mathbf{K}$ and $\mathbf{K'}$ are termed Dirac points. The vectors $\boldsymbol{G_1} = \frac{4\pi}{a\sqrt{3}}\left(\frac{\sqrt{3}}{2}, \frac{1}{2}\right)$ and $\boldsymbol{G_2} = \frac{4\pi}{a\sqrt{3}}(0,1)$ are the primitive reciprocal-lattice vectors.

the Wannier basis, this Hamiltonian is written [Castro Neto *et al.* (2009); Peres (2010); Abergel *et al.* (2010)] as

$$H = t \sum_{\boldsymbol{R}} \{|A, \boldsymbol{R}\rangle \left[\langle B, \boldsymbol{R} + \boldsymbol{\delta}_1| + \langle B, \boldsymbol{R} + \boldsymbol{\delta}_2| + \langle B, \boldsymbol{R} + \boldsymbol{\delta}_3|\right] + \text{H.c.}\} ,$$

(2.36)

where $|A, \boldsymbol{R}\rangle$ is the Wannier state in the carbon atom located at position $\boldsymbol{R}$, in the sub-lattice A, and an equivalent definition holds for $|B, \boldsymbol{R} + \boldsymbol{\delta}_i\rangle$, where $\boldsymbol{\delta}_i$, with $i = 1, 2, 3$, are the vectors connecting a given carbon atom in sub-lattice A with its the three nearest neighbors carbon atoms in sub-lattice B. Here, H.c. stands for the Hermitian-conjugate. Next, we introduce the representation of the Wannier states in terms of the Bloch states, of wavevector $\mathbf{k} = (k_x, k_y)$, in the above Hamiltonian, thereby obtaining

$$H = t \sum_{\mathbf{k}} [|A, \mathbf{k}\rangle\langle B, \mathbf{k}| \phi(\mathbf{k}) + \text{H.c.}] , \quad (2.37)$$

which can also be written as $H = \sum_{\mathbf{k}} \Phi_{\mathbf{k}}^\dagger H_k \Phi_{\mathbf{k}}$, where $\Phi_{\mathbf{k}}^\dagger = [|A, \mathbf{k}\rangle, |B, \mathbf{k}\rangle]$,

$$H_k = \begin{bmatrix} 0 & t\phi(\mathbf{k}) \\ t\phi^*(\mathbf{k}) & 0 \end{bmatrix} , \quad (2.38)$$

and

$$\phi(\mathbf{k}) = e^{ik_y a_0} + e^{ik_x \sqrt{3}a_0/2} e^{-ik_y a_0/2} + e^{-ik_x \sqrt{3}a_0/2} e^{-ik_y a_0/2} , \quad (2.39)$$

where $a_0 = a/\sqrt{3}$ is the carbon-carbon distance as defined in Fig. 2.1. It follows from Eq. (2.38) that the spectrum of the π-electrons is then given by

$$E_{\mathbf{k}} = \pm t|\phi(\mathbf{k})| , \quad (2.40)$$

or, more explicitly,

$$E_{\mathbf{k}} = \pm t\sqrt{3 + 2\cos\left(\sqrt{3}k_x a_0\right) + 4\cos\left(\frac{\sqrt{3}}{2}k_x a_0\right)\cos\left(\frac{3}{2}k_y a_0\right)} . \quad (2.41)$$

Thus, we conclude that the spectrum is particle-hole symmetric. Furthermore, since the system is half-filled, the valence band $E_{\mathbf{k}} = -t|\phi(\mathbf{k})|$ is completely filled and the conduction band $E_{\mathbf{k}} = t|\phi(\mathbf{k})|$ is empty. A plot of the band structure of graphene is given in Fig. 2.2. From this figure it is clear that the valence and conduction bands touch each other at six points in the Brillouin zone. Since each point is shared by three Brillouin zones, we have two independent Dirac points (or Dirac cones, named after the shape of the spectrum in their vicinity; see right panel of Fig. 2.2).

When we zoom-in close to one of the Dirac points, we see that the bands are conical. This hints that the low energy (around the Dirac point) Hamiltonian cannot be described by the usual effective-mass approximation used for semiconductors. Let us now derive the low energy Hamiltonian. We start by writing $\mathbf{k} = \mathbf{K} + \mathbf{q}$, where $\mathbf{K}$ is the wavevector of the Dirac point and $|\mathbf{q}|/|\mathbf{K}| \ll 1$, from where it follows that

$$\phi(\mathbf{k}) = e^{i\mathbf{q}\cdot\boldsymbol{\delta}_1} + e^{i2\pi/3} e^{i\mathbf{q}\cdot\boldsymbol{\delta}_2} + e^{-i2\pi/3} e^{i\mathbf{q}\cdot\boldsymbol{\delta}_3} . \quad (2.42)$$

Expanding $\phi(\mathbf{k})$ to first order in $\mathbf{q}$, we have

$$\phi(\mathbf{k}) \approx i\mathbf{q}\cdot\boldsymbol{\delta}_1 + e^{i2\pi/3} i\mathbf{q}\cdot\boldsymbol{\delta}_2 + e^{-i2\pi/3} i\mathbf{q}\cdot\boldsymbol{\delta}_3 , \quad (2.43)$$

and using the explicit values of $\boldsymbol{\delta}_i$, we thereby obtain

$$\phi(\mathbf{k}) \approx \frac{3}{2} a_0 i q_y - \frac{3}{2} a_0 q_x . \quad (2.44)$$

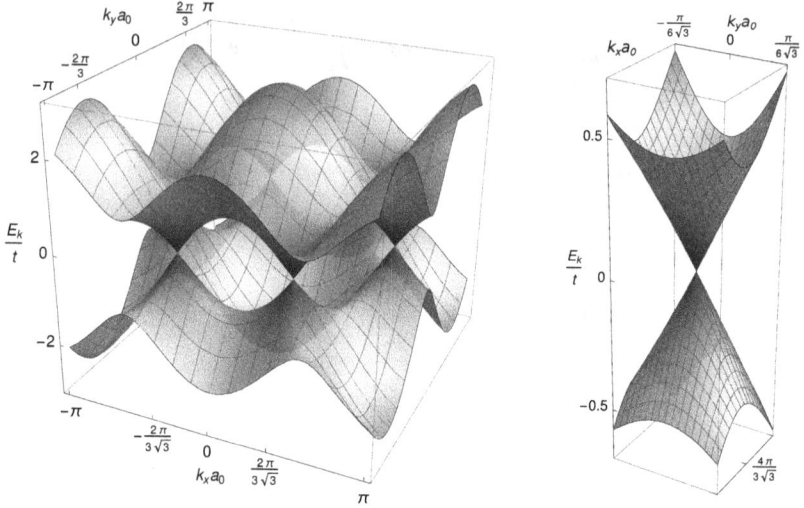

Figure 2.2: Band structure of graphene, as given by $E_{\mathbf{k}} = \pm t |\phi(\mathbf{k})|$ [see Eqs. (2.40) and (2.41)]. Left: note that the valence and conduction bands touch each other at six-points in the Brillouin zone. These particular points of high-symmetry are termed Dirac points and correspond to the momentum values given by $\mathbf{K}$ and $\mathbf{K}'$ (see Fig. 2.1). Right: close-up of the band structure near one of the Dirac points.

With this, the Hamiltonian (2.38) then becomes

$$H_k \approx \frac{3}{2} a_0 t \begin{bmatrix} 0 & -q_x + i q_y \\ -q_x - i q_y & 0 \end{bmatrix}, \qquad (2.45)$$

which leads to the spectrum $E_\pm(\mathbf{q}) \approx \pm \frac{3 t a_0}{2} |\mathbf{q}| = v_F p$ (up to first order in q/K), where $p = \hbar q$, and $v_F \approx c/300$ is the Fermi velocity, where c refers to the speed of light in vacuum. Notice that the electronic spectrum can be written only in terms of q, that is, the momentum measured with respect to the Dirac points. The Hamiltonian H_k is written in momentum space. We now want to write the corresponding Hamiltonian in real space. Since H_k is linear in $\mathbf{q}$, intuition dictates that we should replace the momentum by spatial derivatives. We are then led to write

$$\boxed{H = \frac{3}{2} a_0 t \begin{bmatrix} 0 & i \partial_x + \partial_y \\ i \partial_x - \partial_y & 0 \end{bmatrix} = -\frac{3 t a_0}{2\hbar} (\sigma_x p_x + \sigma_y p_y) = -v_F \boldsymbol{\sigma} \cdot \boldsymbol{p},}$$

$$(2.46)$$

where σ_x and σ_y are the Pauli matrices and the momentum operator is $\boldsymbol{p} = (p_x, p_y)$. Clearly, the Hamiltonian (2.46) resembles the massless Dirac

Hamiltonian in two spatial dimensions. This justifies the terminology used above when referring to $\mathbf{K}$ and $\mathbf{K}'$ as "Dirac points". We can now work backwards and solve the eigenvalue problem $H\psi = E\psi$, where H is given by Eq. (2.46). To this end, we write

$$\psi_{\mathbf{q},\alpha}(\mathbf{r}) = \frac{1}{\sqrt{A}}\begin{pmatrix} a_\alpha \\ b_\alpha \end{pmatrix} e^{i\mathbf{q}\cdot\mathbf{r}} = \frac{1}{\sqrt{A}}\chi_\alpha e^{i\mathbf{q}\cdot\mathbf{r}}, \qquad (2.47)$$

leading to an eigenvalue problem of the form

$$\frac{3}{2}a_0 t \begin{bmatrix} 0 & q_x - iq_y \\ q_x + iq_y & 0 \end{bmatrix}\begin{pmatrix} a_\alpha \\ b_\alpha \end{pmatrix} = E_{\mathbf{q},\alpha}\begin{pmatrix} a_\alpha \\ b_\alpha \end{pmatrix}, \qquad (2.48)$$

whose solution is

$$\chi_\alpha = \frac{1}{\sqrt{2}}\begin{pmatrix} e^{-i\theta_\mathbf{q}/2} \\ \alpha e^{i\theta_\mathbf{q}/2} \end{pmatrix}, \qquad (2.49)$$

with eigenvalues

$$\boxed{E_\pm(\mathbf{q}) = \pm\hbar v_F q}, \qquad (2.50)$$

where in the above equations, A stands for the area of the system, $\alpha = \mathrm{sgn}(E)$ is the band index, and $\theta_\mathbf{q} = \arctan(q_y/q_x)$. As Eq. (2.50) plainly demonstrates, the dispersion of the Dirac fermions in graphene is linear with momentum, in the vicinity of the Dirac points. This result has many important implications, and will also have an impact on the dispersion of plasmons in graphene, as will become clear in later chapters.

As an application of the previous solution, let us compute the transmittance of pristine graphene (a more complex application of the above results is done in Chap. 10). The physical idea is simple: an electron in the valence band is excited to the conduction band by the absorption of a photon of angular frequency ω. Since the conduction band is empty and the system has no energy gap, the electron can absorb a photon of an arbitrary low frequency. The rate per unit area at which the material absorbs energy from the electromagnetic field can be computed from Fermi's golden rule [Zettili (2009)].

The coupling of the electrons to the electromagnetic field is made via the usual minimal coupling. Within this framework, we write the Hamiltonian as

$$H = v_F \boldsymbol{\sigma} \cdot (\boldsymbol{p} + e\boldsymbol{A}), \qquad (2.51)$$

where $\boldsymbol{A}$ is the vector potential and $e > 0$ the elementary charge. We now choose the form of the vector potential as

$$\boldsymbol{A} = \frac{1}{2}(A_0, 0)(e^{i\omega t} + e^{-i\omega t}), \qquad (2.52)$$

which is appropriate for describing a monochromatic plane wave. In the spirit of first-order time-dependent perturbation theory, the perturbation, V, reads

$$V = e v_F \sigma_x \frac{A_0}{2} \, . \tag{2.53}$$

The wavefunctions to be used in the calculation of the matrix element of the perturbation are [cf. Eqs. (2.47) and (2.49)]

$$\psi_-(\mathbf{r}) = \frac{1}{\sqrt{2A}} \begin{pmatrix} e^{-i\theta_\mathbf{q}/2} \\ -e^{i\theta_\mathbf{q}/2} \end{pmatrix} e^{i\mathbf{q} \cdot \mathbf{r}} \, , \tag{2.54}$$

for the valence band and

$$\psi_+(\mathbf{r}) = \frac{1}{\sqrt{2A}} \begin{pmatrix} e^{-i\theta_\mathbf{q}/2} \\ e^{i\theta_\mathbf{q}/2} \end{pmatrix} e^{i\mathbf{q} \cdot \mathbf{r}} \, , \tag{2.55}$$

for the conduction band. Thus, the matrix element of the perturbation can be readily determined,

$$\langle \psi_+ | V | \psi_- \rangle = -A_0 \frac{i v_F e}{2} \sin \theta \, , \tag{2.56}$$

so that the transition rate, as given by Fermi's golden rule, follows:

$$W_{+-}(k) = \frac{2\pi}{\hbar} \frac{v_F^2}{4} e^2 A_0^2 \sin^2 \theta \, \delta(2\hbar v_F k - \hbar\omega) \, . \tag{2.57}$$

The rate of transition, $W_{+-}(k)$, is associated to a single momentum value, $p = \hbar k$. We now have to sum over all the momentum values in the Brillouin zone, and account for the spin degeneracy and for the existence of two Dirac points as well (also dubbed as valley degeneracy). Then, the total transition rate is

$$\frac{1}{\tau} = 4 \int \frac{d\mathbf{k}}{(2\pi)^2} W_{+-}(k) = \frac{e^2 |A_0|^2 \omega}{8\hbar^2} \, , \tag{2.58}$$

where the factor of four is due to the spin and valley degeneracies mentioned above. Therefore, from the previous equation, the power per unit area absorbed by the material reads

$$W_a = \frac{\hbar\omega}{\tau} = \frac{e^2 E_0^2}{8\hbar} \, , \tag{2.59}$$

where we have expressed the amplitude of the vector potential A_0 in terms of the corresponding amplitude of the external electric field E_0 ($A_0 = -i E_0/\omega$). On the other hand, the power flux per unit area contained in the field is given by

$$W_i = \frac{1}{2} c \epsilon_0 E_0^2 \, . \tag{2.60}$$

Having the quantities (2.59) and (2.60), we can obtain the transmittance of graphene via

$$\boxed{T = 1 - \frac{W_a}{W_i} = 1 - \pi\alpha \simeq 97.7\%}, \tag{2.61}$$

where $\alpha = e^2/(4\pi\epsilon_0\hbar c)$ is the fine structure constant of atomic physics. From the previous result, we conclude that the transmittance of graphene is independent of the frequency of the impinging radiation and also independent of any other material's parameters. Naturally, this result holds

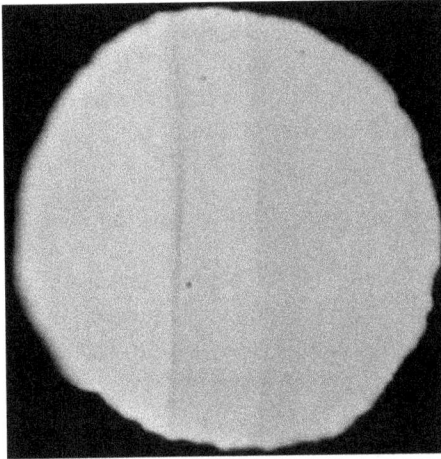

Figure 2.3: A hole drilled in a metallic scaffold is partially covered by graphene and its bilayer. The hole is illuminated by white light from behind. The covering of the hole is, from left to right, air, monolayer graphene, bilayer graphene. We can see by contrast the presence of the single and bilayer graphene sheets, due to the absorption of 2.3%/layer of the impinging radiation. The greyish colour is a consequence of the independence of the transmittance on the frequency of the incoming radiation (image courtesy of A. K. Geim).

true only as long as the momentum expansion around the Dirac point remains valid, and only applies to neutral (undoped) graphene. In Fig. 2.3 we show a graphene sheet (and its bilayer) partially covering a hole drilled in a metallic scaffold. Due to the absorption of 2.3% of the impinging radiation (reflection is negligible) we can see with the bare eye a single graphene sheet.

Interesting enough, the previous result for the transmittance allow us to compute the conductivity of graphene. It is known that the transmittance through a thin conducting film relates to the conductivity through[5]

$$T = 1/|1 + \sigma(\omega)/(2c\epsilon_0)|^2 \,, \tag{2.62}$$

where ϵ_0 is the vacuum permittivity. Since we have computed T we can invert this latter equation and obtain $\sigma(\omega)$. This procedure leads to[6]

$$\boxed{\sigma(\omega) = \pi e^2/2h \equiv \sigma_0}\,, \tag{2.63}$$

that is, a constant value, independent of the material parameters (a situation not very common in condensed matter). If the effect of temperature is included, then the conductivity of graphene has the form

$$\sigma(\omega, T) = \sigma_0 \tanh\left(\frac{\hbar\omega}{4k_B T}\right)\,, \tag{2.64}$$

where the hyperbolic tangent is a consequence of the Fermi distribution function. For doped graphene at finite temperature, the previous expression further transforms to

$$\sigma(\omega, T) = \frac{1}{2}\sigma_0\left[\tanh\left(\frac{\hbar\omega + 2E_F}{4k_B T}\right) + \tanh\left(\frac{\hbar\omega - 2E_F}{4k_B T}\right)\right]\,, \tag{2.65}$$

where E_F is the Fermi energy. This latter expression includes only interband transitions. The intraband transitions can be described by the Drude model. The conductivity of graphene will be discussed ahead in more detail, specially the case of doped graphene, which we have considered here only in passing.

In this book, we shall focus on graphene as being a plasmonic material. As far as graphene plasmonics is concerned, all we need to describe its electromagnetic properties and light-matter interactions is encompassed within graphene's optical conductivity (which in turn could be extended to include non-local effects—more on this in Sec. 2.3.3). For a detailed treatment of graphene physics, please refer to [Geim (2009); Castro Neto *et al.* (2009); Peres (2010)].

2.3.2   *The optical conductivity of graphene*

Along with its linear dispersion, graphene's optical conductivity is the key ingredient in graphene optics and plasmonics, since it contains all the relevant information on the physics governing the electromagnetic interactions

---

[5]See derivation in Appendix N.

[6]The reader should not confuse σ_0 with the quantum of conductance. σ_0 is termed the universal AC conductivity of graphene.

between graphene and external stimuli such as electromagnetic radiation or fast electrons.

In order to better characterize the conductivity, it is useful to split this physical quantity into two distinct contributions: one describing intraband transitions (transitions within the conduction (or valence) band where the momentum is not conserved), and another accounting for interband transitions (vertical transitions from the valence to the conduction band, in which case there is momentum conservation), namely

$$\sigma_g(\omega) = \sigma_{\text{intra}}(\omega) + \sigma_{\text{inter}}(\omega) \ . \tag{2.66}$$

In writing the last expression it is implicit that we will ignore spatial dispersion (i.e. non-local effects) henceforth (later, non-local effects will be introduced). An analytical expression for the dynamical conductivity of graphene can be obtained under the framework of linear response theory and is computed via Kubo's formula [Falkovsky and Varlamov (2007); Falkovsky (2008); Stauber *et al.* (2008); Wunsch *et al.* (2006); Gusynin *et al.* (2009)], using some sort of approximation. Traditionally the bare bubble [Mahan (2000)], which is related to the charge-charge susceptibility (or polarizability), is computed. Within the latter approach we have[7]

$$\sigma_{\text{intra}}(\omega) = \frac{\sigma_0}{\pi} \frac{4}{\hbar\gamma - i\hbar\omega} \left[E_F + 2k_B T \ln\left(1 + e^{-E_F/k_B T}\right) \right] \ , \tag{2.67}$$

characterizing intraband processes, where $\sigma_0 = e^2/(4\hbar)$, k_B is Boltzmann's constant, γ is the relaxation rate (see discussion below), and E_F the Fermi energy, while for interband transitions we have

$$\sigma_{\text{inter}}(\omega) = \sigma_0 \left[G(\hbar\omega/2) + i\frac{4\hbar\omega}{\pi} \int_0^\infty dE \frac{G(E) - G(\hbar\omega/2)}{(\hbar\omega)^2 - 4E^2} \right] \ , \tag{2.68}$$

with

$$G(x) = \frac{\sinh\left(\dfrac{x}{k_B T}\right)}{\cosh\left(\dfrac{E_F}{k_B T}\right) + \cosh\left(\dfrac{x}{k_B T}\right)} \ . \tag{2.69}$$

At zero temperature, or as long as the condition $E_F \gg k_B T$ is fulfilled, the above equations reduce to

$$\sigma_{\text{intra}}(\omega) = \frac{\sigma_0}{\pi} \frac{4E_F}{\hbar\gamma - i\hbar\omega} \ , \tag{2.70}$$

$$\sigma_{\text{inter}}(\omega) = \sigma_0 \left[\Theta(\hbar\omega - 2E_F) + \frac{i}{\pi} \ln\left|\frac{\hbar\omega - 2E_F}{\hbar\omega + 2E_F}\right| \right] \ , \tag{2.71}$$

---

[7]A derivation of Eqs. (2.67) and (2.68) is given in Appendix B.

where $\Theta(x)$ stands for the Heaviside step-function. Notice that for $2E_F > \hbar\omega$ the term describing interband transitions gives a negligible contribution to the conductivity due to Pauli blocking. Therefore, we note that for frequencies in the terahertz (THz) and mid-IR spectral region, at room temperature, and under typical doping levels, graphene's optical conductivity is mostly dominated by the term describing intraband electronic processes, following a Drude-like expression [Stauber *et al.* (2007); Peres *et al.* (2007)]

$$\sigma_g(\omega) \approx \frac{\sigma_0}{\pi} \frac{4E_F}{\hbar\gamma - i\hbar\omega}. \tag{2.72}$$

The results for the (local) optical conductivity of graphene are summarized in Fig. 2.4. We note, however, that in some conditions the previous expressions for the local conductivity are not enough and we are forced to use an expression for graphene's dynamical conductivity including non-local effects[8]. For example, it can be shown that the value of the decay length of surface plasmons as they propagate along a graphene sheet, depends on whether we use the local (homogeneous) or the non-local (non-homogeneous) conductivity. Another important remark regards the interpretation of γ in the previous expressions for the conductivity. From an experimental point of view, γ should be considered as a phenomenological parameter that one uses to fit the experiments and that includes the different pathways by which an excited electron can relax to the ground state. On the other hand, from a theoretical point a view, the calculation of γ from first principles is an important and complex task, but which, in the end, should agree with the experimentally obtained value. Clearly, γ will depend on intrinsic aspects of the physics of graphene, such as electron-impurity, electron-phonon, and electron-electron interactions [Principi *et al.* (2013)], but it will also depend on extrinsic aspects, such as the substrate on which graphene is deposited. Whether the calculated value of γ will agree with its phenomenological counterpart remains an open question.

We now turn to the discussion of the non-local conductivity of graphene, as computed from Kubo's formula. In addition, we will also consider an extension of the bare polarizability that includes the effects of relaxation due to γ.

---

[8]See Appendix C for a derivation of the non-local optical conductivity of graphene in the context of Drude model. We shal use these results in Chap. 10.

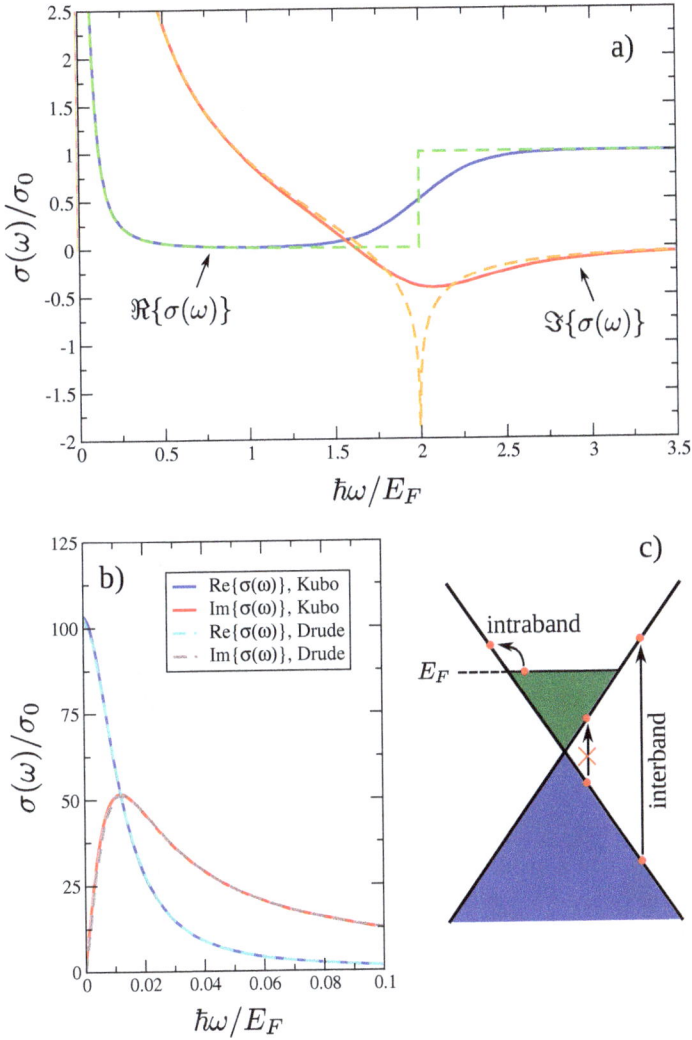

Figure 2.4: Optical conductivity of graphene: a) real and imaginary parts of the conductivity computed using the local (that is, in long-wavelength limit) Kubo expression [Eqs. (2.67) and (2.68)] at $T = 300$ K (solid lines) and at $T = 0$ K (dashed lines); b) graphene's optical conductivity in the THz regime, obtained via Kubo's formula (solid lines) and using Drude's expression (2.72) (dashed); c) schematic illustration of the transition processes in doped graphene. Parameters: $E_F = 0.3$ eV and $\hbar\gamma = 3.7$ meV. Cross over the arrows indicate processes prevented by Pauli blocking.

2.3.3   *Lindhard function: Beyond the local approximation*

In the preceding subsection we have shown that, for typical values of doping ($E_F \sim$0.2-0.5 eV), the conductivity of graphene in the THz and mid-IR is well described by the Drude model expression, given by Eq. (2.72), and ignoring non-local effects. This conclusion is also supported by a number of experimental works [Lee *et al.* (2011); Rouhi *et al.* (2012); Sule *et al.* (2014)], which have reported that in the spectral range spanning 100 GHz – 10 THz the optical response of the material is well described by the Drude model [Sule *et al.* (2014)]. These results do give confidence in the theoretical description of SPP in graphene based on the Drude model (in that window of the electromagnetic spectrum, and up to the mid-IR for samples with higher doping concentrations).

While the simple description of the optical (frequency-dependent) conductivity of graphene suffices for most experimental and theoretical scenarios, one needs to be equipped with a more general theory for situations falling outside the aforementioned regime or in circumstances where q-dependent, that is non-local effects, emerge, such as in the case of graphene surface plasmons (GSPs) with large wavevectors approaching $k_F = E_F/(\hbar v_F)$ and for a quantum emitter in the vicinity of a graphene sheet (see Chap. 10).

One of the most popular models in the nanophotonics community capable of delivering the dependence of the linear response on the entire (q,ω)-space is the Lindhard dielectric function [Lindhard (1954)], which can be obtained from the polarization bubble. The Lindhard dielectric function is also often dubbed as the dielectric function in the random-phase approximation (RPA) [Mahan (2000)]. This theoretical framework follows from a microscopic, quantum mechanical (perturbative) description, whose main assumption is that the electrons respond not to the bare Coulomb potential, V_C, but to an *effective* potential — $V_{eff}(\mathbf{q}, \omega) = V_C(\mathbf{q}, \omega) + V_s(\mathbf{q}, \omega)$ (here written in Fourier space) — comprising both V_C and the screening potential, V_s, produced by the electronic polarization, resulting from the dynamical rearrangement of charges in response to V_C. Note that this model includes electron-electron interactions by virtue of the screening interaction term.

The random-phase approximation can be obtained either by writing the Hamiltonian of the electron gas in the presence of an external potential and solving the equation of motion[9] using a self-consistent method for the

---

[9]The derivation follows along the lines of Appendix F, which provides the calculation of the same quantity for double-layer graphene.

potential [Ehrenreich and Cohen (1959)], or by the diagrammatic technique using Green's functions [Mahan (2000)]. In what follows, we will summarize the procedure towards the calculation of the bare polarizability (or non-interacting dynamical susceptibility) of graphene, which will allow us to compute the RPA dielectric function. The interested reader is thus advised to refer to G. Mahan's textbook on many-body physics [Mahan (2000)] for further details (where the RPA results for the 3D electron gas are given). In addition, a thorough description of the derivation of the 2D polarizability of graphene can be found in [Wunsch *et al.* (2006); Hwang and Sarma (2007)] and in the textbook by Jishi [Jishi (2014)] (see also [Mihaila (2011)]).

Generically, the retarded density-density correlation function may be defined as[10]

$$\chi_{\rho\rho}(\mathbf{r} - \mathbf{r}', t - t') = -\frac{i}{\hbar}\Theta(t - t')\langle[n(\mathbf{r}, t), n(\mathbf{r}', t')]\rangle , \qquad (2.73)$$

together with its momentum space version,

$$\chi_{\rho\rho}(\mathbf{q}, t - t') = -\frac{i}{\hbar}\Theta(t - t')\langle[n(\mathbf{q}, t), n(-\mathbf{q}, t')]\rangle , \qquad (2.74)$$

corresponding to the Fourier transformed response function with respect to the spatial coordinate, with $n(\mathbf{q}, t) = \sum_{\mathbf{k}\sigma} c_{\mathbf{k}\sigma}^{\dagger}(t)c_{\mathbf{k}+\mathbf{q}\sigma}(t)$ being the electronic density operator written in the second quantization formalism[11]. In the language of quantum field theories, Eq. (2.74) simply relates the probability amplitude for an electron-hole pair created at time t' with momentum $\mathbf{q}$ (and spin σ), to be found at a later time t with the same momentum (and spin). During its propagation, the pair can experience a number of scattering processes owing to electron-electron interactions (e.g. represented by self-energies and vertex correction diagrams [Fetter and Walecka (2003); Mahan (2000); Abedinpour *et al.* (2011)]). The random-phase approximation ignores these corrections, and approximates the proper density-density correlation function by the summation of all bare bubble diagrams connected by Coulomb lines. Working from Eq. (2.74) and to lowest order in perturbation theory, one arrives to the 2D polarizability[12] at finite doping, defined by the non-interacting (bare) bubble diagram as [Wunsch *et al.*

---

[10]The retarded "two-point" correlation function preserves causality (the system's response is only triggered *after* the perturbation is switched on, hence the name). We further assume a translational invariant system, so that the response function only depends on the difference between spatial variables, i.e. $\chi_{\rho\rho}(\mathbf{r}, \mathbf{r}', t-t') \equiv \chi_{\rho\rho}(\mathbf{r}-\mathbf{r}', t-t')$.
[11]This form of $n(\mathbf{q}, t)$ holds for the jellium model [Fetter and Walecka (2003)].
[12]An explicit calculation of the integrals in Eq. (2.75) is given in Appendix A.

(2006); Shung (1986); Hwang and Sarma (2007); Jablan *et al.* (2009)]

$$\chi^0_{\rho\rho}(\mathbf{q},\omega) \equiv P(\mathbf{q},\omega)$$

$$= \frac{1}{\pi^2} \int d^2k \sum_{\alpha,\alpha'=\pm 1} F_{\alpha,\alpha'}(\mathbf{k},\mathbf{q}) \frac{f_\alpha(k) - f_{\alpha'}(|\mathbf{k}+\mathbf{q}|)}{\hbar\omega + E_\alpha(k) - E_{\alpha'}(|\mathbf{k}+\mathbf{q}|) + i\eta} \,,$$

$$(2.75)$$

with the energies being $E_\pm(k) = \pm\hbar v_F k - E_F$, and $f_\pm(\mathbf{k}) = \left[e^{E_\pm(k)/k_B T} + 1\right]^{-1}$ expresses the Fermi-Dirac distribution, and where the form factor

$$F_{\alpha,\alpha'}(\mathbf{k},\mathbf{q}) = \frac{1}{2}\left(1 + \alpha\alpha' \frac{k^2 + kq\cos\theta}{k|\mathbf{k}+\mathbf{q}|}\right) \,, \qquad (2.76)$$

comes from the overlap of the wavefunctions of Dirac electrons [see Eq. (2.47)], and θ refers to the angle formed by the wavevectors $\mathbf{k}$ and $\mathbf{q}$. Additionally, we note that in writing Eq. (2.75) we have already included the spin ($g_s = 2$) and valley ($g_v = 2$) degeneracies. At $T = 0$ K the Fermi distributions reduce to step functions, easing the evaluation of the integral in Eq. (2.75). Furthermore, one may apply Sokhotsky's formula[13] to the polarizability (2.75), thereby separating its real and imaginary parts,

$$P(\mathbf{q},\omega) = P_{re}(\mathbf{q},\omega) + iP_{im}(\mathbf{q},\omega) \,, \qquad (2.77)$$

and then computing $P_{re}(q,\omega)$ and $P_{im}(q,\omega)$ separately. After some lengthy algebra (see Appendix A), these read:

$$P_{re}(x,y) = \begin{cases} -\frac{2k_F}{\pi\hbar v_F} + \frac{1}{4\pi}\frac{k_F}{\hbar v_F}\frac{x^2}{\sqrt{y^2-x^2}}\left[C_h\left(\frac{y+2}{x}\right) - C_h\left(\frac{2-y}{x}\right)\right] \text{, in region 1B} \\ -\frac{2k_F}{\pi\hbar v_F} \text{, in region 1A} \\ -\frac{2k_F}{\pi\hbar v_F} + \frac{1}{4\pi}\frac{k_F}{\hbar v_F}\frac{x^2}{\sqrt{y^2-x^2}}C_h\left(\frac{y+2}{x}\right) \text{, in region 2B} \\ -\frac{2k_F}{\pi\hbar v_F} + \frac{1}{4\pi}\frac{k_F}{\hbar v_F}\frac{x^2}{\sqrt{x^2-y^2}}C\left(\frac{2-y}{x}\right) \text{, in region 2A} \end{cases}$$

$$(2.78)$$

for the real part of the polarizability, and

$$P_{im}(x,y) = \begin{cases} 0 \text{, in region 1B} \\ \frac{1}{4\pi}\frac{k_F}{\hbar v_F}\frac{x^2}{\sqrt{x^2-y^2}}\left[C_h\left(\frac{2-y}{x}\right) - C_h\left(\frac{y+2}{x}\right)\right] \text{, in region 1A} \\ \frac{1}{4\pi}\frac{k_F}{\hbar v_F}\frac{x^2}{\sqrt{y^2-x^2}}C\left(\frac{2-y}{x}\right) \text{, in region 2B} \\ -\frac{1}{4\pi}\frac{k_F}{\hbar v_F}\frac{x^2}{\sqrt{x^2-y^2}}C_h\left(\frac{y+2}{x}\right) \text{, in region 2A} \end{cases}$$

$$(2.79)$$

---

[13]Sokhotsky's result: $\lim_{\eta\to 0^+}\frac{1}{x+i\eta} = \mathcal{P}\left(\frac{1}{x}\right) - i\pi\delta(x)$, where $\mathcal{P}$ denotes the Cauchy principal value [Vladimirov (1971)].

for the corresponding imaginary part. Note that both quantities are defined by branches, whose domains are depicted in 2.5. Also, in writing Eqs. (2.78) and (2.79) we have defined $x = q/k_F$, $y = \hbar\omega/E_F$, as well as the (real) auxiliary functions

$$C_h(a) = a\sqrt{a^2 - 1} - \text{arccosh}(a) \ , \tag{2.80}$$

$$C(a) = a\sqrt{1 - a^2} - \arccos(a) \ . \tag{2.81}$$

At this point, let us anticipate that long-lived plasmons can only exist in region 1B, that is, where $\Im\{P(\mathbf{q}, \omega)\} = 0$. Outside this region, the imaginary part of the polarizability (which relates to absorption) is non-zero and the plasmons become severely damped, quickly decaying into excited electron-hole pairs. For this reason, zones 2B, 2A and 1A are usually designated by electron-hole continuum or Landau damping [14] regions (interband Landau damping takes place in region 2B, whilst areas 2A and 1A lie in the intraband Landau damping regime—see Fig. 2.5).

The results contained in Eqs. (2.77)-(2.79) describe the bare polarizability of doped graphene, in the whole (q, ω)-plane depicted in Fig. 2.5. Having finally obtained the response function $P(\mathbf{q}, \omega)$, we find ourselves just a stone's throw away from building the RPA dielectric function and the full, non-local, conductivity. Once in possession of these quantities, we have all the essential ingredients to fully characterize and compute the spectrum of plasmonic excitations in graphene.

The RPA dielectric function can be written in terms of the Lindhard polarization as [Mahan (2000)]

$$\varepsilon^{\text{RPA}}(\mathbf{q}, \omega) = \epsilon_r - \nu_q P(\mathbf{q}, \omega) \ , \tag{2.82}$$

---

[14]This terminology is somewhat loose. Strictly speaking, Landau damping is a reversible process occurring when the phase velocity of an electromagnetic wave (SPP, in our case) matches that of a free electron. The electrons moving slightly slower than the wave are accelerated and remove energy from the wave. The condition for it to occur is $v_{\text{phase}} = \hbar\omega/q \leq v_F$ (where v_F is the Fermi velocity). Thus, the region on the right of the line $\hbar\omega = \hbar q v_F$ is where Landau damping can take place (see regions 1A and 2A in Fig. 2.5). Alternatively, SPPs can lose their energy by creating electron-hole pairs. Contrary to usual electromagnetic wave (free radiation), SPPs have large momenta and can participate in indirect (or non-vertical) interband electron-hole transitions. The minimal energy of an electron-hole pair in doped graphene is $E_{e-h} = 2E_F - \hbar q v_F$, where q is the total momentum of the pair (see Fig. 2.5 – region 2B). Therefore it becomes possible to satisfy the momentum and energy conservation restriction for creation of an electron-hole pairs at the expenses of one SPP when $\hbar\omega > 2E_F - \hbar q v_F$ and $q \leq k_F$. This process is somewhat different from Landau damping. It is just resonant absorption of plasmons by generation of electron-hole pairs. However, and despite this comment, we will keep using the terminology "Landau damping" in a broad sense to conform with its use in the literature on plasmons in graphene.

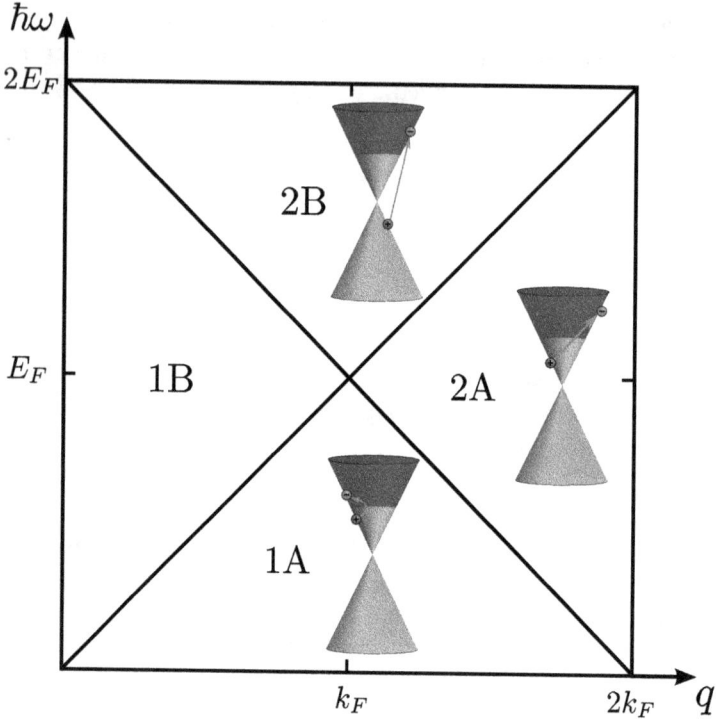

Figure 2.5: Regions of the $(q, \hbar\omega)$-space corresponding to the polarizability as given by Eqs. (2.78) and (2.79). Some of the processes characterizing the polarizability are also portrayed in a pictorial fashion.

where ϵ_r is the relative permittivity of the medium in which graphene is embedded, and $\nu_q = \frac{e^2}{2\epsilon_0 q}$ stands for the Fourier transform of the Coulomb interaction (here ϵ_0 denotes the permittivity of free space). Collective excitations coincide with the zeros of the dynamical dielectric function, that is, they correspond to solutions of the equation $\varepsilon^{\mathrm{RPA}}(\mathbf{q}, \omega) = 0$. With this end in view, it is possible to calculate the dipersion of GSPs via [Wunsch *et al.* (2006); Jablan *et al.* (2009)]

$$\Re\{P(q, \omega)\} = \frac{2\epsilon_r \epsilon_0}{e^2} q \,, \tag{2.83}$$

which is valid for weak damping and has solutions for $\Re\{P(q, \omega)\} > 0$ (fulfilled in regions 1B and 2B) as is the case of doped graphene. However, Eq. (2.83) is only solvable by numerical means. This is an implicit equation which allows the determination of $\omega(q)$.

In parallel, one may find preferable to use the non-local dynamical conductivity instead, to describe the electronic processes and light-matter interactions. This can be achieved by writing the longitudinal conductivity in terms of the density-density response function, through the relation[15]

$$\sigma(q,\omega) = ie^2 \frac{\omega}{q^2} P(q,\omega) . \qquad (2.84)$$

This expression is an extension of the local optical (frequency-dependent only) conductivity, given by Eqs. (2.70) and (2.71), to the case of an arbitrary wavevector. For this reason it is often referred to as the full non-local conductivity.

To conclude this subsection, we note that Eqs. (2.82) and (2.84) encode the fundamental physics describing the optoelectronic interactions in graphene (within the linear response framework outlined above), and, in particular, constitute the pivotal ingredients to investigate plasmon-related phenomena in this material. These closed-form expressions enable us to determine the system's response for any frequency and wavevector in the interval $\frac{q}{k_F}, \frac{\hbar\omega}{E_F} \in [0,2]$, at any doping concentration.

2.3.4 *Lindhard polarization function in relaxation-time approximation: The case of graphene*

In the preceding paragraphs we have introduced the Lindhard polarization in the case where electronic scattering processes have been ignored. Following that description, we then presented the general lines of the derivation of the non-local dielectric function and conductivity [which stem from the polarizability via Eqs. (2.82) and (2.84)] for the particular case of graphene.

Moreover, we have seen that the behavior of the polarizability is strongly dependent on the specific region of the (q,ω)-plane under consideration. This is due to the fact that each of these regions is characterized by different processes contributing to the density-density response function — see, for instance, Fig. 2.5. Notice that in the analysis made above, the only mechanisms which can yield plasmonic losses (affecting the GSPs-lifetime) correspond to processes involved in the so-called Laudau damping; these are, in essence, *intrinsic* processes which mostly depend on the physics governing the material's electronic properties. On the other hand, *extrinsic* processes such as impurity scattering or lattice defects, cannot be accounted by that elementary description[16].

---

[15] A detailed derivation is presented in Appendix D.
[16] In this paragraph, we borrowed the language of [Lin and Liu (2014)]. Notwithstanding, let us remark that a literal interpretation can be misleading: for example, scattering

A possible route to take extrinsic processes into account, inspired on Boltzmann's semi-classical equation for electronic transport, is to employ the relaxation-time approximation [Ashcroft and Mermin (1976)]. Within this approximation, collisions are introduced by a phenomenological parameter $\gamma = \tau^{-1}$, known as scattering rate[17]. This extension of the Lindhard function leads to the same quantity in the so called relaxation-time approximation (RT). Since the naive replacement $\omega \to \omega + i\gamma$ fails to conserve the local electron number[18], one needs to do some extra work in order to eliminate this artifact. To circumvent this defect, Mermin proposed that the collisions relax the electronic density to a local equilibrium distribution, basically an equilibrium distribution with a shifted chemical potential (varying in space and time), in such a way that it satisfies the continuity equation [Mermin (1970)]. Using this prescription, we arrive at the following expression for the 2D polarizability in the RT approximation [Mermin (1970); Jablan *et al.* (2009)]:

$$P_\gamma(q,\omega) = \frac{(1+i\gamma/\omega)\,P(q,\omega+i\gamma)}{1+i\gamma/\omega \cdot P(q,\omega+i\gamma)/P(q,0)} . \qquad (2.85)$$

A derivation of this result for a single band system was given by Das [Das (1975, 1974)] (see also [Nakajima (1955)] for the formalism used by Das). In Appendix E we provide a derivation of Eq. (2.85) for the case of graphene. Naturally, the Lindhard dielectric function and the non-local conductivity in the relaxation-time approximation follows from Eqs. (2.82), (2.84) and Eq. (2.85), upon substitution of $P(q,\omega)$ by $P_\gamma(q,\omega)$, that is:

$$\varepsilon^{\mathrm{RPA-RT}}(q,\omega) = \epsilon_r - \nu_q P_\gamma(q,\omega) , \qquad (2.86)$$

and

$$\sigma(q,\omega) = ie^2 \frac{\omega}{q^2} P_\gamma(q,\omega) , \qquad (2.87)$$

respectively.

A contour plot of graphene's 2D polarizability in the RT approximation is shown in Fig. 2.6 [similar to the one presented in [Wunsch *et al.* (2006)]

---

with intrinsic phonons is, by the above classification, an *intrinsic* process, albeit it could be encapsulated by the *extrinsic* phenomenological scattering parameter γ (see next paragraph).

[17]Note that we have also introduced this parameter earlier in this Chapter – cf. Sec. 2.2. Boltzmann's approach goes beyond the Drude model by considering Fermi-Dirac statistics.

[18]Alternatively, we can show that $P(q,\omega+i\gamma)$ is complex when $\omega \to 0$, when it should be purely real since it corresponds to the static response function.

for the collisionless case, i.e. $P(q, \omega)$]. Note that the imaginary part of the polarizability is always negative, a consequence of $P_\gamma(q, \omega)$ being a retarded response function. The real part of the same quantity has not a fixed sign.

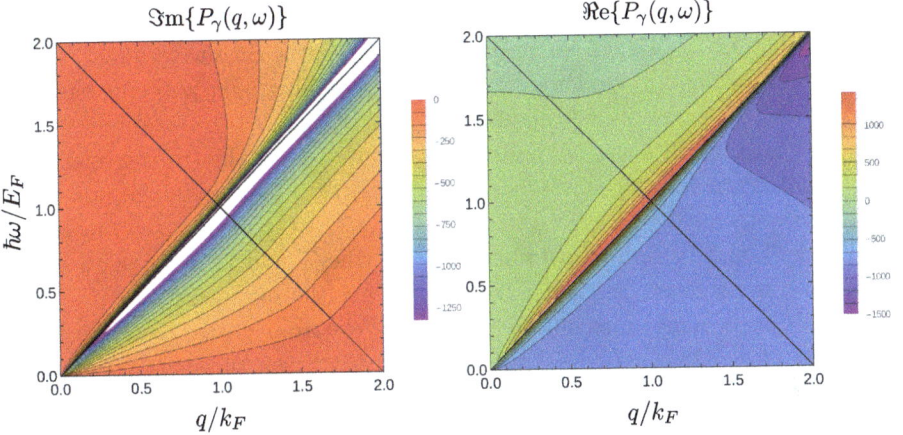

Figure 2.6: Real and imaginary parts of the 2D polarizability of graphene in the RT approximation, as given by Eq. (2.85). Left: contour plot of $\Im m\{P_\gamma(q, \omega)\}$; right: contour plot of $\Re e\{P_\gamma(q, \omega)\}$. We have used $\hbar\gamma = 16.5$ meV which corresponds to 4 THz [Ju *et al.* (2011)].

2.4 The Transfer-Matrix Method and the First Appearance of Plasmons in Graphene

The transfer-matrix method constitutes a fundamental tool in photonic sciences to investigate the optical response of layered media —including dielectrics, conductors, or even combinations of both. The fundamental concept behind this technique is the ability of building a "transfer-matrix" that links the electromagnetic fields at a given point X in terms of the corresponding fields at another point X'. It is particularly useful, for instance, in the computation of the transmission and reflection coefficients (and thus also absorption) of a structured material composed of several layers. In addition, it is easy to implement numerically, even when disorder is included. Another advantage of this technique is the computation of the band-structure of photonic crystals, which are the photonic versions of crystalline solids in solid state physics. Here too —i.e. in condensed matter physics— an electronic version of the transfer-matrix technique is widely employed in the context of electronic transport in solids, namely for determining the probability of transmitting an electron through a given device. In the context of condensed matter, one deals with electronic wavefunctions and the physics is governed by Schrödinger's equation, whereas in the photonics case one works with electromagnetic waves and the underlying dynamics stems from Maxwell's equations (in the classical regime). In both scenarios, infinite periodic structures exhibit electronic/photonic band gaps where propagation is forbidden.

Still, despite the many uses and advantages of the transfer-matrix (T-matrix) method, in this book we are primarily interested in plasmon-related phenomena[19]. Therefore, in what follows, we will first set the basis of the T-matrix formalism while deriving the T-matrix for single-, double- and multi-layer graphene, thereby finding the expressions for the transmittance, reflectance and absorbance of electromagnetic radiation in such systems. We then reveal the connection between the matrix elements of the T-matrix and the spectrum of graphene plasmons via the poles of the Fresnel reflection coefficient.

2.4.1 *Transfer-matrix for a graphene monolayer*

Let us first describe the electromagnetic scattering at a single-layer graphene interface, in which the graphene sheet is sandwiched by two semi-infinite dielectric media characterized by the relative permittivities ϵ_1 and ϵ_2, as depicted in Fig. 2.7. The T-matrix method accounts both for incoming waves (and corresponding reflected and transmitted waves) coming from above (i.e. from medium 1) and from below (i.e. from medium 2) the graphene interface.

We shall consider the electromagnetic fields in the form of transverse magnetic (i.e. TM or p-polarized) plane waves. Therefore, we write the magnetic field as superposition of forward- and backward-traveling waves,

$$\mathbf{B}_y^{(j)}(\mathbf{r}, t) = \left(A_j e^{ik_{j,z}z} + B_j e^{-ik_{j,z}z} \right) e^{i(qx - \omega t)} \, \hat{\mathbf{y}} \,, \qquad (2.88)$$

for waves in the medium j. The corresponding electric field follows from Maxwell's equations [cf. Sec. 2.1.1], for which its x-component reads

$$\mathbf{E}_x^{(j)}(\mathbf{r}, t) = \frac{k_{j,z}c^2}{\omega \epsilon_j} \left(A_j e^{ik_{j,z}z} - B_j e^{-ik_{j,z}z} \right) e^{i(qx - \omega t)} \, \hat{\mathbf{x}} \,, \qquad (2.89)$$

---

[19]For a good introduction to the transfer-matrix technique in photonics, the reader is advised to check the book of [Markos and Soukoulis (2008)]. For the transfer-matrix method in the context of graphene see [Zhan *et al.* (2013)].

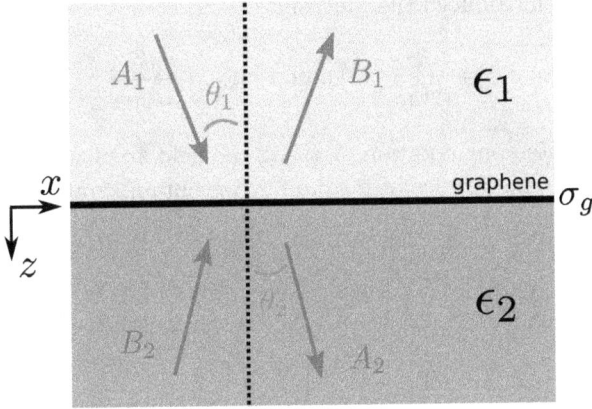

Figure 2.7: Electromagnetic scattering at a single graphene 2D layer. The graphene sheet is located at the plane defined by $z = 0$, and cladded between two semi-infinite dielectric media characterized by the relative permittivities ϵ_1 and ϵ_2 as indicated in the figure. The electromagnetic properties of graphene are encompassed by its conductivity, σ_g.

where the projection of the wavevector along $\hat{\mathbf{z}}$ satisfies

$$k_{j,z}^2 = \epsilon_j \frac{\omega^2}{c^2} - q^2 . \tag{2.90}$$

Having introduced the expressions for the electromagnetic fields, our task will be to find the relations between field amplitudes in adjacent dielectric media. In order to do that, we need to invoke the boundary conditions, expressing both the continuity of the tangential component of the **E**-field across the graphene interface and the discontinuity of $\mathbf{B}_y$ owing to the surface currents in the graphene monolayer, that is,

$$E_x^{(1)}(z = 0) = E_x^{(2)}(z = 0) , \tag{2.91}$$

$$B_y^{(1)}(z = 0) - B_y^{(2)}(z = 0) = \mu_0 \sigma_g E_x^{(1)}(z = 0), \tag{2.92}$$

where σ_g is the graphene optical conductivity. Using Eqs. (2.88) and (2.89) in the above boundary conditions, yields

$$A_1 - B_1 = \frac{\epsilon_1 k_{2,z}}{\epsilon_2 k_{1,z}} (A_2 - B_2) , \tag{2.93}$$

$$A_1 + B_1 = \frac{\sigma_g k_{2,z}}{\omega \epsilon_0 \epsilon_2} (A_2 - B_2) + A_2 + B_2 . \tag{2.94}$$

We now convert this system of linear equations for the amplitudes into matrix form,

$$\begin{pmatrix} 1 & -1 \\ 1 & 1 \end{pmatrix} \begin{pmatrix} A_1 \\ B_1 \end{pmatrix} = \begin{pmatrix} \eta & -\eta \\ 1 + \xi_\sigma & 1 - \xi_\sigma \end{pmatrix} \begin{pmatrix} A_2 \\ B_2 \end{pmatrix} , \tag{2.95}$$

where we have introduced the functions

$$\eta = \frac{\epsilon_1 k_{2,z}}{\epsilon_2 k_{1,z}} \quad \text{and} \quad \xi_\sigma = \frac{\sigma_g k_{2,z}}{\omega \epsilon_0 \epsilon_2} \,, \tag{2.96}$$

in order to lighten our notation. Writing the field amplitudes of medium 1 in terms of the amplitudes of medium 2, we obtain, from Eq. (2.95), the following result

$$\begin{pmatrix} A_1 \\ B_1 \end{pmatrix} = \frac{1}{2} \begin{pmatrix} 1 + \eta + \xi_\sigma & 1 - \eta - \xi_\sigma \\ 1 - \eta + \xi_\sigma & 1 + \eta - \xi_\sigma \end{pmatrix} \begin{pmatrix} A_2 \\ B_2 \end{pmatrix}, \tag{2.97}$$

$$= \mathbf{T}_{1 \to 2} \begin{pmatrix} A_2 \\ B_2 \end{pmatrix}. \tag{2.98}$$

Here, the matrix connecting the field amplitudes in adjacent dielectrics — that is, one in each side of the graphene layer— is the transfer-matrix across a single graphene, namely

$$\boxed{\mathbf{T}_{1 \to 2} = \frac{1}{2} \begin{pmatrix} 1 + \eta + \xi_\sigma & 1 - \eta - \xi_\sigma \\ 1 - \eta + \xi_\sigma & 1 + \eta - \xi_\sigma \end{pmatrix}} \,. \tag{2.99}$$

This result constitutes one of the fundamental building blocks from which we can now compute the optical properties of any complex structure composed by an arbitrary number of graphene layers, or even a periodic configuration of graphene layers separated by insulators akin to a graphene photonic crystal or metamaterial.

The other basic building block is the propagation matrix, which accounts for the free-propagation of waves within a dielectric slab. We shall introduce this matrix later when dealing with the graphene double-layer.

Before moving on, we further note that the result for the T-matrix across an interface between two media in which the graphene sheet is absent, can be fetched from Eq. (2.99) by taking the limit when $\sigma_g \to 0$, producing

$$\mathbf{T}_{1 \to 2}^0 = \frac{1}{2} \begin{pmatrix} 1 + \eta & 1 - \eta \\ 1 - \eta & 1 + \eta \end{pmatrix}. \tag{2.100}$$

In possession of the T-matrix for a graphene monolayer, Eq. (2.99), we can now readily determine the transmission and reflection coefficients of electromagnetic radiation impinging onto a graphene layer, along with the spectrum of plasmonic excitations in that material.

2.4.2 Transmittance, reflectance and absorbance of electro-magnetic radiation by a graphene monolayer

Since the relation between the field amplitudes above and below the graphene sheet are embedded in the T-matrix $\mathbf{T}_{1\to2}$, the expressions for the scattering coefficients can be straightforwardly obtained from the matrix elements[20] of the 2×2 matrix $\mathbf{T}_{1\to2}$; in particular, via

$$t = \frac{1}{T_{11}^{1\to2}} \quad \text{and} \quad r = \frac{T_{21}^{1\to2}}{T_{11}^{1\to2}}, \tag{2.101}$$

respectively for the transmission and reflection coefficients. From these, the expressions for the transmittance and reflectance across the graphene single-layer are given by

$$\mathcal{T}_{SLG} = \frac{\epsilon_1 k_{2,z}}{\epsilon_2 k_{1,z}} \left| \frac{1}{T_{11}^{1\to2}} \right|^2 = \frac{\epsilon_1 k_{2,z}}{\epsilon_2 k_{1,z}} \left| \frac{2}{1 + \eta + \xi_\sigma} \right|^2, \tag{2.102}$$

and

$$\mathcal{R}_{SLG} = \left| \frac{T_{21}^{1\to2}}{T_{11}^{1\to2}} \right|^2 = \left| \frac{1 - \eta + \xi_\sigma}{1 + \eta + \xi_\sigma} \right|^2, \tag{2.103}$$

in the same order. At normal incidence, the above quantities read

$$\mathcal{T}_{SLG}^{\text{normal}} = \sqrt{\frac{\epsilon_1}{\epsilon_2}} \left| \frac{2\sqrt{\epsilon_2}}{\sqrt{\epsilon_1} + \sqrt{\epsilon_2} + \frac{\sigma_g}{c\epsilon_0}} \right|^2, \tag{2.104}$$

$$\mathcal{R}_{SLG}^{\text{normal}} = \left| \frac{\sqrt{\epsilon_2} - \sqrt{\epsilon_1} + \frac{\sigma_g}{c\epsilon_0}}{\sqrt{\epsilon_2} + \sqrt{\epsilon_1} + \frac{\sigma_g}{c\epsilon_0}} \right|^2, \tag{2.105}$$

which is a well-know result [Markos and Soukoulis (2008)]. Figure 2.8 illustrates the behavior of the transmittance, reflectance, and absorbance as a function of the impinging wave's frequency, at normal incidence for suspended doped graphene. The latter quantity follows from the other two via

$$\mathcal{A}_{SLG} = 1 - \mathcal{T}_{SLG} - \mathcal{R}_{SLG}, \tag{2.106}$$

Note that at frequencies in the THz up to the mid-IR spectral range the system's response is dominated by graphene's absorption akin to the so-called Drude peak. However (and rather remarkably), at visible wavelengths, the absorbance is essentially frequency-independent with a constant value of approximately 2.3% —cf. the figure's inset and recall Eq. (2.61). The reader should keep in mind the results of Fig. 2.8 when the moment to compare these with the plasmonic response of graphene arises.

---

[20]We use the notation $\mathbf{T}_{1\to2} = \begin{pmatrix} T_{11}^{1\to2} & T_{12}^{1\to2} \\ T_{21}^{1\to2} & T_{22}^{1\to2} \end{pmatrix}$.

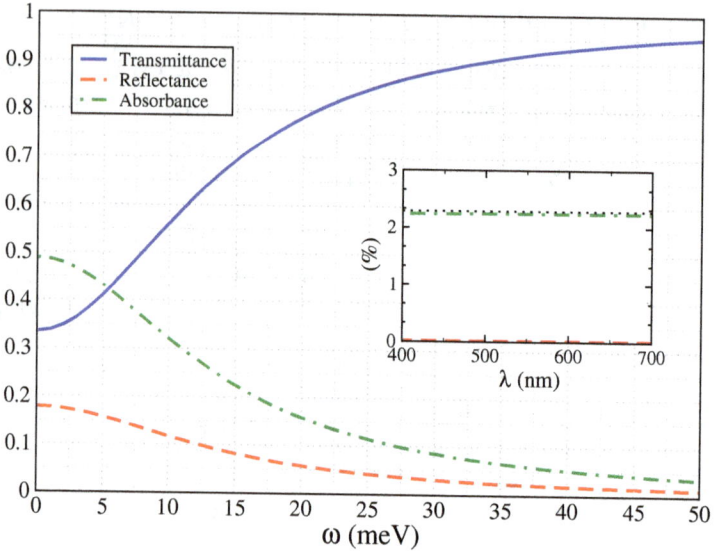

Figure 2.8: Transmittance, reflectance and absorbance of electromagnetic radiation at normal incidence through a suspended ($\epsilon_1 = \epsilon_2 = 1$) graphene monolayer, as a function of the frequency of the impinging radiation. The inset shows the same quantities (in percentage) in the visible region of the electromagnetic spectrum; the black-dotted line indicates the value $\sim \alpha\pi \simeq 2.3\%$, characteristic of the "universal"/frequency-independent absorption of single-layer graphene. Here, graphene's conductivity is modeled by Kubo's formula [Eqs. (2.70) and (2.71)], with $E_F = 0.4$ eV and $\hbar\gamma = 5$ meV.

2.4.3 *Transfer-matrix for a graphene double-layer*

The transfer-matrix method is a very powerful tool in the context of multilayer structures. For instance, after the derivation of the T-matrix ascribed to the propagation of electromagnetic radiation across a single graphene layer, the T-matrix for a graphene double-layer immediately follows from the former. Here, by double-layer graphene we mean two graphene layers separated by a slab of dielectric material with dielectric constant ϵ_1, and clad by two semi-infinite insulators as illustrated in Fig. 4.8. Indeed, and without any further calculations, we can readily write the T-matrix for double-layer graphene as[21]

$$\boxed{\mathbf{T}_{1\to3} = \mathbf{T}_{1\to2} \cdot \mathbf{P}_2(d) \cdot \mathbf{T}_{2\to3}}\,, \qquad (2.107)$$

where the "dot" denotes matrix multiplication, and $\mathbf{P}_2$ is a propagation matrix describing the free wave-propagation along a dielectric layer of length d and characterized by a relative permittivity ϵ_2; explicitly we have

$$\mathbf{P}_2(d) = \begin{pmatrix} e^{-ik_{2,z}d} & \\ 0 & e^{ik_{2,z}d} \end{pmatrix}. \qquad (2.108)$$

This propagation matrix links the amplitudes of the fields at a given z-coordinate, say, $z_0 + d$, with those at the previous point z_0 (that is, a point located at a distance d behind).

Therefore, the structure of Eq. (2.107) is easy to understand. The matrix $\mathbf{T}_{1\to2}$ is just the T-matrix describing the propagation of light across a single graphene sheet (i.e. the same that we have studied in Sec. 2.4.1). After the first graphene layer, the electromagnetic radiation needs to propagate along an insulating slab of thickness d before reaching the second graphene layer; this is accounted for by the matrix $\mathbf{P}_2(d)$. Finally, we just need to consider the effect of the second graphene layer by multiplying the previous matrices from the right with $\mathbf{T}_{2\to3}$, which is the same T-matrix of single-layer graphene, but now describing light's propagation from medium 2 to medium 3. This description is portrayed in Fig. 2.9.

From Eq. (2.107), and as before, the transmission, reflection and absorption probabilities are evaluated using the matrix elements of $\mathbf{T}_{1\to3}$, that is,

$$\mathcal{T}_{DLG} = \frac{\epsilon_1 k_{3,z}}{\epsilon_3 k_{1,z}} \left| \frac{1}{T_{11}^{1\to3}} \right|^2, \qquad (2.109)$$

$$\mathcal{R}_{DLG} = \left| \frac{T_{21}^{1\to3}}{T_{11}^{1\to3}} \right|^2, \qquad (2.110)$$

$$\mathcal{A}_{DLG} = 1 - \mathcal{T}_{DLG} - \mathcal{R}_{DLG}. \qquad (2.111)$$

Notice that these equations have exactly the same form as Eqs. (2.102), (2.103) and (2.106). In fact, this structure is maintained independently of the number of graphene layers and of the type of dielectric slabs. Naturally, for this reason, and also benefiting from the fact that we can construct *any*

---

[21]The functions η and ξ_σ introduced when deriving $\mathbf{T}_{1\to2}$ are straightforward to generalize when writing $\mathbf{T}_{2\to3}$ in full form. More explicitly, one just needs to replace the indices by performing the substitutions $1 \to 2$ and $2 \to 3$ (and so on in other multi-layer configurations).

Figure 2.9: Electromagnetic scattering at a graphene double-layer (here depicted at normal incidence). The two graphene sheets are located at the $z = 0$ and $z = d$ planes, and are separated by a dielectric slab (with ϵ_2) of thickness d. The process leading to the construction of the overall T-matrix for the double-layer configuration is also illustrated on the left, where the individual constitutent matrices are shown.

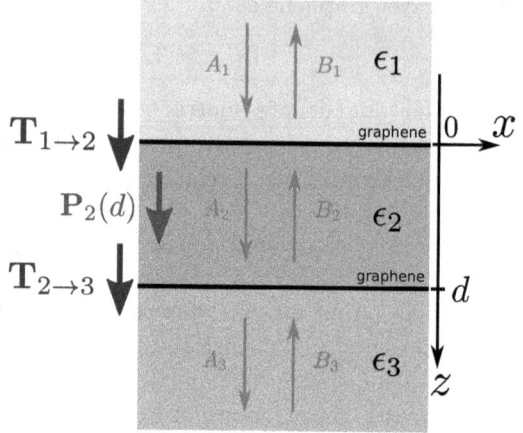

N-layered structure from only two elementary matrices —the T-matrix for the propagation across a graphene layer, $\mathbf{T}_{i \to k}$, and the free-propagation matrix $\mathbf{P}_j(d_j)$—, the potential of the transfer-matrix technique to describe multi-layered photonic systems is, as already noted, obviously enormous.

2.4.4 *Transfer-matrix for multi-layer graphene structures*

As mentioned above, once we have determined the T-matrix for the transmission across one graphene layer, one can now easily extend our study to multi-layer graphene structures and/or graphene-based photonic crystals.

To illustrate this claim, let us consider a layered structure composed by N graphene layers separated by insulating slabs with thickness d and dielectric constant ϵ_2 —see, for instance, Fig. 2.10. The whole system is assumed to be embedded in a homogeneous material with relative permittivity ϵ_1. From the figure, it becomes apparent that the transfer-matrix of the *total* system can be built in terms of the propagation matrices $\mathbf{P}_i$ and $\mathbf{T}_{j \to k}$, yielding

$$\mathbf{T}_{tot} = \mathbf{T}_{1 \to 2} \cdot (\mathbf{T}_{u.c.})^{N'} \cdot \mathbf{P}_2 \cdot \mathbf{T}_{1 \to 2}^{-1} , \qquad (2.112)$$

where the matrix

$$\mathbf{T}_{u.c.} = \mathbf{P}_2 \cdot \mathbf{T}_{2 \to 2} , \qquad (2.113)$$

repeats itself N'-times. We bring to the reader's attention that in writing these expressions the proper handling of the phases related to different

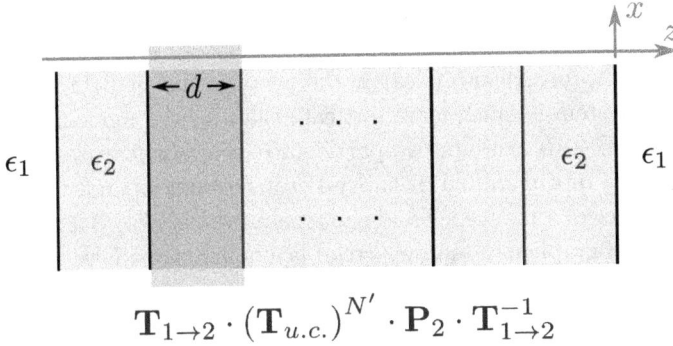

$$\mathbf{T}_{1\to2} \cdot (\mathbf{T}_{u.c.})^{N'} \cdot \mathbf{P}_2 \cdot \mathbf{T}_{1\to2}^{-1}$$

Figure 2.10: Pictorial representation of a composite structure containing N-layers of graphene separated by dielectric slabs with thickness d and dielectric constant ϵ_2. The system is embedded in a homogeneous medium with ϵ_1. Here, N' is defined from $N = N' + 2$. The shaded region highlights the unit-cell (u.c.) of the structure.

positions along the z-axis is implicit[22] (this lightens up the notation). The computation of $(\mathbf{T}_{u.c.})^{N'}$ can be carried out efficiently by performing an eigenvectors decomposition of the matrix, that is,

$$(\mathbf{T}_{u.c.})^{N'} = \mathbf{U} \cdot \begin{pmatrix} \lambda_1^{N'} & 0 \\ 0 & \lambda_2^{N'} \end{pmatrix} \cdot \mathbf{U}^{-1}, \qquad (2.114)$$

where λ_1, λ_2 are the eigenvalues of the matrix (2.113) and $\mathbf{U} = (\vec{v_1}, \vec{v_2})$ is the matrix formed by its eigenvectors (here identified by $\vec{v_1}$ and $\vec{v_2}$). Thus, the scattering probabilities directly follow from the matrix elements of $\mathbf{T}_{tot}$:

$$\mathcal{T}_{tot} = \left| \frac{1}{T_{11}^{tot}} \right|^2 \qquad \text{and} \qquad \mathcal{R}_{tot} = \left| \frac{T_{21}^{tot}}{T_{11}^{tot}} \right|^2, \qquad (2.115)$$

along with, $\mathcal{A}_{tot} = 1 - \mathcal{T}_{tot} - \mathcal{R}_{tot}$ (this expression holds for the case where the optical conductivity of graphene is a complex number). The transmittance and reflectance of electromagnetic radiation (in the THz) across a N-layer graphene system like the one depicted in Fig. 2.10 is shown in Fig. 2.11, for $N = 5$ and $N = 50$ graphene sheets. In the figure's top panel, the photonic band-structure related to an infinite (and periodic) graphene-based photonic crystal is also illustrated (the real part of the conductivity

---

[22]In other words, the matrix $\mathbf{P}_2(d)$ as defined in Eq. (2.108) refers *only* to the propagation from $z = 0$ to $z = d$. The generalization to other regions of the z-axis is straightforward.

has been ignored in this calculation, as we meant to present a simple example of the transfer-matrix method). All results were obtained through the transfer-matrix formalism. Clearly, the propagation of light through the multi-layer system is much more intricate than for the monolayer case (cf. Fig. 2.8). Fig. 2.10 exhibits two particularly relevant features: the first is the emergence of a manifold of Fabry-Perot resonances, and the second is the appearance of a photonic band gap, spanning the $2.5 - 3$ THz region, as the number of graphene (and dielectric) layers increases; here, light's propagation is forbidden (in the same way that electronic motion (current) is null for energies within the electronic band gap [where the density of states equals to zero]). While this suppression of light's transmission is already

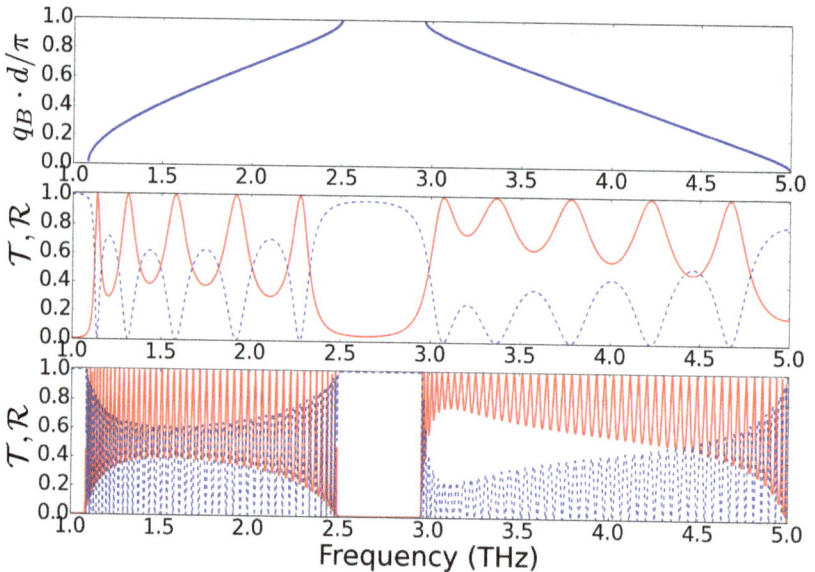

Figure 2.11: Top panel: photonic band-structure for an infinite graphene-based photonic crystal (similar to the one depicted in Fig. 2.10 when $N \to \infty$) in the first Brillouin zone (q_B stands for Bloch's momentum). The remaining panels show the transmittance (solid red) and reflectance (blue; dashed) of THz radiation across a N-layer graphene structure, with $N = 5$ (middle) and $N = 50$ (bottom). Graphene's conductivity is modelled using the local-Kubo formula with negligible damping. Parameters: $d = 30$ μm, $E_F = 0.5$ eV, $\epsilon_1 = 1$ and $\epsilon_2 = 4$ (image courtesy of A. Chaves).

significant in the $N = 5$ case, the effect is particularly noteworthy for the $N = 50$ configuration, for which $\mathcal{T}_{tot} \simeq 0$.

2.4.5 Plasmons: A transfer-matrix approach

Above we have briefly reviewed the transfer-matrix method in the context of light's propagation in single- and multi-layer graphene-based photonic structures.

It turns out that the transfer-matrix technique can also be applied to the description of plasmonic excitations in the aforementioned systems. This should not be entirely surprising, since plasmon resonances can be thought as electromagnetic surface waves which in principle affect light's propagation along the N-layered graphene configuration (provided that the conditions for the excitation of graphene plasmons are met).

Starting from the simplest case of a graphene monolayer clad between two semi-infinite dielectrics, we have seen that the T-matrix for this system is simply the matrix $\mathbf{T}_{1\to 2}$, as given by Eq. (2.99). From this, the Fresnel reflection coefficient immediately follows [cf. Eq. (2.101)],

$$r = \frac{1 - \eta + \xi_\sigma}{1 + \eta + \xi_\sigma} . \tag{2.116}$$

The condition from which the plasmon dispersion is derived corresponds to the poles of the reflection coefficient, and thus the spectrum of graphene plasmons can be determined from it; in particular we have

$$1 + \eta + \xi_\sigma = 0 \quad \Rightarrow \quad 1 + \frac{\epsilon_1 k_{2,z}}{\epsilon_2 k_{1,z}} + \frac{\sigma_g k_{2,z}}{\omega \epsilon_0 \epsilon_2} = 0 . \tag{2.117}$$

Since plasmons correspond to surface waves bounded at the interface (or, for a 2D plasmonic material, localized at the material's plane), then we must have

$$k_{j,z} = i\sqrt{q^2 - \epsilon_j \omega^2 / c^2} \equiv i\kappa_j , \tag{2.118}$$

with $\Re\{\kappa_j\} > 0$, so that Eq. (2.117) becomes

$$1 + \frac{\epsilon_1 \kappa_2}{\epsilon_2 \kappa_1} + \frac{i\sigma_g \kappa_2}{\omega \epsilon_0 \epsilon_2} = 0 , \tag{2.119}$$

$$\Rightarrow \boxed{\frac{\epsilon_1}{\kappa_1} + \frac{\epsilon_2}{\kappa_2} + \frac{i\sigma_g}{\omega \epsilon_0} = 0} , \tag{2.120}$$

which is the condition for the existence of graphene plasmons that we shall obtain in Chapter 4 —cf. Eq. (4.13). Naturally, one can extend this procedure to the graphene double-layer structure, and so on.

A popular way to probe the plasmon dispersion is by producing an intensity plot of the imaginary part of the Fresnel reflection coefficient, whose poles indicate the presence of plasmonic (and other polaritonic) excitations. This approach is illustrated in Fig. 2.12 for a single- and a double-layer graphene configuration (see figure's caption for further details). In the figure, warm colors —i.e. from yellow to red— indicate the region of the phase-space where the condition for the existence of plasmons is fulfilled. A detailed discussion of these results will be given in Chapter 4.

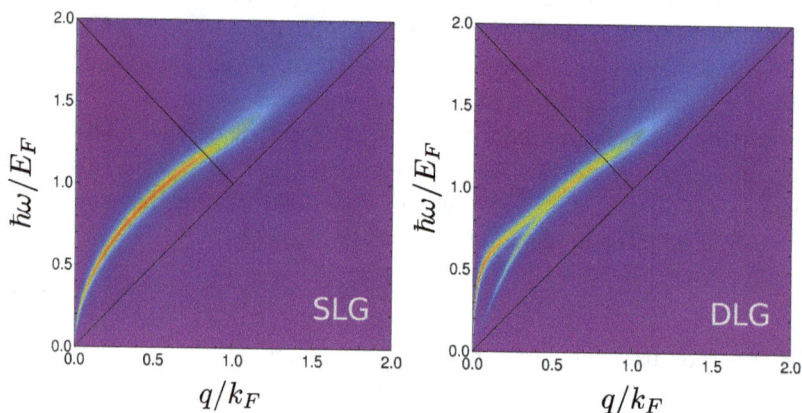

Figure 2.12: Intensity plots of the imaginary part of the Fresnel reflection coefficient, indicating plasmons in single- and double-layer graphene. Left: single-layer graphene (SLG), sandwiched between dielectrics with $\epsilon_1 = 1$ and $\epsilon_2 = 4$. Right: double-layer graphene (DLG), where the insulators possess relative permittivities of $\epsilon_1 = \epsilon_3 = 1$ and $\epsilon_2 = 4$. In both cases, the graphene sheets have $E_F = 0.45$ eV, $\hbar\gamma = 16.5$ meV and the conductivity of the material is modeled by its non-local expression in terms of graphene's irreducible polarizability, using Mermin's prescription to include a (particle-conserving) relaxation-time [see Eq. (2.87)].

Chapter 3

Surface Plasmon-Polaritons at Dielectric-Metal Interfaces

Before discussing plasmonic-related phenomena in graphene structures, it is instructive to review the classical case of light confinement occurring at dielectric-metal interfaces. In this Chapter, we will apply the formalism developed previously to address the question of how the interface between two media with different optical properties can originate surface modes confined to dimensions smaller than the wavelength. This special solution of Maxwell's equations is known as surface plasmon-polaritons.

3.1 Single Dielectric-Metal Interface

We begin by considering the simplest system supporting surface plasmon-polaritons (SPPs): a single (planar) interface between a dielectric and a metal. For convenience, we assume that both media are semi-infinite and isotropic, with the dielectric occupying the half-region defined by $z < 0$, as depicted in Fig. 3.1 (also, note the orientation of the z−axis).

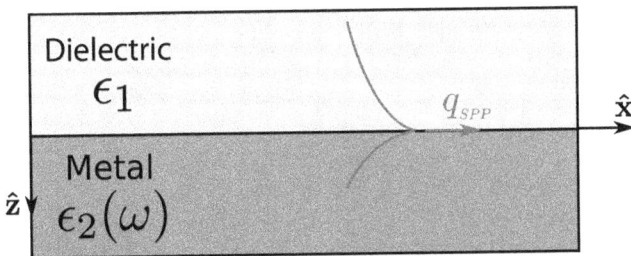

Figure 3.1: Schematic representation of a surface plasmon-polariton at an interface between a dielectric and a metal. The field profile is shown qualitatively for guidance.

The dielectric medium is characterized by a positive real dielectric constant, ϵ_1, whereas the electromagnetic properties of the metal will be taken into account by a complex, frequency-dependent, dielectric function, $\epsilon_2(\omega)$, of the form

$$\epsilon_2(\omega) = 1 - \frac{\omega_p^2}{\omega^2 + i\omega\gamma} \ . \tag{3.1}$$

Recall that equation (3.1) is the dielectric function of the metal within the free electron gas model [cf. equation (2.33)]. This level of approximation suffices for the current purposes as long as the energy of the free space radiation is not high enough to induce inter-band transitions. (Some limitations of the Drude model are discussed in Sec. 3.1.1.2.)

As shown in what follows, in the spectral region where $\Re\{\epsilon_2(\omega)\} < 0$, the dielectric-metal interface supports longitudinal waves that are tightly confined to the interface between the two media (see, for instance, Fig. 3.1). These surface waves are SPP modes.

3.1.1 *Dispersion relation*

In the presence of an interface solutions with longitudinal electric field or magnetic field components are possible. Let us first consider transverse magnetic (TM) solutions. These can be written as

$$\mathbf{E}_j(\mathbf{r}, t) = (E_{j,x}\hat{\mathbf{x}} + E_{j,z}\hat{\mathbf{z}}) \, e^{-\kappa_j|z|} e^{i(qx-\omega t)} \ , \tag{3.2}$$

$$\mathbf{B}_j(\mathbf{r}, t) = B_{j,y}\hat{\mathbf{y}} \, e^{-\kappa_j|z|} e^{i(qx-\omega t)} \ , \tag{3.3}$$

where the index $j = 1, 2$ identifies the medium as in Fig. 3.1. Inserting equations (3.2) and (3.3) into Maxwell's equations (while assuming that the materials are non-magnetic, $\mu_j = \mu_0$)

$$\nabla \times \mathbf{E}_j = -\frac{\partial \mathbf{B}_j}{\partial t} \ , \tag{3.4}$$

$$\nabla \times \mathbf{B}_j = \frac{\epsilon_j}{c^2} \frac{\partial \mathbf{E}_j}{\partial t} \ , \tag{3.5}$$

yields

$$B_{j,y} = -\frac{\omega\epsilon_j}{c^2 q} E_{j,z} \ , \tag{3.6}$$

$$B_{j,y} = -i \, \text{sgn}(z) \frac{\omega\epsilon_j}{c^2 \kappa_j} E_{j,x} \ , \tag{3.7}$$

$$E_{j,x} = -i \, \text{sgn}(z) \frac{\kappa_j}{q} E_{j,z} \ , \tag{3.8}$$

$$\kappa_j = \sqrt{q^2 - \epsilon_j \omega^2 / c^2}. \tag{3.9}$$

Maxwell's equations require the continuity of the tangential components (relative to the interface) of the electric and magnetic fields across the interface (i.e., at $z = 0$),

$$E_{1,x} = E_{2,x} \ , \tag{3.10}$$

$$B_{1,y} = B_{2,y} \ , \tag{3.11}$$

(notice that for non-magnetic media $H_{1,y} = H_{2,y} \Leftrightarrow B_{1,y} = B_{2,y}$). Note that Eq. (3.10) implies that we are not dealing with a perfect conductor (see Sec. 3.3). Making use of equation (3.7) to eliminate the magnetic field amplitudes in the boundary conditions (3.10)-(3.11), we arrive at the following pair of equations:

$$E_{1,x} = E_{2,x} \ , \tag{3.12}$$

$$\frac{\epsilon_1}{\kappa_1} E_{1,x} = -\frac{\epsilon_2(\omega)}{\kappa_2} E_{2,x} \ . \tag{3.13}$$

The linear system composed by equations (3.12) and (3.13) has a non-trivial solution if

$$\boxed{\frac{\epsilon_1}{\kappa_1(q,\omega)} + \frac{\epsilon_2(\omega)}{\kappa_2(q,\omega)} = 0} \ , \tag{3.14}$$

from which the dispersion relation of TM modes can be obtained. Note that in equation (3.14) we emphasize the fact that $\kappa_{1,2}$ depends on the frequency as well as on the wavevector in accordance to equation (3.9). Moreover, in order to have an electromagnetic wave confined to the interface —i.e., $\Re\{\kappa_{1,2}\} > 0$; see equations (3.2) and (3.3)— it follows from equation (3.14) that if $\epsilon_1 > 0$, then we must have $\Re\{\epsilon_2(\omega)\} < 0$. This means that SPPs can only exist at interfaces between materials whose (real parts of the) dielectric permittivities have opposite signs, such as an interface between an insulator and a metal.

We can write a more explicit relation for the SPPs' wavevector, q_{SPP}, as a function of the frequency, by substituting equation (3.9) into equation (3.14), obtaining

$$\boxed{q_{SPP} = \frac{\omega}{c} \sqrt{\frac{\epsilon_1 \epsilon_2(\omega)}{\epsilon_1 + \epsilon_2(\omega)}}} \ . \tag{3.15}$$

We emphasize that equation (3.15) remains valid when $\epsilon_2(\omega)$ is a complex-valued function, that is, in spectral regions where damping cannot be neglected, $\gamma > 0$. To better appreciate the role of the imaginary component of $\epsilon_2(\omega)$, we analyze separately the cases with and without attenuation.

3.1.1.1   *Zero damping ($\gamma = 0$)*

In the case of a lossless conductor, the Drude's dielectric function takes the simplified form

$$\epsilon_2(\omega) = 1 - \frac{\omega_p^2}{\omega^2} \, . \tag{3.16}$$

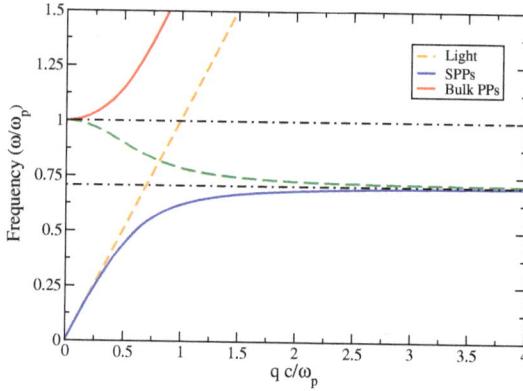

Figure 3.2: Dispersion relation of SPPs at an interface between air and a Drude metal without damping. Three different regions are borne out: an upper branch with radiative modes (red curve); a lower branch with bound SPPs modes (blue curve); and an intermediate region, with $\Re\{q\} = 0$, where propagation is forbidden (green dashed curve).

Figure 3.2 shows the dispersion relation obtained from inserting equation (3.16) into equation (3.15). The continuous curves represent the real part of the plasmon's wavevector and the dashed green curve its imaginary part. It is clear that whenever the value of q is a purely imaginary number, the existence of propagating modes is forbidden. In Fig. 3.2, this corresponds to the region between the two horizontal dashed-dotted lines. Outside this region, we can identify two kinds of fundamentally different plasmons: bulk plasmons (red curve) and SPPs (blue curve). The former consists of collective oscillations of the free electrons within the metal (also called volume plasmons), while the latter corresponds to bound excitations confined at the interface. In addition, in the upper branch of the dispersion relation relative to bulk plasmons, we have $\omega > \omega_p$, so that radiation propagates through metal —radiative modes— with q being real and $\kappa_j \to i\kappa_j$. On the other hand, since SPPs correspond to bound modes with both q and κ_j real, its dispersion curve must lie to the right of the light line in the dielectric (yellow line). Consequently, SPPs cannot be excited by direct illumination, so that one has to come up with a mechanism capable of coupling light to surface plasmons. Examples of such techniques are prism and grating coupling, which will be described in more detail in Chapter 5.

Before moving on to the discussion of the case with a real (lossy) metal, let us take a closer look at the behavior of SPPs in two limiting cases, those of small and large wavevectors. In the regime of small wavevectors, the dispersion curve lies close to the light line, and the SPP resembles an incident light wave tangent to the interface —Sommerfeld-Zenneck wave. Thus, $\kappa_1 \ll 1$ [cf. equation (3.9)] and the electromagnetic field is poorly confined, penetrating deep inside the dielectric. Conversely, for large wavevectors, it is apparent from Fig. 3.2 that the SPPs dispersion curve approaches a limiting frequency as $q \to \infty$. This frequency can be obtained from equation (3.15) by setting $\epsilon_1 + \epsilon_2(\omega) = 0$, yielding

$$\omega_{sp} = \frac{\omega_p}{\sqrt{1 + \epsilon_1}} \, , \qquad (3.17)$$

which is known as the *surface plasmon frequency* and corresponds to the lower horizontal line in Fig. 3.2. Note, however, that in this regime the group velocity $v_g = \frac{\partial \omega}{\partial q} \to 0$. Therefore this is a dispersionless mode with an electrostatic character, dubbed a *surface plasmon*.

3.1.1.2 *Finite damping* ($\gamma \neq 0$)

In real metals, there is often a miscellania of damping mechanisms which contribute to the attenuation of SPPs, ranging from electron-electron interactions to impurity and lattice scattering, just to mention a few. Within the framework of the Drude model, one attemps to account for all these phenomena by introducing a phenomenological parameter, γ, as in equation (3.1). Nevertheless, although the model captures fairly well the dominant features of noble metals at low frequencies, it fails to describe interband transitions at higher frequencies. In order to avoid further complications, we shall use the actual values of the dielectric function taken from experiments. In Figs. 3.3 and 3.4 the dispersion relation of SPPs is plotted using experimental data obtained from [Johnson and Christy (1972)]. The most striking feature is that in this case there is a finite, maximum allowed wavevector for the SPPs, in contrast with the previous undamped case (compare, for instance, Figs. 3.2 and 3.3).

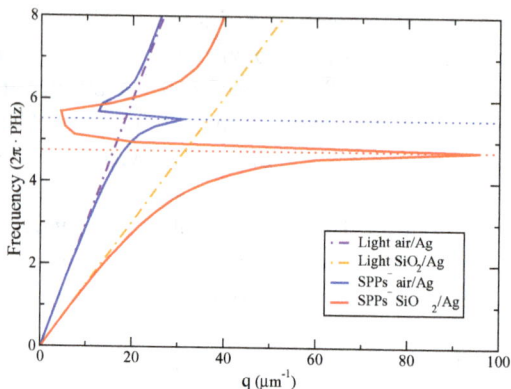

Figure 3.3: Dispersion relation of TM SPPs in air/Ag (blue) and SiO$_2$/Ag (red) interfaces, where we have used experimental data from [Johnson and Christy (1972)] to account for silver's dielectric function. Owing to damping, the wavevector of the bounded SPPs modes is now limited to a maximum value at ω_{sp} (dotted lines).

This not only sets up a lower bound for the SPPs wavelength, given by $\lambda_{SPP} = 2\pi/\Re\{q\}$, but also limits the degree of confinement of the mode. Additionally, due to the non-vanishing γ, we now have the more general situation where the dielectric function of the conductor is a complex-valued function, and, therefore, so it is q, as illustrated in Fig. 3.4. Thus, the SPPs will be attenuated, losing energy as they travel along the system until they eventually fade away.

Finally, note that unlike the case of a lossless metal —in which there is a gap, between ω_{sp} and ω_p, where the existence of modes was prohibited (i.e. with a purely imaginary propagation constant)— we now observe the emergence of *quasibound modes* in the aforementioned region.

Before moving on with our study of the basic features of SPPs, let us take a step back to analyse the case of a s-polarized, transverse electric (TE) surface electromagnetic wave. These modes have the electric field perpendicular to the plane of incidence, so that we will be seeking solutions of the type

$$\mathbf{E}_j(\mathbf{r}, t) = E_{j,y}\hat{\mathbf{y}} \; e^{-\kappa_j|z|}e^{i(qx-\omega t)} \; , \tag{3.18}$$

$$\mathbf{B}_j(\mathbf{r}, t) = (B_{j,x}\hat{\mathbf{x}} + B_{j,z}\hat{\mathbf{z}}) \, e^{-\kappa_j|z|}e^{i(qx-\omega t)} \; . \tag{3.19}$$

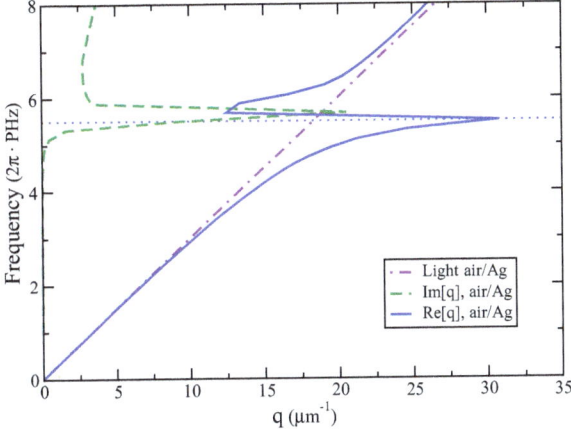

Figure 3.4: Dispersion relation with both real and imaginary parts of the propagation constant, q, of TM SPPs at an interface between air and silver, using experimental data from [Johnson and Christy (1972)].

As before, we substitute the fields (3.19) and (3.18) into Maxwell's equations (3.4) and (3.5), obtaining

$$B_{j,z} = \frac{q}{\omega} E_{j,y} , \tag{3.20}$$

$$B_{j,x} = -i \, \text{sgn}(z) \frac{\kappa_j}{\omega} E_{j,y} , \tag{3.21}$$

$$B_{j,x} = -i \, \text{sgn}(z) \frac{\kappa_j}{q} B_{j,z} , \tag{3.22}$$

$$\kappa_j = \sqrt{q^2 - \epsilon_j \omega^2/c^2}. \tag{3.23}$$

The boundary conditions at $z = 0$, now read

$$E_{1,y} = E_{2,y} , \tag{3.24}$$

$$B_{1,x} = B_{2,x} . \tag{3.25}$$

Introducing equation (3.21) into the second boundary condition (3.25), and further combining equations (3.24) and (3.25), we arrive at the following condition:

$$(\kappa_1 + \kappa_2)E_{1,y} = 0 . \tag{3.26}$$

This condition may be satisfied if either $E_{1,y} = 0$ or $\kappa_1 + \kappa_2 = 0$. Note, however, that we must have $\Re\{\kappa_1\} > 0$ and $\Re\{\kappa_2\} > 0$ since we are looking for solutions confined to the dielectric-metal interface. Thus, the condition given by equation (3.26) can only be fulfilled if $E_{1,y} = E_{2,y} = 0$, that is, TE surface waves cannot exist at interfaces between a dielectric and a conductor[1].

3.1.2 *Propagation length and field confinement*

In the above discussion of TM surface plasmon-polaritons, we claimed that the inherent Ohmic losses in metals originate a complex-valued dielectric function. Thus, from equation (3.15) it can be seen that q will also be complex, $q = q' + iq''$, so that the attenuation of the SPPs solutions will be caracterized by an exponentially decaying term as $\exp(-q''x)$.

In particular, defining $\epsilon_2(\omega) = \epsilon_2' + i\epsilon_2''$ and inserting this into equation (3.15), we obtain

$$q = \frac{\omega}{c}\sqrt{\frac{\epsilon_1(\epsilon_2' + i\epsilon_2'')}{\epsilon_1 + \epsilon_2' + i\epsilon_2''}} \approx \frac{\omega}{c}\sqrt{\frac{\epsilon_1\epsilon_2'}{\epsilon_1 + \epsilon_2'}}\left[1 + i\frac{\epsilon_1\epsilon_2''/\epsilon_2'}{\epsilon_1 + \epsilon_2'}\right]^{1/2}, \qquad (3.27)$$

where we have explicitly assumed that $|\epsilon_2'| > \epsilon_2''$ (condition satisfied from frequencies up to about ω_{sp}, i.e. the SPPs region) and we have also ignored the terms of order $\mathcal{O}\left([\epsilon_2'']^2\right)$. Expanding equation (3.27), yields

$$q \approx \frac{\omega}{c}\sqrt{\frac{\epsilon_1\epsilon_2'}{\epsilon_1 + \epsilon_2'}}\left[1 + \frac{i}{2}\frac{\epsilon_1\epsilon_2'}{\epsilon_1 + \epsilon_2'}\frac{\epsilon_2''}{(\epsilon_2')^2}\right], \qquad (3.28)$$

from which we identify

$$q' = \frac{\omega}{c}\sqrt{\frac{\epsilon_1\epsilon_2'}{\epsilon_1 + \epsilon_2'}} \qquad \text{and} \qquad q'' = \frac{1}{2}\frac{\omega}{c}\left(\frac{\epsilon_1\epsilon_2'}{\epsilon_1 + \epsilon_2'}\right)^{3/2}\frac{\epsilon_2''}{(\epsilon_2')^2}. \qquad (3.29)$$

Since the electromagnetic field intensity of SPP decreases as $\exp(-2q''x)$, its characteristic propagation length is given by

$$\boxed{L_{SPP} = \frac{1}{2\Im\mathrm{m}\{q\}} \approx \frac{1}{2q''} = \frac{c}{\omega}\left(\frac{\epsilon_1 + \epsilon_2'}{\epsilon_1\epsilon_2'}\right)^{3/2}\frac{(\epsilon_2')^2}{\epsilon_2''}}, \qquad (3.30)$$

which defines the distance that the SPP travels until its intensity falls by $1/e$. Figure 3.5 shows the propagation lengths for surface plasmon-polaritons at air/ silver and silica/silver interfaces (blue and red curves, respectively). Note that for wavelengths within the visible region, the propagation distances are of the order of a few micrometers up to 100 μm, whereas

---

[1]Note that plasmon-polaritons, by definition, couple to collective oscillations, so **k** and **E** must be parallel (for such coupling to occur). However, in TE modes, by definition, **k** $\perp$ **E** so the expression "TE plasmon-polaritons", sometimes used, is incorrect. TE should be named "confined TE modes" or even "surface TE modes".

in the infrared they can reach values between 100 and 1000 μm. Naturally, these values depend upon the specific dielectric/metal involved[2]. Additionally, we find that the propagation length attains its minimum value as the frequency approaches ω_{sp}, as we had already anticipated in the previous subsection.

One of the most interesting properties of SPPs is their ability to confine the electromagnetic field, in the direction perpendicular to the interface, down to subwavelength scales beyond the diffraction limit. Since along this direction the fields fall off as $\exp(-\kappa_j |z|)$, we can quantify the "degree of field confinement" by introducing the penetration depth, ζ, defined as

$$\zeta_j = \frac{1}{\Re\{\kappa_j\}} \, .$$ (3.31)

We can obtain an approximate expression for ζ, by substituting Eq. (3.15) into Eq. (3.9), from which we arrive to

$$\zeta_{\text{diel.}} = \frac{c}{\omega} \sqrt{\frac{|\epsilon_2'| - \epsilon_1}{\epsilon_1^2}} \quad \text{and} \quad \zeta_{\text{metal}} = \frac{c}{\omega} \sqrt{\frac{|\epsilon_2'| - \epsilon_1}{(\epsilon_2')^2}}$$ (3.32)

for both the dielectric and the metal, as indicated by the subscripts in Eq. (3.32). In writing these expressions, we explicitly used the fact that ϵ_2' must be negative in order for SPPs to exist [recall our discussion after Eq. (3.14)], and we have neglected ϵ_2''. The behavior of the penetration depths (also called attenuation length) of the electromagnetic field due to SPPs as a function of the corresponding vacuum wavelength is depicted in Fig. 3.6.

In the metal, the penetration depth is roughly constant for wavelengths below the surface plasmon frequency and it can be several orders of magnitude smaller than in the dielectric. In contrast, the penetration depth of the electromagnetic field into the dielectric grows with increasing wavelength. This means that the ability of SPPs to confine the electromagnetic field deteriorates as the wavelength (frequency) increases (decreases).

We end our preliminary discussion of the basic properties of SPPs at single interfaces by noting that while the field confinement is weaker at lower frequencies (as we have just seen), the corresponding propagation

---

[2]We further note that even for the very same metal we can come up with different results depending on the source of empirical data retrieved from the literature - e.g. compare Figs. 3.5a and 3.5b, obtained using data taken from [Johnson and Christy (1972)] and [Palik (1997)], respectively. The reasons for the discrepancies observed among several experimental data available in the literature are still motif of debate and we shall not enter here in such muddy waters.

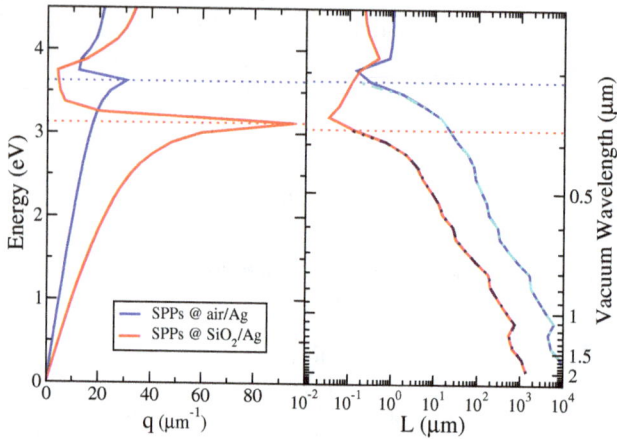

(a) Left panel: dispersion relation of TM SPPs at air/Ag and SiO$_2$/Ag interfaces (similar to Fig. 3.3). Right panel: corresponding propagation lengths. The solid curves were obtained directly from $L = (2\Im m\{q\})^{-1}$, while the dot-dashed curves were computed using the approximate expression indicated by Eq. (3.30). The dielectric data was taken from [Johnson and Christy (1972)].

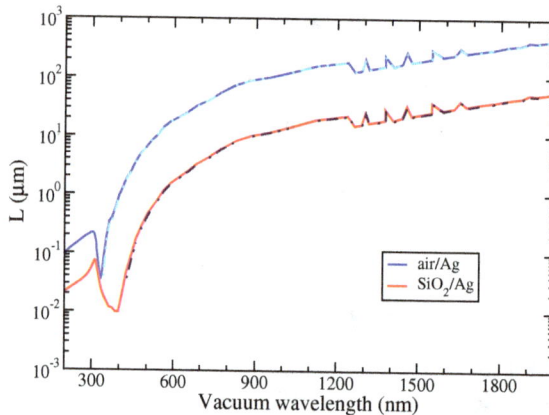

(b) The same as in Fig. 3.5a (right panel) but now using the experimental data from [Palik (1997)].

Figure 3.5: Propagation lengths of SPPs at dielectric/metal interfaces.

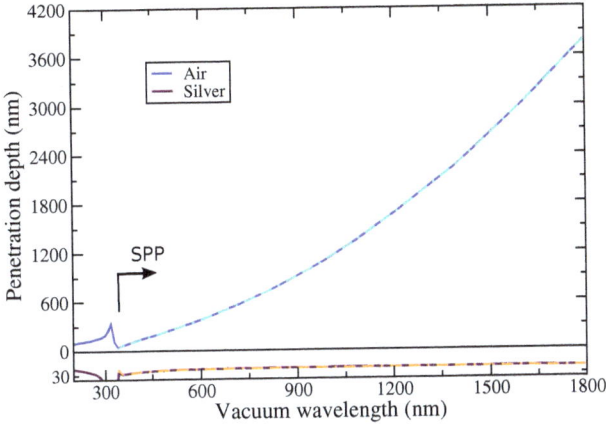

Figure 3.6: Penetration depths of the electromagnetic fields due to SPPs — starting from the arrow and to its right— into air (blue curve) and silver (maroon curve), calculated using the optical constants from [Johnson and Christy (1972)]. Note that we have plotted the penetration into the metal below the $\zeta = 0$ line, and notice the different scale in that region. The dot-dashed curves correspond to the approximate values obtained using expressions (3.32).

length is larger (cf. Fig. 3.5). Thus, higher field localization comes hand in hand with larger losses. This trade-off between field confinement and propagation length is characteristic of plasmonics.

3.2 Multilayer Structures

3.2.1 *Double Interface*

Up to now, we have only treated the simplest case of a single flat dielectric-metal interface. Nevertheless, the landscape encompassing SPPs turns out to be much richer. The natural next step is to look for SPPs solutions in more complex (but still planar) geometries, such as systems containing several interfaces. From our previous study, we expect that these multilayer structures have the capacity to sustain SPPs modes whenever there is an interface between a dielectric and a conductor. However, as we shall see ahead, SPPs in such heterostructures may exhibit interesting phenomena by their own, that is, out of reach of ordinary SPPs at single interfaces. For the sake of clarity and pedagogy, as well as historical significance, here we will consider essentially two different systems: a thin metal film cladded between two semi-infinite dielectric media - dielectric/metal/dielectric (DMD) structure - and its "image", that being a dielectric layer sandwiched between two semi-infinite metallic layers - metal/dielectric/metal (MDM) structure.

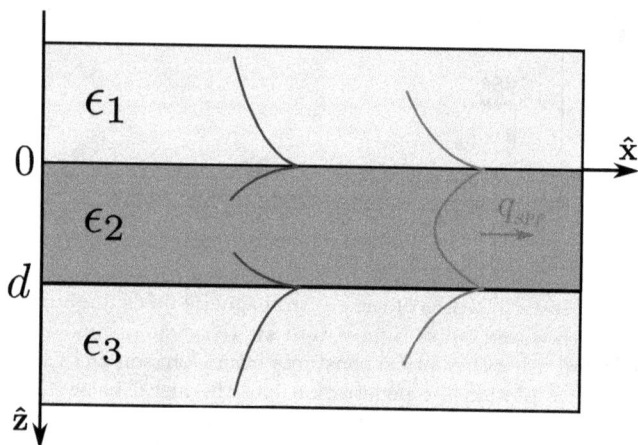

Figure 3.7: Schematic representation of a three-layer heterostructure - either DMD or MDM - capable of supporting surface plasmon-polaritons at each of its dielectric/metal interfaces. For metallic media it is implicit that its corresponding dielectric function is frequency-dependent, i.e. $\epsilon_m(\omega)$, where the subscript stands for "metal". The width of the core layer is caracterized by d, and the cladding media are assumed to be semi-infinite. In addition, the field of both two non-interacting (blue) surface plasmon-polaritons and a hybridized mode (red) are qualitatively illustrated for eye guidance.

The common denominator is that both heterostructures can be regarded as three-layer systems with a double interface. In the case of a thin film, we already antecipated that bound SPPs solutions should exist at each of its interfaces. The novelty emerges when the thickness of the metallic film, d, becomes sufficiently small[3] to enable the communication between the electromagnetic fields of the SPPs modes at opposite interfaces. The result of this interaction will be the splitting of the *retarded* SPP condition, given by equation (3.15), into a set of two distinct new modes.

---

[3]To be more quantitative, by sufficiently small we mean $\kappa_2 d \ll 1$.

3.2.2 Dispersion relation

The procedure to derive the dispersion relation of surface plasmon-polaritons in a double interface will follow, in essence, the same steps as in Sec. 3.1.1 for one dielectric-metal interface. The electromagnetic fields in the top and bottom media have the usual form of a TM surface wave

$$\mathbf{E}_j(\mathbf{r}, t) = (E_{j,x}\hat{\mathbf{x}} + E_{j,z}\hat{\mathbf{z}}) e^{-\kappa_j |z|} e^{i(qx-\omega t)} , \tag{3.33}$$

$$\mathbf{B}_j(\mathbf{r}, t) = B_{j,y}\hat{\mathbf{y}} \, e^{-\kappa_j |z|} e^{i(qx-\omega t)} , \tag{3.34}$$

with

$$B_{j,y} = -\frac{\omega \epsilon_j}{c^2 q} E_{j,z} , \tag{3.35}$$

$$B_{j,y} = -i \operatorname{sgn}(z) \frac{\omega \epsilon_j}{c^2 \kappa_j} E_{j,x} , \tag{3.36}$$

$$E_{j,x} = -i \operatorname{sgn}(z) \frac{\kappa_j}{q} E_{j,z} , \tag{3.37}$$

$$\kappa_j = \sqrt{q^2 - \epsilon_j \omega^2/c^2}. \tag{3.38}$$

now for $j = 1, 3$ only [where $z < 0$ in region 1 and $z > d$ in region 3 [4] (see Fig. 3.7)]. The situation becomes different for medium $j = 2$ (where $0 < z < d$). Here, the fields can be described as a superposition of an exponentially growing wave with an exponentially decaying wave, of the form

$$\mathbf{E}_2(\mathbf{r}, t) = \left\{ \left(E_{2,x}^{(+)}\hat{\mathbf{x}} + E_{2,z}^{(+)}\hat{\mathbf{z}} \right) e^{\kappa_2 z} + \left(E_{2,x}^{(-)}\hat{\mathbf{x}} + E_{2,z}^{(-)}\hat{\mathbf{z}} \right) e^{-\kappa_2 z} \right\} e^{i(qx-\omega t)}, \tag{3.39}$$

$$\mathbf{B}_2(\mathbf{r}, t) = \left(B_{2,y}^{(+)} e^{\kappa_2 z} + B_{2,y}^{(-)} e^{-\kappa_2 z} \right) \hat{\mathbf{y}} e^{i(qx-\omega t)}, \tag{3.40}$$

satisfying the relations

$$B_{2,y}^{(+)} = -\frac{\omega \epsilon_2}{c^2 q} E_{2,z}^{(+)}, \tag{3.41}$$

$$B_{2,y}^{(+)} = i \frac{\omega \epsilon_2}{c^2 \kappa_2} E_{2,x}^{(+)}, \tag{3.42}$$

$$B_{2,y}^{(-)} = -\frac{\omega \epsilon_2}{c^2 q} E_{2,z}^{(-)}, \tag{3.43}$$

$$B_{2,y}^{(-)} = -i \frac{\omega \epsilon_2}{c^2 \kappa_2} E_{2,x}^{(-)}, \tag{3.44}$$

$$\kappa_2 = \sqrt{q^2 - \epsilon_2 \omega^2/c^2}, \tag{3.45}$$

---

[4]Note that in the latter we could also have written the z-dependence of the fields in Eqs. (3.33) and (3.34) as $e^{-\kappa_3(z-d)}$. As written in the text, the factor $e^{\kappa_3 d}$ is taken into account "inside" the amplitudes.

connecting the amplitudes of the magnetic field with the corresponding amplitudes of the electric field. Since now we have two interfaces, we will need to employ two sets of boundary conditions, one at $z = 0$,

$$E_{1,x} = E_{2,x}^{(+)} + E_{2,x}^{(-)} , \tag{3.46}$$

$$B_{1,y} = B_{2,y}^{(+)} + B_{2,y}^{(-)} , \tag{3.47}$$

and the other at $z = d$,

$$E_{2,x}^{(+)} e^{\kappa_2 d} + E_{2,x}^{(-)} e^{-\kappa_2 d} = E_{3,x} e^{-\kappa_3 d} , \tag{3.48}$$

$$B_{2,y}^{(+)} e^{\kappa_2 d} + B_{2,y}^{(-)} e^{-\kappa_2 d} = B_{3,y} e^{-\kappa_3 d} . \tag{3.49}$$

Using expressions (3.35)-(3.38) and (3.41)-(3.45), one arrives at the following homogeneous linear system of equations

$$\begin{pmatrix} -1 & 1 & 1 & 0 \\ -\frac{\epsilon_1}{\kappa_1} & \frac{\epsilon_2}{\kappa_2} & -\frac{\epsilon_2}{\kappa_2} & 0 \\ 0 & e^{\kappa_2 d} & e^{-\kappa_2 d} & -e^{-\kappa_3 d} \\ 0 & \frac{\epsilon_2}{\kappa_2} e^{\kappa_2 d} & -\frac{\epsilon_2}{\kappa_2} e^{-\kappa_2 d} & \frac{\epsilon_3}{\kappa_3} e^{-\kappa_3 d} \end{pmatrix} \begin{pmatrix} E_{1,x} \\ E_{2,x}^{(+)} \\ E_{2,x}^{(-)} \\ E_{3,x} \end{pmatrix} = 0 . \tag{3.50}$$

The dispersion relation for transverse magnetic SPPs in a double interface is then obtained by solving the condition that the determinant of the matrix in (3.50) must be zero in order to exist a non-trivial solution, yielding

$$\boxed{ \left(1 + \frac{\epsilon_2 \kappa_1}{\epsilon_1 \kappa_2} \right) \left(1 + \frac{\epsilon_2 \kappa_3}{\epsilon_3 \kappa_2} \right) = \left(1 - \frac{\epsilon_2 \kappa_1}{\epsilon_1 \kappa_2} \right) \left(1 - \frac{\epsilon_2 \kappa_3}{\epsilon_3 \kappa_2} \right) e^{-2\kappa_2 d} } . \tag{3.51}$$

This expression must be solved numerically, and it is implicit that $\kappa_j \equiv \kappa_j(q, \omega)$ and that $\epsilon_j \equiv \epsilon_j(\omega)$ for $j = 1, 3$ referring to metallic media[5]. Note that in the limit $d \to \infty$ (or $\kappa_2 d \gg 1$), corresponding to thick films in DMD heterostructures or to thick dielectric cores in MDM heterostructures, we get from Eq. (3.51) the conditions

$$\frac{\epsilon_1}{\epsilon_2} + \frac{\kappa_1}{\kappa_2} = 0 \qquad \text{and/or} \qquad \frac{\epsilon_3}{\epsilon_2} + \frac{\kappa_3}{\kappa_2} = 0 , \tag{3.52}$$

which are just the decoupled dispersion relations of the individual SPPs at each interface [cf. (3.14)].

It is interesting to consider the particular case of a symmetrical environment, in which the substrate and the superstrate are made of the same

---

[5]In principle, the dielectric function of the dielectrics could also depend on the frequency, but we will ignore this complication for now, and assume that in the region of interest the dielectric functions of the insulators are constant.

material, and thus $\epsilon_1 = \epsilon_3$ and $\kappa_1 = \kappa_3$. Under these circumstances, the dispersion relation (3.51) can be further simplified, and splits into two modes delivered by the following pair of equations:

$$\omega_+ : \quad \tanh(\kappa_2 d/2) = -\frac{\epsilon_2 \kappa_1}{\epsilon_1 \kappa_2} , \qquad (3.53)$$

and

$$\omega_- : \quad \coth(\kappa_2 d/2) = -\frac{\epsilon_2 \kappa_1}{\epsilon_1 \kappa_2} , \qquad (3.54)$$

corresponding to a high-frequency mode and a low-frequency mode, as indicated by the subscripts. In addition, it is also common to encounter these solutions in the literature classified as an odd mode, where the tangential component of the electric field (E_x) is antisymmetric with respect to the plane $z = d/2$; and an even mode, in which the tangential component of the electric field is symmetric relative to the $z = d/2$ plane. Figure 3.8 shows the dispersion relation of non-radiative[6] SPPs in silver thin films (DMD heterostructures) free-standing in air, computed using Eqs. (3.53) and (3.54) for several different widths. Silver's dielectric function, $\epsilon_2(\omega)$, is modeled as a free electron gas with negligible damping [see, for instance, Eq. (3.16)].

From Fig. 3.8 we see that in the case of the thickest film (orange dot-dashed line), its dispersion curve consists of two almost decoupled, degenerate modes (due to the symmetrical environment) and is indistinguishable to the one of a single air/Ag interface (black curve). However, as the film thickness is decreased, the evanescent tails of the SPPs in each interface overlap and hybridize, modifying the (independent) SPP dispersion. This interaction between the electromagnetic fields on different interfaces act as a perturbation which removes the degeneracy of the spectrum. As a consequence, there is a splitting of the original dispersion relation into two branches, ω_+ and ω_-, corresponding to modes in which the component of the electric field along the propagation direction is an odd and an even function of $z - d/2$, respectively, as Fig. 3.9 illustrates.

Furthermore, for sufficiently thin films, the antisymmetric solution exhibits an obvious maxima at frequencies above ω_{sp} in contrast both to the symmetric solution and to the single-boundary dispersion curve. Recall that for the latter, there is a gap between ω_{sp} and ω_p where propagation is

---

[6]We mention that in this geometry the dispersion relation also returns solutions between ω_p and the curves plotted in Fig. 3.8 (not shown for clarity). These are highly damped, short-lived modes [Dionne *et al.* (2005)] (even when $\gamma \to 0$), so that we shall ignore them onwards.

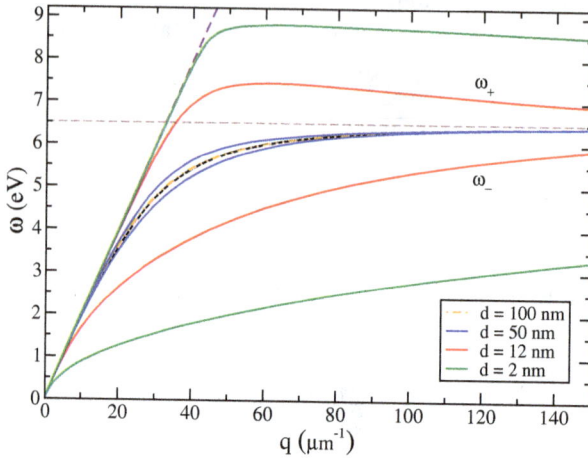

Figure 3.8: Dispersion relation of TM SPPs in a thin film, namely an air/silver/air multilayer structure, showing the hybridized, coupled modes for various mettalic layer thicknesses, d, indicated in the plot. The black curve represents the dispersion curve of SPPs in a single air/silver interface and the dashed violet line the dispersion of light in the dielectric. The broken horizontal line indicates the asymptotic surface plasmon frequency. Here, silver is assumed to be a lossless Drude metal with $\omega_p = 9.2$ eV.

not allowed (cf. Fig. 3.2). Additionally, note that the odd mode approaches the asymptotic surface plasmon frequency from above, with negative group velocity. On the other hand, the even mode has the ability to achieve large wavevectors at frequencies well below ω_{sp}, meaning that the corresponding SPP wavelengths will be extremely short.

The splitting of the SPPs dispersion curve due to the two coupled (bounded) modes in a thin oxydized aluminum film was experimentally demonstrated by Pettit *et al* [Pettit *et al.* (1975)] using electron-loss spectroscopy. Their results are shown in Fig. 3.10, where the typical dispersion of the hybridized ω_+ and ω_- modes can be clearly seen and follows closely the theoretical model.

So far, for this geometry, we have limited our discussion to the ideal case of a lossless conductor. In general, as discussed earlier, the metal's dielectric function will be described by a complex-valued function, which in turn can modify significantly the SPP dispersion relation. Thus, and in parallel with what we have seen in the case of a single interface, we expect that q will approach a maximum value before folding back (eventually

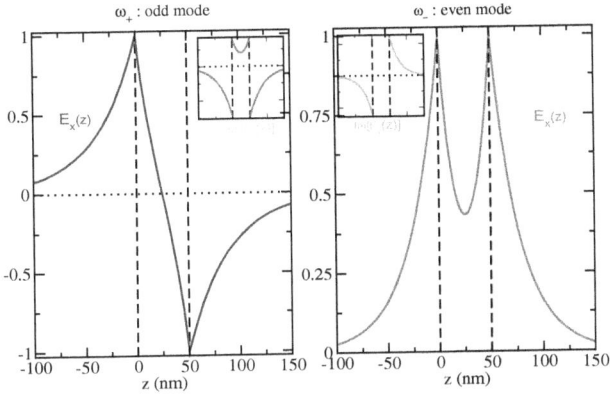

Figure 3.9: Profile of the tangential component of the electric field of SPPs in a double interface: air/silver/air. The insets also show $E_z(z)$ for completeness. Left: antisymmetric (odd) mode. Right: symmetric (even) mode. Parameters: $d = 50$ nm and $\omega = 5.5$ eV (yielding $q_{\text{odd}} = 38~\mu\text{m}^{-1}$ and $q_{\text{even}} = 46~\mu\text{m}^{-1}$).

crossing the light line) to smaller wavevectors. An interesting feature of this type of three-layer system is that losses do not affect the two modes in the same way. For the odd mode, it can be shown that the imaginary part of the propagation constant decreases with decreasing film thickness. According to Eq. (3.30), this results in very long propagation distances and for that reason this mode was named *long-range surface plasmon-polariton* (LRSPP). In fact, multicentimeter propagation lengths were achieved experimentally, thus making it a good candidate for waveguiding applications. This behavior can be understood by noting that the fraction of the field in the metal decreases with decreasing thickness, owing to the presence of a node in the longitudinal electric-field component at the film's mid-plane. The consequence of a smaller portion of the electric field inside the mettalic layer is a weaker interaction of the mode with the dissipative mechanisms of the film. Nevertheless, the reduction in attenuation comes along with a concomitant decrease in field confinement, resembling a plane wave travelling in the homogeneous dielectric media in the limit of a very thin film [Dionne *et al.* (2005); Maier (2007)]. Conversely, the even mode is more and more attenuated as the film width is reduced (due to an increasingly higher fraction of the tangential electric-field component within the conductor), leading to short propagation distances —*short-range surface*

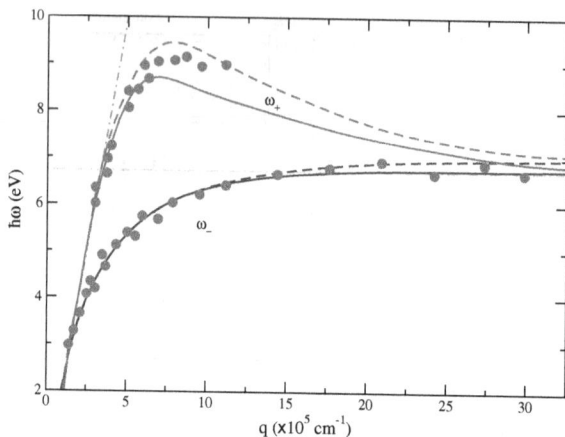

Figure 3.10: Experimental data of the SPPs dispersion in an oxydized 16 nm thin Al foil. The curves correspond to a theoretical fit of the two coupled modes for a 12 nm Al film sandwiched between two oxydized layers with a nominal thickness of 4 nm (continuous curve: α-Al$_2$O$_3$; dashed curve: amorphous aluminum). Adapted from [Pettit *et al.* (1975)].

plasmon-polariton (SRSPP). The dependence of the propagation length of both modes on the thickness of the core conductive layer is summarized in Fig. 3.11.

Throughout this Chapter we have briefly reviewed the fundamental concepts underlying the field of metal plasmonics. A lot more could have been told about this vast subject, like the discussion of the results for MDM heterostructures and also the case with $\epsilon_1 \neq \epsilon_3$, or even the blooming branch of plasmonics in metal-based nanoparticles. However, without these omissions this book would be otherwise too long and, perhaps more importantly, we would be steering away from the topic of graphene plasmonics. A comprehensive description of plasmonics in metals, metal-dielectric interfaces, and metallic nanoparticles has been written by Maier [Maier (2007)].

3.3 A Short Note on Perfect Conductors

Despite what has been said in the previous sections about SPPs on dielectric-metal planar interfaces, it has been pointed out that a planar interface between a dielectric and a perfect metal does not support a surface

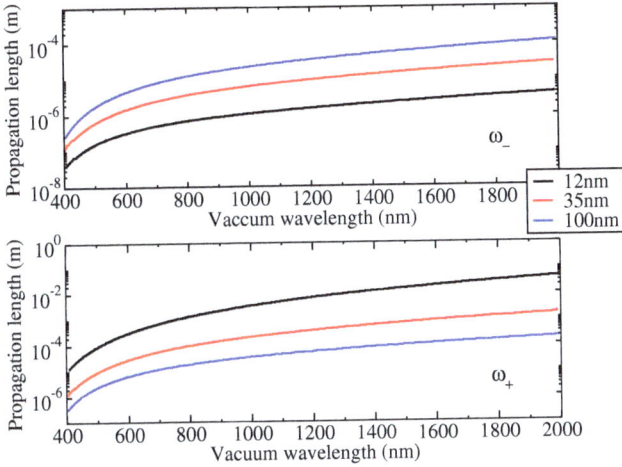

Figure 3.11: Propagation distance of even (top panel) and odd (bottom panel) SPPs modes in SiO₂/silver/SiO₂ for different thickness, d, computed using optical constants retrieved from experiments. Adapted from [Dionne *et al.* (2005)].

electromagnetic wave [Maradudin (2014)]. The proof is straightforward. Let us assume a $p-$polarized electromagnetic wave of frequency ω propagating along the $x-$direction as in Fig. 3.1. The metal occupies the $z > 0$ region (see Fig. 3.1). For the polarization under consideration, only the $y-$component of the magnetic field exists, which obeys the equation

$$\left(\frac{\partial^2}{\partial x^2} + \frac{\partial^2}{\partial z^2} + \frac{\omega^2}{c^2} \right) B_y = 0 \,. \tag{3.55}$$

A solution of the previous equation that propagates along the metal-dielectric interface and decays to zero for $z \to -\infty$ reads

$$B_y = A e^{iqx} e^{\kappa_0 z} \,, \tag{3.56}$$

where κ_0 is defined as

$$\kappa_0 = \sqrt{q^2 - \omega^2/c^2} \,, \tag{3.57}$$

and is real and positive. Since we are dealing with a perfect metal, the boundary condition at the interface is

$$E_x = 0 \Leftrightarrow \frac{\partial B_y}{\partial z} = 0 \,. \tag{3.58}$$

Applying the previous boundary condition to the solution (3.56) it follows that

$$A \kappa_0 B_y = 0 \,. \tag{3.59}$$

A non-trivial solution requires that $A \neq 0$ and $\kappa_0 = 0$, which leads to $\omega = qc$. This solution, however, is not a surface wave, as it implies that

$$B_y = Ae^{i\omega x/c}, \tag{3.60}$$

which does not decay as we move away from the interface. We should however note that in structured surfaces, made of a perfect metal, SPPs can exist [Maradudin (2014)].

Chapter 4

Graphene Surface Plasmons

In recent years we have been witnessing an impressive outburst in the field of plasmonics, with papers on the subject being published everyday at a great pace. On the other hand, graphene has been one of the most active topics in (but not limited to) the condensed matter physics community since its isolation in 2004, and ten years later our interest in the wonder material still thrives as it continues to present new challenges to us. The prediction of graphene surface plasmon-polaritons (GSPs) and their experimental observation led to the emergence of the field of graphene nanoplasmonics, which combines two of the most vibrant topics in physics today, and promises to deliver a huge impact in future technologies ranging from novel optoelectronic devices [Liang *et al.* (2014)] to applications in nanomedicine [Zheng *et al.* (2012)].

The main purpose of this chapter is to describe the physics governing surface plasmon-polaritons in simple graphene-based systems, before moving on to the treatment of more complex structures. In particular, we shall obtain the dispersion relation of GSPs in monolayer and double-layer graphene embedded in dielectric media, and investigate their properties, e.g. propagation length, field confinement, versatility, tunability, etc.

4.1 Monolayer Graphene

We begin our study of surface plasmon-polaritons in graphene by considering a system consisting in a single graphene sheet cladded between two semi-infinite dielectric media, characterized by the real dielectric constants (relative permittivities) ϵ_1 and ϵ_2, referring respectively to the top and bottom dielectrics, as depicted in Fig. 4.1.

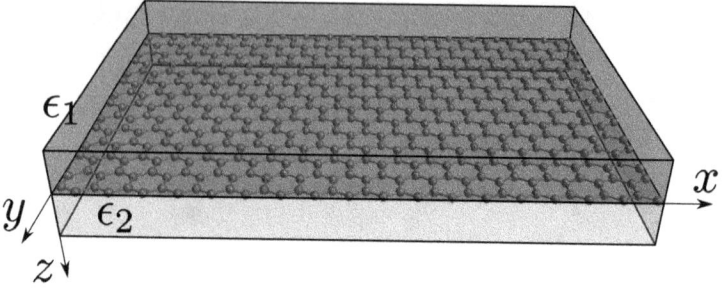

Figure 4.1: Illustration of a single graphene sheet sandwiched between two semi-infinite insulators with dielectric constants ϵ_1 and ϵ_2. Medium 1 occupies the $z < 0$ half-space and the $z > 0$ half-space is occupied by medium 2. The graphene sheet is located at the $z = 0$ plane.

4.1.1 *Spectrum of GSPs*

Let us assume a solution of Maxwell's equations, in the form of a p-polarized electromagnetic wave (TM wave). As in Chapter 3, we use the ansatz

$$\mathbf{E}_j = (E_{j,x}\hat{\mathbf{x}} + E_{j,z}\hat{\mathbf{z}})\, e^{iqx} e^{-\kappa_j|z|} \ , \tag{4.1}$$

$$\mathbf{B}_j = B_{j,y} e^{iqx} e^{-\kappa_j|z|}\hat{\mathbf{y}} \ , \tag{4.2}$$

where $j = 1, 2$ refers to the dielectric media with relative permittivities ϵ_1, ϵ_2. Equations (4.1) and (4.2) describe an electromagnetic surface wave which is confined to the neighborhood of the graphene sheet and propagates along the $\hat{\mathbf{x}}$-direction. Due to the translational invariance of the system, the linear momentum along the propagation direction must be conserved, enabling us to write $q \equiv q_1 = q_2$. Moreover, we note that in writing these equations it should be understood that they correspond to the spatial components of the fields. The time dependence, in what follows, is assumed to be of the typical harmonic form, i.e. $e^{-i\omega t}$.

Introducing Eqs. (4.1) and (4.2) into Maxwell's equations for dielectric media

$$\nabla \times \mathbf{E}_j = -\frac{\partial \mathbf{B}_j}{\partial t} \ , \tag{4.3}$$

$$\nabla \times \mathbf{B}_j = \frac{\epsilon_j}{c^2}\frac{\partial \mathbf{E}_j}{\partial t} \ , \tag{4.4}$$

yields

$$-\operatorname{sgn}(z)\kappa_j E_{j,x} - iq E_{j,z} = i\omega B_{j,y} \ , \tag{4.5}$$

$$\operatorname{sgn}(z)\kappa_j B_{j,y} = -i\omega\epsilon_j/c^2 E_{j,x} \ , \tag{4.6}$$

$$iq B_{j,y} = -i\omega\epsilon_j/c^2 E_{j,z} \ , \tag{4.7}$$

which can be rewritten with respect to the magnetic field amplitudes, while also obtaining an expression for $\kappa_j(q, \omega)$

$$E_{j,x} = i\,\mathrm{sgn}(z)\frac{\kappa_j c^2}{\omega\epsilon_j}B_{j,y}, \tag{4.8}$$

$$E_{j,z} = -\frac{qc^2}{\omega\epsilon_j}B_{j,y}, \tag{4.9}$$

$$\kappa_j^2 = q^2 - \omega^2\epsilon_j/c^2 . \tag{4.10}$$

The boundary conditions linking the electromagnetic fields at $z = 0$ read[1]

$$E_{1,x}(x, z)|_{z=0} = E_{2,x}(x, z)|_{z=0} , \tag{4.11}$$

$$B_{1,y}(x, z)|_{z=0} - B_{2,y}(x, z)|_{z=0} = \mu_0 J_x(x) = \mu_0 \sigma_{xx} E_{2,x}(x, z)|_{z=0} , \tag{4.12}$$

which assures the continuity of the tangential component of the electric field and the discontinuity of the tangential component of the magnetic field across the interface. We emphasize that the conductivity of graphene enters in the boundary condition only. Note that is the case of metal-dielectric interface, the right-hand-side of Eq. (4.12) is zero (see Chap. 3). Additionally, we further assume that graphene is a truly two-dimensional material[2], whose entire electromagnetic properties are accounted in its frequency-dependent conductivity, $\sigma(\omega)$. Letting $\sigma(\omega) \equiv \sigma_{xx} = \sigma_{yy}$, meaning that graphene's conductivity is isotropic (which is true for unstrained graphene) and substituting the fields into the boundary conditions (4.11) and (4.12), we arrive to

$$\boxed{\frac{\epsilon_1}{\kappa_1(q, \omega)} + \frac{\epsilon_2}{\kappa_2(q, \omega)} + i\frac{\sigma(\omega)}{\omega\epsilon_0} = 0} , \tag{4.13}$$

which describes the dispersion relation, $\omega(q)$, of graphene TM surface plasmon-polaritons. The reader should note the difference between Eq. (4.13) and Eq. (3.14). Note that Eq. (4.13) is an implicit equation, in the sense that the conductivity depends on the frequency, and κ_j, given by Eq. (4.10), is a function of both SPP wavenumber, q, and frequency. Hence, despite the simple geometry of the system, the dispersion relation (4.13) does not have an analytical solution and must be solved by numerical means. In addition, it is clear that Eq. (4.13) only has real solutions when the imaginary part of the conductivity is positive and its real part vanishes. On the other hand, if the real part of the conductivity is non-zero,

---

[1]Within the linear response regime. This holds and gives accurate results in most situations.
[2]In this case, the (surface) current density is given by $\mathbf{J}(\mathbf{r}) = \mathbf{J}_s(x, y)\delta(z)$.

then, in general, the solutions will be complex. We shall discuss both cases separately in the next paragraphs.

Recall that, in the spectral region in which we are interested in this monograph —from the THz to the mid-IR regime— the conductivity of graphene is dominated by a Drude-type term[3]

$$\sigma(\omega) = \sigma_0 \frac{4i}{\pi} \frac{E_F}{\hbar\omega + i\hbar\gamma} , \qquad (4.14)$$

provided that $E_F \gg k_B T$ and $E_F \gg \hbar\omega$ (cf. Sec. 2.3.2).

Figure 4.2 shows the spectrum of TM graphene surface plasmon-polaritons, obtained from the numerical solution of Eq. (4.13), where we have neglected absorption (that is, $\gamma = 0$). Note that, once again, the dispersion curves lie to the right of the light lines in the dielectrics (due to the bound nature of SPPs). Thus, as we have seen in Chapter 3 when we studied SPPs in dielectric-metal interfaces, it is not possible to excite GSPs simply by directly shining electromagnetic radiation onto the structure. Moreover, in this case ($\gamma = 0$), for doped graphene, the conductivity is a purely imaginary function of the frequency, so that the wavevector, q, coming out of Eq. (4.13) will be real. Conversely, in the general (and more realistic) case where absorption is present, i.e. for finite γ, the dispersion relation yields a complex-valued wavevector, $q = q' + iq''$, whose imaginary part characterizes the attenuation of the SPP, as we shall see further ahead in the text[4].

In Fig. 4.3 we present the dispersion curve of TM SPPs in graphene, retrieved from Eq. (4.13) using different values for the damping parameter (γ) entering in the conductivity (4.14). From the figure it is apparent that the GSP dispersion relation is shifted toward higher frequencies with respect to the dispersion curve with $\Gamma = \hbar\gamma = 0$, and that this effect is larger

---

[3]We want to stress that the use of the full non-homogeneous conductivity of graphene will originate results quantitatively different from those reported in this and the next section. The option for using here Drude's conductivity relates to the simplicity of the obtained expressions. Qualitatively there are no appreciable changes.

[4]Actually, considering a complex wavevector is not the only possibility. One can also consider a complex frequency and follow an alternative interpretation. If $\gamma \neq 0$, Eq. (4.13) yields two relations, which link ω and q where, generally speaking, both are complex quantities. One can take either of them as a real independent parameter, then Eq. (4.13) determines entirely the real and imaginary parts of the other. If we take q as independent (real) variable, we shall obtain $\omega = \omega_1 + i\omega_2$ and the imaginary part, ω_2, is the inverse of the lifetime of SPP eigenmodes (for a certain q). This is relevant to the situation where SPPs are excited by pulsed radiation. On the other hand, if we take ω as independent variable (continuous wave excitation of SPPs), the dispersion relation (4.13) yields a complex-valued eigenvector.

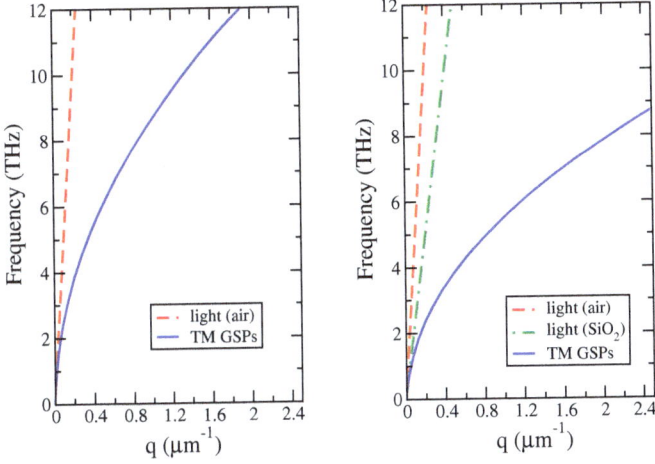

Figure 4.2: Dispersion relation of TM graphene surface plasmon-polaritons (GSPs) in a dielectric/graphene/dielectric heterostructure, as depicted in Fig. 4.1. Left: free-standing graphene in air; right: air/graphene/SiO$_2$. Here graphene's conductivity is assumed to be given by Eq. (4.14) without damping ($\gamma = 0$), so that the conductivity only has an imaginary part. We have used the following parameters: $E_F = 0.45$ eV, $\epsilon_{air} = 1$ and $\epsilon_{SiO_2} = 3.9$.

as Γ increases. The value $\Gamma = 3.7$ meV corresponds to a typical damping parameter obtained from transport measurements [Li *et al.* (2008)], while $\Gamma = 20$ meV corresponds to a rather high value. Nevertheless, taking absorption into account only changes the spectrum significantly in the region of low wavevectors, shown in the inset of Fig. 4.3. Looking at this figure at first glance, one might think that it should be possible to excite GSPs directly using free-propagating electromagnetic beams, since the GSPs dispersion curves cross the light lines at some point. However, that is indeed not the case. Here, the usual reasoning fails because the region plotted in the inset of Fig. 4.3 falls within the *overdamped* regime ($\omega_{SPP}/\gamma < 1$), in which GSPs (and SPPs in general) cannot be sustained[5]. We should stress that although the presence of a non-zero Γ does not significantly alter the aspect of the dispersion curve, it plays a leading role in the attenuation of SPPs, which increases with increasing Γ. Losses are often the main limiting factor in plasmonic-based systems, so that one usually wants a Γ value as small as possible [Yan *et al.* (2013); Khurgin and Boltasseva (2012)].

---

[5]In fact, for the higher $\Gamma = 20$ meV value, the overdamped regime extends beyond the inset of Fig. 4.3.

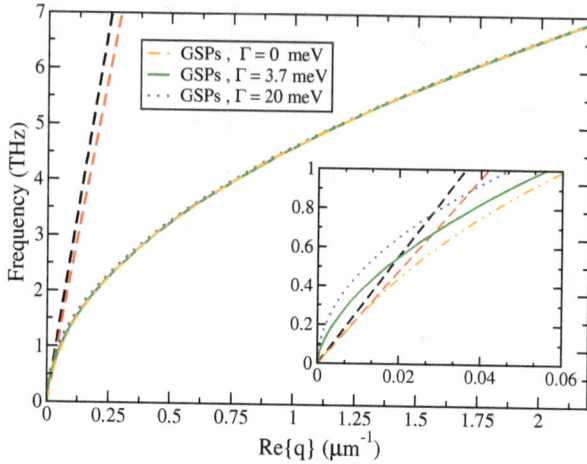

Figure 4.3: TM GSPs dispersion curves obtained from Eq. (4.13) using different values for the damping parameter, $\Gamma = \hbar\gamma$. For comparison, we have also included the dispersion relation with $\Gamma = 0$ (orange dot-dashed curve). The dashed lines represent the light dispersion in the cladding dielectrics. The inset shows a close-up of the low-q' region (the axis labels are the same as in the main figure). Parameters: $\epsilon_1 = 3$ (ion gel), $\epsilon_2 = 4$ (SiO$_2$) and $E_F = 0.45$ eV.

Albeit by now we have essentially all we need in order to describe SPPs in single-layer graphene, in the end we still must rely on numerical means to solve Eq. (4.13). For pratical matters this does not constitute a problem *per se*, since any personal laptop will be able to compute the plasmonic spectrum with little effort. It is instructive, however, to analyze a couple of limits and simplifications where it is possible to obtain approximate expressions in closed-form. This will not only allow the reader to gain a deeper understanding of the theoretical framework, but also to build an intuition on the physics governing SPPs in graphene.

We start by assuming that the graphene sheet is embedded in a symmetrical dielectric environment ($\epsilon \equiv \epsilon_1 = \epsilon_2$), which is the degenerate case of the more general situation where the dielectric constants of each media are replaced by their average value, i.e. $\epsilon = (\epsilon_1 + \epsilon_2)/2$. In this limit, the dispersion relation (4.13) simplifies to

$$1 + i\frac{\sigma(\omega)}{2\omega\epsilon\epsilon_0}\sqrt{q^2 - \epsilon\omega^2/c^2} = 0 , \qquad (4.15)$$

but still needs to be solved numerically. On the other hand, in the non-retarded regime where $q \gg \sqrt{\epsilon}\omega/c$, Eq. (4.15) reduces to

$$q \approx i\frac{2\omega\epsilon\epsilon_0}{\sigma(\omega)} \ . \tag{4.16}$$

Ignoring absorption, graphene's conductivity (4.14) becomes

$$\sigma(\omega) = 4i\alpha\epsilon_0 c\frac{E_F}{\hbar\omega} \ , \tag{4.17}$$

where $\alpha = \frac{e^2}{4\pi\epsilon_0\hbar c}$ is the fine-structure constant, so that expression (4.16) transforms to

$$\hbar\omega_{GSP} \approx \sqrt{\frac{2\alpha}{\epsilon}E_F\hbar cq} \ , \tag{4.18}$$

with the typical $\omega_{GSP} \propto \sqrt{q}$ scaling. Furthermore, since for graphene we have $E_F = \hbar v_F k_F$ with $k_F \propto n_e^{1/2}$, then the GSPs' frequency depends on the electronic density as

$$\omega_{GSP} \propto n_e^{1/4} \ . \tag{4.19}$$

While the plasmon dispersion $\omega_p \propto \sqrt{q}$ is inherent to all two-dimensional electronic systems [Stern (1967)], the behaviour portrayed in Eq. (4.19) is exclusive of graphene and has been confirmed by several experiments [Ju *et al.* (2011); Yan *et al.* (2013)].

We now take one step further in our analysis by including absorption. This is particularly important, since it will allow us to define the GSPs propagation length. Inserting Drude's full expression (4.14) for the conductivity into Eq. (4.15), we obtain

$$\kappa(q,\omega) = \frac{\epsilon\hbar}{2\alpha c}\frac{1}{E_F}\left(\omega^2 + i\gamma\omega\right) \ . \tag{4.20}$$

From Eq. (4.20), we can quantify the degree of confinement of the electromagnetic fields of GSPs in the dielectrics using the penetration depth, defined as

$$\zeta_{GSP} = \frac{1}{\Re e\{\kappa\}} = \frac{2\alpha}{\epsilon}\hbar c\frac{E_F}{(\hbar\omega)^2} \ . \tag{4.21}$$

Note that for the same level of doping, the field confinement (penetration depth) is a monotonically increasing (decreasing) function of the frequency.

As we have mentioned before, for a non-zero damping parameter the dispersion relation returns a complex-valued wavevector (for a real frequency), $q = q' + iq''$, which introduces the exponentially decaying term $\exp(-q''x)$ in the fields, and characterizes the attenuation of SPPs as they propagate along the graphene sheet. In the electrostatic limit ($\kappa \to q$), the dispersion relation (4.20) reduces to

$$q' + iq'' \approx \frac{\epsilon \hbar}{2 \alpha c} \frac{1}{E_F} \left(\omega^2 + i\gamma\omega \right) , \qquad (4.22)$$

(assuming $\gamma \ll \omega$) from which we retrieve

$$q' = \frac{\epsilon}{2 \alpha \hbar c} \frac{(\hbar\omega)^2}{E_F} , \qquad (4.23)$$

and

$$q'' = \frac{\epsilon \hbar}{2 \alpha c} \frac{\gamma\omega}{E_F} . \qquad (4.24)$$

Note that Eqs. (4.23) and (4.24) are only approximate as we used the homogeneous version of the conductivity instead of its non-local version. It follows from Eq. (4.24) that the propagation length of SPPs in single-layer graphene (defined as the distance covered by GSPs when its intensity falls by $1/e$) is given by

$$L_{GSP} = \frac{1}{2q''} = \frac{\alpha}{\epsilon} \frac{\hbar c}{\Gamma} \frac{E_F}{\hbar\omega} , \qquad (4.25)$$

which decreases with increasing Γ (recall that $\Gamma = \hbar\gamma$) as expected. Additionally, notice that expression (4.23) has the same functional form as Eq. (4.18). Using the former, one can relate the GSP wavelength $\lambda_{GSP} = 2\pi/\Re\{q\} \approx 2\pi/q'$ to its corresponding light wavelength in vacuum, λ_0, namely[6]

$$\frac{\lambda_{GSP}}{\lambda_0} = \frac{2\alpha}{\epsilon} \frac{E_F}{\hbar\omega} . \qquad (4.26)$$

Let us now consider an example: taking $\omega/(2\pi) = 10$ THz, $E_F = 0.3$ eV and $\epsilon = 4$, we obtain $\lambda_{GSP}/\lambda_0 \approx 0.026$; that is, the plasmon-polariton wavelength is almost two orders of magnitude shorter than the incident light's wavelength, which is a respectable amount of localization far beyond the diffraction limit. For frequencies around $\omega_F = E_F/\hbar$ (usually in the mid-IR under typical doping concentrations), the subwavelength confinement is even stronger, being of the order of $\alpha \approx 1/137 \approx 0.007$ as compared to λ_0 [Koppens *et al.* (2011); de Abajo (2014)]. This is often the maximum attainable confinement for such a dielectric/graphene/dielectric system since

---

[6]It is worth to point out that it is possible to correct Eq. (4.26) for the case with asymmetric dielectric environments ($\epsilon_1 \neq \epsilon_2$) upon the replacement $\epsilon \to (\epsilon_1 + \epsilon_2)/2$ (but still in the non-retarded regime!). The same holds for relations (4.16) and (4.22)-(4.25).

for frequencies $\omega > \omega_F$ GSPs become severely damped due to the onset of interband Landau damping [Yan *et al.* (2013)].

The ability of GSPs to achieve such high levels of spatial confinement in the THz and mid-IR regimes is especially relevant, since in this spectral range SPPs in noble metals have a light-like behaviour and poor confinement, with wavelengths close to $\lambda_0/\sqrt{\epsilon}$ (at those frequencies the SPPs dispersion curve is indistinguishable from the light line; cf. Fig. 3.3). Moreover, the concentration of charge carriers in graphene can be easily controlled by electrical or chemical doping, allowing the realization of GSPs with tunable properties. Thus, all of the above makes graphene an attractive material for THz plasmonics with tailored properties.

4.1.2 *Field profile of GSPs in single-layer graphene*

Having obtained the dispersion relation of p-polarized SPPs in single-layer graphene, we now want to know in more detail what is the behavior of the electromagnetic fields owing to this type of surface waves. In what follows, we will show how one can work through the equations obtained in the last subsection in order to reconstruct the fields. This exercise will also help us to better visualize, in an explicit fashion, the electromagnetic fields of GSPs. For the sake of clarity, here we will only consider the electric field[7].

With that in mind, let us pick a *characteristic* frequency, say $\omega_c = 2\pi \times 1\text{THz}$; we can then find the corresponding wavenumber q_c by solving numerically the dispersion relation [see, for instance, Eq. (4.13)]. At this point, we only need to set up the expressions for the amplitudes of the fields. Recall that, from Eqs. (4.8) and (4.9), we have

$$E_{j,z} = i\,\mathrm{sgn}(z)\frac{q}{\kappa_j}E_{j,x} \ . \tag{4.27}$$

Note that we can compute the field amplitudes up to a proportionality constant, that is, we have the freedom to normalize the fields as we wish (at least from a theoretical point of view). Normalizing the amplitudes of the fields with respect to $E_{1,x}$, we obtain

$$E_{1,x} = 1 \ , \tag{4.28}$$

$$E_{1,z} = -i\frac{q_c}{\kappa_1} \ , \tag{4.29}$$

$$E_{2,x} = 1 \ , \tag{4.30}$$

$$E_{2,z} = i\frac{q_c}{\kappa_2} \ , \tag{4.31}$$

---

[7]The magnetic field can be obtained in a similar manner. Additionally, one may use the electric-field amplitudes to build the corresponding amplitudes of the magnetic field.

where $E_{1,x} = E_{2,x}$ follows from the first boundary condition (4.11). Once again, in the writing of the above equations, the dependence of $\kappa_{1,2}$ on the frequency (and wavevector) is implicit, i.e. $\kappa_{1,2} \equiv \kappa_{1,2}(q_c, \omega_c)$. Therefore, we now have everything we need to construct the electric field profile, namely a pair (q_c, ω_c) which satisfies the dispersion relation, the parameters $\kappa_{1,2}(q_c, \omega_c)$ and the electric-field amplitudes, given by Eqs. (4.28)-(4.31). Plugging all these into

$$\mathbf{E}_{GSP}\left(\mathbf{r}\right) = E_x(x, z)\,\hat{\mathbf{x}} + E_z(x, z)\,\hat{\mathbf{z}}\ , \tag{4.32}$$

where

$$E_\nu(x, z) = \begin{cases} \Re\left\{E_{1,\nu}e^{iqx}e^{-\kappa_1|z|}\right\}\ , & \text{for } z \leq 0 \\ \Re\left\{E_{2,\nu}e^{iqx}e^{-\kappa_2|z|}\right\}\ , & \text{for } z > 0 \end{cases}\ , \tag{4.33}$$

with $\nu = x, z$ [cf. equation (4.1)], allows us to compute the electric vectorial field and its corresponding magnitude at any given point. The results are summarized in Fig. 4.4. In the top panel we show the (normalized) absolute value of the electric field in the xz-plane, with its respective field lines superimposed. On the other hand, in the central panel we have plotted $\text{sgn}(z)E_z(x, z)$, which enables us to get a clearer picture of the coherent charge-density fluctuations, differentiating the locations with higher density of electrons (blue) and holes (red) [notice the direction of the field lines].

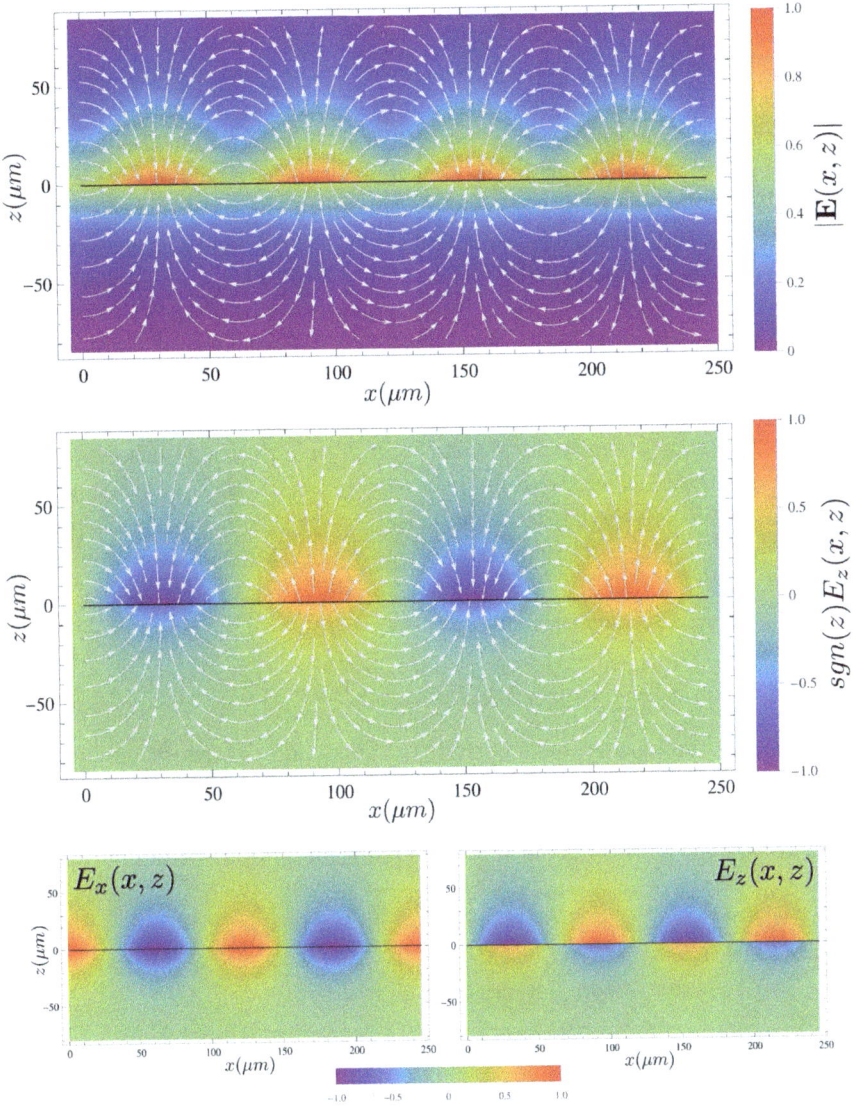

Figure 4.4: Graphical representations of the electric field (a.u.) of TM GSPs in an air/graphene/SiO$_2$ heterostructure, as depicted in Fig. 4.1 (note that here the plots are rotated with respect to Fig. 4.1). The graphene sheet is represented as a black horizontal line. We have used the following parameters: $\omega_c = 2\pi \times 1\,\text{THz}$, $q_c = 0.051\ \mu\text{m}$, $E_F = 0.45$ eV, $\gamma = 0$, $\epsilon_{\text{air}} = 1$ and $\epsilon_{\text{SiO}_2} = 3.9$.

Finally, in the bottom panels we have density plots of the individual x- and z-components of the electric field.

Further, from Fig. 4.4 we can observe the surface-wave character of GSPs, since the magnitude of the field decays exponentially as we move away from the graphene plane (shown as a horizontal black line at $z = 0$ in the figure). Also evident is the collective behavior of these charge-density oscillations, which is particularly evident in the figure's middle panel. On top of that, the figure clearly shows that the electric field attains rather high intensities within a very small region in the vicinity of the graphene sheet. It is then not difficult to conceive that GSPs are indeed interesting candidates to probe small changes in the dielectric environments at the meso- and nano-scales (easily achieved using GSPs with frequencies in the mid-IR). In fact, that is precisely the basic concept which constitutes the main idea behind plasmonic biochemical sensors. Indeed, a graphene-based plasmonic sensor has been experimentally demonstrated in a recent publication, where the sensitivity of the graphene biosensor to the presence of certain proteins (the building blocks of life) utterly outperformed its gold-based counterpart [Rodrigo *et al.* (2015)]. This is undoubtedly a major step forward in the field of graphene plasmonics.

4.1.3   *Plasmon dispersion revisited*

In subsection 4.1.1 we have derived the spectrum of (transverse magnetic) GSPs in extended graphene encapsulated between two insulators, and analyzed the fundamental properties of SPPs in single-layer graphene. In what follows, we will extend our previous study of graphene surface plasmons to the whole (q, ω)-plane (for $\frac{q}{k_F}, \frac{\hbar\omega}{E_F} \in [0, 2]$). With that end in view, we will revisit GSPs' dispersion relation in this extended region using the theoretical tools detailed in Chapter 2.

For the sake of completeness, let us rewrite Eq. (4.13),

$$\frac{\epsilon_1}{\sqrt{q^2 - \epsilon_1 \omega^2/c^2}} + \frac{\epsilon_2}{\sqrt{q^2 - \epsilon_2 \omega^2/c^2}} + i\frac{\sigma_g(q, \omega)}{\omega\epsilon_0} = 0 , \qquad (4.34)$$

which allows us to retrieve the plasmonic spectrum for monolayer graphene. Notice that in writing the previous expression, we have now generalized the system's response by replacing the optical conductivity, $\sigma(\omega)$, by $\sigma_g(q, \omega)$, thereby including non-local effects. In addition, from this equation it is evident that the plasmon dispersion will essentially depend on whatever the conductivity $\sigma_g(q, \omega)$ might be; in other words, we should, in principle, obtain different outcomes if we choose Drude conductivity for graphene or its non-local (Lindhard) counterpart instead. The results ahead clarify when it is permissible to use Drude's conductivity or the non-local optical response instead. The reader should note the difference between Eq. (4.34) and Eq. (3.14) for the metal-dielectric interface case.

In order to assess the sensitivity of the plasmon dispersion to the underlying theoretical framework in which the conductivity is computed, we will solve Eq. (4.34) using (i) graphene's Drude conductivity, $\sigma_D(\omega)$; (ii) local (or homogeneous) Kubo conductivity, $\sigma^{\mathrm{Kubo}}(q \to 0, \omega)$, that is, the dynamical conductivity in the so-called *long-wavelength limit*, and; (iii) the full non-local (or non-homogeneous) conductivity, $\sigma(q, \omega)$. The results are summarized in Fig. 4.5 (see caption for further details). It is clear from the figure that the behavior of the plasmonic band of single-layer graphene seems to be indeed determined by the chosen model for the conductivity. This is particularly true at larger wavevectors (that is, shorter wavelengths) approaching k_F, where the distinction can be rather striking. However, at sufficiently small wavevectors, all models converge to a single, well-defined dispersion curve behaving as $\omega_{GSP} \propto \sqrt{q}$. This validates our previous claim, that one can safely describe the plasmonic spectrum of doped graphene (under typical doping conditions) in the THz up to the lower end of the mid-IR spectral regions, solely by using Drude's expression for the graphene conductivity.

Let us now proceed with our discussion by attempting to discriminate the origins of the differences of the GSPs' dispersion relation observed in Fig. 4.5. We shall withhold our attention to the figure's left panel, since it corresponds to free-standing graphene (that is, graphene in vacuum). In this way we get rid of any influences of the dielectric environment. As noted before, the plasmon dispersion towards the lower left corner of the plot is independent of the model describing graphene's conductivity. The reason for this is two-fold: on the one hand this region corresponds to the long-wavelength regime, so that a local description of the conductivity is sufficient. On the other hand, the contribution coming from the interband part of the conductivity is negligible, and plasmons are also free from interband Landau damping (owing to Pauli blocking). This explains the accuracy of GSPs' descriptions relying on the Drude model in many experimental and theoretical landscapes (as long as they satisfy both of the statements made above). Yet, as we approach the region where plasmons can decay into interband electron-hole excitations (upper inverted triangle in Fig. 4.5), the curve arising from the conductivity modeled by Drude's expression departs

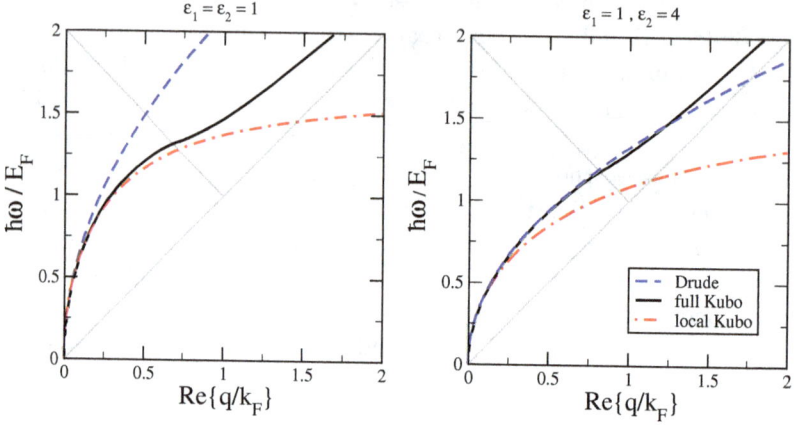

Figure 4.5: Dispersion relation of GSPs using conductivities corresponding to different approximations, in particular: Drude model (blue), as given by Eq. (2.72); local (red) Kubo conductivity, expressed by Eqs. (2.70) and (2.71); and its full, non-local version (black) [via Eqs. (2.84) and (4.34); an equivalent way would be to find the zeros of the RPA dielectric function, i.e. by solving Eq. (2.83) directly from the 2D polarizability of graphene]. All spectra were obtained by employing numerical techniques, since Eq. (4.34) cannot be solved analytically. Left: results for free standing graphene; right: results for graphene cladded between dielectric media with $\epsilon_1 = 1$ and $\epsilon_2 = 4$. Remaining parameters: $E_F = 0.5$ eV and $\hbar\gamma \approx 16$ meV (as reported in [Ju *et al.* (2011)]). All the computations were carried out at zero temperature, which is a very good approximation since $T \ll T_F$, where $T_F = E_F/k_B$. In the right panel, we have also plotted the light line (green), but it is barely indistinguishable from the figure's vertical axis in this scale.

from the others, as it does not account for interband polarization effects. Finally, as we keep moving towards increasingly larger wavevectors, non-local effects become more important and the plasmon dispersion starts to evolve from a square-root scaling to a linear-like behaviour.

Given the resuts dicussed in the preceding paragraphs, it is clear that the plasmon spectrum of monolayer graphene is strongly dependent on the level of approximation we make for the optical condutivity. Hence, depending on the specific region of the (q, ω)-space under investigation the appropriate approximation for the conductivity has to be used.

4.1.4 *Loss function*

Electron energy loss spectroscopies (EELS) have been widely used for many decades to investigate and characterize fundamental properties of various materials. In the early days of plasmonics (in the 1950s-60s [Ritchie (1957); Powell and Swan (1959); Teng and Stern (1967); Raether (1988)]), before the development of a plethora of different techniques to excite and engineer SPPs, EELS was *the method* to probe the physics governing these coherent excitations. More recently, high-resolution reflection EELS was also applied to uncover the dispersion and damping of graphene plasmons in a graphene sheet prepared on a silicon carbide wafer [Liu *et al.* (2008)]. The idea is to expose the sample to a beam of fast electrons with well-defined energies, and then measure the (angle-resolved) energy loss of the scattered electrons, which, namely, carry information on the material's Coulomb interactions and collective excitations, such as plasmons. Within the electron spectroscopies community, it is common to describe the electronic properties of the materials using the dielectric function of the system —usually within the microscopic formalism of the RPA—, from which the *loss function* can be determined; explicitly:

$$L(q,\omega) \propto -\Im\left\{\frac{1}{\varepsilon(q,\omega)}\right\},\tag{4.35}$$

where the longitudinal dielectric function reads,

$$\varepsilon^{\text{RPA}}(q,\omega) = \epsilon_r - \nu_q P(q,\omega),\tag{4.36}$$

as defined in subsection 2.3.3 [see Eq. (2.82)]. By inspection of Eq. (4.35), one can see that this quantity will exhibit peaks corresponding to the poles of $\varepsilon^{-1}(q,\omega)$ [recall that plasmon dispersion may be obtained from $\varepsilon(q,\omega) = 0$]. Therefore, the loss function yields a direct analogy to the EELS experimental spectra.

Still, the usefulness of the loss function extends beyond the explanation of the EELS spectrum, as it can also be used to infer about the collective plasmonic modes excited by optical means. Here, the region of larger wavevectors is reached by coupling light to GSPs in graphene nanostructures and arrays produced using standard lithography techniques, and thus enabling us to surpass the large momentum mismatch between electromagnetic radiation and graphene surface plasmon-polaritons (see, for instance, the right panal of Fig. 4.5, where the dispersion of free-propagating light approaches the vertical axis at $q = 0$; as a consequence, GSPs provide an impressive degree of subwavelength confinement when bechmarked against traditional plasmonic materials such as noble metals).

The loss function of free-standing monolayer graphene is shown in Fig. 4.6, for different doping concentrations. These were computed using the longitudinal dielectric function within the RPA-RT approximation, that is, $L(q, \omega) = -\Im \left\{ \varepsilon^{-1}_{\text{RPA-RT}}(q, \omega) \right\}$. The striking feature in Fig. 4.6 is the presence of a strong plasmon band, owing to the high probability of energy-loss due to the excitation of SPPs in graphene. Note that the GSPs' dispersion obtained from the zeros of $\varepsilon^{\text{RPA}}(q, \omega)$ [continuous curve] is superimposed with the plasmonic band calculated using the loss function, as it should. Once again, in the long-wavelength limit, the loss function is well described by the Drude model (orange dashed curve). Furthermore, notice that the GSPs band becomes fainter until it is utterly washed-out as it enters the interband Landau damping domain [indicated in panel d)]. The reason that is to blame for this latter effect is due to the shutdown of the so-called optical gap sustained by Pauli blocking: by doping graphene, one raises the Fermi level from the Dirac point, leading to the emergence of an optical gap where long-lived plasmons can exist (since $\Im\{ P_\gamma(q, \omega) \} \approx 0$) and decay into electron-hole quasiparticles is forbidden by Pauli's exclusion principle (this is known as the Moss-Burstein effect). This gap has a maximum size of $2E_F$ at $q = 0$ and closes at the Fermi wavevector, $q = k_F$ (for vertical transitions). Hence, the low-loss, stable graphene plasmons which are noteworthy for technological applications will be inside the optical gap where the intensity of the loss function is higher (and the plasmonic band is robust).

While historically it is costumary to compute the loss function from the Lindhard dielectric function, it is also possible to construct an "alternative" loss function by taking the imaginary part of the reflection coefficient (whose poles are indicative of collective excitations). Towards that end, let us write the reflection coefficient for single-layer graphene in vacuum, which reads:

$$r = \left[1 - i \frac{2\omega \epsilon_0}{\sigma(q, \omega) \sqrt{q^2 - \omega^2/c^2}} \right]^{-1}, \qquad (4.37)$$

for TM, or p-polarized, electromagnetic waves (see Appendix N). The general expression for the reflection coefficient of graphene cladded between two dielectrics with arbitrary dielectric constants, ϵ_1 and ϵ_2, is derived in

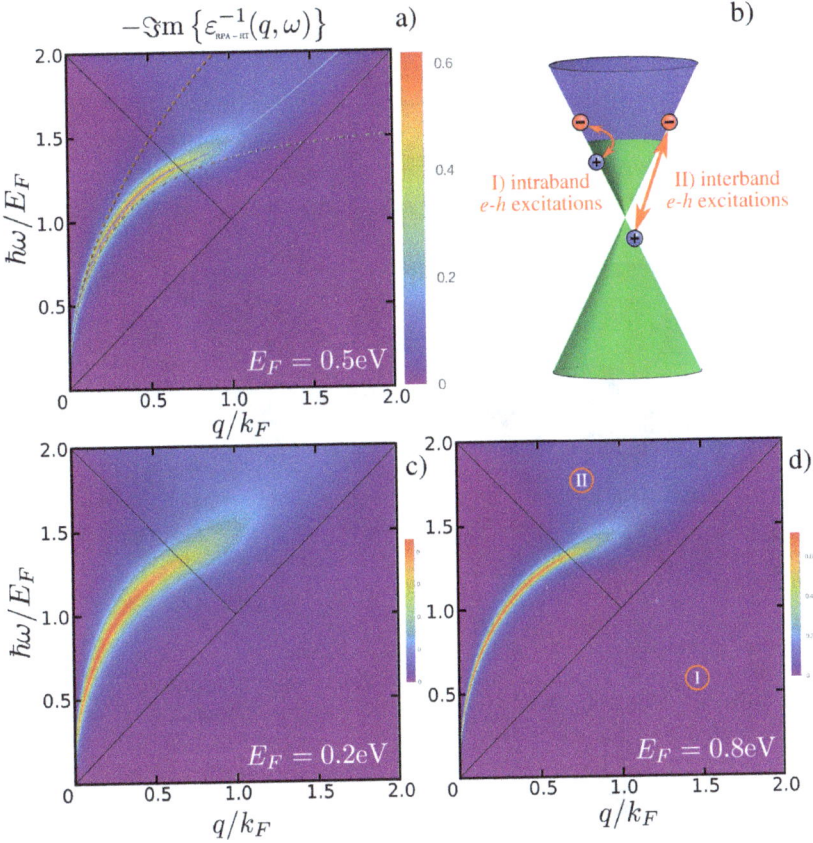

Figure 4.6: Graphene's loss function computed within the RPA-RT framework, $L(q,\omega) = -\Im\left\{\varepsilon_{\text{RPA-RT}}^{-1}(q,\omega)\right\}$, for different doping levels. A strong plasmonic band is clearly visible in the loss function, losing its intensity as it penetrates in the Landau damping regime. The dispersion curves obtained via Drude model (dashed orange), local-RPA (dot-dashed gray) and full RPA (continuous blue) are superimposed with the loss function's intensity plot in panel a). The black lines define the domains corresponding to the regions of Landau damping (electron-hole continuum). In panel b) we have portrayed the electronic processes taking place in those regions, which are labelled accordingly in the last panel. The intensity plots are presented in a $\log_{100}\left[1 + L(q,\omega)\right]$ scale. We have used $\hbar\gamma = 16.5$ meV (as reported in [Ju *et al.* (2011)]). Here, the FWHM of the GSPs band grows with along with the ratio $\hbar\gamma/E_F$.

Sec. N.2. Thus, as noted above, the loss function can also be defined as

$$L_r(q, \omega) = \Im \{r\},$$ (4.38)

where the amplitude r follows from Eq. (4.37). The sole purpose of the subscript in the previous definition is just to emphasize that this latter loss function is written in terms of the reflection coefficient, accounting for both the q- and ω-dependence of the optical excitations in graphene. Figure 4.7

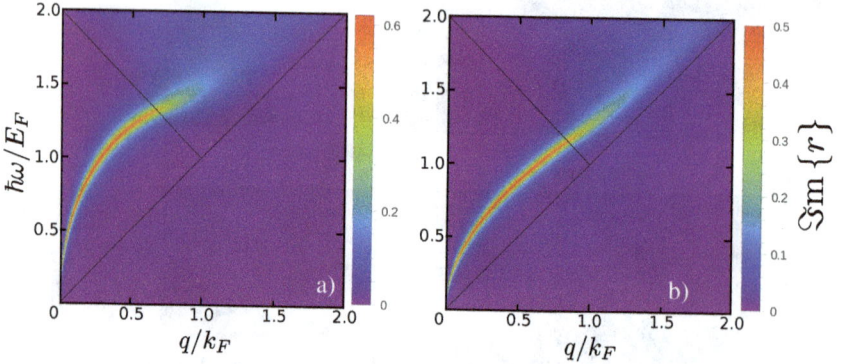

Figure 4.7: Graphene's loss function obtained from the imaginary part of the reflection coefficient, $L(q, \omega) = \Im\{r\}$. The dispersion relation of GSPs can be cleary inferred. Left panel: result for free-standing graphene, i.e. $\epsilon_1 = \epsilon_2 = 1$. Right panel: result for single-layer graphene placed on top of a dielectric with $\epsilon_2 = 4$ (and $\epsilon_1 = 1$). The intensity plots are presented in a $\log_{100}$ scale, and the value of $\hbar\gamma$ is the same as in Fig. 4.6. The Fermi energy was set to $E_F = 0.5$ eV.

depicts the loss function as obtained via Eq. (4.38) in two different cases: for free-standing graphene and for graphene on a substrate with relative permittivity $\epsilon_2 = 4$. Both cases exhibit strong absorption (or energy loss) akin to the excitation of SPPs within the graphene sheet, while rendering the plasmon dispersion in this material. Furthermore, by comparing the panels a) displayed in Figs. 4.7 and 4.6 —which were produced using the very same parameters— it is possible to confirm that the loss function expressed in Eq. (4.38) [using the Fresnel reflection coefficient] is equivalent to the one defined by Eq. (4.35) [in terms of the RPA dielectric function]. This constitutes a clear indication that these two loss functions are indeed equivalent. On the other hand, it should be noted that the regions of high intensity in Fig. 4.7 match the dispersion curves calculated in Fig. 4.5 (black curves, corresponding to the use of the RPA dielectric function), as expected.

In this last two subsections we have provided a further approach to study plasmonic effects in graphene and characterize the plasmon dispersion of GSPs. The major improvements brought forward by these methods —epitomized by Figs. 4.5, 4.6 and 4.7— with respect to the elementary approach outlined in the beginning of this Chapter, is the incorporation of polarization effects and Landau damping, and the extension to a nonlocal response. Although the latter is often negligible in current patterned graphene structures (such as ribbons and disks), whose dimensions are typically larger than Fermi's wavelength, these are predicted to become important as we approach structures with sizes of only a few tens of nanometers [Thongrattanasiri *et al.* (2012b)], which will eventually be within our capabilities taking into consideration the rapid development of fabrication techniques that we have been witnessing during the last couple of decades.

4.1.5 *A word on TE surface waves*

Unlike common metals, graphene has the ability to support transverse electric (TE) surface modes [Mikhailov and Ziegler (2007)]. In the following, we shall obtain the spectrum of TE modes and discuss the necessary conditions for their existence. A direct optical excitation of the transverse electric mode in graphene has recently been achieved using the prism-coupling technique [Menabde *et al.* (2016)]. Experimentally, the excitation of the TE mode leaves its footprint in the form of dips in the reflectance curves (see Fig. 5.5).

Here, we will be looking for solutions of Maxwell's equation of the type

$$\mathbf{E}_j(\mathbf{r},t) = E_{j,y}\hat{\mathbf{y}} \, e^{-\kappa_j|z|}e^{i(qx-\omega t)} \, , \qquad (4.39)$$

$$\mathbf{B}_j(\mathbf{r},t) = (B_{j,x}\hat{\mathbf{x}} + B_{j,z}\hat{\mathbf{z}}) \, e^{-\kappa_j|z|}e^{i(qx-\omega t)} \, , \qquad (4.40)$$

where the relations between the fields' amplitudes are the same as in Eqs. (3.20)-(3.22), and, as it is by now ordinary, $\kappa_j = \sqrt{q^2 - \epsilon_j \omega^2/c^2}$. In this case, the boundary conditions read

$$E_{1,y}(x,z)|_{z=0} = E_{2,y}(x,z)|_{z=0} \, , \qquad (4.41)$$

$$B_{2,x}(x,z)|_{z=0} - B_{1,x}(x,z)|_{z=0} = \mu_0 J_y(x,y) = \mu_0 \sigma_{yy} E_{2,y}(x,z)|_{z=0} \, , \qquad (4.42)$$

which upon substituting the fields (4.39) and (4.40), satisfying the relations (3.20)-(3.22), yields the following equation for the dispersion relation of TE modes in graphene,

$$\boxed{\kappa_1(q,\omega) + \kappa_2(q,\omega) = i\mu_0\omega\sigma(\omega)} \, , \qquad (4.43)$$

where we have once again assumed that the local dynamical conductivity is isotropic. We note that Eq. (4.43) only has solutions if and only if the imaginary part of $\sigma(\omega)$ is negative. A non-zero real part of the conductivity introduces damping by making $\Im m\{q\} \neq 0$. For conventional two-dimensional electron systems, e.g. in GaAs/AlGaAs quantum-well structures, the conductivity follows the Drude model, as long as interband transitions are ignored, so that the imaginary part of the optical conductivity is always positive. Therefore, TE surface modes cannot be realized in 2DEGs. However, graphene's conductivity provides a much richer landscape: besides the Drude-like intraband contribution, there is another term accounting for interband electronic processes, which can be negative. Thus, whenever the net sum of both contributions yields $\Im m\{\sigma\} < 0$, the excitation of TE GSPs is possible —this occurs in the window of frequencies $1.667 < \hbar\omega/E_F < 2$, where interband transitions prevail [Mikhailov and Ziegler (2007); Grigorenko *et al.* (2012); Stauber (2014)].

We further note that the spectrum given by Eq. (4.43) needs to be computed numerically; notwithstanding, since the term on the right-hand side is small —being proportional to α— the TE surface modes dispersion does not depart much from the light line, i.e. $\hbar\omega \lesssim \hbar cq/\sqrt{\epsilon}$, which translates in poor confinement. This last feature makes them less interesting than TM modes, and thus we will consider only TM modes hereafter. Let us add that, in fact, these TE-type "surface modes" are the limiting case of so-called guided waves that can propagate in thin films and exist in both $s-$ and p-polarizations. In contrast with the genuine surface waves, the field profile inside the film is described by a harmonic (and not exponentially decaying) function. In the limit of vanishing film thickness the s-polarized guided waves are transformed into the TE surface waves.

4.2   Double-Layer Graphene

In the last section we investigated the main features of SPPs in a single (continuous) graphene sheet. We have shown that GSPs frequencies typically lie between the THz and the mid-IR regions of the electromagnetic spectrum. In this section we will proceed with our studies of SPPs in graphene, taking one step further on the complexity ladder by considering a system with two graphene layers. This double-layer graphene structure consists in two graphene sheets, separated by a dielectric medium with relative permittivity ϵ_2 and thickness d, enclosed by two semi-infinite top and bottom dielectrics as illustrated in Fig. 4.8. Note that this is somewhat

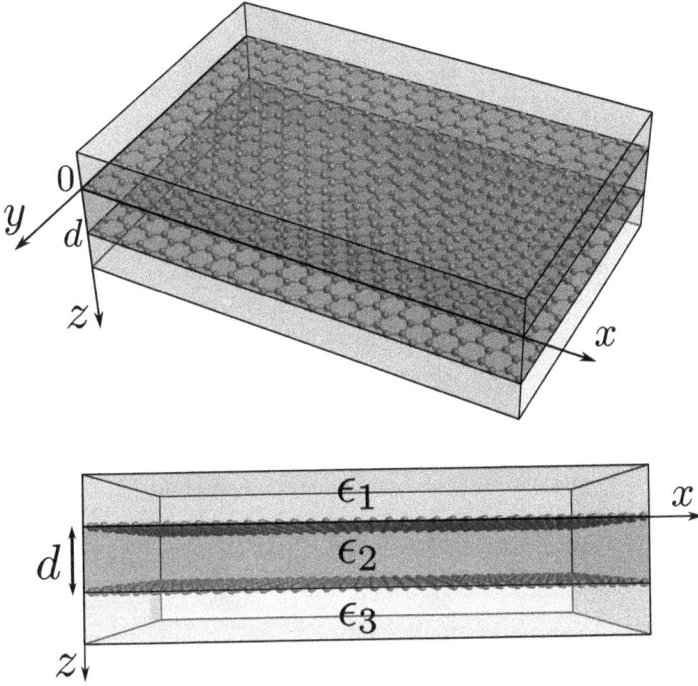

Figure 4.8: Double-layer graphene structure. The graphene sheets (placed at the planes defined by $z = 0$ and $z = d$) are separated by a dielectric slab of thickness d and relative permittivity ϵ_2. Medium 1 occupies the $z < 0$ half-space and the $z > d$ half-space is occupied by the dielectric medium 3.

similar to the case of a dielectric/metal/dielectric heterostructure, in the sense that the system has two interfaces, each capable of sustaining SPPs on their own. Thus, based on what we have seen in subsection 3.2.1, we may already anticipate that when the graphene layers are sufficiently close to one another, the two individual GSPs will interact leading to the emergence of a pair of hybridized GSPs modes. Nonetheless, as in the present case the conductor is (doped) graphene, the corresponding plasmon-polaritons will inherit the remarkable plasmonic properties of graphene that we have outlined in the previous section.

4.2.1 *Dispersion relation*

With the purpose of computing the spectrum of GSPs in double-layer graphene, we assume solutions of Maxwell's equations in the (by now customary) form of,

$$\mathbf{E}_j(\mathbf{r},t) = \left(E_{j,x}\hat{\mathbf{x}} + E_{j,z}\hat{\mathbf{z}}\right) e^{-\kappa_j|z|} e^{i(qx-\omega t)} , \tag{4.44}$$

$$\mathbf{B}_j(\mathbf{r},t) = B_{j,y}\hat{\mathbf{y}}\, e^{-\kappa_j|z|} e^{i(qx-\omega t)} , \tag{4.45}$$

for $j = 1$ ($z < 0$) and $j = 3$ ($z > d$) alone. Here we shall limit ourselves to the study of transverse magnetic modes (such that the electric field lies in the plane of incidence, perpendicular to the interfaces/graphene sheets). For the core dielectric medium $0 < z < d$, we do not need to meet the requirement of vanishing fields at $z \to \pm\infty$, so that, in general, these can be written as a linear combination of exponentially growing "(+)" and exponentially decaying "($-$)" electromagnetic waves, that is

$$\mathbf{E}_2(\mathbf{r},t) = \left\{ \left(E_{2,x}^{(+)}\hat{\mathbf{x}} + E_{2,z}^{(+)}\hat{\mathbf{z}}\right) e^{\kappa_2 z} + \left(E_{2,x}^{(-)}\hat{\mathbf{x}} + E_{2,z}^{(-)}\hat{\mathbf{z}}\right) e^{-\kappa_2 z} \right\} e^{i(qx-\omega t)},$$
$$\tag{4.46}$$

$$\mathbf{B}_2(\mathbf{r},t) = \left(B_{2,y}^{(+)} e^{\kappa_2 z} + B_{2,y}^{(-)} e^{-\kappa_2 z} \right) \hat{\mathbf{y}}\, e^{i(qx-\omega t)} . \tag{4.47}$$

The relations between the amplitudes of the magnetic field with respect to their corresponding electric-field amplitudes, are given by expressions (3.35) and (3.36) for $j = 1, 3$, and Eqs. (3.41)-(3.44) for $j = 2$. Additionally, we have $\kappa_j = \sqrt{q^2 - \epsilon_j \omega^2/c^2}$, as before.

Assuming the same doping concentration in both graphene layers (we shall relax this assumption later), the boundary conditions at $z = 0$ and $z = d$ read

$$E_{1,x} = E_{2,x}^{(+)} + E_{2,x}^{(-)} , \tag{4.48}$$

$$B_{1,y} = B_{2,y}^{(+)} + B_{2,y}^{(-)} + \mu_0\sigma(\omega)E_{1,x} , \tag{4.49}$$

and

$$E_{2,x}^{(+)} e^{\kappa_2 d} + E_{2,x}^{(-)} e^{-\kappa_2 d} = E_{3,x} e^{-\kappa_3 d} , \tag{4.50}$$

$$B_{2,y}^{(+)} e^{\kappa_2 d} + B_{2,y}^{(-)} e^{-\kappa_2 d} - B_{3,y} e^{-\kappa_3 d} = \mu_0\sigma(\omega)E_{3,x} e^{-\kappa_3 d} , \tag{4.51}$$

respectively. Eqs. (4.48)-(4.51) form a linear system, which may be casted in matrix form as

$$\begin{pmatrix} -1 & 1 & 1 & 0 \\ \frac{\epsilon_1}{\kappa_1} + i\frac{\sigma}{\omega\epsilon_0} & -\frac{\epsilon_2}{\kappa_2} & \frac{\epsilon_2}{\kappa_2} & 0 \\ 0 & e^{\kappa_2 d} & e^{-\kappa_2 d} & -e^{-\kappa_3 d} \\ 0 & \frac{\epsilon_2}{\kappa_2} e^{\kappa_2 d} & -\frac{\epsilon_2}{\kappa_2} e^{-\kappa_2 d} & \left(\frac{\epsilon_3}{\kappa_3} + i\frac{\sigma}{\omega\epsilon_0}\right)e^{-\kappa_3 d} \end{pmatrix} \begin{pmatrix} E_{1,x} \\ E_{2,x}^{(+)} \\ E_{2,x}^{(-)} \\ E_{3,x} \end{pmatrix} = 0. \tag{4.52}$$

The dispersion relation of GSPs in a double-layer graphene structure can be obtained by solving $\det(\mathbf{M}) = 0$, where $\mathbf{M}$ is the matrix figuring in Eq. (4.52). After some algebra, one arrives to

$$e^{\kappa_2 d}\left(\frac{\epsilon_3}{\kappa_3} + i\frac{\sigma}{\omega\epsilon_0} + \frac{\epsilon_2}{\kappa_2}\right)\left(\frac{\epsilon_1}{\kappa_1} + i\frac{\sigma}{\omega\epsilon_0} + \frac{\epsilon_2}{\kappa_2}\right) =$$
$$e^{-\kappa_2 d}\left(\frac{\epsilon_3}{\kappa_3} + i\frac{\sigma}{\omega\epsilon_0} - \frac{\epsilon_2}{\kappa_2}\right)\left(\frac{\epsilon_1}{\kappa_1} + i\frac{\sigma}{\omega\epsilon_0} - \frac{\epsilon_2}{\kappa_2}\right). \quad (4.53)$$

The reader should check the differences and similarities between Eq. (3.51) for a metal thin film and Eq. (4.53) for a graphene double-layer. The plasmonic spectrum rendered by Eq. (4.53) is plotted in Fig. 4.9, and it clearly shows the two hybridized modes arising from the interaction among GSPs in opposite graphene sheets (provided that d is small enough to allow the "communication" between the evanescent tails of each individual GSP). This inter-layer electromagnetic interaction splits the spectrum into two branches: one solution lying above the dispersion curve of SPPs in single-layer graphene, corresponding to the *optical mode*; and a lower frequency solution, corresponding to the *acoustic mode*. For the former, the electric-field component along the propagation direction, $E_x(z)$, is symmetric with respect to the $z = d/2$ plane (the charges in both layers oscillate in phase), while for the latter it is antisymmetric with respect to the same plane (the charge-density oscillations are in antiphase)[8]. Moreover, we note that the splitting of the spectrum is stronger for smaller values of d, as Fig. 4.9 illustrates. Even further, it has been shown that the double-layer structure can be optimized in order to obtain low attenuation values for the two modes [Lin and Liu (2014)]. This optimization depends on the value of d.

As before, in the limit $\kappa_2 d \to \infty$, we recover the expressions for the dispersion relations corresponding to the individual, non-interacting GSPs in each graphene layer,

$$\left(\frac{\epsilon_3}{\kappa_3} + \frac{\epsilon_2}{\kappa_2} + i\frac{\sigma}{\omega\epsilon_0}\right)\left(\frac{\epsilon_1}{\kappa_1} + \frac{\epsilon_2}{\kappa_2} + i\frac{\sigma}{\omega\epsilon_0}\right) = 0, \quad (4.54)$$

which are equivalent to Eq. (4.13).

An interesting situation is that of a symmetric environment with $\epsilon_1 = \epsilon_3$ (and, likewise, $\kappa_1 = \kappa_3$). In this case it is possible to find simpler expressions for each solution, namely

$$\omega_{opt}: \quad \frac{\epsilon_2}{\kappa_2}\tanh(\kappa_2 d/2) + \frac{\epsilon_1}{\kappa_1} + i\frac{\sigma}{\omega\epsilon_0} = 0, \quad (4.55)$$

[8]At least in the case of a symmetrical dielectric environment, $\epsilon_1 = \epsilon_3$. If $\epsilon_1 \neq \epsilon_3$ phase-matching may be prevented in some circumstances.

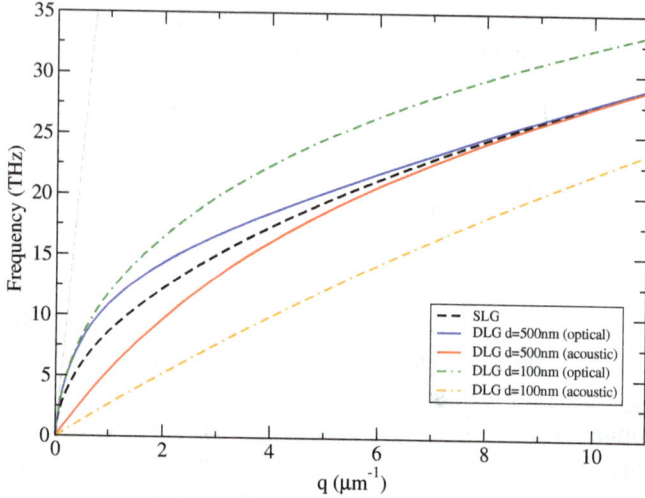

Figure 4.9: Spectrum of TM GSPs in a double-layer graphene structure. The interaction between SPPs in different graphene sheets splits the dispersion relation into two solutions: a higher frequency optical mode and a lower frequency acoustic mode. The dispersion curve of GSPs in single-layer graphene is shown as a black dashed curve for eye guidance. Parameters: $\epsilon_1 = \epsilon_2 = \epsilon_3 = 1$, $E_F = 0.45$ eV and $\gamma = 0$. Note that, for a given frequency, the acoustic mode has a much larger wavenumber, therefore exhibiting a much larger degree of spatial localization.

for the optical mode, and

$$\omega_{ac}: \quad \frac{\epsilon_2}{\kappa_2}\coth(\kappa_2 d/2) + \frac{\epsilon_1}{\kappa_1} + i\frac{\sigma}{\omega\epsilon_0} = 0 \,, \qquad (4.56)$$

for the acoustic one.

Taking the non-retarded limit and assuming that the dielectrics are all made of the same material, i.e. $\epsilon \equiv \epsilon_1 = \epsilon_2 = \epsilon_3$, or within a naive "effective medium" approach, Eq. (4.53) reduces to

$$2\epsilon + i\frac{q\sigma}{\omega\epsilon_0}\left(1 \pm e^{-qd}\right) = 0 \,, \qquad (4.57)$$

where the upper and lower signs correspond to the optical (symmetric) and acoustic (antisymmetric) solutions, respectively. The above expression could also have been obtained from Eqs. (4.55) and (4.56). If we consider that graphene's conductivity is roughly given by Drude's contribution with negligible damping, we can write an explicit, closed-form formula for the GSPs frequency in double-layer graphene:

$$(\hbar\omega_{GSP})^2 \approx \frac{2\alpha}{\epsilon}E_F\hbar cq\left(1 \pm e^{-qd}\right) \,. \qquad (4.58)$$

Naturally, for $qd \to \infty$ we recover the relation for the dispersion of GSPs in single-layer graphene [cf. Eqs. (4.58) and (4.18)].

Supplemented with the treatment outlined above, one could continue to stack as many graphener layers (separated by insulator slabs) as one wants [Kaipa *et al.* (2012); Zhan *et al.* (2013)]. In general, for an arbitrary N-layer graphene configuration, the plasmon spectrum will consist in one optical branch and $N-1$ acoustic ones [Zhu *et al.* (2013b)].

4.2.2 *Field profile of GSPs in double-layer graphene*

Analogously to what we did before in subsection 4.1.2, it is worthwhile to represent the full vectorial electromagnetic fields akin to GSPs modes in double-layer graphene. This will endow us with further insight on the characteristics of each one of the modes. To this end, we set $q_c = 1\ \mu\text{m}$ and then solve the dispersion relation (4.53) numerically, from which we end up with the pair of frequencies $\omega_{\text{opt}}(q_c)$ and $\omega_{\text{ac}}(q_c)$, corresponding to the optical and acoustic modes, respectively. Afterwards, one only needs to reconstruct the fields (4.44)-(4.47) accordingly, in a similar fashion to the case described in subsection 4.1.2 for monolayer graphene.

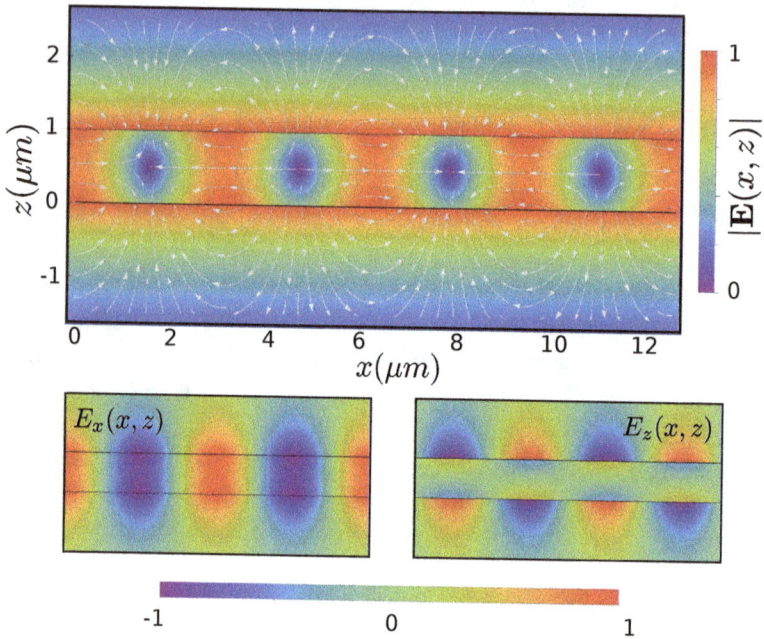

Figure 4.10: Illustration of the computed (vectorial) electric field corresponding to the optical mode, $\omega_{opt}(q_c)$. The top panel shows the respective field lines and the color indicates the field's normalized magnitude. The panels at the bottom show density plots for each component separately (same xz-region as in the main figure). Parameters: $\epsilon_1 = \epsilon_3 = 1$, $\epsilon_2 = 4$, $d = 1$ μm, $E_F = 0.45$ eV and $\gamma = 0$ (for the sake of simplicity; in the case of finite γ, the only difference would be that the intensity of the fields would drop gradually as the GSPs travel along the x-direction).

Figure 4.10 shows the resulting GSPs' electric field relative to the optical mode, whereas Fig. 4.11 displays the electric field corresponding to the acoustic mode. Clearly, the optical mode is made up of in-phase collective charge-density oscillations of the Dirac fermions in each graphene layer, while for the acoustic mode the charge-carriers in each graphene sheet oscillate out-of-phase[9]. Moreover, looking at the panels depicting the

---

[9]This is exactly the opposite to what happens in lattice vibrations. Moreover, in a N-layer structure we should have one "optical" SPP mode and $N - 1$ acoustic modes, again in contrast with the case of phonons. This terminology concerning SPP modes may be confusing because it is contradiction with the well established terminology about phonons. Such a hassle is justified though, since the "acoustic plasmon" exhibits a linear dispersion for small wavevectors, exactly in the same way as "acoustic phonons" do.

electric-field component tangential to the graphene's planes, $E_x(x, z)$, we realize that, for the former, the value of the x-component of the field is always the identical at both graphene sheets, at any given x (at least in this particular case where $\epsilon_1 = \epsilon_3 = 1$); conversely, for the acoustic mode, the value of $E_x(x, z)$ at any given x in one graphene layer has the opposite sign with respect to its value on the other layer. As a consequence, this

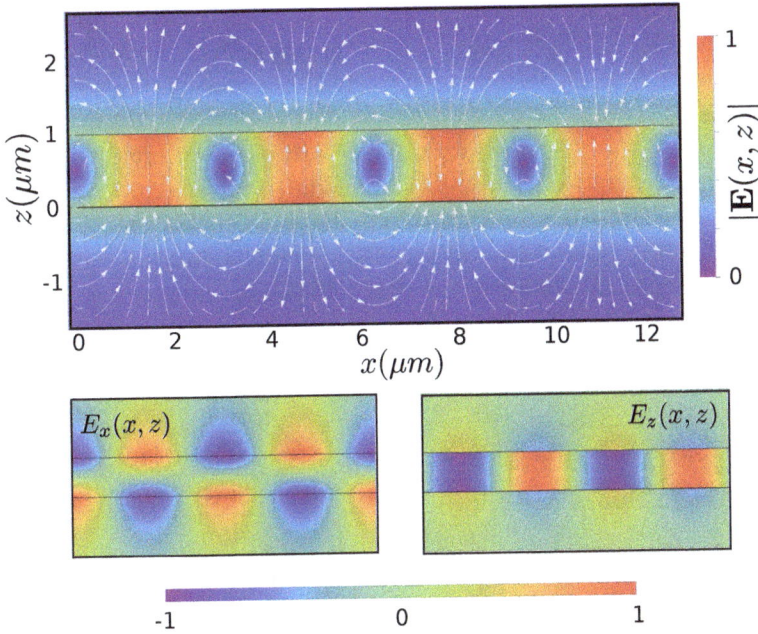

Figure 4.11: Same as in Fig. 4.10, but now relative to the acoustic mode, $\omega_{\text{ac}}(q_c)$. Parameters: $\epsilon_1 = \epsilon_3 = 1$, $\epsilon_2 = 4$, $d = 1$ μm, $E_F = 0.45$ eV and $\gamma = 0$.

determines whether the collective charge-density perturbations oscillate in-phase or out-of-phase. The spatial configurations of the charges akin to both modes are also represented in a pictorial fashion in the figures for eye guidance.

Curiously, notice that in the case of the acoustic branch the inner medium experiences a large polarization, so that the charge densities on the two layers resemble a set of capacitors with alternated polarity.

4.2.3 Plasmon dispersion beyond the Drude model approximation

As we have seen in subsection 4.1.3, the Drude model provides a good description of the plasmonic excitations in graphene in the THz regime up to the lower end of the mid-IR region of the electromagnetic spectrum. The study of the plasmon dispersion in double-layer graphene within Drude's theory was conducted in subsection 4.2.1. Notwithstanding, non-local phenomena and interband polarization effects become increasingly important as we approach higher frequencies and correspondingly larger wavevectors. These were studied in detail for monolayer graphene in the above-mentioned subsection 4.1.3. Here, we shall briefly extend our previous investigation to the double-layer case.

Perhaps the most straightforward approach based on what we have already studied in the double-layer case, would be just to employ the non-local Kubo conductivity —obtained via the graphene's 2D polarizability function [cf. Eq. (2.84)]— instead of the Drude's conductivity, in the equation from which the dispersion relation of GSPs in double-layer graphene is obtained. Within this approach, Eq. (4.53) now becomes:

$$
\begin{aligned}
e^{\kappa_2 d} \left[\frac{\epsilon_3}{\kappa_3} + \frac{\epsilon_2}{\kappa_2} + i\frac{\sigma_2(q,\omega)}{\omega\epsilon_0} \right] \left[\frac{\epsilon_1}{\kappa_1} + \frac{\epsilon_2}{\kappa_2} + i\frac{\sigma_1(q,\omega)}{\omega\epsilon_0} \right] = \\
e^{-\kappa_2 d} \left[\frac{\epsilon_3}{\kappa_3} - \frac{\epsilon_2}{\kappa_2} + i\frac{\sigma_2(q,\omega)}{\omega\epsilon_0} \right] \left[\frac{\epsilon_1}{\kappa_1} - \frac{\epsilon_2}{\kappa_2} + i\frac{\sigma_1(q,\omega)}{\omega\epsilon_0} \right] ,
\end{aligned}
\tag{4.59}
$$

where $\sigma_j(q,\omega)$ refers to the full non-local conductivity of the top and bottom graphene sheets in Fig. 4.8 [respectively, $j = 1$ for the one at $z = 0$ (top) and $j = 2$ for the other at $z = d$ (bottom)].

Another possible route to describe the spectrum of the collective excitations arising in this structure, is to extend the framework of the random-phase approximation (RPA) to the case of a graphene double-layer. In this system, the RPA dielectric function reads [Flensberg et al. (1995); Hwang and Sarma (2007)]:

$$
\begin{aligned}
\varepsilon_{\mathrm{DLG}}^{\mathrm{RPA}}(q,\omega) = [1 - \nu_1(q)P_1(q,\omega)] [1 - \nu_2(q)P_2(q,\omega)] \\
- \nu_{12}(q)\nu_{21}(q)P_1(q,\omega)P_2(q,\omega) ,
\end{aligned}
\tag{4.60}
$$

a result that also appears in the context of Coulomb drag between two Fermi gases [Flensberg et al. (1995); Jauho (1998)]. A derivation of the previous equation is given in Appendix F. Here $P_1(q,\omega)$ is the irreducible polarization function of the top graphene layer, while $P_2(q,\omega)$ is the same quantity for the bottom layer. Additionally, $\nu_1(q) = \nu_2(q) = \frac{e^2}{2\epsilon_0 q}$ and $\nu_{12}(q) = \nu_{21}(q) = \frac{e^2}{2\epsilon_0 q}e^{-qd}$ are, respectively, the Fourier transform of the intra- and inter-layer Coulomb interaction in the elementary case with $\epsilon_1 = \epsilon_2 = \epsilon_3 = 1$ [10]. The main idea behind the model that leads to Eq. (4.60), is to arrive to an effective response function that accounts for the Coulomb interaction not only within the same sheet, but also amongst the two graphene layers, while these are also at the same time coupled by an external scalar potential[11].

The plasmon dispersion in the graphene double-layer structure can now be obtained by finding the zeros of the longitudinal RPA dielectric function of the entire system (recall the discussion at the end of Sec. 2.2),

$$\epsilon_{\text{DLG}}^{\text{RPA}}(q,\omega) = 0 . \tag{4.61}$$

The plasmonic spectra yielded from the numerical solution of Eq. (4.61), for different values of the inter-layer separation, d, is plotted in Fig. 4.12. The figure clearly shows a rather dramatic splitting of the plasmonic band corresponding to the single-layer system (black dot-dashed line), into a pair of distinct hybridized modes: an optical mode (solid lines), where the collective charge-density oscillations in each layer oscillate in-phase; and an acoustic mode (dashed lines), in which the charge-carriers in opposite layers oscillate out-of-phase. These two modes originate from the coupling between the collective excitations in each graphene sheet, and are shared between them, propagating along the structure as a single, composite excitation (this can also be seen in Figs. 4.10 and 4.11). The splitting of the plasmon dispersion observed in the figure is strongly dependent on the inter-layer separation, d, being more prominent as d diminishes[12]. This is direct consequence of the larger coupling between the two sheets for smaller values of d, where each layer "*feels*" the presence of the other more substantially, which can be understood on the basis of $\nu_{\text{inter}}(q) \equiv \nu_{12}(q) = \nu_{21}(q)$ (inter-layer Coulomb interaction) becoming comparable to $\nu_{\text{intra}}(q) = \nu_1(q) = \nu_2(q)$ (intra-layer Coulomb interaction).

This behavior is completely similar to the one we have described in Sec. 4.2.1 (cf. Figs. 4.9 and 4.12). In that section, we have limited ourselves to the study of GSPs in double-layer graphene at THz frequencies, where Drude theory suffices. Here, we extended our previous analysis to a larger region of the (q,ω)-space, which includes non-local and polarization effects,

---

[10]The general case with arbitrary relative permittivities (that of different dielectrics) is considered in Appendix G.

[11]The interested reader is advised to consult Appendix F for further details.

[12]Actually, as the product qd diminishes, to be more rigorous.

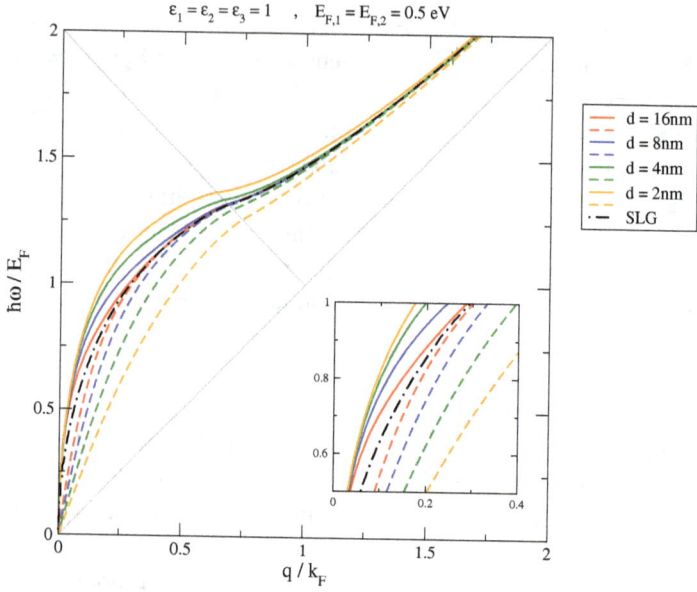

Figure 4.12: Spectra of the plasmonic excitations in double-layer graphene in the phase-space contained in the interval $\{\hbar\omega/E_F, q/k_F\} \in [0,2]$, for several values of d (separation between graphene layers) —see legend box. The dispersion relation, obtained from $\varepsilon_{\mathrm{DLG}}^{\mathrm{RPA}}(q,\omega) = 0$, clearly exhibits two distinct modes: one optical (solid lines) and one acoustic (dashed lines) for each value of d. Both layers posses the same doping ($E_F = 0.5$ eV) and are assumed to be embedded in air. Also shown is the single plasmonic band for the case of monolayer graphene (or, else, for $qd \to \infty$ where the modes are decoupled), which is depicted as a black dot-dashed curve for reference.

and is particularly relevant when studying mid-IR plasmons in double-layer graphene.

In parallel to what we have done for the single-layer case, let us introduce the loss function corresponding to the double-layer structure, defined as

$$L_{\mathrm{DLG}}(q,\omega) = -\Im\mathfrak{m}\left\{ \frac{1}{\varepsilon_{\mathrm{DLG}}^{\mathrm{RPA}}(q,\omega)} \right\} , \qquad (4.62)$$

where the RPA dielectric function is given by Eq. (4.60). This loss function is represented in Fig. 4.13 for a fixed value of d and graphene sheets with balanced doping. Again, we identify two clearly separated branches, one mode with higher frequency (optical) and another one with lower frequency

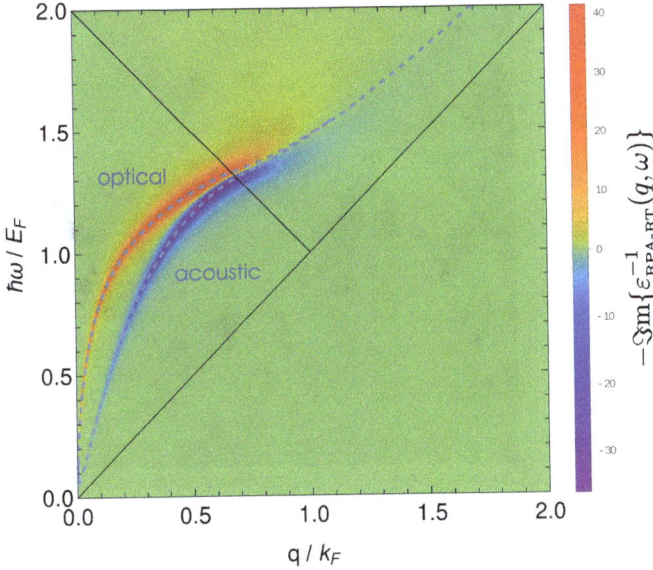

Figure 4.13: Loss function for double-layer graphene embedded in air, exhibiting a branched plasmonic spectrum with two modes which are identified in the figure. The superimposed dashed curves represent the dispersion relation obtained alternatively from Eq. (4.61). We have assumed that both graphene sheets are equally doped and have $E_F = 0.5$ eV, and that they possess the same relaxation-time, $\tau = 2\pi/\gamma$, where $\hbar\gamma = 16.5$ meV. The inter-layer separation was set to $d = 5$ nm.

(acoustic) –when compared to the decoupled modes (Fig. 4.12 black dot-dashed line)—, for a given wavevector. These high-intensity bands are in accordance with the dispersion relation rendered by the zeros of the RPA dielectric function for the double-layer, Eq. (4.61), as they should be. The latter solutions are shown as dashed lines in the figure. Not surprisingly, the GSPs modes lose their strength as they enter the regime dominated by Landau damping. Before ending the discussion of Fig. 4.13, let us remark that the fact that the loss function changes sign does not constitute a violation of causality, since the only physical meaning is ascribed to its zeros, irrespective of the loss function's sign in their neighborhood.

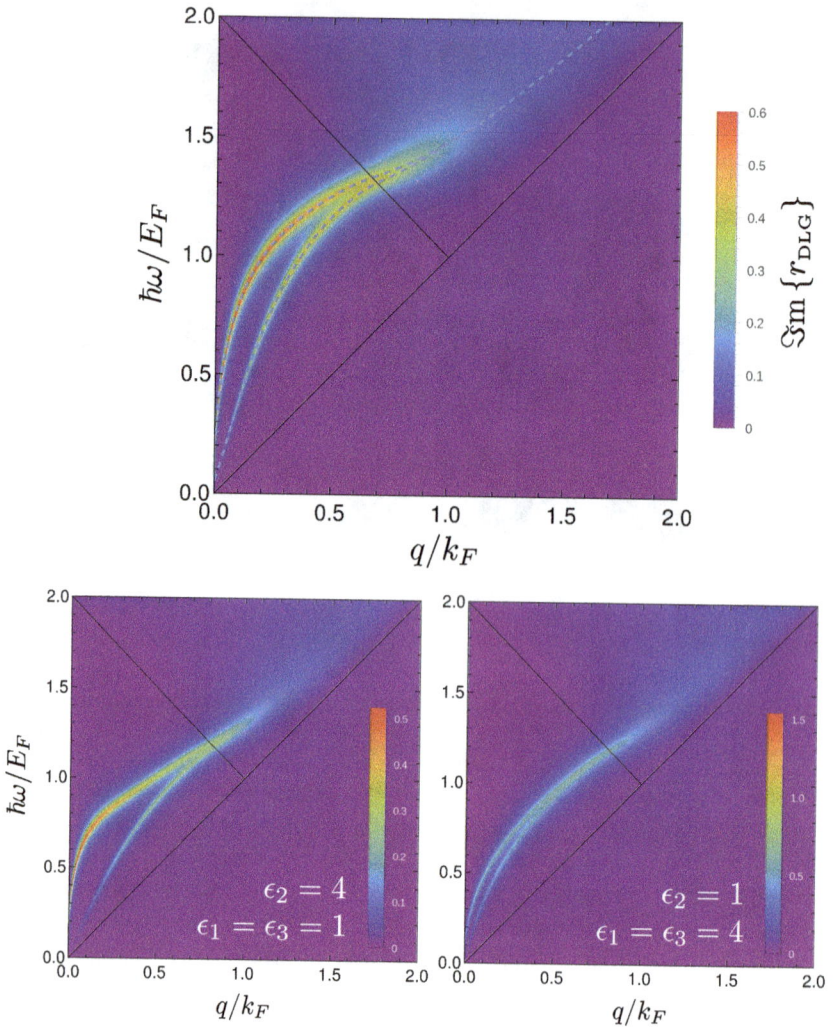

Figure 4.14: Top panel: loss function for double-layer graphene embedded in air, obtained from the imaginary part of the Fresnel reflection coefficient, i.e. via Eq. (4.63) [in logarithmic scale, namely, $\log_{100}(1 + \Im m\, r_{\mathrm{DLG}})$]. The two optical and acoustic hybrid modes appear as well-defined branches of high intensity, following the same dispersion as obtained from the zeros of the RPA dielectric function of the system (dashed line). The bottom panels correspond to configurations with more than one type of dielectric, as indicated in each panel. In all panels we have considered graphene sheets with balanced doping, having set $E_F = 0.5$ eV; the damping was assumed to be $\hbar\gamma = 16.5$ meV (in both sheets) and the inter-layer separation was set to $d = 5$ nm.

In subsection 4.1.4 we have introduced yet another loss function, defined by the imaginary part of the Fresnel reflection coefficient. The same line of reasoning can also be applied to the double-layer system, where the loss function will now be given in terms of the reflection amplitude for the present geometry, that is,

$$L_r(q,\omega) = \Im\left\{r_{\mathrm{DLG}}(q,\omega)\right\} , \qquad (4.63)$$

where r_{DLG} is determined by solving the scattering problem in the double-layer graphene structure. Naturally, at least as far as the dispersion relation is concerned, Eq. (4.63) does indeed produce an outcome equivalent to the ones we have already considered above —that is, the ones determined from either Eq. (4.59), (4.61) or (4.62)—, as Fig. 4.14 plainly demonstrates (e.g. compare, for instance, Figs. 4.14 and 4.13). As in Fig. 4.13, the spectrum contains the same pair of well-defined, coupled hybrid modes, whose position in the (q,ω)-space coincides with the result determined by the solution of Eq. (4.61), which is depicted in Fig. 4.14 as a dashed line. Also shown are alternative configurations involving more than one type of dielectric material, which in turn illustrates that these can be used to alter and diversify the plasmon dispersion in increasingly complex heterostructures.

Note that, up to this point, we have assumed that both graphene layers were equally doped. However, that of course does not necessarily have to be the case. Therefore, before closing our study of the plasmonic spectrum in double-layer graphene, we shall relax this assumption and consider the situation in which the top and bottom graphene sheets possess different values of the Fermi energy. Such a case is portrayed in Fig. 4.15. Once again, we easily recognize the existence of one optical and one acoustic mode, that in the figure appear as two bands of enhanced intensity. We note that in general each mode is affected differently depending on the specific region of the (q,ω)-space where it lives. For instance, a region could be free from Landau damping ascribed to one graphene layer, but suffer from Landau damping arising from the other (e.g. this is the case within the area encompassed by smaller inverted triangle displayed in Fig. 4.15).

Finally, as a way of supplementing the information delivered by the intensity plot, we have also plotted the plasmon dispersion obtained from the numerical solution of Eq. (4.59), using the non-local Kubo conductivity of graphene. Here, the dispersion of the optical plasmon is represented as a dashed line, whereas the dispersion of the acoustic plasmon is shown as a dot-dashed curve. These dispersion curves clearly match the corresponding

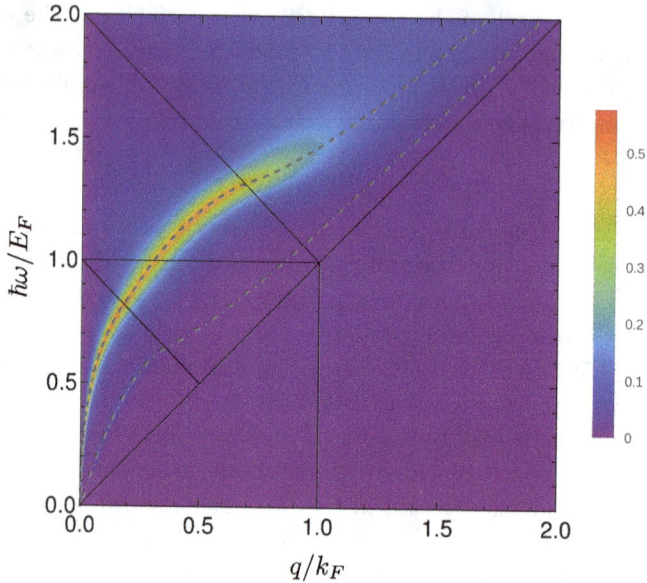

Figure 4.15: Loss function $[\log_{100}(1 + \Im m r_{\mathrm{DLG}})]$ for double-layer graphene embedded in air, in the case where the graphene sheets have different Fermi energies, namely $E_F \equiv E_{F,1} = 0.4$ eV and $E_{F,2} = E_{F,1}/2 = 0.2$ eV. We have assumed a scattering rate of $\hbar\gamma = 16.5$ meV in both sheets and that $d = 5$ nm. The plasmon dispersion as obtained from Eq. (4.59) is also shown, where the dashed and dot-dashed lines represent the optical and acoustic plasmon, respectively. In the making of the plot we have employed graphene's non-local Kubo conductivity, as given by Eq. (2.87).

plasmonic branches seen in the color plot. Furthermore, notice that in the present case, even when the modes decouple (at large qd), their dispersion does not converge into a single line, owing to their different Fermi energies.

To conclude, we bring to the reader's attention that, in principle, nothing prevents us from stacking even more graphene layers into multi-layered graphene structures [Yan *et al.* (2012a); Zhu *et al.* (2013b); Wachsmuth *et al.* (2014)]. We also remark that a theoretical description of such systems could be easily obtained by following the study outlined in this subsection (and corresponding Appendices), since the physics is, in essence, the same.

4.3   Surface Plasmon-Phonon-Polaritons in Graphene

In this section, we will turn our attention to the interaction between graphene surface plasmon-polaritons and surface phonons arising when graphene is deposited (or grown) on a polar substrate, e.g. SiO_2 or hBN.

A material is said to be *polar* if its crystalline structure contains atoms with different electronegativities[13], in other words, when a material chemical bonds are ionic, at least partially. The simplest case are binary solids such as hBN, NaCl or GaAs. Phonons are collective vibrational modes in a solid, and can be categorized as: optical, when the spectrum in the limit $q \to 0$ tends to a finite value; and acoustic, the spectrum in the limit $q \to 0$ tends to zero [14,15]. The latter exhibit a linear dispersion at low wavevectors, with slope equal to material's sound velocity, whereas the optical phonons are essentially dispersionless in the long-wavelength limit. Only optical phonons can directly couple to electromagnetic waves, hence the name (for this reason, optical phonons are dubbed IR-active[16], since their resonant frequencies typically lie in the IR spectral range).

For the sake of simplicity, let us consider an ionic crystal arranged as a sequence of two alternating atomic species, that is, a positively charged ion (cation) and its negatively charged companion (anion). When a lattice mode corresponding to an LO optical phonon reverberates throughout the crystal, the two oppositely charged neighbors (ions) oscillate in antiphase, much like an oscillating electric dipole. The electric field created by these "dipoles" can scatter electrons via this long-range Coulomb interaction, which is known as Fröhlich interaction (between electrons and optical phonons) or simply referred to as polar coupling [Mahan (2000)]. Transverse optical (TO) phonons also interact with electrons through a short-range mechanism called Optical Deformation Potential (ODP). Contrary

---

[13]Thus, mono-elemental crystals cannot be polar.

[14]We bring to the reader's attention that in GSPs both the optical and the acoustic modes in double-layer graphene tend to zero in the limit $q \to 0$. This should cause no confusion.

[15]The classification in acoustic and optical phonons can also be defined in terms of the relative vibration of the atoms. In the limit $q \to 0$ the atoms in the acoustic branch vibrate in phase, whereas the optical phonons are out-of-phase in the same limit.

[16]However, not all optical phonons are IR active. In a simple (partially) ionic crystal like NaCl or CdTe only TO phonons are IR active, in the sense that they can couple directly to EM radiation. In more complex crystals even transverse modes may not be IR active. Being active or not depends on their symmetry. A mode is IR active when its irreducible representation is compatible with that of a polar vector in the point symmetry group of the crystal.

to the Fröhlich interaction, the latter exists also in non-polar semiconductors such as Si.

The coupling between plasmons and phonons was first reported in the sixties [Mooradian and Wright (1966)], in the context of Raman scattering in n-type GaAs samples. Strong plasmon-phonon coupling in graphene has been experimentally observed both using HREELS techniques [Liu and Willis (2010); Koch *et al.* (2010)] (in extended graphene) and by optical excitation (in periodic arrays of patterned graphene [Luxmoore *et al.* (2014); Zhu *et al.* (2014); Yan *et al.* (2013)] and in graphene nanoresonators [Brar *et al.* (2013, 2014)]). As we shall see further ahead, the interaction between graphene surface plasmon-polaritons and the surface optical (SO) phonons of the (polar) substrate reshapes the dispersion of GSPs, and leads to the emergence of new hybrid modes that have been labeled as surface plasmon-phonon-polaritons (SPPhPs).

The aim of this section will be the study of these hybrid modes in a system analogous to the one depicted in Fig. 4.1 —that of single graphene sheet between two insulating media—, but now in the case where the underlying substrate is a polar material. The most straightforward manner to account for this interaction is to relax our initial assumption of a constant dielectric function, to a generalized, frequency-dependent complex dielectric function of the form[17] [Ashcroft and Mermin (1976); Fox (2010)]

$$\epsilon_{\text{subs.}}(\omega) = \epsilon_\infty + \sum_{j=1}^{N} f_j \frac{\omega_{\text{TO},j}^2}{\omega_{\text{TO},j}^2 - \omega^2 - i\omega\Gamma_{\text{TO},j}} , \qquad (4.64)$$

akin to a Lorentz oscillator model with N modes. Here, $\omega_{\text{TO},j}$ corresponds to the frequency of the material's j-th bulk transverse optical (TO) phonon[18], which, for a single phonon frequency, is related to the corresponding Fuchs-Kliewer SO phonon [Fuchs and Kliewer (1965)] via $\omega_{\text{SO}} = \omega_{\text{TO}}\sqrt{\frac{1+\epsilon_s}{1+\epsilon_\infty}}$ [19], where ϵ_s and ϵ_∞ denote the static and high-frequency dielectric constants of the insulator, respectively. Similarly to

[17]The alternative way would be to start from the Fröhlich Hamiltonian describing the coupling amongst graphene's charge carriers and bulk LO phonons of a polar medium [Mahan (2000); Fratini and Guinea (2008); Amorim *et al.* (2012); Schiefele *et al.* (2012)], and to add this interaction to the RPA dielectric function, thus constructing a generalized dielectric function of the entire system. However, this is an unnecessary hassle, as it really does not enlighten us more on the affairs here considered.
[18]The frequency of the bulk longitudinal phonon (LO) is connected to the frequency of the corresponding TO phonon by the Lyddane-Sachs-Teller relation, i.e. $\frac{\omega_{\text{LO}}}{\omega_{\text{TO}}} = \sqrt{\frac{\epsilon_s}{\epsilon_\infty}}$ (in the case of a single mode).
[19]In the case of more than one oscillator (phononic mode), and when the different

the electronic scattering rate, γ, introduced in the previous Chapters, the parameter Γ_{TO} figuring in Eq. (4.64) stands for the TO phonon scattering rate. The dimensionless quantity denoted by f_j characterizes the oscillator strength, that is, it gauges the relative contribution of each mode to the total screening properties of the substrate, and it satisfies the sum rule $\sum_j f_j = \epsilon_s - \epsilon_\infty$. The values of each $f_j = \epsilon_{(j-1)} - \epsilon_{(j)}$ (with $\epsilon_{(0)} \equiv \epsilon_s$ and $\epsilon_{(N)} \equiv \epsilon_\infty$) can be inferred from experiment [Fischetti *et al.* (2001)], where $\epsilon_{(j)}$ are the "intermediate" dielectric constants of the ionic substrate.

The frequencies corresponding to the SO phonons (Fuchs-Kliewer excitations) arising at the interface can also be determined from [Mönch (2010)]

$$\epsilon_{\text{subs.}}(\omega) + 1 = 0 \ , \tag{4.65}$$

(in the absence of graphene and neglecting damping) where we have assumed that the other dielectric is air.

Having outlined the elementary optical properties of polarizable substrates, we are now in possession of all the necessary ingredients to describe surface plasmon-phonon-polaritons in graphene. In the light of the framework introduced above, the expression for the dispersion relation of these surface modes becomes

$$\boxed{\frac{\epsilon_1}{\sqrt{q^2 - \epsilon_1 \omega^2/c^2}} + \frac{\epsilon_{\text{subs.}}(\omega)}{\sqrt{q^2 - \epsilon_{\text{subs.}}(\omega)\omega^2/c^2}} + i\frac{\sigma_g(q,\omega)}{\omega\epsilon_0} = 0} \ , \tag{4.66}$$

which is just Eq. (4.13) with the permittivity of the bottom dielectric replaced by the dynamical dielectric function of a polar substrate, $\epsilon_{\text{subs.}}(\omega)$. Also note that we have allowed for a momentum dependence of the conductivity. Naturally, nothing stops us to consider that the top dielectric is also polar; in that case, we would proceed in the same fashion, that is, performing the replacement $\epsilon_1 \to \epsilon_1(\omega)$ as well.

In what follows, we are going to study the emergence of these hybrid plasmon-phonon modes in monolayer graphene resting on different polar substrates that are typically used in experiments, such as SiC and SiO_2. We shall consider these two cases separately.

---

phonon modes are well separated in frequency, this relation transforms into $\omega_{\text{SO},j} \approx \omega_{\text{TO},j}\sqrt{[1 + \epsilon_{(j-1)}]/[1 + \epsilon_{(j)}]}$, where $\epsilon_{(j)}$ are the "intermediate" relative permittivities of the polar medium (inferred from experiments) [Fischetti *et al.* (2001)]. In general, a given $\omega_{\text{SO},j}$ is a function of all the others $\omega_{\text{TO},j}$.

4.3.1 *Graphene on SiC*

The dynamics of coupled surface plasmon-phonon modes has been experimentally investigated in graphene/silicon carbide heterostructures as "early" as 2010, using REELS methodologies [Liu and Willis (2010); Koch *et al.* (2010)]. The case of SiC has the peculiarity that, apart from being relevant from the experimental point of view, it can also be doped, as reported in [Koch *et al.* (2010)]. Therefore, we also need to include a contribution to the SiC's dielectric function coming from the concentration of (free) charge-carriers induced by doping, which we will do by adding to Eq. (4.64) a term representing the dielectric function of a free electron gas, thereby yielding the following *total* dielectric function for SiC:

$$\epsilon_{\mathrm{SiC}}(\omega) = \epsilon_\infty + \frac{(\epsilon_s - \epsilon_\infty)\omega_{\mathrm{TO}}^2}{\omega_{\mathrm{TO}}^2 - \omega^2 - i\omega\Gamma_{\mathrm{TO}}} - \frac{\omega_p^2}{\omega^2 + i\omega\gamma_e}, \qquad (4.67)$$

where $\omega_p^2 = \frac{e^2 n_e}{m_e \epsilon_0}$ is the plasma frequency. Notice that this expression contains both contributions: a phononic one with a single mode, as in Eq. (4.64), and another accounting for the free electron plasma, given by Eq. (2.33). The spectrum of SPPhPs is then obtained numerically from Eq. (4.66), by taking $\epsilon_{\mathrm{subs.}}(\omega) = \epsilon_{\mathrm{SiC}}(\omega)$.

The resulting dispersion relation of coupled surface plasmon-phonon modes in graphene/SiC is illustrated in Fig. 4.16, indicated by the solid lines (in blue). The key feature visible in the figure is the complete reshaping of the typical $\omega_{GSP} \propto \sqrt{q}$ dispersion of bare GSPs (black dashed curve), to a two-branched dispersion in the vicinity of the frequency of the SiC's optical phonon. This behaviour is suggestive of strong coupling between Fuchs-Kliewer SO phonons arising at the graphene/SiC interface and the collective excitations of charge-carriers in graphene, culminating with the subsequent emergence of hybrid plasmon-phonon coupled modes which, as the name suggests, possess mixed plasmonic and phononic content. The amount of plasmon-like or phonon-like content can be *qualitatively* inferred by studying the spectrum. Looking at Fig. 4.16, we note that for short wavevectors the lower branch is clearly plasmonic-like, since it closely follows the dispersion of uncoupled GSPs (black dashed line). On the other hand, in the same limit, the upper branch seems to posses more phononic content, approaching ω_{SO} as $q \to 0$. Conversely, at larger wavevectors this behaviour is reversed, with the lower branch being phonon-like and essentially dispersionless —resembling a flat band with growing q, at frequency ω_{TO}—, whereas the upper branch of the dispersion relation of SPPhPs converges to the dispersion of bare GSPs modes. This anti-crossing behaviour

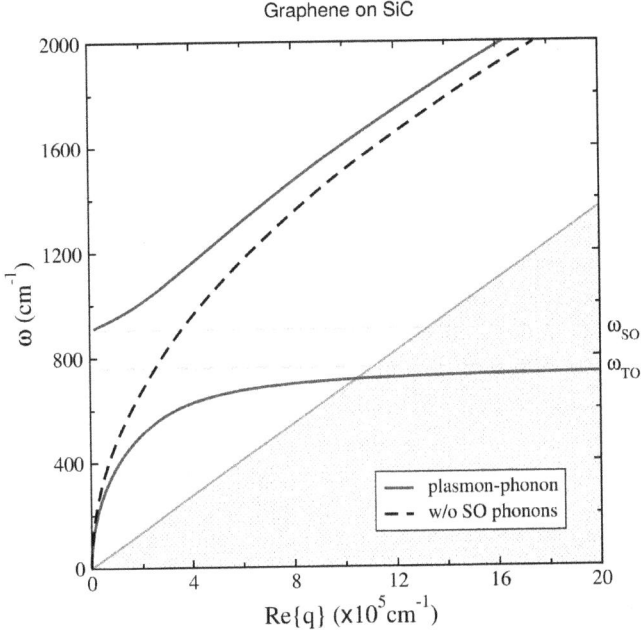

Figure 4.16: Spectrum of surface plasmon-phonon-polaritons for graphene on SiC ionic substrate, which has a Fuchs-Kliewer SO phonon at frequency $\omega_{SO} = 907$ cm^{-1}. The presence of the polar coupling, via Fröhlich interaction, reshapes the plasmonic spectrum of bare (i.e. uncoupled) SPPs in graphene, now exhibiting two branches separated around the frequency of the dispersionless (in the long-wavelength limit) optical phonon. We have modeled graphene's conductivity using Drude's theory, with $E_F = 0.45$ eV and $\hbar\gamma = 8.6$ meV, which is a good approximation within the (q, ω)-region considered in this plot. For the lightly doped 6H-SiC(0001) substrate, we have assumed a frequency-dependent dielectric function with one phononic resonance, as given by Eq. (4.67), parameterized with $\epsilon_\infty = 6.4$, $\epsilon_s = 9.54$, $\omega_p = 16$ cm^{-1}, $\gamma_e = 11$ cm^{-1}, $\omega_{TO} = 760$ cm^{-1} and $\Gamma_{TO} = 3$ cm^{-1} [Koch et al. (2010)]. The brownish shaded area indicates the region of intraband Landau damping for single-particle excitations in graphene.

is indeed symptomatic of strong plasmon-phonon coupling, and has been recognized in experiments with graphene on SiC [Liu and Willis (2010); Koch et al. (2010)], SiO$_2$ [Brar et al. (2013, 2014); Luxmoore et al. (2014); Zhu et al. (2014); Yan et al. (2013)] and hBN [Brar et al. (2014); Woessner et al. (2015)]. Furthermore, one of them [Koch et al. (2010)] reported the observation of quenching of silicon carbide's SO phonons, which was ascribed to the strong plasmon-phonon coupling [20].

4.3.2 *Graphene on SiO$_2$*

Silicon dioxide (SiO$_2$) is perhaps the most common substrate in experiments involving graphene and graphene-based structures[21]. Aside from this fact, it turns out that SiO$_2$ is also polar, so that the analysis of the polar coupling between graphene's electrons (or quasiparticles) and the optical phonons of SiO$_2$ is of the utmost importance in nanophotonics experiments with GSPs (at least whenever the probed region of frequencies is in the neighborhood of the phononic resonances).

As noted above in the beginning of the present section, these phononic resonances contribute to the dielectric function of the material, owing to the polarization field created by the oscillating dipoles (arising from the collective vibrations of the ionic lattice, i.e. optical phonons). Thus, in order to study the plasmon-phonon coupling in a graphene/SiO$_2$ heterosystem, it is compulsory to know the dielectric function of silicon dioxide. This quantity, in the region of the electromagnetic spectrum encopassing the THz and the mid-IR regimes, is plotted in Fig. 4.17. From the figure, one can observe three distinct resonances of phononic nature, which are ascribed to the optical phonons of SiO$_2$. Thus, in this particular case, the dielectric function of the substrate can be written from Eq. (4.64) with $N = 3$ oscillators, that is:

$$\epsilon_{\mathrm{SiO}_2}(\omega) = \epsilon_\infty + \sum_{j=1}^{3} f_j \, \frac{\omega_{\mathrm{TO},j}^2}{\omega_{\mathrm{TO},j}^2 - \omega^2 - i\omega\Gamma_{\mathrm{TO},j}} \, , \qquad (4.68)$$

in terms of the frequencies of TO phonons of the bulk silicon dioxide, weighted by the f_j's (and with linewidths proportional to the damping, $\Gamma_{\mathrm{TO},j}$). The exact value of these parameters vary from experiment to experiment (within a reasonable margin); thereafter, they are often determined for each specific experiment from the corresponding measured data.

As before, the dispersion relation of the plasmonic excitations of the system is determined from Eq. (4.66) using the suitable dielectric function of the substrate, which in the current case is given by Eq. (4.68). The

---

[20]We note that the so called Hopfield coefficients (see, e.g., [Hopfield (1958); Tokunaga *et al.* (2001)]) are a quantitative measure of the fraction of plasmon and phonon in the mixed excitation. They can be expressed through the dispersion curves of the coupled and uncoupled excitations.

[21]In fact, the existence of one-atom-thick graphene was empirically determined for the first time on a SiO$_2$/Si substrate [Novoselov *et al.* (2004)].

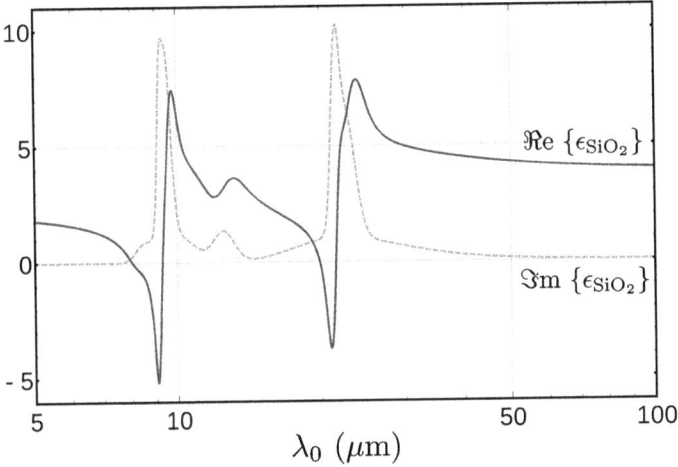

Figure 4.17: Dielectric function of SiO$_2$ as a function of the vacuum wavelength, from the THz up to the mid-IR spectral range. The solid blue curve represents the real part of the dielectric function of the silicon dioxide, while its imaginary part is represented by the dashed orange line. This plot was obtained using the parameterization given in [Kitamura *et al.* (2007)] for silica glass.

solution of the aforementioned transcendental equation is depicted in Fig. 4.18. The phenomenological parameters entering the dielectric function of SiO$_2$ are taken from [Constant *et al.* (2015); Luxmoore *et al.* (2014)] (refer to the figure's caption for further details). The figure plainly shows that the polaritonic spectrum in a graphene/SiO$_2$ heterosystem is much more complex than the one in the case where graphene sits on a non-polar dielectric. To begin with, the dispersion relation is not a single curve growing proportionally to $\sqrt{q}$ (in this region of the phase-space), but rather a set of well-defined branches exhibiting an anti-crossing behaviour, each of them separated by the horizontal lines at the SO phonons' frequencies. Note that this is exactly the same behavior that we have seen in the previous subsection 4.3.1; the difference is that in the present case we have three distinct phononic resonances, but the underlying physics is the same: the optical phonons of the polar crystal interact with graphene's electrons (or holes) through the long-range Fröhlich interaction. This electron-phonon coupling is the origin of the plethora of new hybrid surface plasmon-phonon modes visible in the figure. The degree of plasmonic/phononic content of each mode, strongly depends on its spectral position relative to the frequencies of the substrate's optical phonons, with the SPPhPs modes becoming

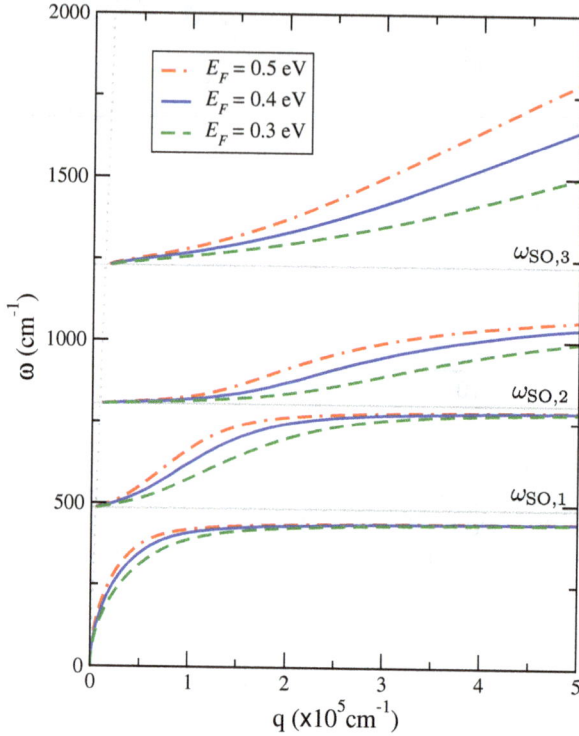

Figure 4.18: Polaritonic spectrum of surface plasmon-phonon-polaritons in graphene placed on a polar SiO_2 substrate, for different levels of doping. In the vicinity of the resonances akin to the optical phonons of SiO_2, these hybridize with graphene's surface plasmons, thus originating strongly coupled plasmon-phonon surface modes which display anti-crossing behavior. We have used the following parameters: $\omega_{TO,j} = \{448.3, 792.2, 1128.8\}$ cm^{-1}, $f_j = \{0.7514, 0.1503, 0.6011\}$ and $\epsilon_\infty = 2.4$ [Constant *et al.* (2015); Luxmoore *et al.* (2014)]. Here, damping is assumed to be negligible and $\sigma_g(q,\omega) \to \sigma_g^{\text{Drude}}(\omega)$. Light's dispersion is also shown as a dotted gray line.

more phonon-like (and dispersionless) as they approach any of the SO frequencies. On the contrary —away from $\omega_{SO,j}$— the SPPhPs modes tend to become essentially plasmon-like. Finally, notice that the transition between modes which are plasmon-like, to modes that are phonon-like, tends to be more abrupt as we move towards higher doping concentrations. Also, and perhaps not surprisingly, one can see that as we increase the Fermi level, the spectra shift towards higher energies as for bare GSPs modes. Therefore,

SPPhPs in grahene carry the same tunability as uncoupled graphene sur-face plasmons, while suffering from lower losses [Woessner *et al.* (2015); Ja-cob (2014)]; these two characteristics, together with strong subwavelength confinement, are indeed very much desirable for technological applications. Strong plasmon-phonon coupling between GSPs and SiO_2's optical phonons has been demonstrated in recent experiments [Brar *et al.* (2013, 2014); Lux-moore *et al.* (2014); Zhu *et al.* (2014); Yan *et al.* (2013)], whose reported behavior is consistent with the results we have just discussed.

In a sort of conclusion, let us close this section with a practical example which sums up all the relevant phenomena that we have been discussing so far, which is demonstrated in Fig. 4.19. This figure condenses all the theoretical framework outlined in the preceding lines, applied to the specific experimental conditions as reported by [Luxmoore *et al.* (2014); Constant *et al.* (2015)][22]. The most eye-catching element in the figure is the color-plot referring to the imaginary part of the Fresnel reflection coefficient for *p*-polarized light —loss function (cf. subsection 4.1.4)— which gives in-formation on the optical excitations in the system, namely, SPPhP bands. These can be clearly seen in the plot as four high-intensity bands displaying the same anti-crossing behavior akin to the strong coupling between surface plasmons in graphene and the optical phonons in SiO_2. The width of these hybrid plasmon-phonon bands goes hand in hand with both electronic and phononic damping (that is, higher damping corresponds to broader bands). Furthermore, the numerical solution of Eq. (4.66) is shown as gray dots; since now the damping is finite, for each real frequency, Eq. (4.66) renders a complex-valued wavevector. Thus, the horizontal axis actually represents $\Re\{q\}$ when referring to the dots in gray. Another side-effect of $\gamma \neq 0$ is the back-folding of the dispersion curves near the transitions between each band (or, equivalently, near $\omega_{SO,j}$). This is to be expected, in the light of what we have studied in subsection 3.1.1 when we considered the case of a dielectric-metal interface with a lossy conductor. In addition, notice that the numerically computed solutions of Eq. (4.66) are superimposed with the regions of higher intensity of the color-plot, as they should be. In contrast, the classic $\omega_{GSP} \propto \sqrt{q}$ spectrum of bare GSPs is portrayed as a dashed curve in white color. The difference between this line and the

---

[22]Please see the caption of the figure for details. The only phenomenological parameter not taken from experimental data was the electronic scattering rate, that we have set to $\gamma = 8$ meV. This does not alter our analysis in any way, with the exception of the widths of the SPPhP bands (refer to the main text below).

SPPhP bands provides compelling evidence of the strong plasmon-phonon coupling in this system, thus revealing that polarizable substrates such as SiO_2, SiC or BN influence the carrier dynamics and collective excitations in graphene.

In recent times, the investigation of these tunable surface plasmon-phonon coupled modes in graphene and graphene-based structures has been picking up momentum as it starts to unfold new possible routes of exploration in plasmonics and nanophotonics, which may be potentially relevant for future photonic technologies and devices.

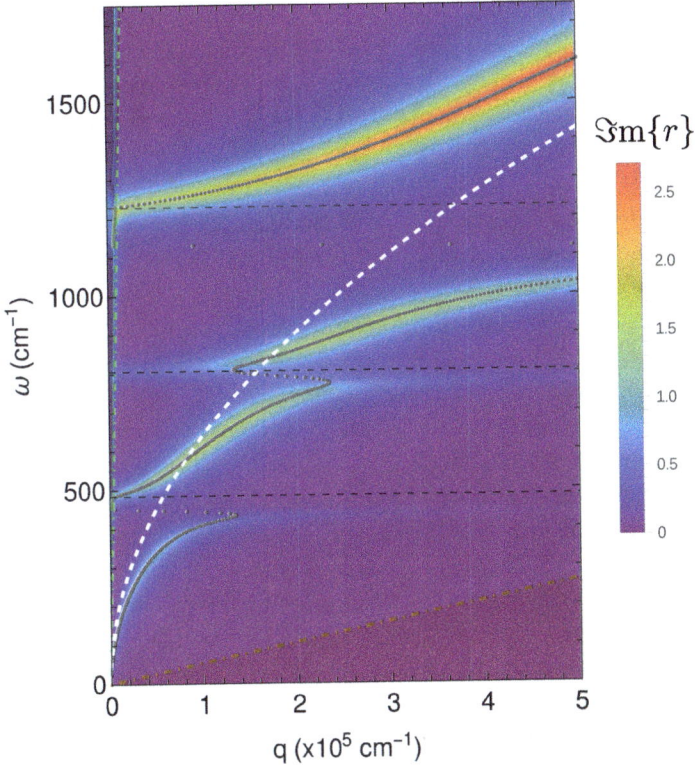

Figure 4.19: Loss function (in logarithmic scale) for graphene on SiO$_2$, exhibiting hybrid surface plasmon-phonon-polaritons modes resulting from the long-range Fröhlich interaction amongst GSPs and optical phonons of the silicon dioxide. The dielectric function of the latter is parameterized as in Fig. 4.18, but now in the presence of phonon damping, with $\hbar\Gamma_{TO} = \{26.7, 42.4, 42.4\}$ cm^{-1} [Constant *et al.* (2015); Luxmoore *et al.* (2014)], and graphene's conductivity is modeled using Drude's theory, which is accurate in the represented (q, ω)-space, where we have assumed $E_F = 0.37$ eV [Constant *et al.* (2015); Luxmoore *et al.* (2014)] and $\hbar\gamma = 8$ meV. The dispersion of coupled plasmon-phonon modes, obtained via the numerical solution of Eq. (4.66) [using Eq. (4.68)] is shown as gray dots [with coordinates $(\Re\{q\}, \omega)$, since q is complex-valued in the case of finite damping]. The observed folding of the dispersion curve is a result of the non-zero damping. The bare GSPs' dispersion ($\propto \sqrt{q}$) is plotted as a dashed curve (in white), and the $\omega_{SO,j}$ frequencies are illustrated as horizontal dashed lines (in black). The shaded areas in the lower right and in the upper left depict the region of intraband Landau damping (in brown) and the dispersion of light in the SiO$_2$ (in green), respectively.

4.4 Magneto-Plasmons in Monolayer Graphene

It is well-known that the introduction of a magnetic field influences the dynamics of charged particles. One of the most elementary examples of such an effect that any undergraduate student in physics or engineering encounters is the Hall effect [Hall (1879); Ashcroft and Mermin (1976)], whose classical description directly stems from the Lorentz force law applied to a free charged particle. In fact, the study and exploitation of magnetic fields interacting with matter is a broad subject by itself, encopassing several areas of human knowledge: from the astrophysics of interstellar space and plasmas, electrodynamics, magneto-optics, or even as an aid in diagnostic medicine, with magnetic resonance imaging (MRI) being the most prominent example.

Also in what plasmonics is concerned, the interaction of plasmons with applied magnetic fields offers yet another dimension to be explored. The goal of the present section is, however, more humble, since the number of topics that one can possibly cover in one book is naturally limited. Nonetheless, it is important, at least, to discuss briefly the area of magneto-plasmonics in graphene. In what follows, we provide a derivation of the dispersion relation of surface magneto-plasmon-polaritons (MPPs) in single-layer graphene, when this material is subjected to a static external magnetic field perpenducilar to the graphene sheet (see Fig. 4.20). First, we will characterize the polaritonic spectrum of the hybrid MPPs within a semi-

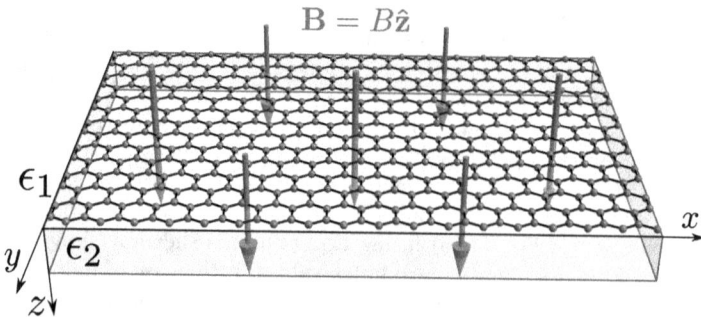

Figure 4.20: Pictorial representation of monolayer graphene permeated by an uniform **B** field perpendicular to its plane. The graphene sheet (at $z = 0$) is encapsulated by two semi-infinite dielectrics with relative permittivities ϵ_1 and ϵ_2, respectively occupying the half-planes defined by $z < 0$ and $z > 0$.

classical approach (and discuss its regime of validity); from this we shall obtain an analytical expression in closed-form for the spectrum, which can be obtained by making some (not too drastic) approximations. Finally, we end this section by comparing those results with the ones obtained using a quantum mechanical approach that takes Landau level quantization into account.

4.4.1 *Derivation of the spectrum's condition*

When an external magnetic field is present, the boundary conditions at the interface change because a Hall current is produced. As such, we cannot simply use the form of the fields from subsections 4.1.1 and 4.1.5, so that we have to start from the general solution instead:

$$\mathbf{E}_j = (E_{j,x}, E_{j,y}, E_{j,z})e^{i(q_j x - \omega t)}e^{-\kappa_j |z|} , \qquad (4.69)$$

$$\mathbf{B}_j = (B_{j,x}, B_{j,y}, B_{j,z})e^{i(q_j x - \omega t)}e^{-\kappa_j |z|} , \qquad (4.70)$$

where $q_j, \kappa_j \in \mathbb{C}$, with $\Re\{\kappa_j\} \geq 0$ and $\Re\{q_j\}/\Im\{q_j\} \geq 0$, and j labels the media in accordance with Fig. 4.20. Note that these solutions are no longer TM or TE waves in general.

Maxwell's equations in the absence of volume charges and currents lead to the relations,

$$B_{j,x} = -i\,\mathrm{sgn}(z)\frac{\kappa_j}{\omega}E_{j,y} , \qquad (4.71)$$

$$B_{j,y} = i\,\mathrm{sgn}(z)\frac{\kappa_j}{\omega}E_{j,x} - \frac{q_j}{\omega}E_{j,y} , \qquad (4.72)$$

$$B_{j,z} = \frac{q_j}{\omega}E_{j,y} , \qquad (4.73)$$

from the one involving the curl of the electric field, and

$$B_{j,x} = -i\,\mathrm{sgn}(z)\frac{q_j}{\kappa_j}B_{j,z} + i\,\mathrm{sgn}(z)\frac{\omega\epsilon_j\mu_j}{c^2\kappa_j}E_{j,y} , \qquad (4.74)$$

$$B_{j,y} = -i\,\mathrm{sgn}(z)\frac{\omega\epsilon_j\mu_j}{c^2\kappa_j}E_{j,x} , \qquad (4.75)$$

$$B_{j,y} = -\frac{\omega\epsilon_j\mu_j}{c^2 q_j}E_{j,z} , \qquad (4.76)$$

from the other relating the curl of the magnetic field with the electric-field time derivative. As in the above-mentioned subsections, these equations lead to the same relation between wavevector components found with the TM/TE ansatz, i.e. that q_j and κ_j are connected through the relation

$$q_j^2 = \kappa_j^2 + \epsilon_j\mu_j\omega^2/c^2 , \qquad (4.77)$$

where ϵ_j and μ_j refer to the dielectric constant and relative (magnetic) permeability of the medium j, respectively. As for the boundary conditions (cf. subsection 2.1.2), they have some extra terms due to the extra field components. These give:

$$E_{1,x} = E_{2,x} \wedge E_{1,y} = E_{2,y}\,, \qquad (4.78)$$

$$q \equiv q_1 = q_2\,, \qquad (4.79)$$

$$\mu_1^{-1}B_{1,y} - \mu_2^{-1}B_{2,y} = \mu_0\left[\sigma_{xx}E_{1,x} - \sigma_{xy}E_{1,y}\right]\,, \qquad (4.80)$$

$$\mu_2^{-1}B_{2,x} - \mu_1^{-1}B_{1,x} = \mu_0\left[\sigma_{yy}E_{1,y} - \sigma_{yx}E_{1,x}\right]\,. \qquad (4.81)$$

The boundary condition for the normal component of the electric displacement vector was not written since it gives no additional information.

Let us study the condition relating the fields expressed by Eq. (4.80). We shall focus on the isotropic case, where $\sigma_{xx} = \sigma_{yy}$ and $\sigma_{yx} = -\sigma_{xy}$. Using the same procedure as in subsections 4.1.1 and 4.1.5 [namely, with the aid of Eqs. (4.75) and (4.78)], we arrive at

$$\frac{1}{\mu_0\mu_1}\left(1 + \frac{\epsilon_2\kappa_1}{\epsilon_1\kappa_2}\right)B_{1,y} = \sigma_{xx}\left(\frac{c^2\kappa_1}{i\omega\epsilon_1\mu_1}\right)B_{1,y} - \sigma_{xy}E_{1,y}\,. \qquad (4.82)$$

Notice that, contrary to the case with $\sigma_{xy} = 0$, here we cannot simply make $B_{j,x} = B_{j,z} = E_{j,y} = 0$ in order to obtain a TM mode, since in that case Eq. (4.81) would imply $E_{1,x} = 0$, and thus $B_{1,y} = 0$ [cf. Eq. (4.75)]; i.e. a wave would not exist! As for the last boundary condition, Eq. (4.81), we end up with

$$\frac{1}{i\omega\mu_0}\left(\frac{\kappa_1}{\mu_1} + \frac{\kappa_2}{\mu_2}\right)E_{1,y} = \sigma_{xx}E_{1,y} + \sigma_{xy}E_{1,x}\,, \qquad (4.83)$$

where we have proceeded in a similar fashion as for the other condition. Also, if one now attempts to obtain a TE mode by setting $E_{j,x} = B_{j,y} = E_{j,z} = 0$ when $\sigma_{xy} \neq 0$, this would lead to all components being null: no wave. A wave solution still can be worked out, provided that all the components are non-zero.

Let us now obtain the condition from which the spectrum of magnetoplasmons in graphene stems. By making use of the relation (4.75), Eq. (4.80) becomes

$$\left(\frac{\epsilon_1}{\kappa_1} + \frac{\epsilon_2}{\kappa_2} + i\frac{\sigma_{xx}}{\omega\epsilon_0}\right)E_{1,x} = -i\frac{\sigma_{xy}}{\omega\epsilon_0}E_{1,y}\,, \qquad (4.84)$$

whereas Eq. (4.83) implies that

$$E_{1,x} = \frac{i\epsilon_0 c^2}{\omega \sigma_{xy}} \left(\frac{\kappa_1}{\mu_1} + \frac{\kappa_2}{\mu_2} - i\omega\mu_0\sigma_{xx} \right) E_{1,y}, \qquad (4.85)$$

and thus, from the combination of the two above equations, we finally obtain the spectrum's condition akin to MPPs in graphene [Ferreira *et al.* (2012); Chiu and Quinn (1974)]:

$$\boxed{\left[\frac{\epsilon_1}{\kappa_1} + \frac{\epsilon_2}{\kappa_2} + i\frac{\sigma(\omega)}{\omega\epsilon_0} \right] \left[\frac{\kappa_1}{\mu_1} + \frac{\kappa_2}{\mu_2} - i\omega\mu_0\sigma(\omega) \right] = -Z_0^2\sigma_H^2(\omega)}, \qquad (4.86)$$

where $Z_0 = \sqrt{\mu_0/\epsilon_0}$ is the vacuum impedance and we have introduced the following notation for the components of the : $\sigma(\omega) \equiv \sigma_{xx}(\omega) = \sigma_{yy}(\omega)$, describing the longitudinal homogeneous dynamical conductivity of graphene, and $\sigma_H(\omega) \equiv \sigma_{xy}(\omega) = -\sigma_{yx}(\omega)$ for graphene's Hall conductivity. Moreover, note that the first term between brackets on the left-hand side of the previous equation corresponds to the expression for the spectrum of TM graphene surface plasmons, while the second term can be identified as the condition from which the spectrum of graphene's TE surface plasmons follows. Clearly, when the magnetic field is zero, so it is σ_H, and Eq. 4.86 factorizes into a product of two terms: one akin to TM GSPs and the other to TE GSPs. However, if the system is subjected to a non-vanishing magnetic field, a Hall current is produced and the dispersion relation of magneto-plasmons in monolayer graphene becomes more intricate, obeying Eq. (4.86).

In the case where the two insulating media are equal, the spectrum is obtained by solving a simpler transcendental equation, that is

$$\boxed{\left[\frac{2\epsilon}{\kappa} + i\frac{\sigma(\omega)}{\omega\epsilon_0} \right] \left[\frac{2\kappa}{\mu} - i\omega\mu_0\sigma(\omega) \right] = -Z_0^2\sigma_H^2(\omega)}. \qquad (4.87)$$

4.4.2 *Solution of the semi-classical spectrum's condition in a special limit*

As we have just seen, the dispersion relation of MPPs in single-layer graphene follows from the condition (4.86). This is a transcendental equation that needs to be solved numerically in order to obtain the spectrum of MPPs; furthermore, it is not self evident, just by looking at Eq. (4.86), how the dispersion relation of GSPs is modified when an external (uniform) magnetic field is applied. Therefore, our only hope to get a glimpse of the effect of the magnetic field on the GSPs' dispersion —apart from a plotting

the numerical solution—, is to solve Eq. (4.86) in a particular limit, where
some kind of approximation could be employed (and which hopefully retain
most of the physics).

For the sake of simplicity, let us assume that $\mu_1 = \mu_2 = 1$. Neglecting
retardation effects (which is often an accurate approximation in graphene
plasmonics due to the high subwavelength confinement owing to GSPs),
that is, taking $\kappa_1, \kappa_2 \to q$, Eq. (4.86) translates to

$$\left[\frac{2\bar{\epsilon}}{q} + i\frac{\sigma(\omega)}{\omega\epsilon_0}\right][2q + i\mu_0\omega\sigma(\omega)] = -Z_0^2\sigma_H^2(\omega) , \qquad (4.88)$$

where $\bar{\epsilon} = (\epsilon_1 + \epsilon_2)/2$. Solving this equation for q, yields

$$q(\omega) = \frac{H(\omega) \pm \sqrt{H^2(\omega) + 16\bar{\epsilon}F(\omega)G(\omega)}}{4F(\omega)} , \qquad (4.89)$$

where we have defined the following auxiliary functions:

$$H(\omega) = F(\omega)G(\omega) - Z_0^2\sigma_H^2(\omega) - 4\bar{\epsilon} , \qquad (4.90)$$

$$F(\omega) = \frac{i\sigma(\omega)}{\omega\epsilon_0} , \qquad (4.91)$$

$$G(\omega) = i\mu_0\omega\sigma(\omega) . \qquad (4.92)$$

Naturally, the fundamental ingredient in the present calculation is of course
the magneto-optical conductivity tensor. In the present discussion, we shall
neglect the role of optical phonons and consider only the effect of intraband
transitions driven by the applied fields, in the long-wavelength (local) limit.
The latter is justified for small wavevectors, that is $q \ll k_F, l_B$, where
$l_B = \sqrt{\frac{\hbar}{eB}}$ denotes the *magnetic length*. Hence, within the semi-classical
approach of Boltzmann's transport theory, the components of graphene's
conductivity tensor read [23]:

$$\sigma_{xx}(\omega) = \sigma_0\frac{4|E_F|}{\pi}\frac{(\Gamma - i\hbar\omega)}{(\Gamma - i\hbar\omega)^2 + \Omega_c^2} \equiv \sigma(\omega) , \qquad (4.93)$$

$$\sigma_{xy}(\omega) = -\sigma_0\frac{4|E_F|}{\pi}\frac{\Omega_c}{(\Gamma - i\hbar\omega)^2 + \Omega_c^2} \equiv \sigma_H(\omega) , \qquad (4.94)$$

$$\sigma_{yx}(\omega) = -\sigma_{xy}(\omega) , \qquad (4.95)$$

$$\sigma_{xx}(\omega) = \sigma_{yy}(\omega) , \qquad (4.96)$$

where $\Gamma = \hbar\gamma$ and $\Omega_c/\hbar = \omega_c = ev_F^2B/|E_F|$ is termed *cyclotron frequency*.
Notice that, in fact, the conductivity is also a function of the magnetic field,

---

[23] A complete derivation of these expressions for the components of the magneto-optical
conductivity tensor of graphene is presented in Appendix H.

although it is not explicitly written in the above formulae. In order to have an idea of the order of magnitude of the cyclotron energy, we note that for $E_F = 0.4$ eV and $B = 5$ T (both typical values found in experiments [Yan *et al.* (2012b)]), its value is $\Omega_c \simeq 8.2$ meV. This semi-classical Boltzmann approach should be reliable for standard (moderate) doping levels or higher, as well as for sufficiently small magnetic fields, in the spectral regions where the optical weight is dominated by intraband transitions.

Now, a rather crucial point in the approximation towards a closed-form expression for the MPPs' spectrum is to note that, for frequencies around the THz and for small damping, the product $F(\omega)G(\omega) \ll H^2(\omega)$, so that we can approximate expression (4.89) as

$$q(\omega) \approx \frac{1}{2}\frac{H(\omega)}{F(\omega)} = \frac{1}{2}\left[G(\omega) - \frac{Z_0^2\sigma_H^2(\omega)}{F(\omega)} - \frac{4\bar{\epsilon}}{F(\omega)}\right], \qquad (4.97)$$

where we have chosen the "+" solution since the other would give $q \simeq 0$. In order to proceed further in our computation, it is now necessary to expand the components of the conductivity tensor for small γ/ω and small ω_c/ω, that is, $\Gamma, \Omega_c \ll \hbar\omega$. In this regime, after performing a series expansion on Eqs. (4.93) and (4.94), one obtains

$$\sigma(\omega) \simeq \frac{\sigma_0}{\pi}\frac{4E_F}{\hbar\omega}\left[\frac{\gamma}{\omega} + i\left(1 - \frac{\gamma^2}{\omega^2} + \frac{\omega_c^2}{\omega^2}\right)\right], \qquad (4.98)$$

$$\sigma_H(\omega) \simeq \frac{\sigma_0}{\pi}\frac{4E_F}{\hbar\omega}\left[\frac{\omega_c}{\omega} - i\frac{2\gamma\omega_c}{\omega^2}\right], \qquad (4.99)$$

up to quadratic terms in the expansion variables (and corresponding crossed terms). In writing these equations we have assumed, without loss of generality, that $E_F > 0$, and we shall do so henceforth. A numerical estimation of the three terms figuring in Eq. (4.97), reveals that the last term dominates provided that $\hbar\omega/E_F \gg \alpha \approx 0.01$ (where α is the fine-structure constant). Thus, we can further approximate Eq. (4.97) by

$$q(\omega) \approx -\frac{2\bar{\epsilon}}{F(\omega)} = i\frac{2\bar{\epsilon}\epsilon_0\omega}{\sigma(\omega)}, \qquad (4.100)$$

that is,

$$q(\omega) = i\frac{\bar{\epsilon}}{2\alpha}\frac{(\hbar\omega)^2}{E_F\hbar c}\left[\frac{\gamma}{\omega} + i\left(1 - \frac{\gamma^2}{\omega^2} + \frac{\omega_c^2}{\omega^2}\right)\right]^{-1}, \qquad (4.101)$$

$$\approx \frac{\bar{\epsilon}}{2\alpha}\frac{(\hbar\omega)^2}{E_F\hbar c}\left(1 - \frac{\omega_c^2}{\omega^2} + i\frac{\gamma}{\omega}\right), \qquad (4.102)$$

where in the last step we have performed another expansion on the small parameters $\gamma/\omega, \omega_c/\omega$ (again, up to quadratic order). Splitting the real and imaginary parts by writing the wavevector as $q = q' + iq''$, from Eq. (4.102) we learn that:

$$\begin{cases} q' = \dfrac{\bar{\epsilon}}{2\alpha} \dfrac{(\hbar\omega)^2}{E_F \hbar c} \left(1 - \dfrac{\omega_c^2}{\omega^2} \right) \\[2ex] q'' = \dfrac{\bar{\epsilon}}{2\alpha} \dfrac{(\hbar\omega)^2}{E_F \hbar c} \dfrac{\gamma}{\omega} \end{cases}, \qquad (4.103)$$

where in order to have a weakly damped plasmon we need $q'' \ll q'$, which is satisfied whenever $\gamma\omega + \omega_c^2 \ll \omega^2$. In addition, let us remark that from the previous expressions we can also define the magneto-plasmon's field confinement and propagation length [via Eqs. (4.21) and (4.25)]. It is now straightforward to invert the first expression in Eq. (4.103), which after some simple algebra and expansions similar to the ones above, yields the dispersion relation of MPPs in graphene in the long-wavelength limit, reading [Goerbig (2011); Ferreira *et al.* (2012)]

$$\boxed{\hbar\omega_{MPPs}(q', B) \approx \sqrt{\Omega_{GSPs}^2(q') + \Omega_c^2(B)}}, \qquad (4.104)$$

where $\Omega_{GSPs}(q') = \sqrt{\dfrac{4\alpha}{\epsilon_1 + \epsilon_2} E_F \hbar c q'}$ is the well-known dispersion of GSPs in a zero magnetic field that we have already seen in subsection 4.1.1 [notice that we have also restored $\bar{\epsilon} \to (\epsilon_1 + \epsilon_2)/2$]. We have thus arrived to a closed-form expression for the spectrum of MPPs in graphene, from which we can readily infer that the presence of the magnetic field pushes the frequency of the hybrid MPPs towards higher frequencies when compared to the GSPs' resonant frequency for $B = 0$. Consistently, for $B = 0$ we have $\Omega_c = 0$ and we recover the dispersion of graphene surface plasmons in the absence of a magnetic field. For ordinary 2DEGs, the result is similar to Eq. (4.104), but with Ω_{GSPs} replaced by its parabolic free electron gas counterpart [Fedorych *et al.* (2009); Baskin *et al.* (2011)].

Having derived Eq. (4.104), let us compare this analytical (approximate) expression with the numerical solution obtained through Eq. (4.86), which employs the full semi-classical (local) magneto-optical conductivity of graphene and includes retardation effects. The resulting spectra of MPPs in graphene are plotted in Fig. 4.21, along with the plasmon dispersion in a vanishing magnetic field to serve as reference. Firstly, it is apparent from the figure that the analytical approximation is indeed valid for the chosen parameters, as it is nearly indistinguishable from the numerical solution, with the exception of the spectral region defined around $\omega \lesssim \omega_c$ (that is, where the above-made approximations are no longer valid). Furthermore, above that region we see that the dispersion of graphene's hybrid

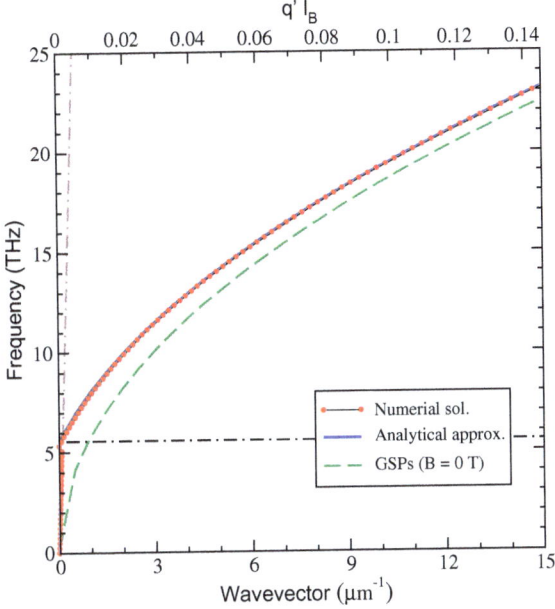

Figure 4.21: Semi-classical dispersion of MPPs in single-layer graphene: approximate analytical expression, $\hbar\omega_{MPPs}\,(q',B)$ (blue solid line); numerical solution of the spectrum's condition, Eq. (4.86), (red dots). The plasmon dispersion in a zero magnetic field, $\Omega_{GSPs}\,(q')$, is shown as a dashed green line to serve as benchmark. Also plotted is the dispersion of light (brownish dashed-dotted line). The horizontal line marks the cyclotron frequency, ω_c. Parameters: $E_F = 0.2$ eV, $B = 7$ T, $\gamma = 8$ meV and $\epsilon_1 = \epsilon_2 = 1$. Note that for consistency with the local approximation for the magneto-optical conductivity, we have restricted the plot to the $ql_B \ll 1$ region (and, naturally, $q \ll k_F$; but in the current case, l_B^{-1} gives a stronger bound on locality).

MPPs is shifted toward higher frequencies with respect to the corresponding GSPs' dispersion for $B = 0$, as we have anticipated in the lines following Eq. (4.104). This is consequence of the hybridization between graphene's 2D plasmons and cyclotronic excitations (with frequency ω_c) driven by the static magnetic field. Secondly, we note that for frequencies below the cyclotron's frequency, the MPPs modes are light-like. We can bring some more insight into this result by recalling the reader that, in subsections 4.1.1 and 4.1.5, we have seen that TM GSPs can only exist if $\Im m\{\sigma\} > 0$, whereas TE GSPs can only be sustained provided that $\Im m\{\sigma\} < 0$. In the current case, owing to $B \neq 0$ the imaginary part of the diagonal components of the conductivity tensor, $\Im m\{\sigma\}$, are negative for frequencies below ω_c, and positive for frequencies above ω_c. Thus, in the latter spectral region, hybrid MPPs and TM GSPs behave alike, while in the former spectral range the modes resemble TE GSPs and for this reason they have been dubbed as quasi-transverse electric (QTE) modes [Ferreira *et al.* (2012)].

4.4.3 *Spectrum of magneto-plasmons in the quantum regime: Landau quantization*

Above we have studied the dispersion of magneto-plasmon-polaritons in graphene subjected to a static magnetic field in the semi-classical regime. We shall end this section by treating the same problem in another regime —that we will dub "quantum regime"—, in which the emergence of Landau levels owing to the integer quantum Hall effect in graphene will disrupt the smoothly varying plasmonic band seen in the semi-classical case.

In the present case, the single-particle Hamiltonian in the Dirac-cone approximation can be written as (where the applied fields are included via minimal coupling/Peierls substitution)

$$H = H_0 + H_{LM} + ev_F \boldsymbol{\sigma} \cdot \mathbf{A}(\mathbf{r}) , \qquad (4.105)$$

where $H_0 = \hbar v_F q$ is the free Dirac Hamiltonian, and $H_{LM} = ev_F \boldsymbol{\sigma} \cdot \mathbf{A}_{EM}(\mathbf{r})$ accounts for the light-matter interaction between a linearly polarized electromagnetic wave oscillating at frequency ω and graphene's Dirac fermions. The last term describes the interaction of the electrons in graphene with the external static $\mathbf{B}$ field. The eigenstates of the Hamiltonian (4.105) are found to be (in the gauge where $\mathbf{A}(\mathbf{r}) = Bx\,\hat{\mathbf{y}}$) [Ferreira *et al.* (2011); Goerbig (2011)]

$$\psi_{n,k_y}(\mathbf{r}) = \frac{C_n}{\sqrt{L}} \begin{pmatrix} \phi_{|n|-1}(x) \\ i\,\mathrm{sgn}(n)\phi_{|n|}(x) \end{pmatrix} e^{ik_y y} , \qquad (4.106)$$

and the eigenenergies of the charge-carriers become quantized according to [Ferreira *et al.* (2011); Goerbig (2011)]

$$E_n = \mathrm{sgn}(n) \frac{\hbar v_F}{l_B} \sqrt{2|n|} \quad , \quad n = 0, \pm 1, \pm 2, \dots \; , \qquad (4.107)$$

where $\phi_n(x) = e^{-\xi^2/2} H_n(\xi)/(2^n n! \sqrt{\pi} l_B)^{1/2}$, H_n is the Hermite polynomial of degree $n \geq 0$, $\phi_{-1}(x) = 0$, and $\xi(x) = l_B k_y + x/l_B$ is a dimensionless variable. The parameter L denotes the linear dimension of the system and C_n is a normalization constant that discriminates the zero-energy level from the remaining Landau levels (LLs), evaluating $C_0 = 1$ and $C_n = 2^{-\frac{1}{2}}$ for

$n \neq 0$. Naturally, we should expect that the emergence of Landau level quantization in the electronic spectrum, summarized by Eq. (4.107), will deliver an impact on the plasmonic excitations supported by the system. In practice, one anticipates that the density of electronic states will be enhanced in the whereabouts of the LLs[24], which in turn will lead to the possibility of electronic transitions. The details of this new physics can be obtained by computing the dynamical quantum magneto-optical conductivity by solving the appropriate equations of motion. This is a lengthy calculation, and it can be found in [Ferreira *et al.* (2011)] where the formalism is discussed with some detail (the discussion of MPPs in monolayer graphene in this framework is also discussed in [Ferreira *et al.* (2012)]). For this reason, we will simply call the result[25] from [Ferreira *et al.* (2011)] and discuss its implications in a pictorial fashion, with the aid of Fig. 4.22, whose parameters correspond to a feasible experimental situation. The figure shows the magneto-optical conductivity as a function of the frequency, for doped graphene and taking Landau level quantization into consideration (the semi-classical conductivity is also plotted as dashed lines). Focusing on the longitudinal part of the homogeneous conductivity (larger panel), one can discern multiple peaks in the real part of the conductivity: a prominent peak at lower energies (associated with intraband transitions) and a set of smaller peaks at higher frequencies (owing to interband transitions). These are highlighted by the vertical dashed lines and are ascribed to transitions between different LLs (we note that these must obey some selection rules —see figure's caption for more details)—, as indicated by complementary energy level diagram depicted in the right panel of that figure. This provides a direct and intuitive way to probe the electronic spectrum of graphene in the presence of a static magnetic field, while endowing us some physical insight on the problem, which will be instructive for our discussion of the plasmonic excitations in such regime.

In order to do so, we need to solve the spectrum's condition conveyed by Eq. (4.86) using the quantum magneto-optical conductivity as given in [Ferreira *et al.* (2011)]. The outcome of that numerical solution is portrayed in the left panel of Fig. 4.23. The resulting dispersion relation of MPPs in this quantum regime is strikingly different from its semi-classical analogue, particularly at higher frequencies where interband transitions between distinct LLs dominate the optical response, thus breaking the continuously

---

[24]In the absence of disorder the density of states is an comb of delta-functions.

[25]This is, perhaps, the only occasion that we do such thing in this book, and so we hope that the reader may kindly bear with us.

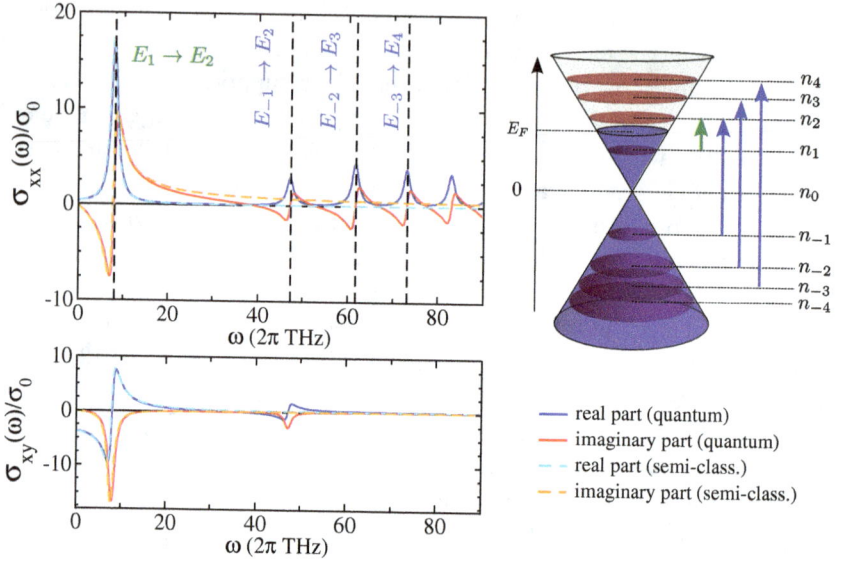

Figure 4.22: Dynamical magneto-optical conductivity of graphene in the quantum regime (with the corresponding semi-classical expressions, Eqs. (4.93) and (4.94) indicted by the dashed lines). The peaks appearing in the real part of σ_{xx} are located at the frequencies defined by $\hbar\omega = E_2 - E_1$, $E_2 - E_{-1}$, $E_3 - E_{-2}$, $E_4 - E_{-3}$, ..., where the E_n's are given by Eq. (4.107), and therefore correspond to transitions between different LLs (cf. with the diagram on the right). These transitions obey the selection rule which says that only transitions between LLs with $|n| - |m| = \pm 1$ can occur (provided that they are not Pauli blocked). Parameters: $E_F = 0.1$ eV, $\Gamma = 3.7$ meV and $B = 5$ T.

varying plasmonic band seen in the semi-classical regime (black-dashed line in the figure) into a set of plasmon branches which fold-back when approaching (from below) a spectral domain where the imaginary part of the conductivity is negative, $\Im m\ \sigma_{xx} < 0$ (shaded regions), as the panel on the right-hand side illustrates. As mentioned in the text following Fig. 4.21, TM-like magneto-plasmons can not be sustained in these regions and the modes are termed QTE, due to their light-like behavior. On the other hand, in the frequency windows where $\Im m\ \sigma_{xx} > 0$, hybrid MPPs modes resembling TM GSPs can be sustained.

Having swiftly reviewed the basic physics behind magneto-plasmon-polaritons in graphene, it is clear that Landau levels quantization may significantly alter the dispersion of MPPs, and thus it is important to be aware

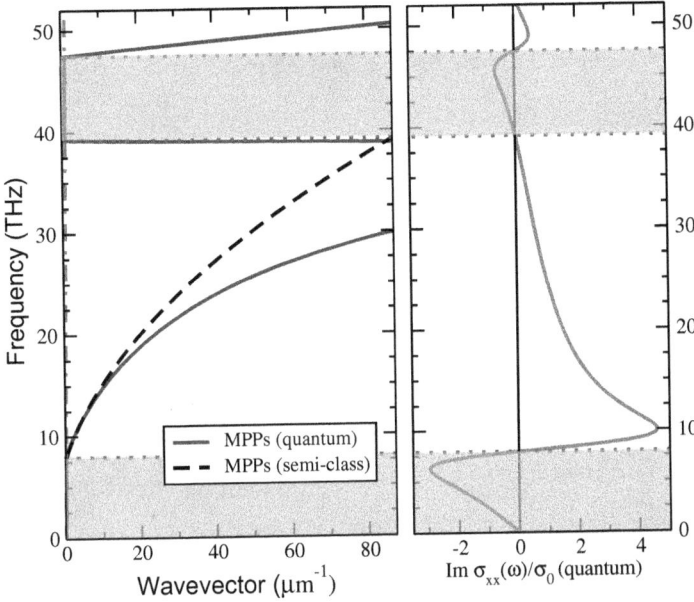

Figure 4.23: Left: spectrum of graphene's MPPs in the quantum regime, where Landau level quantization and interband transitions play a key role. Right: quantum (local) magneto-optical conductivity of graphene as a function of the frequency, with the spectral regions where $\Im m\, \sigma_{xx} < 0$ shaded. These indicate the regions where TM-like plasmons can not be sustained, as the left panel clearly demonstrates. Instead, in such frequency domains the excitations are light-like QTE modes. Conversely, outside the shaded regions, TM-like MPPs can be supported by the system. Parameters: $E_F = 0.1$ eV, $\Gamma = 8$ meV and $B = 5$ T. Since we have taken the local limit, we have restricted the plot of the spectrum to wavevectors in the interval $ql_B \in [0, 1]$.

in which regime one should be working when considering such a problem. It has been realized that graphene is an alluring playground for magneto-optics; examples include the study of Faraday rotation both in extended graphene [Ferreira *et al.* (2011)] and in engineered structures exploiting MPPs to enhance the Faraday rotation [Tymchenko *et al.* (2013)], as well as for graphene placed in an optical cavity to boost even further its already rather high Faraday rotation, considering that this is a two-dimensional material with atomic thickness [Ferreira *et al.* (2011)]. Also on the experimental side, tunable Dirac magneto-plasmons have been observed in arrays of graphene disks using infrared spectroscopy [Yan *et al.* (2012b)], and a scheme to achieve unidirectional excitation of GSPs in graphene placed on top of a magneto-optical substrate [Liu *et al.* (2015a)] has been proposed, which may be interesting for routing plasmons in future optoelectronic circuitry.

4.5   A Detour: Surface Phonon-Polaritons in hBN

Hexagonal boron nitride (hBN) is an uniaxial hyperbolic material [Jacob (2014)]. Being uniaxial means that its dielectric function is actually a tensor, characterized by two different dielectric functions: one along the crystal symmetry axis, $\epsilon_z(\omega)$, and another in the basal plane, $\epsilon_\perp(\omega)$. Hyberbolicity refers to the existence of a spectral range where the product $\epsilon_z(\omega)\epsilon_\perp(\omega)$ is negative. In hBN there are two such regions, denoted by hyperbolicity of type I and II, depending on whether $\epsilon_\perp(\omega)$ is positive (type I) or negative (type II). The two dielectric functions can be condensed into a single formula, reading

$$\epsilon_m(\omega) = \epsilon_{m,\infty} + \epsilon_{m,\infty} \frac{\omega_{\mathrm{LO},m}^2 - \omega_{\mathrm{TO},m}^2}{\omega_{\mathrm{TO},m}^2 - \omega^2 - i\omega\Gamma_m}, \qquad (4.108)$$

m	$\epsilon_{m,\infty}$	$\omega_{\mathrm{TO},m}$ (cm^{-1})	$\omega_{\mathrm{LO},m}$ (cm^{-1})	Γ_m (cm^{-1})
z	2.95	780	830	4
$\perp$	4.87	1370	1610	5

Table 4.1:   Parameters for the dielectric functions of hBN (data taken from [Kumar *et al.* (2015)]). The parameters with $m = z$ are for out-of-plane phonon modes, whereas $m =\perp$ refers to in-plane ones.

where the different parameters are given in Table 4.1. By making a plot of $\epsilon_m(\omega)$ we can easily identify the two different hyperbolic regions, as can be seen in Fig. 4.24. The region within the vertical dashed lines in the bottom left panel of this figure indicates the spectral range where the dielectric function of hBN presents hyperbolic behavior of type II. This will be the region of interest of our study, highlighted in subsection 4.5.2. Within this spectral region, a film of hBN of finite thickness sustains surface phonon-polaritons, as we will detail in subsection 4.5.2. These modes are evanescent outside the hBN film, and resemble a standing wave within the film [Kumar *et al.* (2015)]. Interesting enough, these modes can be excited by the decaying of an excited quantum emitter to the ground state, leading to the Purcell effect (more on this in Chapter 10). This effect

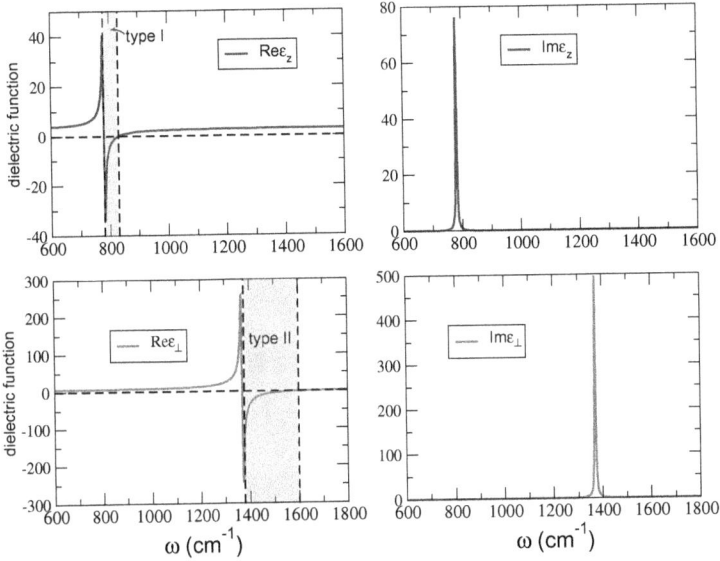

Figure 4.24: Dielectric functions of hBN. The two top panels represent the real and the imaginary parts of $\epsilon_z(\omega)$. The bottom panels represent the same quantities for $\epsilon_\perp(\omega)$. In the left top (bottom) panel, the hyperbolic region of type I (type II) is identified.

refers to the change of the radiative decay rate of an emitter (spontaneous emission) due to the proximity of a system that can support both losses (as either a metal or graphene) and/or surface modes (as, for example, a metal, graphene, or hBN); or, in other words, whenever the emitter can decay through additional channels which do not occur when it is isolated in vacuum (basically, the density of final states is enhanced with respect to the same quantity in vacuum). In the case we are discussing, the additional decay channel is via the excitation of *surface phonon-polaritons*.

4.5.1 Solution of the electromagnetic problem: Fresnel co-efficients

Let us consider an uniaxial crystal with the crystal's axis along the z-direction, as illustrated in Fig. 4.25. We want to find the reflection and transmission coefficients in the situation where the basal plane of the crystal is parallel to the air-hBN interface. To that end we have to solve the electromagnetic problem in air and in hBN. The solution in air is elementary and it is worked out in Appendix P, in a different context. The solution in hBN is slightly different, since the electric field $\mathbf{E}_t$ and the electric displacement field $\mathbf{D}_t$ are not collinear ($\mathbf{E}_t$ and $\mathbf{D}_t$ refer to fields in hBN). We want to find the normal mode of the electric field in hBN, that is, we want to find the wavevector that renders the field in the crystal a plane wave. Let us consider a p-polarized wave impinging on an air-hBN interface, as illustrated in Fig. 4.25. We assume that the time dependence of the fields

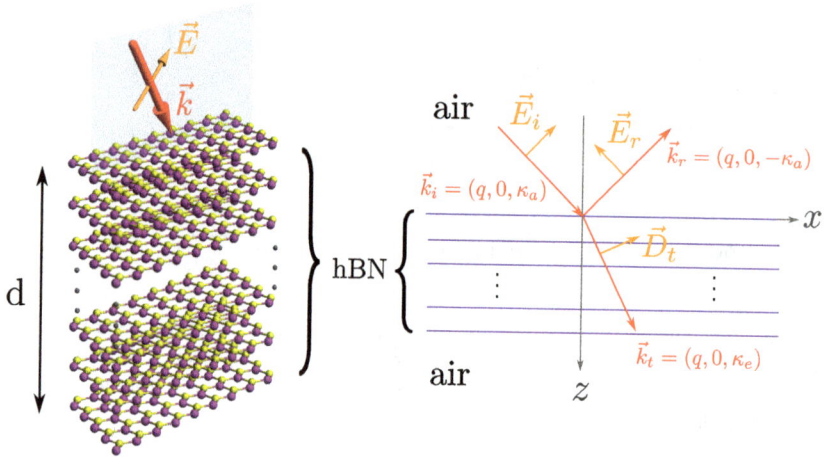

Figure 4.25: Schematics of the scattering of a p-polarized electromagnetic plane wave impinging on a slab of hBN crystal, with its axis of symmetry oriented along the z-direction. The crystal as a finite thickness, d. The dielectric function along the crystal's axis is termed ϵ_z, whereas in the basal plane it is dubbed $\epsilon_\perp$ (hBN is classified as an uniaxial crystal).

has the usual form, $e^{-i\omega t}$, and that the spatial dependence has the form $e^{iqx}e^{\pm i\kappa_e/a z}$, where the $\pm$ sign refers to waves propagating along the positive $(+)$ or negative $(-)$ direction of the z-axis. The curl of the fields in Maxwell's equations read

$$\nabla \times \mathbf{E} = i\omega \mathbf{B} \,, \tag{4.109}$$

$$\nabla \times \mathbf{B} = -i\omega \mathbf{D} = -i\omega \epsilon_0 \boldsymbol{\epsilon} \cdot \mathbf{E} \,, \tag{4.110}$$

where the dielectric tensor $\boldsymbol{\epsilon}$ is given by

$$\boldsymbol{\epsilon} = \begin{bmatrix} \epsilon_\perp & 0 & 0 \\ 0 & \epsilon_\perp & 0 \\ 0 & 0 & \epsilon_z \end{bmatrix} \,. \tag{4.111}$$

Computing the product $\boldsymbol{\epsilon} \cdot \mathbf{E}$ in the region where we have hBN, we obtain $\boldsymbol{\epsilon} \cdot \mathbf{E}_t = \epsilon_\perp E_{t,x}\hat{\mathbf{x}} + \epsilon_z E_{t,z}\hat{\mathbf{z}}$, where $\mathbf{E}_t$ is the transmitted field (i.e. the field in hBN). From the two above Maxwell's equations , it is straightforward to obtain

$$\kappa_e B_y = \mu_0 \epsilon_0 \epsilon_\perp \omega E_{t,x} , \tag{4.112}$$

$$q B_y = -\mu_0 \epsilon_0 \epsilon_z \omega E_{t,z} , \tag{4.113}$$

$$\omega B_y = \kappa_e E_{t,x} - q E_{t,z} . \tag{4.114}$$

Combining the last three equations we have

$$\kappa_e = \sqrt{\epsilon_\perp \frac{\omega^2}{c^2} - \frac{\epsilon_\perp}{\epsilon_z} q^2} , \tag{4.115}$$

where the subscript e is a reference to the extraordinary ray typical of uniaxial crystals. Having determined the value of κ_e, we proceed to the calculation of the Fresnel coefficients. The boundary conditions at the air-hBN interface are the ones discussed in Chapter 2, namely:

$$E_{t,x} = E_{i,x} - E_{r,x} , \tag{4.116}$$

$$B_{t,y} = B_{i,y} + B_{r,y} . \tag{4.117}$$

Relying again on Maxwell's equations, it follows that

$$B_{i,y} = \frac{\epsilon_0 \omega}{\kappa_a c^2} E_{i,x} , \tag{4.118}$$

$$B_{r,y} = -\frac{\epsilon_0 \omega}{\kappa_a c^2} E_{r,x} , \tag{4.119}$$

$$B_{t,y} = \frac{\epsilon_0 \epsilon_\perp \omega}{\kappa_e c^2} E_{t,x} , \tag{4.120}$$

where $\kappa_a = \sqrt{\omega^2/c^2 - q^2}$, and the subscript a refers to the wavevector in air. By plugging in the last three equations into the boundary conditions, we arrive at

$$\frac{E_i}{\kappa_a} + \frac{E_r}{\kappa_a} = \frac{\epsilon_\perp}{\kappa_e}(E_i - E_r) , \tag{4.121}$$

from where it follows the Fresnel reflection coefficient at the air-hBN interface:

$$r_{p,a \to b} = \frac{E_r}{E_i} = -\frac{\kappa_e/\epsilon_\perp - \kappa_a}{\kappa_e/\epsilon_\perp + \kappa_a} , \tag{4.122}$$

a result that agrees with the one found in [Lekner (1991)]. The corresponding transmission amplitude can be computed by following the same steps as before, but it will not be needed for what follows.

4.5.2   Guided phonon-polariton modes

We have almost all we need to compute the dispersion relation of the phonon-polariton modes in hBN, in the region where the material displays hyperbolic behavior. Here, we shall consider the case of type II hyperbolicity, which for hBN it is located in the region of the electromagnetic spectrum spanned by the frequencies $\omega \in [1400, 1600]$ cm^{-1}. Since we are considering a finite crystal of hBN of width d, we need the reflection amplitude of the air-hBN-air heterostructure. Fortunately, this quantity can be determined solely by Eq. (4.122) and the width d. Indeed, the reflection amplitude of a film of width d is given by [Saleh and Teich (2012)]

$$r_{p,a-b-a} = \frac{r_{p,a\to b} + r_{p,b\to a}e^{i2\kappa_e d}}{1 + r_{p,a\to b}r_{p,b\to a}e^{i2\kappa_e d}} , \qquad (4.123)$$

where $r_{p,b\to a} = -r_{p,a\to b}$ is the reflection amplitude of the hBN-air interface (the lower interface in Fig. 4.25). In the spirit of subsection 4.1.4, the dispersion relation of the surface phonon-polaritons guided modes (see discussion ahead) can be determined from the poles of $r_{p,a-b-a}$. Alternatively, we plot the loss function so that the dispersion relation appears as a set of bright lines in the plotted (q, ω)-plane. This approach has been pursued in the making of Fig. 4.26; the obtained results are consistent to those found in [Kumar *et al.* (2015)]. It can be shown that the modes found in Fig. 4.26 are evanescent in air, with decaying wavevector $i\kappa_a = -\sqrt{q^2 - \omega^2/c^2}$, as they stand to the right of the light line, which can barely be seen in the figure. Conversely, in hBN, the modes have the form of standing waves within the hBN slab. The high energy mode has no node inside the hBN film, the second highest energy mode has one node, and so forth [Kumar *et al.* (2015)]. Alternatively, we can understand this heterostructure as a waveguide for the propagation of surface phonon-polaritons, whose electric field decays exponentially into the adjacent medium (made up of air in the present case). Although we have considered a rather idealized situation, there is no qualitative change if we had considered the more realistic case of a hBN crystal on a silicon oxide substrate. As discussed before, surface phonon-polaritons resemble the surface plasmon-polaritons at a metal-dielectric interface (see Chapter 3). This is natural, since due to the hyperbolic nature of the crystal in the considered spectral range, $\epsilon_\perp$ is negative, whereas ϵ_z is positive (thus, the *reststrahlen* band basically mimicks a metal-dielectric interface). It is also clear from Fig. 4.26 that the width of the modes is not large, which suggest that these surface waves are weakly damped (the phonon damping rate can be one or two orders of magnitude smaller than that for electrons

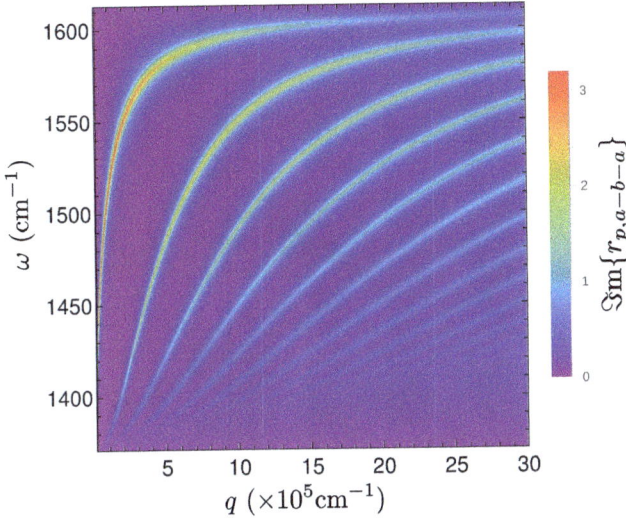

Figure 4.26: Guided surface phonon-polariton modes in a finite thickness hBN crystal, free-standing in air. The thickness is $d = 50$ nm. The intensity plot refers to the loss function [in logarithmic units, or, explicitly $\ln(1 + \Im m\{r_{p,a-b-a}\})$], calculated using Eqs. (4.38) and (4.123).

in a metal [Jacob (2014); Khurgin (2015); Woessner *et al.* (2015)]). Finally, it is not difficult to imagine that by depositing graphene on hBN [Dai *et al.* (2015)], one can create surface plasmon-phonon-polariton coupled modes, which can be tunned by electrostatic gating [Woessner *et al.* (2015)]. This possibility opens new avenues in graphene plasmonics and graphene-based nanophotonics. Further, if a magnetic field is added to the problem, then even more complex types of excitations can arise.

Chapter 5

Excitation of Graphene Surface Plasmons

In this chapter we will briefly review several techniques for the excitation of GSPs. Recall from our previous discussion that, in general, it is not possible to couple an incident electromagnetic beam to a SPP in planar, extended graphene, due to a mismatch between the momentum of the impinging radiation and the GSP's momentum —the dispersion curve of the latter always lies to the right of the light line. This limitation is shared amongst graphene and plasmonic metals (and, of course, metals in general). Thus, it should not come as a surprise that the very same mechanisms for exciting SPPs in metals also apply to the case of graphene.

5.1 Grating Coupling

The excitation of SPPs in graphene via grating coupling has been one of the most popular methods for exciting graphene's plasmon-polaritons [Ju *et al.* (2011); Yan *et al.* (2013); Brar *et al.* (2014); Strait *et al.* (2013); Bludov *et al.* (2012a); Peres *et al.* (2013); Nikitin *et al.* (2012a)]. The basic idea is to take advantage of the general principle from elementary optics that when a beam of light hits a grating structure, it gives rise to set of diffraction orders which carry "extra" momentum contributions, in multiples of $G = 2\pi/R$, defined by the grating period — see Fig. 5.1. Whenever a wavevector akin to a component of the diffracted light coincides to a certain SPP wavevector, the phase-matching condition[1]

$$q_{SPP} = k\sin\theta \pm nG \qquad (5.1)$$

is fulfilled, where $k = \sqrt{\epsilon}\omega/c$ and $n \in \mathbb{Z}$, and excitation takes place. This condition is reciprocal, that is, when a propagating SPP enters in a grating-like region it can (re)couple to the radiation continuum.

---

[1]Note that this 1D condition can be straightforwardly generalized for two dimensions.

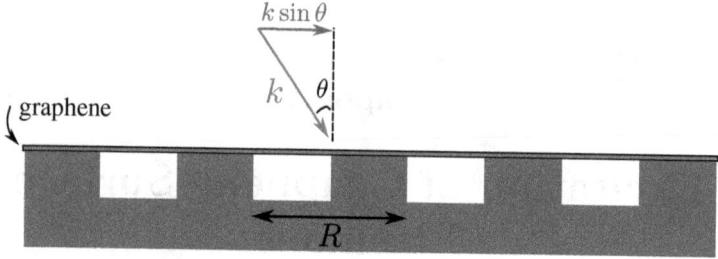

Figure 5.1: Example of a grating coupling scheme for exciting GSPs, in which a graphene sheet in placed on top of a patterned dielectric with period R.

We note that the configuration depicted in Fig. 5.1 is only one of many possible schemes for achieving grating coupling. That particular case was investigated analytically in [Peres *et al.* (2013)]. In Chapter 7 we will consider yet another configuration, in which the grating is formed by graphene itself, patterned into an array of graphene ribbons. Hence, a detailed description of the coupling of electromagnetic radiation to GSPs using gratings can be found in the aforementioned chapter (and also in Chapter 9).

5.1.1 Graphene on Gallium-Arsenide: The role of SO phonons

Let us give an example and consider graphene resting on a square grating made of alternating slabs of air (trenches) and SiO_2 (teeth), with a flat substrate made of Gallium-Arsenide (GaAs) underneath (similar to Fig. 5.1, but in the case where the substrate is made from a different material than the teeth). To make things more interesting, we shall consider a spectral range that includes the optical phonons of this polar material,[2] which appear around 33.9 meV (8.3 THz). To that end, we need the dielectric function of GaAs in that spectral range. This can be parametrized as follows:

$$\epsilon(\omega) = \epsilon_\infty \left(1 + \frac{\omega_{LO}^2 - \omega_{TO}^2}{\omega_{TO}^2 - \omega^2 - i\omega\Gamma_{TO}} \right), \tag{5.2}$$

where ω_{TO} and ω_{LO} are the transverse and longitudinal frequencies of the bulk optical phonons, respectively, and Γ_{TO} is the phonon damping. The parameters are $\omega_{TO} = 270$ cm^{-1}, $\omega_{LO} = 295$ cm^{-1}, $\Gamma_{TO} = 4-5$ cm^{-1}, and $\epsilon_\infty = 10.9$.

---

[2]The reader unfamiliar with these may want to check Sec. 4.3.

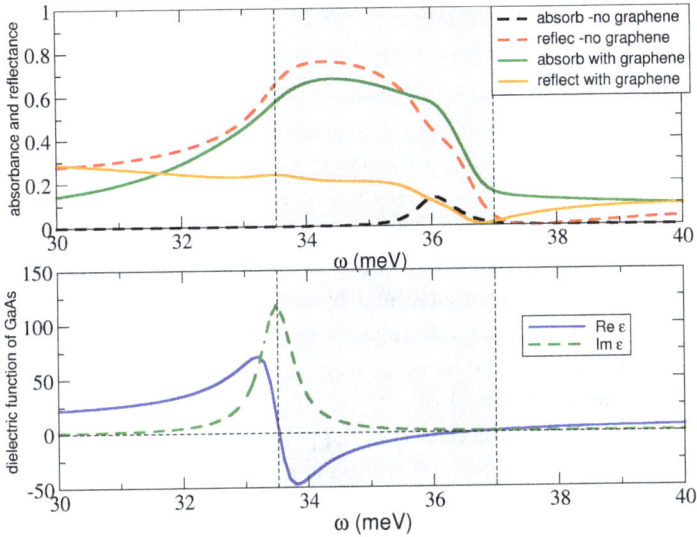

Figure 5.2: Absorbance and reflectance of graphene on a square grating on a flat Gallium-Arsenide substrate. The parameters are: $E_F = 0.4$ eV, the period of the grating is $d = 3.3$ μm, the depth of the grating is $h = 1.7d$, and $\Gamma = 4.1$ meV (1 THz). The grating is made of alternating tooth and trenches of silicon dioxide (assumed to have a dielectric constant of 3.9) and air, respectively. Top: Absorbance and reflectance. Bottom: Real and imaginary parts of the dielectric function of Gallium-Arsenide.

The quantity of interest to be considered here will be the absorbance of radiation by graphene. The result of this calculation is illustrated in Fig. 5.2. It should be appreciate the dramatic changes occurring in the behavior of the grating when graphene is deposited on it. Without graphene, the grating absorbs almost no radiation (with the exception of the little peak around 36.3 meV, which corresponds to the excitation of SO phonons in GaAs) and reflects up to 80% around the energy of 33.9 meV. When graphene is deposited on the grating the absorbance rises up to 70% whereas the reflectance is significantly suppressed. The enhanced absorption seen in the grey window of Fig. 5.2 is due to the negative value of the dielectric function of GaAs, which enhances the excitation of SPP's in graphene (note that the trenches are not made of GaAs but of SiO_2). A similar effect has been seen in a system composed of graphene on a metal grating for frequencies below the plasma frequency of the metal [Ferreira and Peres (2012)].

Thus, the existence phonons in the substrate and in the right spectral range can be explored as a way of enhancing the absorption of graphene.

5.2 Prism Coupling

Another commom approach to overcome the kinematic limitation and to couple (*p*-polarized) photons to SPPs has been the excitation of SPPs via tunneling of evanescent waves originating from total internal reflection of light in a prism made with a higher-index dielectric (Kretschmann and Otto configurations). This can be attained by meeting the conditions: (i) ϵ_{prism} must be the higher than the one relative to the two dielectrics cladding the graphene layer; and (ii) the incident angle must be larger than the critical angle for total internal reflection, $\theta > \theta_c = \arcsin\left(\epsilon_{diel.}/\epsilon_{prism}\right)$. Therefore, the wavevector of light is increased in the prism and it can now afford SPPs' excitation, whenever its in-plane component

$$q = \frac{\omega}{c}\sqrt{\epsilon_{prism}}\sin\theta \, , \qquad (5.3)$$

matches the SPP wavevector. Figure 5.3 shows a diagrammatic representation of this technique[3]. Notice that depending on the impinging angle, θ, different wavevectors can be achieved within the shaded region between the light lines of the dielectric (here assumed to be the same, above and below graphene, for the sake of clarity) and the prism. Under this conditions SPP excitation is detected by observing the presence of a sharp minimum in reflectance —attenuated total reflection (ATR). An example of the geometry in given in the inset of Fig. 5.3 and results are given in Fig. 5.5 (please see Sec. 5.2.1 for a detailed derivation).

5.2.1 *Otto configuration: Single-layer graphene*

Having described briefly the main concept behind the excitation of surface plasmon-polaritons in monolayer graphene via prism coupling, let us now work out the equations for the reflection amplitude, in which the observation of a dip heralds the excitation of GSPs. We shall consider the Otto's approach to achieve light-GSPs coupling (in fact, the Kretschmann configuration does not work for single-layer graphene owing to its intrinsic two dimensionality). The scheme is illustrated in Fig. 5.4. Note the in order to attain plasmon excitation one must guarantee that $\epsilon_p \equiv \epsilon_3 > \epsilon_j$ with

---

[3]Neglecting the influence of the dielectric environment of the prism on the resonant condition/dispersion of GSPs.

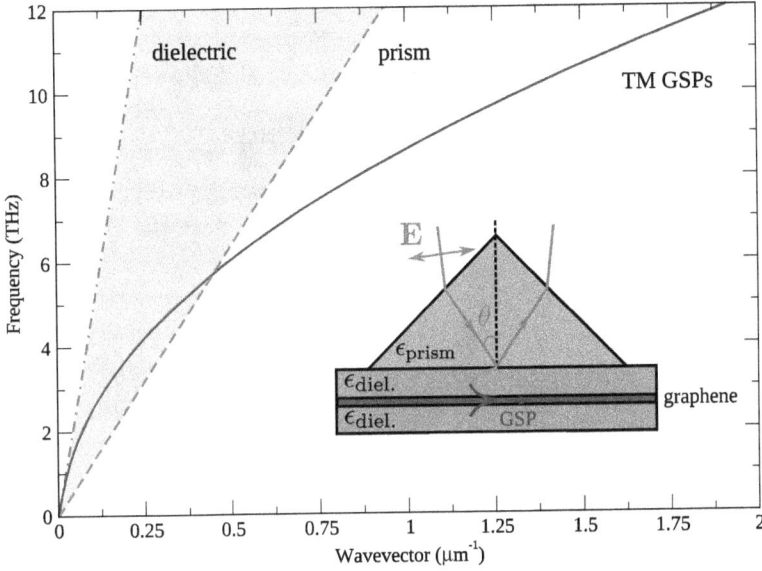

Figure 5.3: Schematic diagram depicting the principle behind the prism coupling technique for exciting GSPs. The inset depicts the coupling of light to GSPs, in the ATR regime, by evanescent tunneling. Doped graphene ($E_F = 0.45$ eV) is assumed to be cladded between one dielectric layer (top) and another semi-infinite dielectric (bottom) with $\epsilon_{diel.} = 1$. We further assumed that $\epsilon_{prism} = 14$.

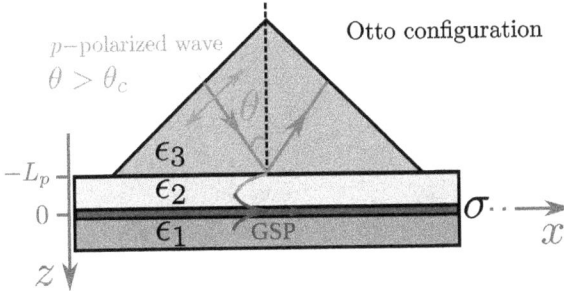

Figure 5.4: Otto configuration with single-layer graphene for exciting GSPs. In the calculations we shall assume that the top medium (prism) is semi-infine, as well as the bottom (medium with ϵ_1).

$j = 1, 2$ (cf. Fig. 5.4) and also to meet the condition for ATR, that is, $\theta > \theta_c = \arcsin(\epsilon_2/\epsilon_3)$.

In the ATR regime, the p-polarized electromagnetic fields in the three dielectrics —i.e. ϵ_3, ϵ_2 and ϵ_1— are of the form:

<u>Medium 3</u> (prism; $-L_p > z$):

$$\mathbf{E}_3(\mathbf{r}) = \left(E_{3,x}^{(i)}, 0, E_{3,z}^{(i)} \right) e^{i\mathbf{k}_i \cdot \mathbf{r}} + \left(E_{3,x}^{(r)}, 0, E_{3,z}^{(r)} \right) e^{-i\mathbf{k}_i \cdot \mathbf{r}} , \tag{5.4}$$

$$\mathbf{B}_3(\mathbf{r}) = \left(0, B_{3,y}^{(i)}, 0 \right) e^{i\mathbf{k}_i \cdot \mathbf{r}} + \left(0, B_{3,y}^{(r)}, 0 \right) e^{-i\mathbf{k}_i \cdot \mathbf{r}} , \tag{5.5}$$

with $\mathbf{k}_i = \sqrt{\epsilon_3}\omega/c\,(\sin\theta, 0, \cos\theta)$ being the impinging wavevector and $\mathbf{k}_r = \sqrt{\epsilon_3}\omega/c\,(\sin\theta, 0, -\cos\theta)$ the reflected wavevector (note the change of sign in the z-component).

<u>Medium 2</u> $(-L_p < z < 0)$:

$$\mathbf{E}_2(\mathbf{r}) = \left\{ \left(E_{2,x}^{(+)}, 0, E_{2,z}^{(+)} \right) e^{\kappa_2 |z|} + \left(E_{2,x}^{(-)}, 0, E_{2,z}^{(-)} \right) e^{-\kappa_2 |z|} \right\} e^{iqx} , \tag{5.6}$$

$$\mathbf{B}_2(\mathbf{r}) = \left\{ \left(0, B_{2,y}^{(+)}, 0 \right) e^{\kappa_2 |z|} + \left(0, B_{2,y}^{(-)}, 0 \right) e^{-\kappa_2 |z|} \right\} e^{iqx} , \tag{5.7}$$

with the propagation constant (named in this way because it reflects the conservation of the in-plane momenta) $q = \sqrt{\epsilon_3}\omega/c\sin\theta$, and where $\kappa_2 = \sqrt{q^2 - \epsilon_2\omega^2/c^2}$.

<u>Medium 1</u> $(z > 0)$:

$$\mathbf{E}_1(\mathbf{r}) = (E_{1,x}, 0, E_{1,z})\, e^{-\kappa_1 |z|} e^{iqx} , \tag{5.8}$$

$$\mathbf{B}_1(\mathbf{r}) = (0, B_{1,y}, 0)\, e^{-\kappa_1 |z|} e^{iqx} , \tag{5.9}$$

where $\kappa_1 = \sqrt{q^2 - \epsilon_1\omega^2/c^2}$. Naturally, an harmonic time dependence of the form $e^{-i\omega t}$ is implicit and it will be assumed to be so throughout.

The next ingredient we need towards an expression for the Fresnel reflection amplitude of the (total) system are the boundary conditions at $z = -L_p$ and $z = 0$, which give

$$E_{3,x}^{(i)} e^{-ik_{3,z} L_p} + E_{3,x}^{(r)} e^{ik_{3,z} L_p} = E_{2,x}^{(+)} e^{\kappa_2 L_p} + E_{2,x}^{(-)} e^{-\kappa_2 L_p} , \tag{5.10}$$

$$B_{3,y}^{(i)} e^{-ik_{3,z} L_p} + B_{3,y}^{(r)} e^{ik_{3,z} L_p} = B_{2,y}^{(+)} e^{\kappa_2 L_p} + B_{2,y}^{(-)} e^{-\kappa_2 L_p} , \tag{5.11}$$

and

$$E_{2,x}^{(+)} + E_{2,x}^{(-)} = E_{1,x} , \tag{5.12}$$

$$B_{2,y}^{(+)} + B_{2,y}^{(-)} = B_{1,y} + \mu_0 \sigma(\omega) E_{1,x} , \tag{5.13}$$

respectively. In the above equations, $k_{3,z} = \sqrt{\epsilon_3}\omega c \cos\theta$ and $\sigma(\omega)$ is the longitudinal dynamical conductivity of graphene. By making use of Maxwell's equations (cf. Sec. 2.1.1), one can express the amplitudes $B_{j,y}$ in terms of the corresponding electric-field components $E_{j,x}$ (or $E_{j,x}^{(\pm)}$ when applicable). Such procedure leads to a linear system of four equations, which may be casted as a standard $\mathbf{Ax = b}$ matrix equation, namely

$$\mathbf{M}_{SLG}^{Otto} \cdot \begin{pmatrix} E_{3,x}^{(r)}/E_{3,x}^{(i)} \\ E_{2,x}^{(+)}/E_{3,x}^{(i)} \\ E_{2,x}^{(-)}/E_{3,x}^{(i)} \\ E_{1,x}/E_{3,x}^{(i)} \end{pmatrix} = \begin{pmatrix} e^{-ik_{3,z}L_p} \\ -\frac{\epsilon_3}{k_{3,z}}e^{-ik_{3,z}L_p} \\ 0 \\ 0 \end{pmatrix}, \tag{5.14}$$

where the matrix $\mathbf{M}_{SLG}^{Otto}$ is given by

$$\mathbf{M}_{SLG}^{Otto} = \begin{pmatrix} -e^{ik_{3,z}L_p} & e^{\kappa_2 L_p} & e^{-\kappa_2 L_p} & 0 \\ -\frac{\epsilon_3}{k_{3,z}}e^{ik_{3,z}L_p} & i\frac{\epsilon_2}{\kappa_2}e^{\kappa_2 L_p} & -i\frac{\epsilon_2}{\kappa_2}e^{-\kappa_2 L_p} & 0 \\ 0 & -1 & -1 & 1 \\ 0 & -\frac{\epsilon_2}{\kappa_2} & \frac{\epsilon_2}{\kappa_2} & \frac{\epsilon_1}{\kappa_1} + i\frac{\sigma(\omega)}{\omega\epsilon_0} \end{pmatrix}. \tag{5.15}$$

Notice that the first entry of the vector in the left-hand-side of Eq. (5.14) [that is, the vector with the unknowns] is exactly the sought reflection coefficient. This entry can be easily computed by using Cramer's rule, which states that the reflection amplitude can be fetched from the following quotient of determinants

$$r = \frac{E_{3,x}^{(r)}}{E_{3,x}^{(i)}} = \frac{\det\left(\mathbf{m}_{SLG}^{Otto}\right)}{\det\left(\mathbf{M}_{SLG}^{Otto}\right)}, \tag{5.16}$$

where the matrix $\mathbf{m}_{SLG}^{Otto}$ is built by replacing the first column of $\mathbf{M}_{SLG}^{Otto}$ by the column vector $(e^{-ik_{3,z}L_p}, -\frac{\epsilon_3}{k_{3,z}}e^{-ik_{3,z}L_p}, 0, 0)^T$, that is, the vector on the right-hand-side of Eq. (5.14). In possession of Eq. (5.16), the reflectance is just one stone's throw away, since it is just the modulus squared of the reflection coefficient. More concretely, after some algebra one arrives to the following expression for the former:

$$\mathcal{R}_{SLG}^{Otto} = \left| \frac{\Xi^* \Lambda^- + \Xi \Lambda^+ e^{2\kappa_2 L_p}}{\Xi \Lambda^- + \Xi^* \Lambda^+ e^{2\kappa_2 L_p}} \right|^2, \tag{5.17}$$

where the "star" denotes the complex conjugate and we have defined the functions

$$\Xi = \epsilon_3 \kappa_2 + i\epsilon_2 k_{3,z}, \tag{5.18}$$

$$\Lambda^\pm = \epsilon_2 \kappa_1 \pm \epsilon_1 \kappa_2 \pm \kappa_1 \kappa_2 \frac{i\sigma}{\omega\epsilon_0}. \tag{5.19}$$

At this point it is worth to note that if $\Re\{\sigma(\omega)\} = 0$, that is, if graphene's conductivity is purely imaginary, then the functions $\Lambda^\pm$ are real, so that the denominator in Eq. (5.17) is simply the complex conjugate of the numerator, and therefore one would obtain $\mathcal{R}_{SLG}^{Otto} = 1$, meaning that all the impinging electromagnetic radiation is reflected. This should not constitute a surprise, since in this regime the graphene sheet would act as a perfect metal, which is effectively capable of shielding all electromagnetic radiation by screening it[4]. Naturally, in empirical scenarios the conductivity of graphene has always a finite-valued real part, either owing to a finite electronic scattering rate and/or due to a contribution from interband effects (or even thermal broadening).

An example of excitation and detection of graphene surface plasmon-polaritons by means of prism coupling in Otto's configuration is depicted in Fig. 5.5. Under such scheme, the observation of sharp dip in the reflectance spectrum constitutes a hallmark of GSP-excitation. This exercise can be accomplished either by fixing the impinging angle and varying the frequency of the incident electromagnetic radiation, or by sweeping the impinging angle at a fixed frequency. Both approaches are illustrated in the panels a) and c) of Fig. 5.5, respectively. In the first case [panel a)], one can see that the resonant frequency moves towards higher values as θ goes up, which is direct consequence of a higher in-plane momentum (which is proportional to $\sin\theta$), and thus the corresponding excitation frequency is also larger (as it grows with $\sqrt{q}$). The same reasoning also explains what we observe in panel c), i.e. for more energetic radiation the phase-matching condition is met and correspondingly higher angles (notice that we have only plotted for angles above the critical angle for total internal reflection, $\theta_c = 32.3°$ in this case). Finally, the dependence on the doping concentration on the reflectance spectra is shown in panel b), whereas panel d) summarizes both the angle and frequency dependence of the reflectance, $\mathcal{R}_{SLG}^{Otto}(\omega, \theta)$, as given by Eq. (5.17). Clearly seen in the density plot of this figure, there is a region in the $(\frac{\omega}{2\pi}, \theta)$-plane where the reflection is strongly attenuated (purple and blue region). This region corresponds to an efficient excitation of surface plasmon-polaritons on graphene.

To conclude, let us stress that the ATR technique with graphene plasmons has also been used as suitable biochemical sensor [Song *et al.* (2010); Wu *et al.* (2010); Zhu and Cubukcu (2015); Wijaya *et al.* (2011)] —SPR graphene sensor. The general principle is that when the local dielectric environment near the graphene sheet is modified, either by the binding of

---

[4]Within the Drude model, this corresponds to the limit of infinite relaxation-time $\tau \to \infty$, or else, zero losses $\gamma = 0$.

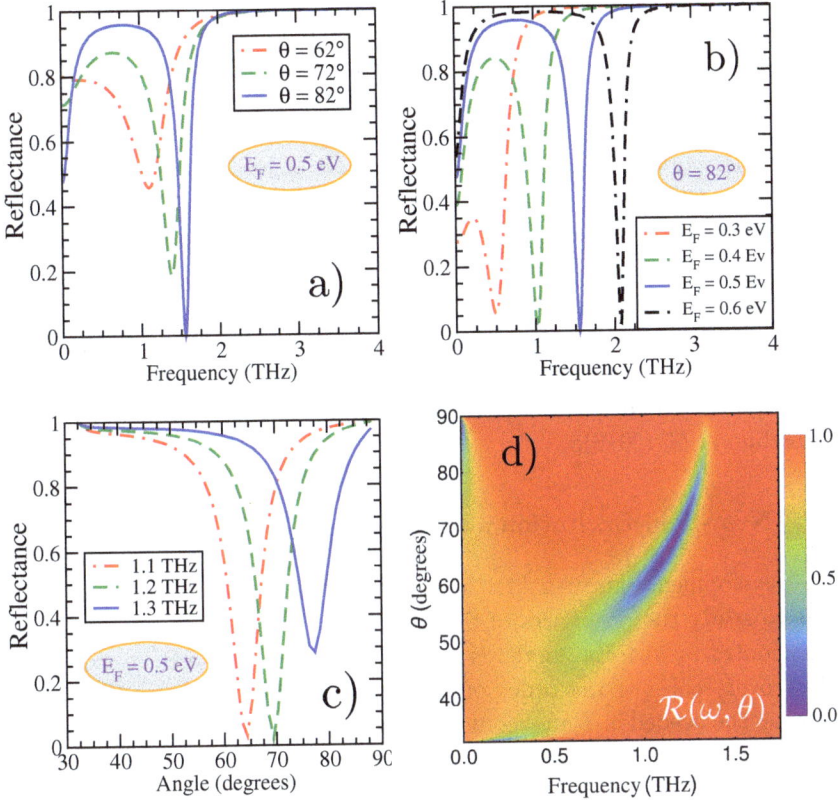

Figure 5.5: Excitation and detection of GSPs using the prism coupling technique in Otto's configuration. Panel a): detection of GSPs' excitation by sweeping the light's frequency at a fixed impinging angle (to different curves correspond different θ-values). The minimum in reflectance, as given by Eq. (5.17), pinpoints the GSPs' resonant frequency. Panel b): dependence of the $\mathcal{R}_{SLG}^{Otto}$ on graphene's Fermi energy as a function of frequency for a fixed incident angle, $\theta = 82°$. In panels a) and b) the remaining parameters are: $\epsilon_{prism} \equiv \epsilon_3 = 14$, $\epsilon_2 = 1$, $\epsilon_1 = 4$, $L_p = 2.5~\mu$m and $\hbar\gamma = 0.1$ meV. Panel c): detection of excitation of GSPs by varying the impinging angle at different fixed frequencies (see legend); as before, the emergence of a reflectance minimum indicates the plasmon's resonant excitation. Panel d): two-dimensional color plot of the reflectance as a function of both the radiation's frequency and the incident angle (we have assumed $E_F = 0.5$ eV). In panels c) and d) the following parameters were used: $\epsilon_{prism} = \epsilon_3 = 14$, $\epsilon_2 = 4$, $\epsilon_1 = 4$, $L_p = 20~\mu$m and $\hbar\gamma = 0.1$ meV.

a molecule or an anti-body, or by a generalized alteration of the concentration of a solution (e.g. flowing in a microfluidic channel adjacent to graphene) containing an analyte of interest, the position of the reflection minimum undergoes a shift which can be used to detect the presence and/or concentration of the biochemical material.

Furthermore, on the light of what we have learned in Sec. 4.4, one can straightforwardly extend the above method for exciting GSPs in Otto's configuration when a dc magnetic field is applied in the transverse direction of the graphene sheet. The reader interested in this exercise is advised to check also [Bludov *et al.* (2013)].

Finally, we note that the excitation of TE surface waves in graphene have recently been experimentally accomplished using the attenuate total reflection method in which a prism in the Otto configuration has been used [Menabde *et al.* (2016)].

5.3   Near-field Excitation and Imaging of GSPs

The development of near-field optical microscopies opened up new possibilities to study the properties of SPPs directly, with nanometer resolution. This contrasts with the methods mentioned above, where excitation takes place over scales of the order of the illuminated spot size[5]. Under this scheme, the needed in-plane momentum is provided by the near-field evanescent components of light coming out from a very sharp tip of a SNOM (scanning near-field optical microscope) with aperture radius $r \ll \lambda_0$. Therefore, by bringing a graphene layer into the vicinity of the SNOM fiber tip, the excitation of GSPs with wavevectors up to $q \sim r^{-1}$ is possible via near-field coupling to the diffracted components of the light leaving the subwavelength aperture. These *locally* excited GSPs will propagate radially away from the source tip.

Recently, two independent research groups performed ground-breaking experiments on graphene plasmonics using a similar concept (though not exactly equivalent) [Fei *et al.* (2012); Chen *et al.* (2012)]. In their pioneering work, the experimentalists have been able not only to excite GSPs, but also to detect and image them. In this case, access to large wavevectors

---

[5]Which can be fairly macroscopic, or at least with dimensions comparable to the wavelength of the incoming radiation propagating in free space (i.e. diffraction-limited).

was achieved by illuminating the (sharp) oscillating tip of an atomic force microscope (AFM) with a focused infrared beam with $\lambda_0 = 11.2$ μm [Fei et al. (2012)], thereby making the conducive tip behave as an antenna[6]. This scattering-type SNOM launches concentric ripples of GSPs propagating (while being damped) along a patch of gated graphene, until they are reflected at the sample edges (or defects). Thus, when the AFM apex is located at a distance $(n + 1/2)\lambda_p/2$ away from an edge, the reflected plasmon waves interfere constructively with the lauched plasmon's waves, forming fringes separated by half the plasmon wavelength, $\lambda_p/2$. The resulting standing (plasmon) waves constitute a convenient way to directly measure the GSP's wavelength. Experimental measurements rendered plasmon wavelengths as small as $\lambda_p \approx 200$ nm [Fei et al. (2012)], thus spatially concentrating electromagnetic radiation by a factor of $\lambda_0/\lambda_p = 56$. By tuning graphene's plasmonic properties using a back-gate, figures of merit in the interval $\lambda_0/\lambda_p = 50 - 60$ were reported [Fei et al. (2012)].

Apart from the excitation capability, the other key feature that makes this technique particularly interesting is its ability to directly resolve and image GSPs in real space (using the very same instrument, which in turn also takes topographic data simultaneously). The so-called nano-imaging works by collecting the fields scattered by the tip. Since the back-scattered signal depends strongly on the distance between the tip shaft and the sample, the background and plasmonic contributions can be isolated by making the tip vibrate vertically at a given frequency (tappering mode), and then demodulating the detected signal. A schematic representation of the setup and nano-imaging technique is shown in Fig. 5.6. This approach has recently been used to study GSPs when the graphene sheet is sitting on a ferroelectric substrate [Goldflam et al. (2015)], where the latter was shown to modify the plasmonic response of graphene through the application of a transient voltage. The same technique was also used to study the plasmonic excitations in graphene bilayers (Bernal stacking) and graphene double-layers, with randomly oriented layers [Fei et al. (2015)]. It was shown that the degree of confinement of plasmons in bilayers is larger than in single- or double-layer graphene, and that interlayer tunneling effects are important in bilayers but negligible in double layers.

---

[6]Or, in other words, it induces an electric dipole at the AFM's tip.

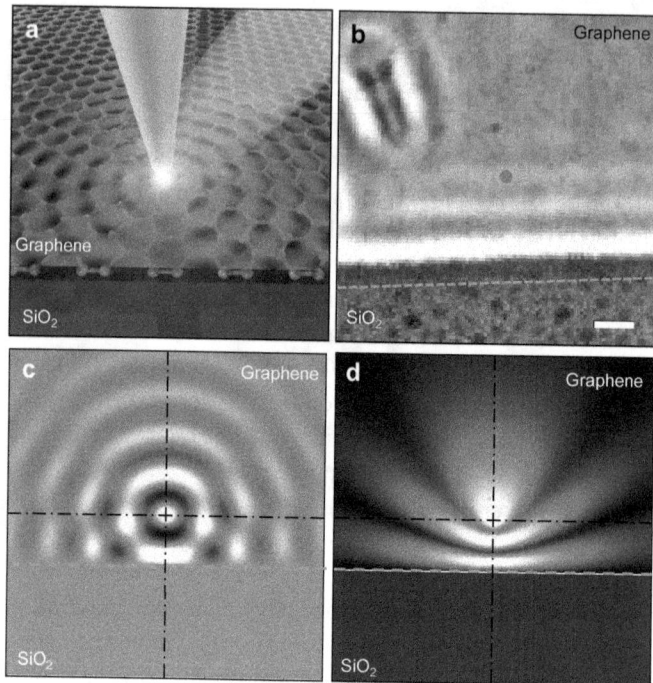

Figure 5.6: Infrared nano-imaging of GSPs. Panel a) illustration of the nano-imaging experimental setup. The launched GSPs are depicted as concentric red ripples propagating away from the AFM's apex upon illumination with IR radiation. Panel b) representative sample, obtained using the nano-imaging technique, of a graphene sheet on silicon oxide exhibiting bright fringes near the edge of the sample due to the interference of the counter-propagating graphene surface plasmons (the green dashed line indicates a defect); Panels c) and (d) depict the real part, $\Re\{E_z\}$, and absolute value, $|E_z|$, of the field profiles (simulations), respectively, when the SNOM tip is located at the red dot of indicated in panel (b). Similar results can be found in [Fei *et al.* (2012)] and [Chen *et al.* (2012)] (these images are courtesy of Z. Fei and D. Basov).

5.4   Others

In the above we presented and reviewed the most recurrent mechanisms to effectively excite GSPs. Nonetheless, it should be noted that these are not the only available techniques.

For instance, surface plasmons can be readily excited using electron beams [Vesseur *et al.* (2012)] by means of electron energy loss spectroscopy

(EELS) . In fact, the first evidences of surface plasmons (in 3D metals) were obtained from electron beam experiments in the 1950s and 1960s [Ritchie (1957); Powell and Swan (1959); Teng and Stern (1967)]. Excitation takes places whenever one of the beam's electrons transfers energy, via inelastic scattering, to a resonant plasmon mode. By measuring the energy loss of the electron transmitted (or reflected) through the sample, the energy $\hbar\omega_{SPP}$ can be determined[7]. Moreover, if the scattered angle is also measured (angle-resolved EELS), one can combine this information to map the corresponding dispersion relation.

In graphene, many EELS-based experiments also reported the excitation of graphene surface plasmons (both in single- and multi-layer graphene) [Liu and Willis (2010); Koch *et al.* (2010); Wachsmuth *et al.* (2014)] and also plasmon-phonons in graphene [Liu and Willis (2010); Koch *et al.* (2010)].

Other examples are the excitation of SPPs by light impinging on a random rough surface and on a defect with the right dimensions, and surface features, among others [Raether (1988); Zayats *et al.* (2005); Maier (2007)].

5.5 Excitation of SPP's by a Moving Line of Charge

In this section we consider in detail another SPP's excitation mechanism based on the propagation of electrons in the vicinity of a graphene sheet. The study of the motion of an electron beam was quite popular in the 1960's in the context of Čerenkov radiation [Pratesi *et al.* (1962); Bass and Yakovenko (1965)]. In particular, the motion of electron beams near a dielectric-metal interface was considered by [Danos (1955)]. In this section we apply those ideas to a dielectric-graphene-dielectric system. We note in passing that a resurgence of these ideas in the context of graphene have recently emerged [Zhan *et al.* (2014); Liu *et al.* (2014b); Tao *et al.* (2015); Kaminer *et al.* (2015)].

In the chapters to follow we will discuss in detail some of the methods described in the previous sections in this chapter. Here we want to give a detailed account of still another possible method: excitation of SPP's by a line of charge moving parallel to the graphene sheet. This regime would roughly correspond to the case of EELS at grazing incidence. Let us assume the configuration of Fig. 5.7. In addition, consider a line of charge, propagating with velocity $\mathbf{v} = u_0\hat{\mathbf{x}}$, traveling in the dielectric ϵ_1, and extending over the graphene sheet along the y−direction, at a distance

---

[7]Also, along with the SPP peak, another peak due to bulk plamons appears at higher energies in the EELS spectrum of traditional metals.

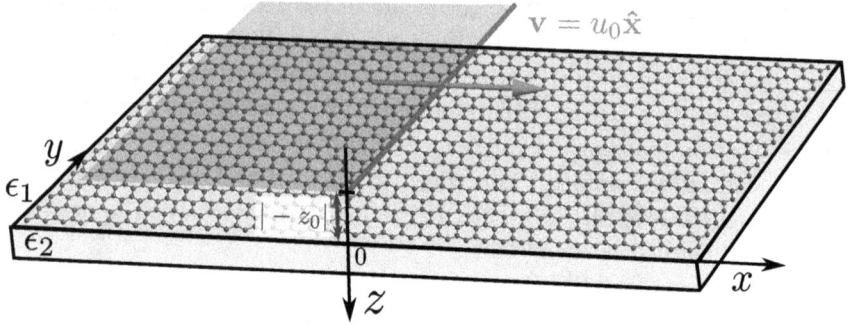

Figure 5.7: An electron beam travelling with constant velocity at a distance z_0 above a graphene monolayer. The beam direction is perpendicular to the beam's travelling velocity, $\mathbf{v} = u_0\hat{\mathbf{x}}$.

$z = -z_0$ from the sheet, as depicted in Fig. 5.7. That is, we are modeling the electron beam (in fact, and electron bunch) by a moving line of charge, with linear charge density $\lambda < 0$. The volume charge density is given by

$$\rho(x,z,t) = \lambda\delta(z+z_0)\delta(x-u_0t) = \frac{\lambda}{u_0}\delta(z+z_0)\delta(x/u_0 - t)\,, \qquad (5.20)$$

and the charge current density (along the x-direction) reads

$$J_x(x,z,t) = u_0\rho(x,z,t) = \lambda\delta(z+z_0)\delta(x/u_0 - t)\,. \qquad (5.21)$$

In terms of the scalar potential, $\phi(\mathbf{r},t)$, and the vector potential, $\mathbf{A}(\mathbf{r},t)$, the wave equations read[Liu *et al.* (2009)]

$$\nabla^2\phi(\mathbf{r},t) - \epsilon_0\mu_0\frac{\partial^2\phi(\mathbf{r},t)}{\partial t^2} = -\frac{\rho(x,z,t)}{\epsilon_1}\,, \qquad (5.22)$$

$$\nabla^2\mathbf{A}(\mathbf{r},t) - \epsilon_0\mu_0\frac{\partial^2\mathbf{A}(\mathbf{r},t)}{\partial t^2} = -\mu_0 J_x(x,z,t)\hat{\mathbf{u}}_x\,, \qquad (5.23)$$

where we have used the Lorentz gauge:

$$\nabla\cdot\mathbf{A}(\mathbf{r},t) + \frac{1}{v_1^2}\frac{\partial\phi(\mathbf{r},t)}{\partial t} = 0\,, \qquad (5.24)$$

where v_1 is the speed of light in medium 1. Since $\mathbf{A}(\mathbf{r},t) = A_x(\mathbf{r},t)\hat{\mathbf{u}}_x$ it follows from comparing the two wave equations that

$$A_x(\mathbf{r},t) = \frac{u_0}{v_1^2}\phi(\mathbf{r},t)\,. \qquad (5.25)$$

Thus, we only have to solve the wave equation for the scalar potential. Let us introduce the Fourier representation of the δ-functions and of the scalar potential:

$$\rho(x,z,t) = \frac{\lambda}{u_0} \int \frac{dk_z}{2\pi} \int \frac{d\omega}{2\pi} e^{ik_z(z+z_0)} e^{i\omega x/u_0} e^{-i\omega t} , \qquad (5.26)$$

and

$$\phi(\mathbf{r},t) = \phi(x,z,t) = \int \frac{dk_z}{2\pi} \int \frac{d\omega}{2\pi} e^{ik_z(z+z_0)} e^{i\omega x/u_0} e^{-i\omega t} \phi(k_z,\omega) . \qquad (5.27)$$

Note that the progatation of the line of charge with speed u_0 leads to the absence in the above Fourier transforms of the k_x wavenumber. Using the previous two Fourier representations in the wave equation we obtain

$$\phi(k_z,\omega) = \frac{\lambda}{\epsilon_1 u_0} \frac{1}{k_z^2 + \omega^2/u_0^2 - \omega^2/c^2} , \qquad (5.28)$$

or

$$\phi(x,z,\omega) = \frac{\lambda}{\epsilon_1 u_0} \int \frac{dk_z}{2\pi} \frac{e^{ik_z(z+z_0)} e^{i\omega x/u_0}}{k_z^2 + \omega^2/u_0^2 - \omega^2/c^2} . \qquad (5.29)$$

The integration over k_z is done in the complex plane, following

$$\phi(x,z,\omega) = \frac{\lambda}{2\epsilon_1 u_0} \frac{e^{-\sqrt{\omega^2/u_0^2 - \omega^2/c^2}|z+z_0|} e^{i\omega x/u_0}}{\sqrt{\omega^2/u_0^2 - \omega^2/c^2}} . \qquad (5.30)$$

The electric field follows from combining [8]

$$\mathbf{E}(x,z,t) = -\nabla\phi(x,z,t) - \frac{\partial \mathbf{A}(x,y,t)}{\partial t} , \qquad (5.31)$$

with the Lorentz condition, which gives [Schieber and Schächter (2001); Liu et al. (2009)] for E_x

$$E_x(x,z,\omega) = -(1 - u_0^2/v_1^2)\frac{\partial \phi(x,z,\omega)}{\partial x} ,$$

$$= -(1 - u_0^2/v_1^2)\frac{\lambda}{2\epsilon_1 u_0} \frac{i\omega}{u_0} \frac{e^{-\sqrt{\omega^2/u_0^2 - \omega^2/c^2}|z+z_0|} e^{i\omega x/u_0}}{\sqrt{\omega^2/u_0^2 - \omega^2/c^2}} . \qquad (5.32)$$

The magnetic field is obtained from

$$\mathbf{B}(x,z,\omega) = \nabla \times \mathbf{A}(x,z,\omega) = \hat{\mathbf{y}}\frac{\partial A_x(x,z,\omega)}{\partial z} , \qquad (5.33)$$

---

[8]This equation follows directly from Maxwell's equations when recasted in terms of the scalar and vector potentials.

that is (for $z_0 < z < 0$)

$$B_y(x, z, \omega) = -\frac{1}{v_1^2} \frac{\lambda}{2\epsilon_1} e^{-\sqrt{\omega^2/u_0^2 - \omega^2/c^2}(z+z_0)} e^{i\omega x/u_0}$$

$$= -i \frac{u_0}{v_1^2} \frac{\sqrt{1 - u_0^2/c^2}}{1 - u_0^2/v_1^2} E_x(x, z, \omega).$$

$$= -i \frac{u_0}{c^2} \frac{1}{\sqrt{1 - u_0^2/c^2}} E_x(x, z, \omega), \qquad (5.34)$$

where in the last line we have specialized to the realistic case where medium 1 in the air and therefore $\epsilon_1 = 1$.

Equations (5.32) and (5.34), along with the remaining $E_y(x, z, \omega)$ field-component (not given), constitute the primary fields, that is, the fields created by the line of charge in motion. Next, we seek solution's of Maxwell's equations where the primary fields couple to SPP's. To that end, we recall the results of Sec. 4.1.1 and the boundary conditions given in Sec. 2.1.2. If we define the SPP's magnetic field in regions 1 and 2 by B_{jy} and the SPP's electric field by E_{jx} (with $j = 1, 2$) and assume that the SPP's propagate with the same wave number (due to translation invariance), $q = \omega/u_0$, as the primary field, the boundary conditions at the graphene interface are

$$E_x + E_{1x} = E_{2x}, \qquad (5.35)$$

$$B_y + B_{1y} - B_{2y} = \mu_0 \sigma_{xx} E_{2x}, \qquad (5.36)$$

where E_x and B_y represent the fields (5.32) and (5.34), respectively. The second boundary condition can be written as (note that $\epsilon_1 = 1$)

$$B_y + i \frac{\omega}{c^2 \kappa_1} E_{1x} + i \frac{\omega \epsilon_2}{c^2 \kappa_2} E_{2x} = \mu_0 \sigma_{xx} E_{2x}, \qquad (5.37)$$

where Eq. (4.6) has been used and

$$\kappa_j = q^2 - \omega^2 \epsilon_j / c^2. \qquad (5.38)$$

Using Eq. (5.34) we can solve for E_{jx} as function of E_x, obtaining for E_{1x}

$$E_{1x} = E_x \frac{-iu_0/\sqrt{1 - u_0^2/c^2} - \sigma_{xx}/\epsilon_0 + i\omega\epsilon_2/\kappa_2}{\sigma_{xx}/\epsilon_0 - i\omega(1/\kappa_1 + \epsilon_2/\kappa_2)}$$

$$= -E_x \frac{\epsilon_2/\kappa_2 - 1/\kappa_1 + i\sigma_{xx}/(\omega\epsilon_0)}{\epsilon_2/\kappa_2 + 1/\kappa_1 + i\sigma_{xx}/(\omega\epsilon_0)}, \qquad (5.39)$$

a result that agrees with that of [Gong *et al.* (2014)] if we take $\sigma_{xx} = 0$. It is clear that the SPP's field E_{1x} has a resonance when

$$\frac{\epsilon_2}{\kappa_2} + \frac{1}{\kappa_1} + i\frac{\sigma_{xx}}{\epsilon_0 \omega} = 0, \qquad (5.40)$$

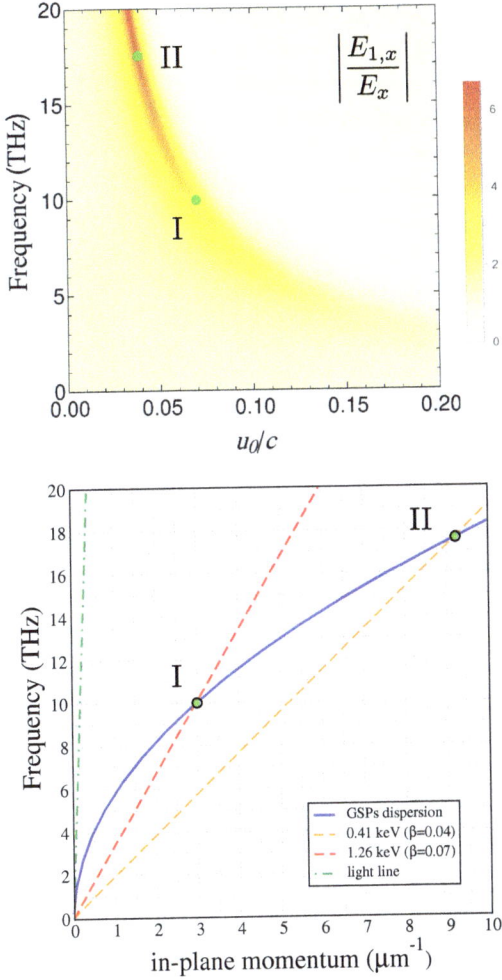

Figure 5.8: Excitation of GSPs by a parallel electron beam, as depicted in Fig. 5.7. Top panel: density plot of the (normalized) electric-field amplitude $|E_{1,x}/E_x|$ for varying frequency and beam's velocity (in units of c). Bottom: Spectrum of graphene plasmons and phase-matching condition for GSPs' excitation via a moving electron beam, with two distinct velocities defined by different parameters $\beta = u_0/c$. The resonant excitations marked by "I" and "II" are also shown in the top panel, for the same $\beta = 0.04$ and $\beta = 0.07$ values. Parameters (both panels): $E_F = 0.5$ eV, $\hbar\gamma = 5$ meV, $\epsilon_1 = 1$ and $\epsilon_2 = 4$.

which is nothing but Eq. (4.13), that is, the implicit dispersion relation for the graphene SPP's. Thus, for a given kinetic energy of the electron beam, parametrized by the velocity u_0, there is a resonant frequency, ω_{SPP}, which corresponds to the excitation of a SPP. Furthermore, note that the prefactor of E_x in Eq. (5.39) is nothing but the reflection coefficient of a $p-$polarized electromagnetic wave at dielectric-graphene interface, as can be confirmed by comparing with Eq. (N.54). This result makes sense, as the total field is constituted by the primary field plus the reflected one. In Fig. 5.8 we represent in its top panel the intensity of the plasmonic field induced by the electron beam as function of frequency and the ratio u_0/c (the ratio $|E_{1,x}/E_x|$ is represented as an intensity plot). In the same panel, two particular points (I and II) are signaled. They correspond to the interception of the $\omega = u_0 k_{\text{in-plane}}$ lines with the plasmonic dispertion relation, that is, when the resonant condition for the excitation of an SPP is attained.

Chapter 6

Launching Plasmons Using a Metallic Antenna

One way of exciting graphene surface plasmon-polaritons (or SPPs in general) is to use a metallic contact placed on top of a graphene sheet. The presence of the contact will scatter an impinging electromagnetic wave and induce plasmons in the nearby graphene. Recently, applying a similar reasoning, Alonso-González *et al.* have experimentally demonstrated the excitation of GSPs using resonant metal antennas, consisting in micrometer-sized gold bars covering a graphene layer [Alonso-González *et al.* (2014)]. When illuminated with resonant light, these behave like an optical dipole which in turn drives the launch of plasmon-polaritons propagating through graphene.

Here we will consider a slightly different system, composed by an infinitely long metal stripe with width L, positioned on top of monolayer graphene. Following closely Satou and Mikhailov [Satou and Mikhailov (2007)], we will develop a theoretical approach that aims at accounting for the system's response to an incoming electromagnetic wave and to describe the assisted launch of GSPs by the metallic contact.

6.1 Theoretical Model and Integral Equation

The system under consideration is depicted in Fig. 6.1. The graphene layer lies in the $z = 0$ plane and the metallic stripe is located on top of it, encompassing the $|x| < L/2$ region and centered at $x = 0$. We shall assume that the thickness of the metallic contact is much smaller than the wavelength of the incident radiation, thus allowing us to treat the contact as a two-dimensional metal. Additionally, we further assume the limit of a highly conductive stripe, in which the system can be perceived as two disjoint graphene sheets linked through the metallic contact. Let us now

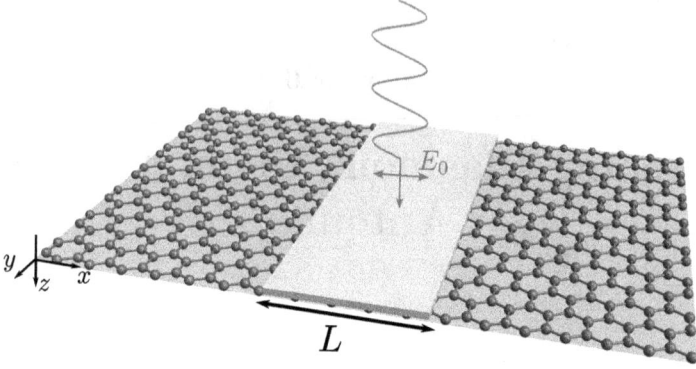

Figure 6.1: Scheme of a metallic contact on graphene. The stripe has infinite length and width L. The whole system is illuminated with electromagnetic radiation linearly polarized along the width of the contact.

suppose that the structure is subjected to a frequency-dependent external electric field, in the form of a monochromatic electromagnetic wave which propagates along the positive z-direction, and has its polarization in the direction perpendicular to the metallic contact,

$$\mathbf{E}^{\text{ext}} = E_0 e^{i(k_z z - \omega t)} \hat{\mathbf{x}} \, , \tag{6.1}$$

with $k_z = \sqrt{\epsilon} \omega / c$, where ϵ refers to the relative permittivity of the embedding (homogeneous) medium. Then, under the assumptions outlined above, the linear response of the whole system can be written as

$$J_x(x) = \sigma_g(\omega) E_x(x,0) + [\sigma_m(\omega) - \sigma_g(\omega)] \Theta \left(L/2 - |x| \right) E_x(x,0) \, , \tag{6.2}$$

where we take the conductivities of graphene, σ_g, and of the metal, σ_m, as being independent of each other. Here, E_x stands for the x-component of the *total* electric field,

$$E_x = E_x^{\text{ext}} + E_x^{\text{ind}} \, , \tag{6.3}$$

which is the sum of the incident field with an induced field, arising from the system's response to the external perturbation. Note that the total field is a function of both x and z. Due to the symmetry of the system, we will seek for solutions of the type $\mathbf{E} = (E_x, 0, E_z)$ and $\mathbf{B} = (0, B_y, 0)$. Moreover, since the structure possesses translational invariance along y-direction, we may write the x-component of the induced electric field as

$$E_x^{\text{ind}}(x, z) = \int_{-\infty}^{\infty} E_x^{\text{ind}}(q, z) e^{iqx - i\omega t} dq \, , \tag{6.4}$$

and similar relations apply for the other components, namely E_z^{ind}, B_y^{ind} and also J_x. Hereafter, we will drop the $e^{-i\omega t}$ term in order to make the forthcoming equations simpler, but it should be noted that the harmonic time dependence is implicit. From Maxwell's equations,

$$\nabla \times \mathbf{E}^{\text{ind}} = -\frac{\partial \mathbf{B}^{\text{ind}}}{\partial t} , \qquad (6.5)$$

$$\nabla \times \mathbf{B}^{\text{ind}} = \frac{\epsilon}{c^2}\frac{\partial \mathbf{E}^{\text{ind}}}{\partial t} + \mathbf{J} , \qquad (6.6)$$

we obtain

$$B_y^{\text{ind}}(q,z) = i\frac{\omega\epsilon}{c^2 k_q^2}\partial_z E_x^{\text{ind}}(q,z) , \qquad (6.7)$$

and

$$\partial_z^2 E_x^{\text{ind}}(q,z) - k_q^2 E_x^{\text{ind}}(q,z) = i\frac{k_q^2}{\omega\epsilon_0\epsilon}J_x(q)\delta(z) , \qquad (6.8)$$

where [1] $k_q^2 = q^2 - \epsilon\omega^2/c^2$ and ϵ_0 is the vacuum permittivity. Upon integration, Eq. (6.8) implies

$$\partial_z E_x^{\text{ind}}(q,0^+) - \partial_z E_x^{\text{ind}}(q,0^-) = i\frac{k_q^2}{\omega\epsilon_0\epsilon}J_x(q) , \qquad (6.9)$$

provided that E_x^{ind} is continuous[2] across the boundary at $z = 0$. Furthermore, Eqs. (6.8) and (6.9) have solutions of the form[3]

$$E_x^{\text{ind}}(q,z) = -i\frac{k_q}{2\omega\epsilon_0\epsilon}J_x(q)e^{-k_q|z|} , \qquad (6.10)$$

where k_q must be chosen in accordance with the following prescription

$$\begin{cases} k_q = \sqrt{q^2 - \omega^2\epsilon/c^2} & \text{for} \quad q^2 > \omega^2\epsilon/c^2 \\ k_q = -i\sqrt{\omega^2\epsilon/c^2 - q^2} & \text{for} \quad q^2 < \omega^2\epsilon/c^2 \end{cases} , \qquad (6.11)$$

describing both evanescent waves and scattered outgoing (propagating) waves, respectively. We now introduce the Fourier transform of the current,

$$J_x(q) = \int_{-\infty}^{\infty}\frac{dx}{2\pi}J_x(x)e^{-iqx} , \qquad (6.12)$$

---

[1] Here we introduce a different notation from that used in Appendices N and P, but no confusion should arise.

[2] Notice that this, along with Eq. (6.9) are equivalent to the usual boundary conditions we have seen in previous chapters.

[3] We Fourier transform Eq. (6.8) in the z-coordinate thus obtaining a algebraic equation for $E_x^{\text{ind}}(q,k_z)$. Solving for the latter and Fourier transforming back to the z-coordinate we obtain the desired result. The last Fourier transform is done in the complex plane. An example of this method is depicted in Fig. G.1.

which together with Eq. (6.10) enables us to write Eq. (6.4) as

$$E_x^{\text{ind}}(x,z) = -\frac{i}{4\pi\epsilon_0\epsilon\omega} \int_{-\infty}^{\infty} dq k_q e^{iqx - k_q|z|} \int_{-\infty}^{\infty} dx' J_x(x') e^{-iqx'} . \quad (6.13)$$

Recall that $J_x(x)$ is a function of the total electric field at $z = 0$, i.e. $E_x(x,0) = E_0 + E_x^{\text{ind}}(x,0)$, so that the computation of the induced field in Eq. (6.13) requires previous knowledge of $E_x^{\text{ind}}(x,0)$. Thus, in order to determine $E_x^{\text{ind}}(x,z)$, one needs find $E_x^{\text{ind}}(x,0)$ beforehand. To this end, plugging Eq. (6.2) into the definition (6.12), we obtain

$$J_x(q) = \sigma_g \left[E_0\delta(q) + E_x^{\text{ind}}(q,0) \right] + (\sigma_m - \sigma_g)\mathcal{I}(L,q) , \quad (6.14)$$

with

$$\mathcal{I}(L,q) = \int_{-L/2}^{L/2} \frac{dx'}{2\pi} E_x(x',0) e^{-iqx'} . \quad (6.15)$$

Using the result (6.14) in Eq. (6.10) for $z = 0$, we arrive at

$$E_x^{\text{ind}}(q, z = 0) = -i\frac{k_q}{2\omega\epsilon_0\epsilon} \left[\sigma_g E_0\delta(q) + \sigma_g E_x^{\text{ind}}(q,0) + (\sigma_m - \sigma_g)\mathcal{I}(L,q) \right] , \quad (6.16)$$

which after some algebra reduces to

$$E_x^{\text{ind}}(q,0) = \frac{1-\xi_q}{\xi_q} \left[E_0\delta(q) + \left(\frac{\sigma_m}{\sigma_g} - 1 \right) \mathcal{I}(L,q) \right] , \quad (6.17)$$

where we have introduced the function[4]

$$\xi_q = 1 + i\frac{\sigma_g k_q}{2\omega\epsilon_0\epsilon} . \quad (6.18)$$

Finally, transforming the expression for the induced field back to real space, while noting that $E_x^{\text{ind}}(q,z) = E_x^{\text{ind}}(q,0)e^{-k_q|z|}$, leads to

$$E_x^{\text{ind}}(x,z) = \frac{1-\xi_0}{\xi_0} E_0 e^{ik_z|z|} + \left(\frac{\sigma_m}{\sigma_g} - 1 \right) \int_{-\infty}^{\infty} dq \frac{1-\xi_q}{\xi_q} \mathcal{I}(L,q) e^{iqx - k_q|z|} . \quad (6.19)$$

As before, notice that to compute of $E_x^{\text{ind}}(x,z)$ we must have determined $\mathcal{I}(L,q)$ [which in turn depends on $E_x^{\text{ind}}(x,0)$]. We shall work towards that goal in what follows.

---

[4]Note that $1/\xi_q$ is nothing more than the transmittance amplitude (N.9) introduced in Appendix N for the p-polarized wave.

6.2 Approximate Solution for $E_x^{ind}(x,0)$ via Fourier Expansion

Let us now take a step back and write the x-component of the *total* electric field at the $z = 0$ plane, that is

$$E_x(x,0) = E_0 + E_x^{ind}(x,0)$$

$$= \frac{1}{\xi_0} E_0 + \left(\frac{\sigma_m}{\sigma_g} - 1\right) \int_{-\infty}^{\infty} dq \frac{e^{iqx}}{\xi_q} \int_{-L/2}^{L/2} \frac{dx'}{2\pi} E_x(x',0) e^{-iqx'}$$

$$+ \left(1 - \frac{\sigma_m}{\sigma_g}\right) \int_{-L/2}^{L/2} dx' E_x(x',0) \underbrace{\int_{-\infty}^{\infty} \frac{dq}{2\pi} e^{iq(x-x')}}_{\delta(x-x')}. \qquad (6.20)$$

Since $-L/2 < x' < L/2$, then the last term in Eq. (6.20) is non-zero only for $x = x' \in [-L/2, L/2]$, owing to the delta function. Therefore, from now on we will treat the electric field inside and outside the metallic contact separately. For the latter, the delta function does not contribute and we have

$$E_x^{out}(x,0) = \frac{1}{\xi_0} E_0 + \left(\frac{\sigma_m}{\sigma_g} - 1\right) \int_{-\infty}^{\infty} dq \frac{e^{iqx}}{\xi_q} \int_{-L/2}^{L/2} \frac{dx'}{2\pi} E_x^{in}(x',0) e^{-iqx'}$$

$$\text{with } |x| > L/2 , \qquad (6.21)$$

and for the field within the metal stripe we find[5]

$$E_x^{in}(x,0) = \frac{\sigma_g}{\sigma_m} \frac{E_0}{\xi_0} + \left(1 - \frac{\sigma_g}{\sigma_m}\right) \int_{-\infty}^{\infty} dq \frac{e^{iqx}}{\xi_q} \int_{-L/2}^{L/2} \frac{dx'}{2\pi} E_x^{in}(x',0) e^{-iqx'}$$

$$\text{with } |x| < L/2 . \qquad (6.22)$$

Notice that in writing expressions (6.21) and (6.22) we have included the superscript "in" in the term $E_x^{in}(x',0)$ to emphasize that it corresponds to the field inside the metallic contact, since the integration takes place only for $|x'| < L/2$. Additionally, note that Eq. (6.22) represents an integral equation for the unknown function $E_x^{in}(x,0)$. On the other hand, we should stress that we can only determine E_x^{out} once we have calculated the electric field inside the metal stripe. Thus, in order to proceed, we must solve Eq.

---

[5]Note the resemblance with the problem of scattering of electrons by an impurity, as described by the Lippmann-Schwinger equation. In the impurity problem, we first solve for the wave function inside the region where the potential is finite and then use this result to determine the wave function outside the region of influence of the impurity.

(6.22). With that in mind, we assume that $E_x^{\text{in}}(x,0)$ can be written as a Fourier expansion,

$$E_x^{\text{in}}(x,0) = \sum_{n=0}^{\infty} A_n \cos\left(\frac{2n\pi x}{L}\right) , \qquad (6.23)$$

where we have chosen the cosines as basis functions due to the even symmetry of the problem[6]. Inserting this representation into Eq. (6.22), we obtain

$$\sum_{n=0}^{\infty} A_n \cos\left(\frac{2n\pi x}{L}\right) = \frac{\sigma_g}{\sigma_m} \frac{E_0}{\xi_0}$$

$$+ \left(1 - \frac{\sigma_g}{\sigma_m}\right) \int_{-\infty}^{\infty} dq \frac{e^{iqx}}{\xi_q} \int_{-L/2}^{L/2} \frac{dx'}{2\pi} e^{-iqx'} \sum_{n=0}^{\infty} A_n \cos\left(\frac{2n\pi x'}{L}\right) . \qquad (6.24)$$

Multiplying both sides of Eq. (6.24) by $\cos\left(\frac{2m\pi x}{L}\right)$ and integrating over $-L/2 \le x \le L/2$, results in (for $m = 0$)

$$LA_0 = L\frac{\sigma_g}{\sigma_m}\frac{E_0}{\xi_0} + \left(1 - \frac{\sigma_g}{\sigma_m}\right) \times$$

$$\times \sum_{n=0}^{\infty} A_n \int_{-\infty}^{\infty} dq \frac{e^{iqx}}{\xi_q} \int_{-L/2}^{L/2} dx \int_{-L/2}^{L/2} \frac{dx'}{2\pi} e^{-iqx'} \cos\left(\frac{2n\pi x'}{L}\right) , \qquad (6.25)$$

and (for $m \ne 0$)

$$\frac{L}{2} \sum_{n=0}^{\infty} A_n \delta_{n,m} = \left(1 - \frac{\sigma_g}{\sigma_m}\right) \times$$

$$\times \sum_{n=0}^{\infty} A_n \int_{-\infty}^{\infty} dq \frac{e^{iqx}}{\xi_q} \int_{-L/2}^{L/2} dx \int_{-L/2}^{L/2} \frac{dx'}{2\pi} e^{-iqx'} \cos\left(\frac{2n\pi x'}{L}\right) \cos\left(\frac{2m\pi x}{L}\right) , \qquad (6.26)$$

where we have used the integral identities arising from the orthogonality of this complete set of basis functions[7]. Performing the integrations over dx' and changing variables to the dimensionless variable $u = qL$, Eqs. (6.25) and (6.26) translate to

$$\frac{\sigma_g}{\sigma_m}\frac{E_0}{\xi_0} = A_0 \left(1 - \frac{4}{\pi}\left[1 - \frac{\sigma_g}{\sigma_m}\right] \int_0^{\infty} \frac{du}{\xi_u} \frac{\sin^2(u/2)}{u^2}\right)$$

$$+ \frac{4}{\pi}\left(1 - \frac{\sigma_g}{\sigma_m}\right) \sum_{n=1}^{\infty} (-1)^{n+1} A_n \int_0^{\infty} \frac{du}{\xi_u} \frac{\sin^2(u/2)}{u^2 - 4n^2\pi^2} , \qquad (6.27)$$

---

[6]Strictly speaking, the even symmetry of the problem only guaranties that the potential is either even or odd. We have chosen to analyze the even solution.

[7]Explicitly, $\int_{-L/2}^{L/2} \cos\left(\frac{2n\pi x}{L}\right) \cos\left(\frac{2m\pi x}{L}\right) dx = \begin{cases} \frac{L}{2}\delta_{n,m} & , \text{ if } n, m \ne 0 \\ L & , \text{ if } m = 0 \end{cases}$.

and

$$
A_m \left(1 - \frac{8}{\pi} \left[1 - \frac{\sigma_g}{\sigma_m}\right] \int_0^\infty du \frac{u^2}{\xi_u} \frac{\sin^2(u/2)}{[u^2 - 4m^2\pi^2]^2}\right)
$$
$$
+ \frac{8}{\pi} \left(1 - \frac{\sigma_g}{\sigma_m}\right) \sum_{\substack{n=0 \\ n \neq m}}^\infty (-1)^{n+m+1} A_n \int_0^\infty du \frac{u^2}{\xi_u} \frac{\sin^2(u/2)}{[u^2 - 4n^2\pi^2][u^2 - 4m^2\pi^2]} = 0 ,
$$

$$(6.28)$$

respectively; here the function ξ_u now reads

$$
\xi_u = 1 - \frac{E_F}{\hbar\omega} \frac{2\alpha\Omega_L}{\hbar\omega + i\hbar\gamma} \sqrt{u^2 - a^2} ,
\tag{6.29}
$$

with the definitions $a = \sqrt{\epsilon \frac{\hbar\omega}{\Omega_L}}$ and $\Omega_L = \hbar c/L$, and where we have assumed a Drude-like conductivity for graphene, as given by expression (2.72). Together, Eqs. (6.27) and (6.28) form a linear system of equations in terms of the Fourier coefficients, A_n. In other words, the above-mentioned expressions can be recast as a matricial equation of the (typical) form $\mathbf{Ax} = \mathbf{b}$; that is, running the sums up to N, we have

$$
\begin{pmatrix}
C_0 & F_1 & F_2 & \cdots & \cdots & F_n \\
H_{0,1} & G_1 & H_{2,1} & \cdots & \cdots & H_{N,1} \\
\vdots & & \ddots & & & \vdots \\
H_{0,j} & \cdots & \cdots & G_j & \cdots & H_{N,j} \\
\vdots & & & & \ddots & \vdots \\
H_{0,N} & H_{1,N} & \cdots & \cdots & H_{N-1,N} & G_N
\end{pmatrix}
\begin{pmatrix}
A_0 \\
A_1 \\
\vdots \\
A_j \\
\vdots \\
A_N
\end{pmatrix}
=
\begin{pmatrix}
\frac{\sigma_g}{\sigma_m} \frac{E_0}{\xi_0} \\
0 \\
\vdots \\
0 \\
\vdots \\
0
\end{pmatrix} .
\tag{6.30}
$$

with

$$
C_0 = 1 - \frac{4}{\pi} \left(1 - \frac{\sigma_g}{\sigma_m}\right) \int_0^\infty \frac{du}{\xi_u} \frac{\sin^2(u/2)}{u^2} ,
\tag{6.31}
$$

$$
F_n = \frac{4}{\pi} \left(1 - \frac{\sigma_g}{\sigma_m}\right) (-1)^{n+1} \int_0^\infty \frac{du}{\xi_u} \frac{\sin^2(u/2)}{u^2 - 4n^2\pi^2} ,
\tag{6.32}
$$

$$
G_m = 1 - \frac{8}{\pi} \left(1 - \frac{\sigma_g}{\sigma_m}\right) \int_0^\infty du \frac{u^2}{\xi_u} \frac{\sin^2(u/2)}{[u^2 - 4m^2\pi^2]^2} ,
\tag{6.33}
$$

$$
H_{n,m} = \frac{8}{\pi} \left(1 - \frac{\sigma_g}{\sigma_m}\right) (-1)^{n+m+1} \int_0^\infty du \frac{u^2}{\xi_u} \frac{\sin^2(u/2)}{[u^2 - 4n^2\pi^2][u^2 - 4m^2\pi^2]} .
$$

$$(6.34)$$

Once the A_n's are determined from the solution of the matrix Eq. (6.30), we are able to construct the field inside the contact via

$$
E_x^{in}(x,0) = \sum_{n=0}^N A_n \cos\left(\frac{2n\pi x}{L}\right) ,
\tag{6.35}
$$

[recall Eq. (6.23)] and then the field in graphene (outside the metal stripe) through Eq. (6.21), yielding

$$E_x^{\text{out}}(x,0) = \frac{1}{\xi_0}E_0 + \frac{2\eta_\sigma}{\pi}\sum_{n=0}^{N}(-1)^n A_n \int_0^\infty du \frac{\cos(ux/L)}{\xi_u}\frac{u\sin(u/2)}{u^2 - 4n^2\pi^2}.$$

(6.36)

where we have defined $\eta_\sigma = \sigma_m/\sigma_g - 1$ for condensing the writing. The first term in the previous equation represents the contribution coming from the uniform part of the structure. In the case where the contact is absent, this would be the only term[8]. Conversely, the second term in Eq. (6.36) accounts for the modification of the electric field induced by the metallic stripe. Also, note that the integral has poles whenever $\xi_u = 0$, corresponding to the excitation of plasmon-polaritons in graphene.

6.3   Results and Discussion

Having the expressions for the fields, given by Eqs. (6.35) and (6.36), we are now in position to investigate the behavior of the in-plane electric field (at $z = 0$) in the system under consideration. Figure 6.2 shows the results obtained for both $E_x^{\text{in}}(x,0)$ and $E_x^{\text{out}}(x,0)$. Here we have plotted the fields for different contact widths, L, while keeping graphene's Fermi level constant — as well as the ratio $\sigma_m/\sigma_g \gg 1$ (highly conductive stripe) — for a fixed incident wavelength, λ_0. In the top panel we have represented the x-component of the electric field in graphene (i.e. outside the contact), which encompasses GSPs fields and also a contribution from the uniform part of the structure [described by the first term in equation (6.36)]. It is apparent from the figure that the system emits plasmon-polaritons travelling away from the metallic stripe exhibiting the characteristic subwavelength confinement and high field enhancement. For instance, in the case with $L = \lambda_0/2$ the GSP wavelength is about 19 times smaller than the wavelength of the impinging electromagnetic wave ($\lambda_p \simeq 0.052\lambda_0$), which is already a fair degree of confinement. Naturally, the plasmonic field weakens with increasing x/L due to the Ohmic losses within graphene. Thus, at large distances away from the contact the field becomes indistinguishable from the response of the homogeneous system, so that $E_x^{\text{out}}(x,0)/E_0 \to \xi_0^{-1}$ (depicted in the figure as a black dotted line). Albeit not particularly enlightening, the field inside the stripe is portrayed in the two panels —b) and c)— at lower left.

---

[8] Notice that if one replaces the contact by a patch of graphene ($\sigma_m \to \sigma_g$), restoring the system's homogeneity, the second term gives a null contribution.

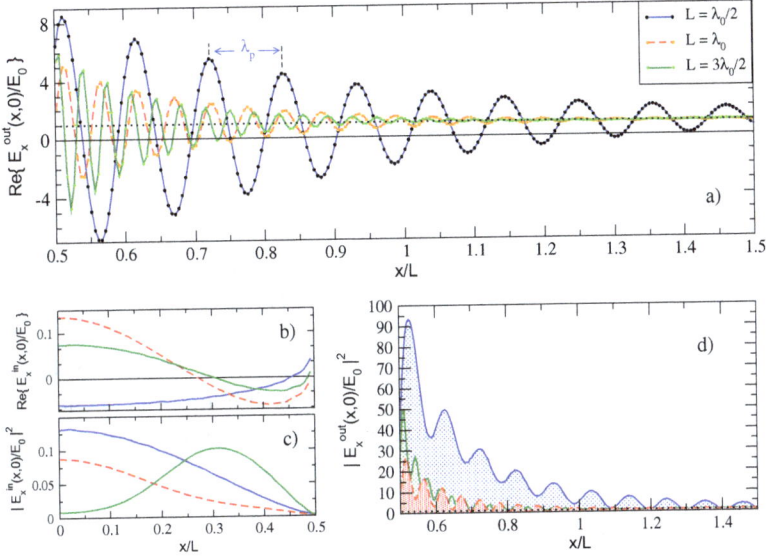

Figure 6.2: In-plane electric fields, both within the metal stripe, $E_x^{in}(x,0)$, and in graphene, $E_x^{out}(x,0)$, as a function of x/L for different contact widths. All fields were normalized with respect to the amplitude of the incident electromagnetic wave. The top panel shows the real part of $E_x^{out}(x,0)/E_0$ as given by Eq. (6.36), while the bottom right panel shows its absolute value squared. The fields inside the contact are plotted in the two panels at lower left. We used the following parameters: $\epsilon = 1$, $\lambda_0 = 10.2$ μm (corresponding to $\hbar\omega = 0.12$ eV or $\omega = 2\pi \times 30$ THz), $E_F = 0.44$ eV, $\hbar\gamma = 5$ meV and $\sigma_g/\sigma_m = 0.01$ (highly conductive stripe).

Every panel in Fig. 6.2 shows a strong dependence on the ratio L/λ_0, which is expected since the width of the contact is of the order of the impinging wavelength. Indeed, for the given parameters in Fig. 6.2 we see that the highest fields are obtained when $L = \lambda_0/2$. For this reason, hereafter we will focus our attention on this particular case.

We now consider not only the in-plane field's fraction, but rather the induced electric field in the whole space. Therefore, using Eqs. (6.19) and (6.10) and proceeding in the same fashion as before, the x-component of the induced field reads

$$E_x^{\text{ind}}(x, z) = \frac{1 - \xi_0}{\xi_0} E_0 e^{ik_z |z|}$$

$$+ \frac{2\eta_\sigma}{\pi} \sum_{n=0}^{\infty} (-1)^n A_n \int_0^\infty du \frac{1 - \xi_u}{\xi_u} \frac{u \sin(u/2)}{u^2 - 4n^2\pi^2} \cos(ux/L) e^{-k_u |z|/L} \ .$$

$$(6.37)$$

Once more, we observe that the field expressed in Eq. (6.37) can be split into two distinct contributions. The first term describes the scattering of the incident electromagnetic wave by the uniform system, i.e. without the contact; thus, it is independent of the x coordinate and represents scattered waves propagating along the z-direction: with reflected waves for $z < 0$ and transmitted waves for $z > 0$. On the other hand, the last term characterizes the screening and the system's response to the inhomogeneity introduced by the metal stripe. For this reason, we shall denote the second term by E_x^{2D}, i.e.

$$E_x^{\text{2D}}(x, z) \equiv \frac{2\eta_\sigma}{\pi} \sum_{n=0}^{\infty} (-1)^n A_n \int_0^\infty du \frac{1 - \xi_u}{\xi_u} \frac{u \sin(u/2)}{u^2 - 4n^2\pi^2} \cos(ux/L) e^{-k_u |z|/L}.$$

$$(6.38)$$

In addition, we note that the above integral contains two different kinds of fields: those in the form of evanescent waves confined near the graphene sheet, when $u > a$; and those corresponding to propagating waves, when $u < a$ (the surface plasmons consist in the former type of fields).

At this point the only item missing in our list is the component of the induced field perpendicular to the structure, that is to say, $E_z^{\text{ind}}(x, z)$. We can retrieve this last piece of information working through Maxwell's equations, which link E_x^{ind}, B_y^{ind} and E_z^{ind}, namely

$$E_z^{\text{ind}}(x, z) = i \frac{c^2}{\omega\epsilon} \partial_x B_y^{\text{ind}}(x, z) \ , \qquad (6.39)$$

and

$$B_y^{\text{ind}}(q, z) = i \frac{\omega\epsilon}{c^2 k_q^2} \partial_z E_x^{\text{ind}}(q, z) \qquad (6.40)$$

$$= -\frac{\text{sgn}(z)}{2\epsilon_0 c^2} J_x(q) e^{-k_q |z|} \ , \qquad (6.41)$$

where in the last step we used Eq. (6.10). Taking the inverse Fourier transform and inserting the previous expression into Eq. (6.39) yields

$$E_z^{\text{ind}}(x, z) = \frac{\text{sgn}(z)}{2\omega\epsilon_0\epsilon} \int_{-\infty}^{\infty} q J_x(q) e^{iqx - k_q |z|} dq \ . \qquad (6.42)$$

Finally, using Eq. (6.14) for $J_x(q)$ and changing variables yet again to $u = qL$, one obtains (after some algebra):

$$E_z^{\text{ind}}(x, z) = -\operatorname{sgn}(z)\frac{2\eta_\sigma}{\pi}\sum_{n=0}^{\infty}(-1)^n A_n \int_0^{\infty} du \frac{1 - \xi_u}{\xi_u}\frac{\sin(ux/L)}{k_u}\frac{u^2 \sin(u/2)}{u^2 - 4n^2\pi^2}$$

$$\times\, e^{-k_u|z|/L}$$

$$\equiv E_z^{\text{2D}}(x, z) \,. \tag{6.43}$$

Notice that for the z-component of the induced field we have $E_z^{\text{ind}} = E_z^{\text{2D}}$, since we are restricting ourselves to the particular case of a normally incident wave with p-polarization. For the sake of completeness, let us make the remark that in the more general situation with an arbitrary angle of incidence, E_z^{ind} would have an extra contribution (analogous to the one in E_x^{ind}) describing scattered waves.

At this point we have a complete knowledge of the induced field, and hence also of $\mathbf{E}^{\text{2D}} = \left(E_x^{\text{2D}}, 0, E_z^{\text{2D}}\right)$. The components of the latter as a function of the spatial coordinates are shown in Fig. 6.3. Recall that this corresponds to the two-dimensional electric field in which GSPs are included. Furthermore, we note that the results depicted in Fig. 6.3 refer to the case where the width of the metallic contact is $\lambda_0/2$, which corresponds to the configuration capable of achieving the highest efficiency to excite GSPs. In panel a) we have represented a two-dimensional plot of $\Re\{E_x^{\text{2D}}/E_0\}$ in the $z = 0$ plane. Owing to the system's translational invariance along the y-direction, we observe a set of high- and low-intensity fringes parallel to the edge of the stripe, gradually decaying as we move away from it. Note that for the chosen value of γ (which is rather reasonable), the decay experienced by the GSPs is rather weak as they can propagate for several plasmon-polaritons wavelengths. Also shown in the figure is the dependence of $\Re\{E_x^{\text{2D}}/E_0\}$ and $\Re\{E_z^{\text{2D}}/E_0\}$ on the (x, z)-coordinates. Both exhibith the typical plasmonic behavior, with strong fields highly localized near the graphene layer and subwavelength confinement. In particular, from the numerical results we have obtained $\lambda_p = 0.052\lambda_0$ for the GSPs' wavelength. At the same time, from the figure we can see that the field at $z = 0.025L = 1.25 \times 10^{-2}\lambda_0$ in the dielectric is already about half of its maximum value. Together, these lead to a large concentration of electromagnetic energy within a very small volume.

While Fig. 6.3 depicts E_x^{2D} and E_z^{2D} separately, it is instructive to gather all this information into a vectorial representation of $\mathbf{E}^{\text{2D}}$. Taking this into consideration, we show in Fig. 6.4 the electric field lines corresponding to $\mathbf{E}^{\text{2D}}(x, z)$. These illustrate well the nature of plasmon-polaritons as collective oscillations in the density of charges, and support

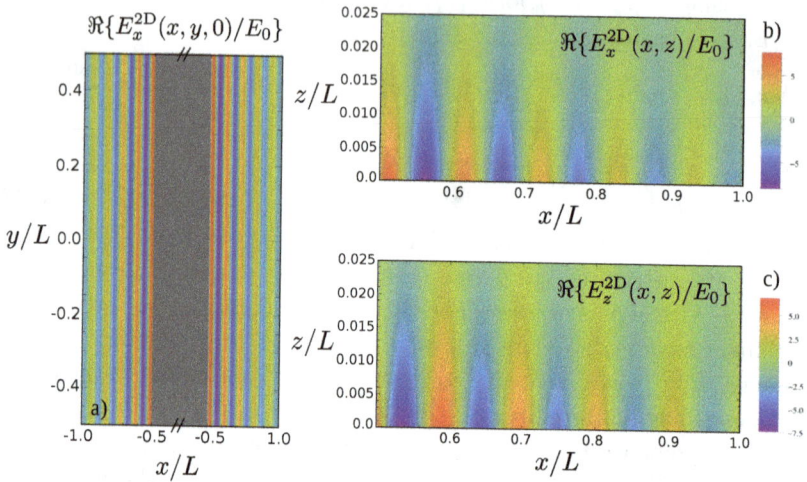

Figure 6.3: Real part of E_x^{2D} and E_z^{2D} normalized with respect to the incoming wave's amplitude. In panel a), the rectangle in gray represents the contact (notice the break on the x-axis). In all panels we considered $L = \lambda_0/2$, that is the condition for the first resonance of the antenna-like metallic stripe. The remaining parameters were the same as in Fig. 6.2.

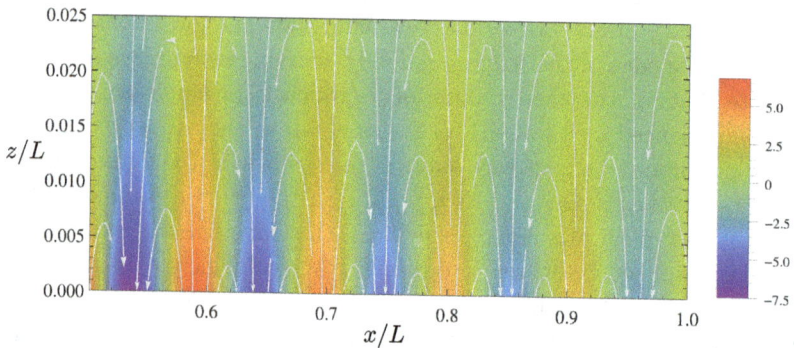

Figure 6.4: Representation of vectorial field $\mathbf{E}^{2D}(x, z)$ in terms of its field lines. These give us a hint of the regions where the concentration of charge carriers is higher (at a given time). The background and respective colorbar refers to the value of $\Re\{E_z^{2D}/E_0\}$ [cf. panel c) in Fig. 6.3].

our interpretation of the problem confirming that this is indeed essentially a plasmonic field.

Thus, our calculations demonstrate the ability to lauch GSPs using light and a highly conducive contact on graphene, endowing us with another method to excite plasmon-polaritons beyond the costumary techniques like prism or grating coupling.

Chapter 7

Plasmonics in Periodic Arrays of Graphene Ribbons

The recent developments in nanofabrication techniques and theoretical/computational modelling have brought plasmonics to the nanoscale, leading to the birth of nanoplasmonics. Recently, the realization of surface plasmon-polaritons in graphene [Ju *et al.* (2011)] has drawn a lot of attention. Since then, research in graphene plasmonics is intense and likely to remain one of the hottest topics in nanophotonics in the forthcoming years. Particular interest has been given to the excitation of GSPs in periodic graphene-based structures, including corrugated graphene [Bludov *et al.* (2013); Slipchenko *et al.* (2013a)], diffraction gratings, graphene plasmonic crystals and metamaterials [Ju *et al.* (2011); Yan *et al.* (2013); Nikitin *et al.* (2012a)], and also conductivity gratings (in which graphene's conductivity is periodically modulated) [Bludov *et al.* (2012a)]. Among these are included arrays of graphene ribbons arranged periodically [Ju *et al.* (2011); Yan *et al.* (2013); Nikitin *et al.* (2012a)], which will be the main subject throughout this Chapter. This problem has also been analyzed from the point of view of circuit theory [Khavasi and Rejaei (2014); Khavasi (2015)].

7.1 A Seminal Paper

As already mentioned in the introduction of this book, by the end of 2011 a seminal paper on surface plasmon-polaritons (SPP) in graphene was published [Ju *et al.* (2011)]. The research conducted in that letter showed the possibility of exciting GSPs, in the THz spectral range, by an array of micro-ribbons, and launched the roots for experimental graphene plasmonics (a representation of the experimental setup is given in Fig. 7.8). Many papers have followed inspired by that reference. In what follows, we summarize the main findings of some of those papers.

In the beginning of 2012 a theoretical work followed [Christensen *et al.* (2012)], studying the manipulation of SPP in graphene nano-ribbons. Motivated by the result of [Ju *et al.* (2011)] other theoretical works were published [Nikitin *et al.* (2012a); Slipchenko *et al.* (2013a)], which, however, made no attempt to simulate the experimental data (something that we will do in the sections ahead). The effect of a magnetic field in THz plasmons was experimentally investigated still in 2012 [Crassee *et al.* (2012)]. In this latter work, large area graphene sheets grown by epitaxial methods were exploited and have shown to support magneto-plasmons, which were also by the same time theoretically investigated [Ferreira *et al.* (2012)].

A natural follow up of the work of [Ju *et al.* (2011)] is the investigation of graphene-based vertical structures, with particular emphasis on stacks of graphene/insulator layers. This line of research was pursued by the IBM group at New York [Yan *et al.* (2012a)], showing that one could manipulate mid- and far-infrared (THz) radiation using such structures, obtaining an enhanced plasmonic response with many interesting applications, such as an almost perfect radiation absorber. They have also shown that these vertical heterostructures can act as linear polarizers [Yan *et al.* (2012a)]. The theoretical foundations of a SPP graphene-based polarizer were investigated at about the same time [Bludov *et al.* (2012b)]. Still in 2012, a mechanism based on a graphene/metamaterial grating to make graphene a perfect absorber, at specific THz frequencies, was proposed [Ferreira and Peres (2012)]; also, an optical switch that would work in the THz spectral range, by exploiting the physics of SPP in graphene, had already been introduced [Bludov *et al.* (2010)]. Moreover, the physics of a polaritonic crystal exploiting GSPs was investigated as early as 2012 [Bludov *et al.* (2012a)].

Excitation of surface plasmons in a metamaterial constituted by a graphene sheet deposited on a layer of hexagonal meta-atoms was considered in [Lee *et al.* (2012)]. A 47% of modulation of the amplitude in the transmitted wave was achieved. A review of different kinds of metamaterials, including those that are graphene-based, can be found in [Zheludev and Kivshar (2012)]. Another review article [Grigorenko *et al.* (2012)], although not really focused on SPPs in graphene, discusses the nature of the collective charge excitations in this material. Moreover, a theoretical consideration of biperiodic meta-surfaces for exciting SPP was given in [Fallahi and Perruisseau-Carrier (2012)], which predicted that the method should work from microwave region of the spectrum up to the infrared. A generalization of the method developed by [Ju *et al.* (2011)], now considering a bilayer of micro-graphene ribbons and working in the far infrared was also proposed. Furthermore, it was shown that an amplitude modulation of the order of 100% was possible to achieve, up to tens of THz [Sensale-Rodriguez *et al.* (2012)]. The geometrical arrangement of two graphene micro-ribbons separated by an insulating gap was considered in [Zhao *et al.* (2013)]. The authors also showed that a single ribbon can effectively work as the double ribbon case, due to electrostatic interactions with the mirror images (in the substrate) of the charges of the single ribbon.

By 2013, based on the geometry studied in [Ju *et al.* (2011)], a biosensor was proposed in [Ishikawa and Tanaka (2013)] and the same idea discussed in another publication [Vasić *et al.* (2013)]. Here, the physical principle is to excite confined SPPs, whose localized electric field exhibits an enhanced intensity, to probe small changes in the local dielectric environment near the graphene's surface due to the adsorption of biomolecules. A further, and natural, generalization of the ideas brought about by [Ju *et al.* (2011)] was to consider a metamaterial made of multi-layer graphene/insulator micro-ribbons, which was demonstrated to have a complex transmission spectrum [Padooru *et al.* (2013)]. This system, however, seems fairly challenging to fabricate without inducing a deterioration in the mobility experienced by the charge-carriers in the graphene ribbons, at least using the current technologies. The ideas described in [Ishikawa and Tanaka (2013); Vasić *et al.* (2013)] are much more feasible from the micro-fabrication point of view, although the results of [Padooru *et al.* (2013)] do hold some interesting promises.

The experimental observation of GSPs in monolayer graphene resting on a two-dimensional grating was first achieved in 2013 [Zhu *et al.* (2013d)], where the authors had in view optoelectronic applications in the infrared spectral range. Since for THz applications one has to use CVD-grown graphene, which is known to have grain boundaries, the issue of SPP reflectance inside the graphene sheet by such type of defects is an important question that was addressed in [Freitag *et al.* (2013)]. Coming back to the ribbon-geometry developed in foundational paper by L. Ju *et al* [Ju *et al.* (2011)], it was demonstrated that such a geometry could be used for photo-detection, depending on the polarization of the impinging radiation [Freitag *et al.* (2013)]. In particular, the coupling of the GSPs to the surface optical-phonons of the silicon dioxide substrate played an important role.

The very same setup —that of a grating of periodically arraged graphene ribbons— was revisited in [Strait *et al.* (2013)], which confirmed the results of the original paper and introduced a numerical model for the calculation of the specrum of GSPs. In addition, the setup proposed in [Peres *et al.* (2013)] was built in [Gao *et al.* (2013)] and the authors demonstrated that, indeed, the transmittance showed minima as function of the infrared frequency owing to the excitation of graphene surface plasmons. The investigation of THz excitation of SPPs in mesas and ribbons grown out of epitaxial graphene was considered for the first time in [Mitrofanov *et al.* (2013)]. An extension of the work done in [Bludov *et al.* (2013)] was considered in [Slipchenko *et al.* (2013a)], by developing an analytical solution for sine-wave graphene-based grating. Presumably, building on the problem posed by [Freitag *et al.* (2013)], an experimental study of the grain-boundaries in CVD-grown graphene used this material's SPPs to map such type of defects [Fei *et al.* (2013)], by taking advantage of the scattering of GSPs at grain-boundaries. The authors also suggested using this kind of boundaries to build SPPs reflectors in graphene. In addition, a two-dimensional array of micro-ribbons was considered, delivering similar qualitative results [Cheng *et al.* (2013)] to its one-dimensional version.

Complementary to the excitation of GSPs in patterned graphene gratings, an intriguing idea was published by two independent groups, proposing the excitation of plasmons in graphene via surface acoustic waves. The underlying principle is to use a sound wave to induce a spatial wave-like displacement of the graphene sheet, which can act effectively as a grating [Schiefele *et al.* (2013); Farhat *et al.* (2011)] (see Chapter 9 for a similar idea). Another proposed approach towards the achievement of plasmonic excitations in graphene, was to explore the possibility of locally doping graphene via molecular adsorption [Peres *et al.* (2012)]. This can be done in a controlled way by imposing specific geometric arrangements. In the end, one can once again produce a periodic conductivity pattern that couples an incident electromagnetic wave to GSPs. This approach was further explored in [Cheng *et al.* (2014)], where ab-initio and electromagnetic simulations were conducted. Experimentally, the idea of selectively reducing insulating fluorinated graphene to regions of conducting material was investigated in [Withers *et al.* (2011)]. This technique can be used to create periodic conductivity patterns without breaking the continuous graphene sheet. A similar concept consists in creating micro-ribbons by hydrogenating graphene —thereby creating graphane— using a mask, thus creating a periodic lattice of graphene/graphane ribbons [Sun *et al.* (2011b)].

Admittedly, the paper of [Ju *et al.* (2011)] sprung a wealth of scientific research and can very well be considered a foundational work. Next, we move to the theoretical description of the setup implemented by [Ju *et al.* (2011)]. Later we will compare the results of our model with the experimental data.

7.2 Theoretical Model

We will begin our theoretical analysis with a concise description of a framework applicable to the most general case of a graphene sheet with a periodically modulated conductivity on a dielectric substrate. As discussed in the previous section, this can be achieved in several ways, such as: placing adsorbed atoms or molecules which modify the LDOS [Silveiro *et al.* (2013); Peres *et al.* (2012)], via spatially-periodic electrical gating, inducing strain, and/or patterning an initially homogeneous graphene layer by employing nanofabrication methods. Here we will investigate the coupling of electromagnetic radiation to GSPs in a periodic array of graphene ribbons [Bludov *et al.* (2013); Ju *et al.* (2011)], by solving the scattering problem and computing the transmittance, reflectance, and absorbance of plane waves impinging upon the aforementioned system. These results will then be compared against experimental results [Ju *et al.* (2011)].

We stress that although the theoretical model we develop in the following section will ultimately be applied to a periodic grating of graphene ribbons, the very same formalism is applicable to any arbitrary (2D) system with periodically modulated conductivity, as for example one created by chemical means [Withers *et al.* (2011); Sun *et al.* (2011b)].

7.2.1 *Setting up the model*

Let us consider a generic two-dimensional system with a spatially-dependent periodic conductivity (see also [Peres *et al.* (2012)]), satisfying $\sigma(x) = \sigma(x + R)$. Here R defines the period of the structure, which can be seen as a one-dimensional (1D) crystal with lattice constant R, allowing us to define the corresponding reciprocal lattice vector, $G = 2\pi/R$. In resemblance to other periodic systems (e.g. an electron subjected to a periodic potential in a solid), the electromagnetic fields obey Bloch's theorem. Therefore, we will seek for transverse magnetic SPP Bloch waves in the

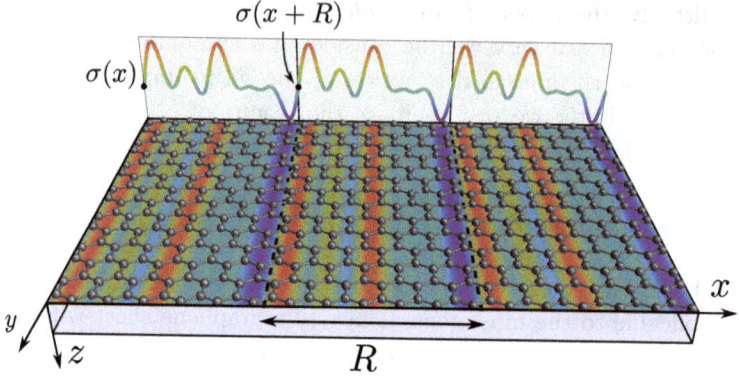

Figure 7.1: Schematic representation of a graphene sheet with an arbitrary spatially-dependent periodic conductivity, such that $\sigma(x) = \sigma(x + R)$, where R is the period of the system. The curve depicts an arbitrary periodic profile for the conductivity, while the colours reflect its corresponding value (a.u.).

form of a Fourier-Floquet series, i.e.

$$\mathcal{E}_x^{(j)}(x, z) = \sum_{n=-\infty}^{\infty} E_{x,n}^{(j)} e^{i(q+nG)x} e^{-\kappa_{j,n}|z|} , \tag{7.1}$$

$$\mathcal{E}_z^{(j)}(x, z) = \sum_{n=-\infty}^{\infty} E_{z,n}^{(j)} e^{i(q+nG)x} e^{-\kappa_{j,n}|z|} , \tag{7.2}$$

$$\mathcal{B}_y^{(j)}(x, z) = \sum_{n=-\infty}^{\infty} B_{y,n}^{(j)} e^{i(q+nG)x} e^{-\kappa_{j,n}|z|} , \tag{7.3}$$

where the index $j = 1$ refers to the substrate ($z > 0$), and $j = 2$ to the top dielectric media ($z < 0$), whose electromagnetic properties are fully characterized by the relative permittivities ϵ_1 and ϵ_2, respectively. If we take the system as being linear, then each Fourier component is independent of the others, so that the following relations can be obtained from Maxwell's equations:

$$B_{y,n}^{(j)} = -i \operatorname{sgn}(z) \frac{\epsilon_j \omega}{c^2 \kappa_{j,n}} E_{x,n}^{(j)} , \tag{7.4}$$

$$B_{y,n}^{(j)} = -\frac{\epsilon_j \omega / c^2}{q + nG} E_{z,n}^{(j)} , \tag{7.5}$$

$$E_{z,n}^{(j)} = i \operatorname{sgn}(z) \frac{q + nG}{\kappa_{j,n}} E_{x,n}^{(j)} , \tag{7.6}$$

$$\kappa_{j,n} = \sqrt{(q + nG)^2 - \epsilon_j \omega^2 c^2} . \tag{7.7}$$

Notice the similarity between equations (7.4)-(7.7) and (4.8)-(4.10), where $q \to q + nG$ for each corresponding diffraction order. This is a natural consequence of the system's periodicity, since now the momentum can be defined up to a reciprocal lattice vector[1]. We proceed by introducing the boundary conditions of the problem (at $z = 0$), which require the continuity of the tangential component of the electric field

$$\mathcal{E}_x^{(2)}(x,0) - \mathcal{E}_x^{(1)}(x,0) = 0$$
$$\Leftrightarrow \sum_n \left[E_{x,n}^{(2)} - E_{x,n}^{(1)} \right] e^{inGx} = 0 , \quad (7.8)$$

and the discontinuity of the magnetic field due to the presence of a surface charge density,

$$\mathcal{B}_y^{(2)}(x,0) - \mathcal{B}_y^{(1)}(x,0) = \mu_0 \sigma(x) \mathcal{E}_x^{(1)}(x,0)$$
$$\Leftrightarrow \sum_n \left[B_{y,n}^{(2)} - B_{y,n}^{(1)} \right] e^{inGx} = \mu_0 \sum_{m,l} \tilde{\sigma}_m E_{x,l}^{(1)} e^{i(m+l)Gx} , \quad (7.9)$$

where the sums run for integers $\in \,]-\infty, \infty[$, and we have written the spatially modulated conductivity as a Fourier series,

$$\sigma(x) = \sum_m \tilde{\sigma}_m e^{imGx} , \quad (7.10)$$

with

$$\tilde{\sigma}_m = \frac{1}{R} \int_0^R \sigma(x) e^{-imGx} dx , \quad (7.11)$$

related to the periodicity of the system. For what follows, it is convenient to redefine the index in the RHS of equation (7.9) as $m = n - l$. Then, equations (7.8) and (7.9) can be rewritten as

$$E_{x,n}^{(2)} - E_{x,n}^{(1)} = 0 , \quad (7.12)$$
$$B_{y,n}^{(2)} - B_{y,n}^{(1)} = \mu_0 \sum_l \tilde{\sigma}_{n-l} E_{x,l}^{(1)} . \quad (7.13)$$

Combining these expressions and making use of relation (7.4), we arrive at

$$\left(\frac{\epsilon_1}{\kappa_{1,n}} + \frac{\epsilon_2}{\kappa_{2,n}} \right) E_{x,n}^{(1)} + \frac{i}{\omega\epsilon_0} \sum_l \tilde{\sigma}_{n-l} E_{x,l}^{(1)} = 0 , \quad (7.14)$$

which is a non-linear eigenvalue problem, whose eigenvalue is the frequency, ω. The solution of equation (7.14) yields the band structure of the polaritonic spectrum in a system where graphene's conductivity is periodic.

[1] Note that a photon, an electron, etc, placed at a given q will *see* the same landscape as its counterpart at $q + nG$, for any $n \in \mathbb{Z}$.

Unfortunately, for the configuration of interest in this Chapter – that of a periodic array of graphene ribbons – the numerical solution of the eigenvalue problem is poorly convergent, due to the sharp discontinuity of the conductivity across the edges of the ribbons. However, for systems with continuous variations of the conductivity, the plasmonic spectrum can be readily obtained; the case of a graphene sheet with cosine-modulated conductivity was adressed in [Bludov *et al.* (2013); Peres *et al.* (2012)] using this technique.

7.2.2 *The scattering problem*

In contrast to the case of a homogeneous system, here the momentum is defined up to a reciprocal lattice vector [recall our discussion after equations (7.4)-(7.7) and (4.8)-(4.10)]; thus, the system's periodicity provides the extra momentum contribution needed to excite SPPs, as their momentum is larger than the one carried by the incident light beam.

Let us now consider the scattering of electromagnetic radiation impinging on a graphene sheet with periodically modulated conductivity and its coupling to GSPs in the form of Bloch-type surface waves. Since the system is periodic, its translational symmetry allows us to use Bloch's theorem to write the electromagnetic fields. For transverse magnetic (*p*-polarized) waves these can be written as

$$\mathcal{E}_x^{(j)}(x,z) = E_x^{inc} e^{iqx} e^{ik_z z} \delta_{j,2} + \sum_n E_{x,n}^{(j)} e^{i(q+nG)x} e^{-\kappa_{j,n}|z|} \,, \tag{7.15}$$

$$\mathcal{E}_z^{(j)}(x,z) = E_z^{inc} e^{iqx} e^{ik_z z} \delta_{j,2} + \sum_n E_{z,n}^{(j)} e^{i(q+nG)x} e^{-\kappa_{j,n}|z|} \,, \tag{7.16}$$

$$\mathcal{B}_y^{(j)}(x,z) = B_y^{inc} e^{iqx} e^{ik_z z} \delta_{j,2} + \sum_n B_{y,n}^{(j)} e^{i(q+nG)x} e^{-\kappa_{j,n}|z|} \,, \tag{7.17}$$

where $q = k\sin\theta$, $k_z = k\cos\theta$ and $\kappa_{j,n} = \sqrt{(q+nG)^2 - \epsilon_j \omega^2/c^2}$ with $k = \sqrt{\epsilon_2}\omega/c$ and $G = 2\pi/R$; note that when the argument of the square root in $\kappa_{j,n}$ is negative, we must write $\kappa_{j,n} = i\sqrt{\epsilon_j \omega^2/c^2 - (q+nG)^2}$. The subscript $j = 1, 2$ refer to the substrate underneath graphene and to the top dielectric medium, respectively, with dielectric constants ϵ_1 and ϵ_2. We consider that light comes from medium 2. Note that whenever $(q+nG)^2 > \epsilon_j \omega^2/c^2$ we have an electromagnetic wave confined to the interface (surface electromagnetic wave), while for $(q+nG)^2 < \epsilon_j \omega^2/c^2$ we have a propagating wave. The relations between the amplitudes of the *n*-th component of the electric and magnetic fields are once again given by expressions (7.4)-(7.7).

In order to meet the boundary conditions, we must impose

$$E_x^{inc} + E_{x,0}^{(2)} = E_{x,0}^{(1)} \quad ; \quad \text{for } n = 0 , \tag{7.18}$$

$$E_{x,n}^{(2)} = E_{x,n}^{(1)} \quad ; \quad \text{for } n \neq 0 , \tag{7.19}$$

and

$$B_{y,0}^{(1)} - B_y^{inc} - B_{y,0}^{(2)} + \mu_0 \sum_l \tilde{\sigma}_{-l} E_{x,l}^{(1)} = 0 \quad ; \quad \text{for } n = 0 , \tag{7.20}$$

$$B_{y,n}^{(1)} - B_{y,n}^{(2)} + \mu_0 \sum_l \tilde{\sigma}_{n-l} E_{x,l}^{(1)} = 0 \quad ; \quad \text{for } n \neq 0 , \tag{7.21}$$

where the Fourier components of the conductivity follow the same prescription as before [see, for instance, equation (7.10)]. Equations (7.18)-(7.21), after some algebra and indentifications, reduce to the following system of equations

$$\left(\frac{\epsilon_1}{\kappa_{1,0}} + \frac{\epsilon_2}{\kappa_{2,0}} \right) E_{x,0}^{(1)} + \frac{i}{\omega\epsilon_0} \sum_l \tilde{\sigma}_{-l} E_{x,l}^{(1)} = i \frac{2\epsilon_2}{k_z} E_x^{inc} , \tag{7.22}$$

$$\left(\frac{\epsilon_1}{\kappa_{1,n}} + \frac{\epsilon_2}{\kappa_{2,n}} \right) E_{x,n}^{(1)} + \frac{i}{\omega\epsilon_0} \sum_l \tilde{\sigma}_{n-l} E_{x,l}^{(1)} = 0 . \tag{7.23}$$

We note that this system can be casted as a matrix equation, whose solutions (for a given ω) are just the amplitudes of the electric field[2]. These shall be obtained numerically using LAPACK's implemented routines for linear algebra[3]. Having computed the fields, we are now able to calculate the transmittance and reflectance of light by the structure, which read

$$\mathcal{T} = \sum_n \mathcal{T}_n \quad \text{with} \quad \mathcal{T}_n = \frac{\epsilon_1}{\epsilon_2} \frac{k_z}{|\kappa_{1,n}|} \left| \frac{E_{x,n}^{(1)}}{E_x^{inc}} \right|^2 , \tag{7.24}$$

and

$$\mathcal{R} = \sum_n \mathcal{R}_n \quad \text{with} \quad \mathcal{R}_n = \frac{k_z}{|\kappa_{2,n}|} \left| \frac{E_{x,n}^{(2)}}{E_x^{inc}} \right|^2 , \tag{7.25}$$

respectively[4]. From these, we define the absorbance as

$$\mathcal{A} = 1 - \mathcal{T} - \mathcal{R} , \tag{7.26}$$

where we should stress that in this expressions *only* the diffraction orders corresponding to propagating modes should be taken into account.

---

[2] For numerical purposes, it is useful to normalize the field amplitudes in the system constituted by equations (7.22) and (7.23) with respect to E_x^{inc}.

[3] We note that the convergence of the numerical solution for the scattering problem is good, even for the rather "extreme" case of a grating of graphene ribbons, in constrast to the computation of the polaritonic bandstructure via equation 7.14.

[4] For a detailed derivation of the expressions for $\mathcal{T}$ and $\mathcal{R}$, please refer to section's I.1 of Appendix I.

7.3 Applications and Results

In the previous section we developed a theoretical model describing the interaction of electromagnetic radiation with single-layer graphene whose conductivity is a periodic function. Here, we will apply this formalism to the specific case of a periodic arrangement of graphene micro-ribbons, and compare our results with experimental data from [Ju *et al.* (2011)].

7.3.1 *Periodic array of graphene ribbons*

The first experimental demonstration (using optical means) of the existence of surface plasmon-polaritons in graphene was achieved by Ju *et al* in 2011, using a periodic array of graphene micro ribbons, dubbed as a metamaterial by the authors [Ju *et al.* (2011)]. Shorty after that, a theoretical account for the experiment was given [Nikitin *et al.* (2012a)], and since then an increasing number of publications followed [Yan *et al.* (2013); Slipchenko *et al.* (2013a); Bludov *et al.* (2013); Strait *et al.* (2013)]. However, the results found in the literature are somewhat sparse and most of the theoretical work describing real experiments was based on finite element or FDTD numerical solutions obtained from commercial codes. Although these usually yield good results, they are computationally demanding, time-consuming and it is harder to get a *"feeling"* on the physics of the problem.

In what follows, we aim to fill this gap by employing the analytical theory discussed in the previous section, to the case of a periodic set of graphene ribbons forming a grating-like structure as depicted in Figure 7.2.

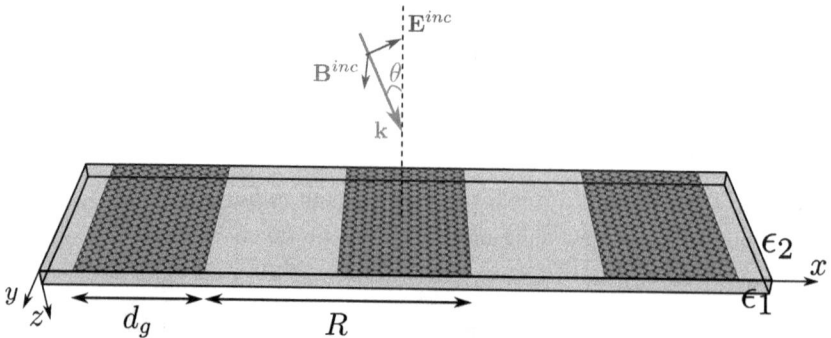

Figure 7.2: Illustration of a *p*-polarized plane wave impinging on a periodic array of graphene ribbons. The size of the unit cell is R and the width of the ribbons d_g. The substrate and capping layer are assumed to be semi-infinite dielectrics.

Notice that the formalism we have built in Section 7.2 is completely general, so that now we only need to specify what are the Fourier coefficients of the conductivity, which enter in the linear system composed by the following pair of equations [cf. expressions (7.22) and (7.23)]

$$\left(\frac{\epsilon_1}{\kappa_{1,0}} + \frac{\epsilon_2}{\kappa_{2,0}} \right) E_{x,0}^{(1)} + \frac{i}{\omega\epsilon_0} \sum_l \tilde{\sigma}_{-l} E_{x,l}^{(1)} = i\frac{2\epsilon_2}{k_z} E_x^{inc} , \qquad (7.27)$$

$$\left(\frac{\epsilon_1}{\kappa_{1,n}} + \frac{\epsilon_2}{\kappa_{2,n}} \right) E_{x,n}^{(1)} + \frac{i}{\omega\epsilon_0} \sum_l \tilde{\sigma}_{n-l} E_{x,l}^{(1)} = 0 . \qquad (7.28)$$

Recalling equation (7.11), we have

$$\tilde{\sigma}_l = \frac{\sigma_g}{R} \int_0^R s(x) e^{-ilGx} dx , \qquad (7.29)$$

where $s(x)$ is determined by the spatial configuration of the system and we shall assume a bulk-like conductivity for the graphene ribbons[5]. As noted in subsection 2.3.2, in the terahertz and mid-IR spectral range and for standard levels of doping ($E_F \gg k_B T$ and $2E_F > \hbar\omega$), graphene's conductivity reduces to Drude's contribution, (2.72). Thus, hereafter we will assume that $\sigma_g = \sigma_g^{Drude}$ unless otherwise stated.

Let us consider the unit cell of our periodic system, which may be divided into one region where graphene is absent, with width d_0, and another region of size d_g occupied by the graphene ribbon —see, for instance, Fig. 7.3. Then, the geometrical factor $s(x)$ can be written in terms of the Heaviside step function,

$$s(x) = \Theta(x - d_0) , \qquad (7.30)$$

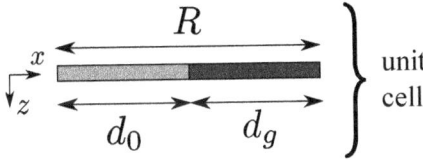

Figure 7.3: Unit cell of the structure portrayed in Figure 7.2 —side view.

[5]This should be a valid approximation (at least) as long as $d_g k_F \gg 1$ and/or $d_g \gg l_{mfp}$, where $k_F = E_F/(\hbar v_F)$ is graphene's Fermi momentum and l_{mfp} the mean free path of the charge carriers. In addition, a study [Thongrattanasiri *et al.* (2012b)] suggested that quantum finite-sized and edge effects in graphene plasmons are only significant for ribbons smaller than 20 nm, which is well below the width of the structures here considered.

describing the sharp discontinuity in the conductivity. Hence, from equation (7.29) we obtain

$$\tilde{\sigma}_0 = \frac{\sigma_g}{R} \int_0^R \Theta(x - d_0) dx = \sigma_g \frac{d_g}{R} \ , \qquad (7.31)$$

for the zero-th coefficient, and

$$\tilde{\sigma}_l = \frac{\sigma_g}{R} \int_0^R \Theta(x - d_0) e^{-ilGx} dx$$

$$= \frac{\sigma_g}{l\pi} \sin\left(l\pi d_g/R\right) e^{il\pi d_g/R} \ , \qquad (7.32)$$

for the l-th Fourier component ($l \neq 0$). In possession of the coefficients for the conductivity, we are now able to solve the linear system formed by equations (7.27) and (7.28). From the corresponding numerical solution, we then compute the transmittance, reflectance and absorbance of electromagnetic radiation through the structure using expressions (7.24)-(7.26). Figure 7.4 shows these quantities for different values of the damping parameter (see also Fig. 2.8). The resulting spectra consist in a set of well-defined GSPs resonances, corresponding to different Bragg vectors (diffraction orders). These GSP-assisted effects yield dramatic changes in the spectra when compared with the case of a pristine (unpatterned) graphene sheet. For instance, it is apparent from the figure that for a grating of graphene ribbons the absorbance at resonant frequencies is higher than in the case of a continuous graphene layer. As it shall be clearer further ahead, this enhancement in absorbance corresponds to the coupling of electromagnetic radiation to Bloch GSPs modes —GSP-induced absorption. Additionally, notice that peaks in absorbance (and reflectance) go hand in hand with dips in transmittance. Also, not surprisingly, small values of Γ render sharp peaks in absorbance, with these becoming successively broader and less pronounced as Γ increases.

Let us now explore the dependence of the absorption spectra on the different parameters of the problem. We begin by studying how the GSPs resonances change with varying carrier density, n_e. The results are displayed in Figure 7.5. From the left panel, it is clear that for larger electronic densities, the resonances are stronger and shift towards higher frequencies. In order to be more quantitative, in the right panel of Figure 7.5 we have plotted the GSP frequency (corresponding to the first resonance) as a function of the doping level, to which we have fitted a function of the type $f(n_e) \propto n_e^b$, having obtained $b = 0.249 \simeq 1/4$ for the exponent (fitting parameter)[6]. This therefore demonstrates that the observed resonances scale

---

[6]The fitting function was obtained using the least-squares method.

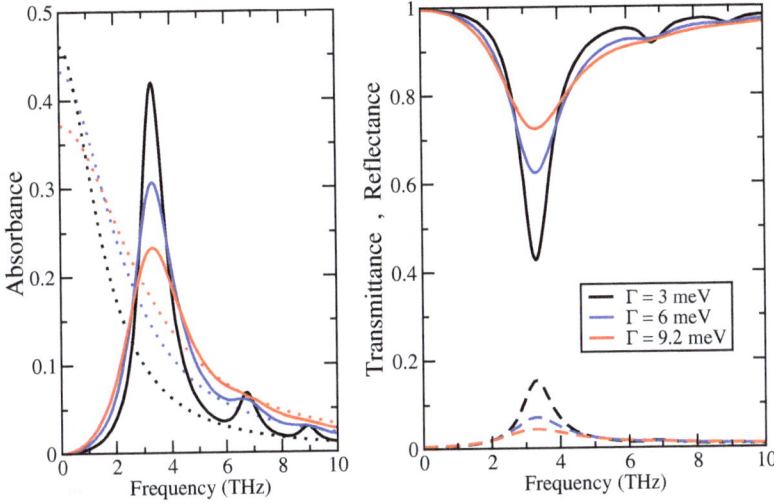

Figure 7.4: Absorbance, transmittance and reflectance of a p-polarized wave through a periodic structure of graphene ribbons for varying $\Gamma = \hbar\gamma$. Left panel: absorbance for a grid of graphene ribbons (continuous curve) and for uniform graphene (dotted curve). Right panel: transmittance (continuous) and reflectance (dashed curve). The remaining parameters are: $E_F = 0.45$ eV, $d_g = 4$ μm, $R = 8$ μm, $\epsilon_1 = 4$, $\epsilon_2 = 3$, $\theta = 0$ (normal incidence). The reader should compare these results with those of Fig. 2.8.

with the density of charge carriers as

$$\omega_{\text{res}} \propto n_e^{1/4} \, , \tag{7.33}$$

which is a specific signature of GSPs. Furthermore, Figure 7.6 (left panel) presents absorbance curves for arrays of graphene ribbons with different periods, R, while keeping the ratio $d_g/R = 0.5$ constant. Here, we observe a blueshift on the GSPs resonances for decreasing values of R. This can be understood by noting that the plasmon-polariton wavevector is approximately given by $q_{GSP} = (2m - 1)\pi/d_g$, with $m = 1, 2, ...$, due to the symmetry of the incident field[7] [Mikhailov and Savostianova (2005); Ju *et al.* (2011); Yan *et al.* (2013)]. Hence, the first, stronger resonance corresponds to the excitation of a localized plasmon with $q = \pi/d_g$, followed by a set of weaker higher-order peaks at higher frequencies, with $q = (2m - 1)\pi/d_g$ for $m \geq 2$. Moreover, the panel at the right in Figure 7.6 clearly demonstrates that the GSP resonances follow the $\sqrt{q}$ dispersion, in agreement with experimental results on plasmons in graphene ribbons [Ju *et al.* (2011); Yan

---

[7]For a thorough discussion, please refer to Section I.2 of Appendix I.

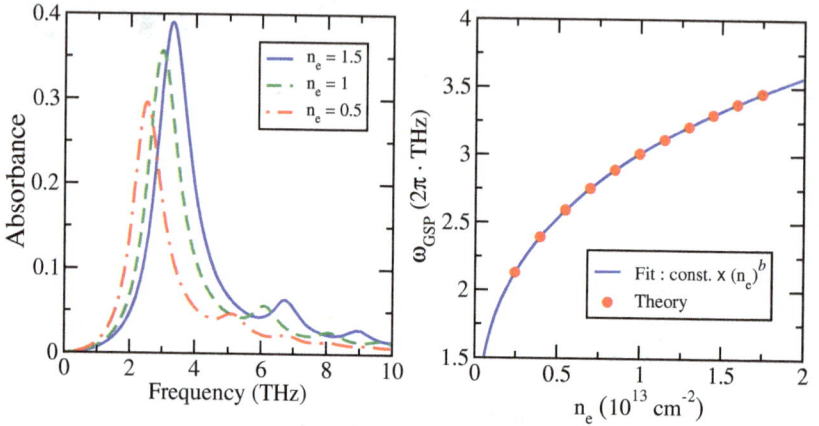

Figure 7.5: Dependence of the GSP frequency on the electronic density. Left panel: Absorbance spectrum for different values of n_e; the legend gives n_e in units of 10^{13} cm^{-2}. Right panel: resonant frequencies —retrieved from our analytic theory (points)— for several values of n_e, and corresponding fitting function, $f(n_e) \propto n_e^b$ with $b = 0.249 \simeq 1/4$. Parameters: $\Gamma = 3.7$ meV, $d_g = 4$ μm, $R = 8$ μm, $\epsilon_1 = 4$, $\epsilon_2 = 3$, $\theta = 0$.

et al. (2013)]. Note that this is somewhat expected, since we have already shown in Chapter 4 that in the non-retarded regime (satisfied as long as $q \approx G \gg k$), we have

$$\hbar\omega_{GSP} \approx \sqrt{\frac{4\alpha}{\epsilon_1 + \epsilon_2} E_F \hbar c q} \, , \qquad (7.34)$$

for homogeneous graphene. In fact, as an exercise, let us consider the parameters corresponding to the blue curve in Figure 7.6 to estimate the plasmon frequency via equation (7.34):

$$\hbar\omega_{GSP} \approx \sqrt{\frac{4}{137 \times 7} 0.3 \times 0.2 \times \frac{2\pi}{4}} \simeq 19.8 \text{ meV} \to 4.8 \text{ THz} \, . \qquad (7.35)$$

Thus, we find that this expression overestimates the GSP resonant frequency, since the fundamental mode in Figure 7.6 appears at 3.9 THz instead. A better estimation can made after realizing that the effective density in the unit cell is $n_{eff.} = n_e/2$, where n_e is the electronic density in the ribbon. In this case, we can define an effective Fermi energy given by

$$E_F^{eff.} = \hbar v_F \sqrt{\pi n_{eff.}} = \hbar v_F \sqrt{\pi \frac{n_e}{2}} \, . \qquad (7.36)$$

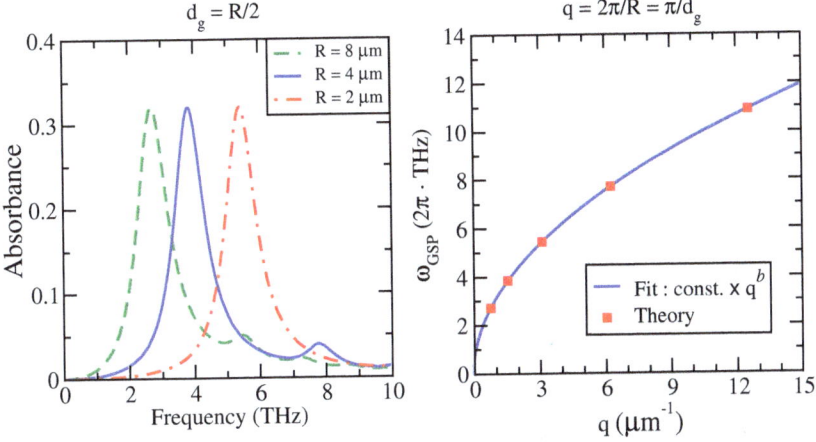

Figure 7.6: Dependence of the GSP frequency on the periodicity of the system and plasmon dispersion for the first resonance. Left panel: Absorbance spectrum for different values of grating period, (or, alternatively, for different widths of the graphene ribbons) R. Right panel: resonant frequencies —obtained using our analytic theory (squares)— as a function of the wavevector $q = 2\pi/R = \pi/d_g$ (fundamental plasmon mode); the curve represents the function fitting the data, namely $\omega_{GSP} \propto q^b$ with $b = 0.5001 \simeq 1/2$. Parameters: $d_g = R/2$, $E_F = 0.3$ eV, $\Gamma = 3.7$ meV, $\epsilon_1 = 4$, $\epsilon_2 = 3$, $\theta = 0$.

Then, Eq. (7.34) is modified to

$$\hbar\omega_{GSP} \approx \sqrt{\frac{4\alpha}{\epsilon_1 + \epsilon_2} E_F^{eff} \cdot \hbar c q} = 2^{-1/4}\sqrt{\frac{4\alpha}{\epsilon_1 + \epsilon_2} E_F \hbar c q}. \qquad (7.37)$$

Plugging in the numbers in Eq. (7.37) we obtain $\hbar\omega_{GSP} \approx 16.6$ meV, which implies $f_{GSP} = \omega_{GSP}/(2\pi) \approx 4.06$ THz, a rather good agreement with the position of the fundamental mode at 3.9 THz. Finally, for the sake of completeness, we have summarized in Fig. 7.7 the effect of different filling factors, d_g/R, and varying incident angle, on the absorbance spectra for an array of graphene ribbons. Notice that the plasmonic resonances undergo a redshift as the filling factor approaches unity. Here, the absorbance spectrum presents itself as a featureless Drude peak related to Ohmic losses —dotted black curve. Further, larger d_g/R ratios attain higher absorbance values owing to the increase of the effective area covered by graphene within the unit cell. With regards to the angle of incidence, Fig. 7.7 shows a weak dependence on this parameter, that being substantial only for large incident angles (i.e. at grazing incidence). Likewise, the position of the resonances remains essentially unchanged, inasmuch as $q = |k\sin\theta + nG| \approx |nG|$ (for odd integers n).

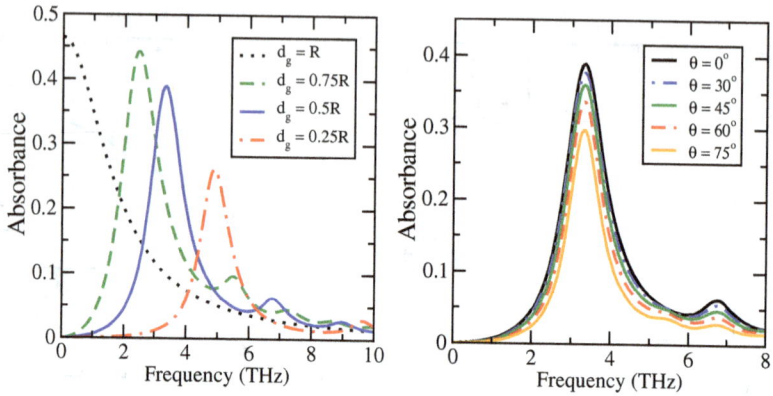

Figure 7.7: Left panel: Absorbance (at normal incidence) of a grid of graphene ribbons for different filling factors, d_g/R, with $R = 8$ μm; the black dotted line corresponds to the case of an unpatterned graphene sheet. Right panel: dependence on the incident angle, for fixed $R = d_g/2 = 8$ μm. The remaining parameters are: $E_F = 0.45$ eV, $\Gamma = 3.7$ meV, $\epsilon_1 = 4$, $\epsilon_2 = 3$.

7.3.2 *Theory* versus *experiment*

So far we have covered virtually every aspect of our theoretical model: we set up the formalism and explored its results within a vast parameter space. In addition, we have demonstrated that our theory correctly describes the excitation of GSPs in graphene ribbons, yielding results consistent with those found in the literature [Ju *et al.* (2011); Nikitin *et al.* (2012a); Yan *et al.* (2013); Bludov *et al.* (2013); de Abajo (2014)]. Still, the ultimate test for any theory must be the direct comparison of its results against actual experimental data. With that in mind, here we will present several spectra computed under the same conditions found in the experiments performed by Ju *et al* [Ju *et al.* (2011)]. This will show to what extent our theory is capable to account for the experimental results.

The experimental setup —consisting in a series of graphene micro-ribbons in a grating-like configuration— is portrayed in Fig. 7.8, along with atomic force microscopy (AFM) images of the fabricated samples. As shown in the figure, these structures have widths of 4, 2 and 1 micrometers, and the ratio $d_g = 0.5R$ was kept constant. It should be noted that the spectra in [Ju *et al.* (2011)] represent the change in transmittance relative to the one taken at the "charge neutral point" (CNP), that is $-\Delta T/T_{CNP}$ with $\Delta T = T - T_{CNP}$. At the CNP, the chemical potential, μ, is zero; thus,

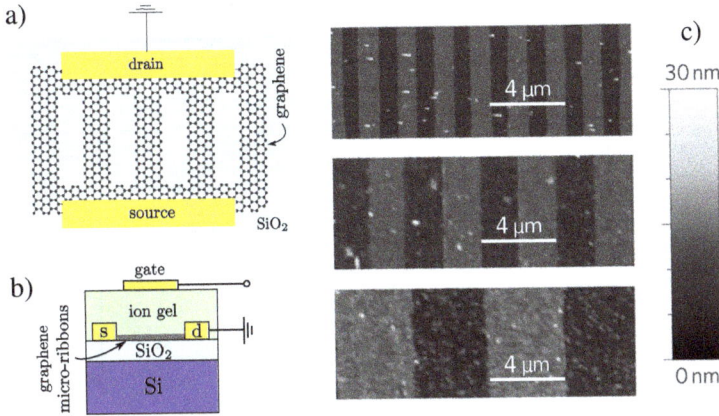

Figure 7.8: Panels a) and b): pictorial representation of the experimental setup, consisting in a periodic array of graphene micro-ribbons on a SiO_2/Si substrate. The carrier concentration can be controlled using a top gate on ion gel (adapted from [Ju *et al.* (2011)]). Panel c): AFM measurements of experimentally fabricated samples, with $d_g = R/2 = 4$ μm, 2 μm and 1 μm (right panel image is courtesy of Feng Wang).

in order to effectively compare the theoretical results with the measured spectra, we must also acknowledge interband processes in the expression for the (local) conductivity of graphene, which we will do by using Kubo's formula [cf. equations (2.67) and (2.68)]), since both conditions $|\mu| \gg k_B T$ and $2|\mu| > \hbar\omega$ break down at the CNP. Hereafter, we shall assume $T = 300$ K and a scattering rate of 4 THz, as indicated in the reference. Also, at this point it is worth to mention that GSPs can interact with surface optical (SO) phonons on polar substrates, such as SiO_2 or hBN [Yan *et al.* (2013); Brar *et al.* (2014)]. This can be modelled by describing their dielectric functions as Lorentz oscillators (instead of a real constants), using the respective SO phonon resonant frequencies. However, for graphene on SiO_2, plasmon-phonon hybridization is only important at frequencies $\omega \gtrsim \omega_{sop} \approx 15$ THz (lowest frequency mode), well above the frequency range contemplated in this work, thereby allowing us to safely set $\epsilon_{SiO_2} = 4$.

Let us first consider the sample with $d_g = 4$ μm. In Fig. 7.9a we compare the theoretical curves, obtained using parameters matching the experimental conditions (see caption for details), with experimental data extracted from [Ju *et al.* (2011)]. For completeness, we have also plotted, along with the transmittance relative to the CNP, $-\Delta T/T_{CNP}$, the absorbance

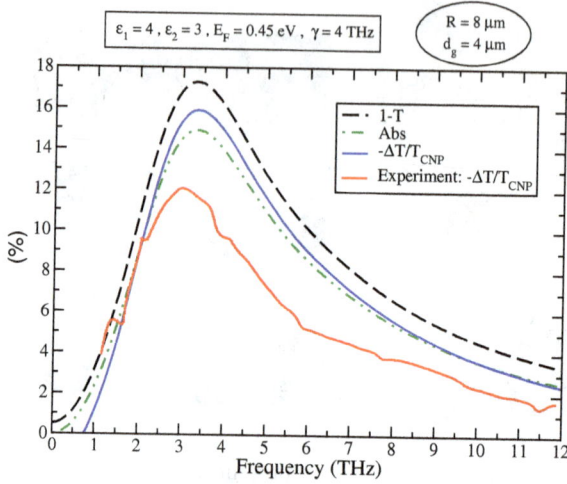

(a) Results for $\epsilon_1 = 4$ (SiO$_2$) and $\epsilon_2 = 3$ (ion gel).

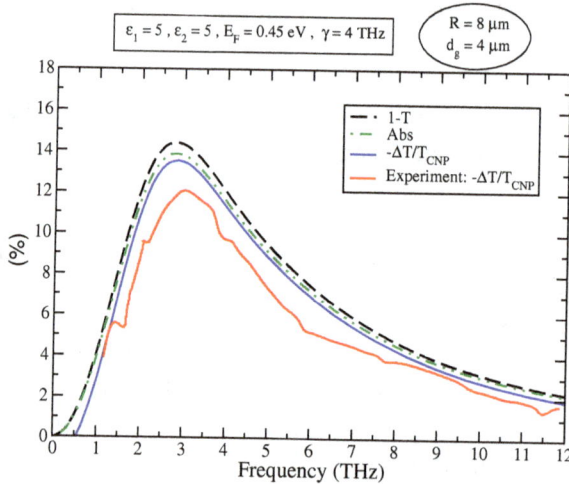

(b) Results obtained by adopting the effective medium approach with $\epsilon = 5$.

Figure 7.9: Transmittance change spectrum, $-\Delta T/T_{CNP} = 1 - T/T_{CNP}$, for an array of graphene micro-ribbons with width $d_g = 4$ μm (and period $R = 8$ μm): theory (blue curve) and experimental data (red curve) [Ju *et al.* (2011)]. Also plotted are the extinction (1-T) and absorbance (Abs) spectra. As in the experiment, we take $\gamma = 4$ THz, $\theta = 0$ (normal incidence) and $n_e = 1.5 \times 10^{13}$ cm^{-2}, which corresponds to $E_F = \hbar v_F \sqrt{\pi n_e} \approx 0.45$ eV.

and extinction spectrum. The figure shows a rather good agreement between theory and experiment, with a prominent plasmon resonance around 3 THz. Explicitly, our calculations rendered $\omega_{GSP}^{\text{theo}} \approx 3.3$ THz, whereas the measured peak is centered at $\omega_{GSP}^{\text{exp}} \approx 3$ THz. Therefore, the only perceptible difference concerns the resonance strength, with theory predicting a peak value about 4 % higher. There are a couple of reasons that may contribute to these slight deviations. First, we note that our model assumes that both dielectrics are semi-infinite —we took $\epsilon_1 = 4$ (SiO$_2$) and $\epsilon_2 = 3$ (ion gel)— neglecting the influence of the Si and also eventual multiple reflections within the finite-sized SiO$_2$ layer[8]. On the other hand, on the experimental side, the authors of the experiment described in [Ju *et al.* (2011)] mention that there was a small part of the beam transmitted through unpatterned graphene in the outskirts of the structure, adding a small background contribution owing to free carrier absorption. The authors state that this extra-contribution was subtracted from the measured data, based on fittings from numerical calculations employing finite element analysis. However, the degree of fidelity of this procedure remains unclear, and perhaps it may have overestimated (or not) the contribution from free carrier absorption. Furthermore, the above-mentioned computational simulations suggested that an effective dielectric environment with $\epsilon = 5$ could be used in order to account for the electromagnetic properties of the different dielectrics[9]. The results corresponding to this case are presented in Fig. 7.9b, in an attempt to encompass not only the effects of the heterogeneous SiO$_2$/Si substrate, but also the intricate nature of the optical constants of ion liquids [Yamamoto *et al.* (2007)]. The figure clearly shows a fair improvement with respect to the previous case (which was already quite good), demonstrating an excellent agreement between our analytical results and experimental data.

Finally, the spectrum of the normalized[10] change in transmittance relative to the CNP, for the different micro-ribbon widths considered in the

---

[8]Taking $q_{GSP} \approx \pi/d_g$, we can estimate the penetration in the dielectrics, in the non-retarded regime, as $\zeta = \Re\{\kappa\}^{-1} \sim q^{-1} \approx 1$ μm. This is about three times larger than thickness of the SiO$_2$ layer, so that some influence due to the Si part of the substrate is to be expected.

[9]This seems to be reasonable, given that the dielectric constants of ion gel, SiO$_2$ and Si are 2-4 [Yamamoto *et al.* (2007)], 4 and 11 [Palik (1997)], respectively (in the THz spectral range).

[10]Although the spectra are normalized, as presented by [Ju *et al.* (2011)], we expect a similar behavior as in Fig. 7.9b.

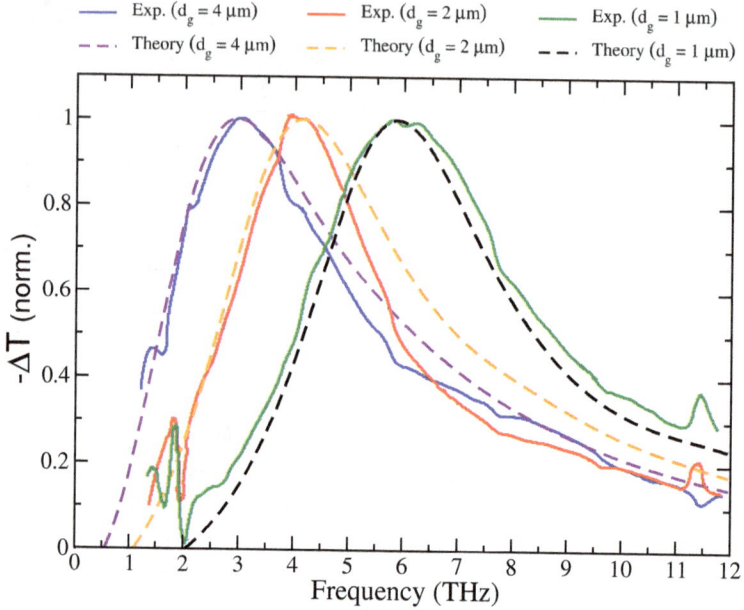

Figure 7.10: Normalized spectra of $-\Delta T$ for arrays of ribbons with different widths: 4, 2 and 1 μm. The computed analytical (dashed) curves are superimposed to the spectra obtained experimentally (continuous curves) for adequate comparison. We adopted $\epsilon = 5$, with the remaining parameters matching the conditions of the experiment, namely $\gamma = 4$ THz, $\theta = 0$ and $n_e = 1.5 \times 10^{13}$ cm^{-2} $\Rightarrow$ $E_F = \hbar v_F \sqrt{\pi n_e} \approx 0.497$ eV (with $v_F = 1.1 \times 10^6$ m/s).

experimental work, is depicted in Fig. 7.10. Once again, we have assumed $\epsilon = 5$ in our calculations[11]. The figure provides further evidence of the outstanding agreement between theory (dashed curves) and experiment (continuous curves), with the former following closely the experimental data. The measured (computed) GSPs resonant frequencies are at 3 (2.9), 4.1 (4) and 6 (5.6) THz for arrays with ribbon widths of 4, 2 and 1 μm, respectively, exhibiting the predicted $\omega_{GSP} \propto d_g^{-1/2}$ scaling behaviour.

---

[11]In addition, we note that in Fig. 7.10 we have used $v_F \approx 1.1 \times 10^6$ m/s, as extracted from experimental data taken from [Ju *et al.* (2011)], to determine the Fermi energy from the carrier concentration, n_e. This value for the velocity of massless Dirac fermions is reported elsewhere [Li *et al.* (2008)], and is a signature of band renormalization owing to many-body interactions [Kotov *et al.* (2012)].

Our results and subsequent analysis testify the ability of our analytical theory to explain the experimental results, without the need to use lengthy, computationally demanding full-wave numerical simulations based either on finite element methods or FDTD. Moreover, the theoretical framework developed over this Chapter constitutes a promising playground for engineering plasmonic micro/nanostructures and metamaterials with tailored plasmonic properties, taking advantage of graphene's tunability and strong light-matter interactions. We expect an even stronger GSP-induced absorption for samples with lower electron scattering rates, which will eventually become available with the continuous improvement of fabrication techniques.

7.4 A THz Polarizer

We would like to end this chapter by discussing a recent experimental work [Sorger *et al.* (2015)] where it was demonstrated that large area, epitaxially grown (on SiC), patterned graphene can act as a grating producing highly polarized THz radiation. The periodicity of the grating was substantially smaller than the free-space THz wavelength. One advantage of the graphene/SiC system is that it allows to combine in one sample different graphene species, for instance, monolayer and quasi-freestanding bilayer graphene, as depicted in Fig. 7.11. Experimentally, the THz response was probed using both a continuous wave THz beam and terahertz time-domain spectroscopy (TDS). The authors of [Sorger *et al.* (2015)] defined the visibility as

$$C = \frac{P_p - P_s}{P_p + P_s},\qquad(7.38)$$

where P_i ($i = s-, p-$polarization) is the fraction of the transmitted power due to impinging radiation in the system associated with a given polarization state (s or p). The authors of that reference observed a visibility up to 20%. Using the latter quantity it was possible to pin down the plasmonic resonance, located at a frequency $f = 2.3 \pm 0.3$ THz for the system depicted in Fig. 7.11. In practice, the studied system works as wire grid polarizer and different configurations were studied in [Sorger *et al.* (2015)]. An interesting aspect of these systems is the existence of a charge modulation, which enhances the contrast and produces an additional periodic structure in the system. This, combined with the physical periodic structure, enhances the plasmonic response of the system. The contrast can be further enhanced by a complete remotion of graphene in a periodic patern,

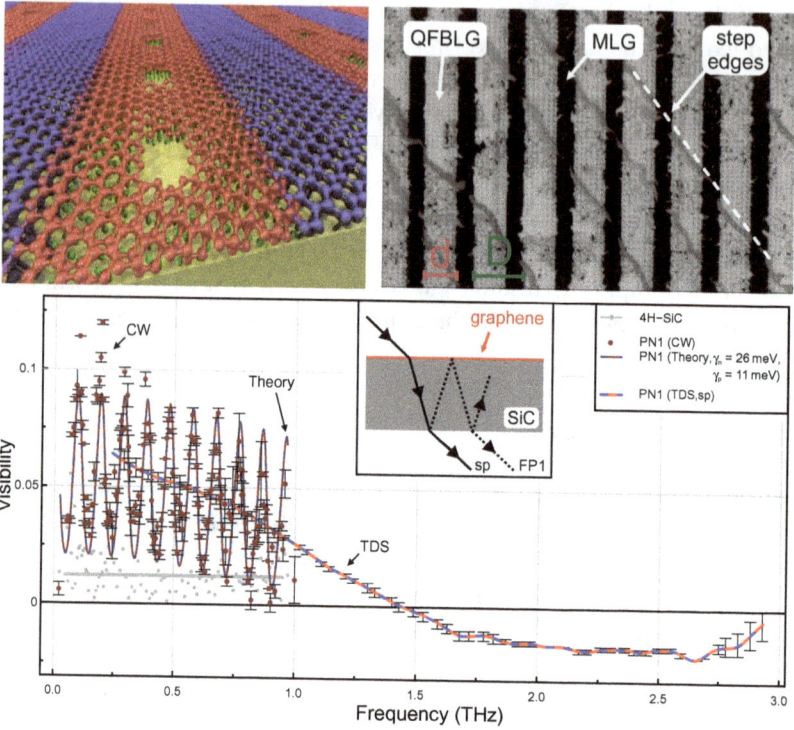

Figure 7.11: Periodic grating of monolayer/bilayer on SiC. Panel (a): Artist's view of a stripe-pattern of charge-modulation-doping within the epitaxial graphene layer on SiC. The n-type monolayer (MLG) graphene-sheet is drawn in blue, with the buffer layer (black) underneath. Local insertion of hydrogen (green) through predefined voids results in p-type bilayer (QFBLG) graphene (red). Panel (b): Scanning electron micrographs depicting an alternating stripe-pattern of QFBLG and MLG. The substrate step edges are oriented at a different angle (dashed line). Bottom panel: experimental results for the visibility. The inset shows the a schematic representation of the system, depicting the Fabry-Perot multiple reflections. The minimum of the visibility at $f = 2.3 \pm 0.3$ THz is due to a resonant excitation of plasmons.

as done in [Lee *et al.* (2012)]. In Fig. 7.12, we plot the experimental data and the theory curves for a periodic system composed of graphene ribbons. A minimum in the visibility around $f = 1.6$ THz is clearly visible which corresponds, according to the simulation, to the excitation of plasmons in the grid system. The shift of the simulations relatively to the experimental data can presumably be attributed to a difference between the effective width of the ribbons and their physical width. There is no reason why they

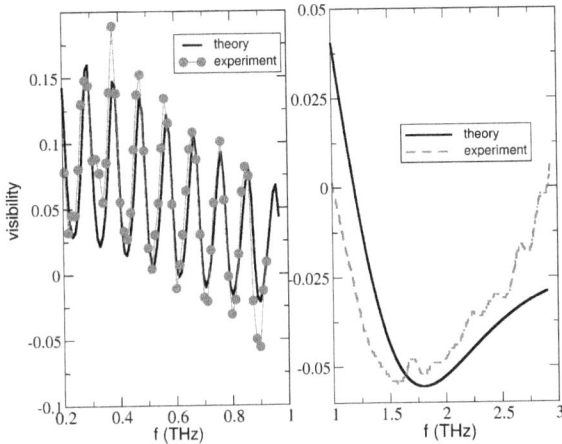

Figure 7.12: Excitation of SPP's in a graphene-based grid polarizer. Left panel: Visibility in the frequency range $f = 0.2-0.9$ THz was measured using a continous wave (CW) beam. Right panel: Visibility in the frequency range $f = 1-3$ THz measured using terahertz time-domain spectroscopy (TDS). The difference in the position of the plasmonic resonance between the experimental data and the model may be due to limitations of the latter.

should coincide, since the modeling assumes a constant conductivity over the full ribbon width, which, admittedly can be considered a first approximation to problem. Given this limitation of the theory, the agreement is quite good. Had the authors chosen to take the width of the ribbons as a fitting parameter the agreement between the experimental data and the theory would be better. Note that the visibility is negative around $f = 1.6$ THz. That is because $P_p < P_s$ due to the excitation of the plasmons. The modeling used in [Sorger *et al.* (2015)] was the same we used in the previous sections of this chapter. The oscillations seen in the left panel of Fig. 7.12 are due to Fabry-Perot multiple reflections in the highly polished SiC substrate.

7.5 Scattering From a Periodic Grid in the Regime $kw < 1$: Analytical Results

Let us revisit the scattering of electromagnetic radiation by a periodic grid of graphene ribbons of width w. In what follows, we shall derive an analytical expression for the reflectance[12] in the regime $kw < 1$ (where $w = d_g$ in the previous sections), with $k = \sqrt{\epsilon_1}\omega/c$. We assume a TM-polarized wave impinging on a grid of graphene ribbons, as illustrated in Fig. 7.13. We further assume that the grid rests in the $y = 0$ plane and that the incoming radiation is propagating along the positive y–direction —cf. Fig. 7.13. The relative permittivity in the medium defined by $y < 0$

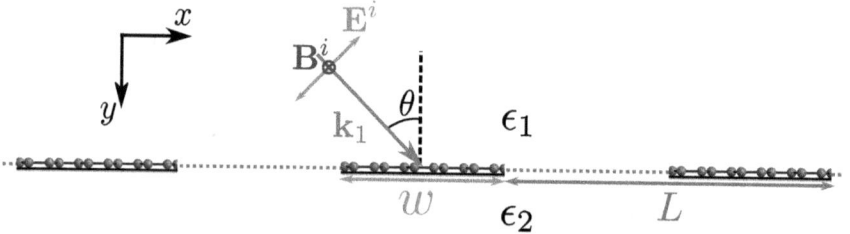

Figure 7.13: Monochromatic p–polarized plane-wave impinging on a grid of graphene ribbons arranged in a grating-like configuration. The graphene stripes are sitting in the plane defined by $y = 0$. The structure is periodic, with period L. The width of the graphene ribbons is constant and defined by w. The system is encapsulated by a top dielectric medium with relative permittivity ϵ_1 (for $y < 0$), and by a bottom dielectric with relative permittivity ϵ_2 (for $y > 0$).

is ϵ_1, while in the medium defined by $y > 0$ the relative dielectric constant is ϵ_2. For this polarization, the incident magnetic field reads

$$\mathbf{B}^i = \hat{\mathbf{z}} B_0^i e^{i(\mathbf{k}_1 \cdot \mathbf{r} - \omega t)}, \tag{7.39}$$

where

$$\mathbf{k}_1 = \frac{\omega}{v_1}(\sin\theta\,\hat{\mathbf{x}} + \cos\theta\,\hat{\mathbf{y}}) = k_x\hat{\mathbf{x}} + k_y\hat{\mathbf{y}}, \tag{7.40}$$

and $v_1 = c/\sqrt{\epsilon_1}$ is the speed of light in the medium characterized by ϵ_1. The reflected magnetic field is written, as before, in the form of a Bloch-sum

$$\mathbf{B}^r = \hat{\mathbf{z}}\sum_{n=-\infty}^{\infty} A_n e^{i(k_{xn}^- x - k_{yn}^- y)}, \tag{7.41}$$

where an implicit time dependence of the form $e^{-i\omega t}$ is assumed, and

$$k_{xn}^- = k_x + \frac{2n\pi}{L}, \tag{7.42}$$

$$\frac{\omega^2}{v_1^2} = (k_{xn}^-)^2 + (k_{yn}^-)^2, \tag{7.43}$$

---

[12] Of course, one could do the same for the transmittance and absorbance (the latter can be fetched from the $\mathcal{A} = 1 - \mathcal{T} - \mathcal{R}$ condition).

where L is the period of the grating ($L = R$ in the previous sections). From Maxwell's curl equation $\nabla \times \mathbf{B} = -i\omega\mu_0\epsilon_0\epsilon_1\mathbf{E}$, it follows that the reflected electric field has the form

$$\mathbf{E}^r = \frac{c^2}{\epsilon_1\omega} \sum_{n=-\infty}^{\infty} A_n e^{i(k_{xn}^- x - k_{yn}^- y)} \left(k_{yn}^- \hat{\mathbf{x}} + k_{xn}^- \hat{\mathbf{y}}\right) . \tag{7.44}$$

Both the impinging electric field and the reflected one add together to produce a current in the graphene ribbons given by

$$\mathbf{J} = J(x)\hat{\mathbf{x}} = \hat{\mathbf{x}}\sigma(x)(E_x^i + E_x^r)|_{y=0} , \tag{7.45}$$

where E_x^r is the x−component of the reflected electric field, $\sigma(x)$ is the position-dependent optical conductivity of the system —which can be simply written as $\sigma(x) = \sigma(w)\Theta(w/2 - |x|)$ in the unit cell, where $\Theta(x)$ is the Heaviside function—, and where E_x^i (the x−component of the incident electric field) reads

$$E_x^i = -\frac{c^2 k_y}{\epsilon_1\omega} B_0^i e^{i(k_x x + k_y y)} . \tag{7.46}$$

As in the case of the reflected magnetic field, the transmitted one can also be written as a Bloch-sum:

$$\mathbf{B}^t = \hat{\mathbf{z}} \sum_{n=-\infty}^{\infty} C_n e^{i(k_{xn}^+ x + k_{yn}^+ y)} . \tag{7.47}$$

from where it follows, together with the above-mentioned Maxwell's equation, that the x−component of the transmitted electric field is

$$E_x^t = -\frac{c^2}{\epsilon_2\omega} \sum_{n=-\infty}^{\infty} C_n k_{yn}^+ e^{i(k_{xn}^+ x + k_{yn}^+ y)} , \tag{7.48}$$

with

$$k_{xn}^+ = k_{xn}^- , \tag{7.49}$$

$$k_{yn}^+ = \sqrt{\frac{\omega^2}{v_2^2} - (k_{xn}^+)^2} , \tag{7.50}$$

where v_2 denotes the speed of light in the medium with relative permittivity ϵ_2. Due to the continuity of the x−component of the electric field above and below the graphene grid (i.e. boundary condition), the equation for the current can be written in two alternative forms as (note that $y = 0$)

$$J(x) = -\sigma(x)\frac{c^2}{\epsilon_1\omega}\left(k_y B_0^i e^{ik_x x} - \sum_{n=-\infty}^{\infty} A_n k_{yn}^- e^{ik_{xn}^- x}\right) , \tag{7.51}$$

$$= -\sigma(x)\frac{c^2}{\epsilon_2\omega} \sum_{n=-\infty}^{\infty} C_n k_{yn}^+ e^{ik_{xn}^+ x} . \tag{7.52}$$

We now introduce the central assumption of this section: the *edge condition* [Rothwell and Cloud (2009)]; this condition states that the current perpendicular to a sharp edge is proportional to the square-root of the distance, ρ, to the edge, that is, $J_x(\rho) \propto \sqrt{\rho}$. Our assumption is that in the regime where $kw < 1$, we can interpolate the current by an expression that includes the edge condition at both edges, $x = \pm w/2$, simultaneously. This assumption allow us to write the current $J_x(x)$ in a ribbon as

$$J_x(x) = \chi e^{ik_x x} \sqrt{w^2/4 - x^2} , \qquad (7.53)$$

where χ is a constant to be determined. We have verified numerically, using the methods presented in Sec. 7.3, that our assumption holds at every ribbon. In order to determine χ, we multiply Eq. (7.52) by a basis function, $e^{-ik_{xm}^+ x}$, and integrate over the unit cell, thus obtaining

$$\chi \frac{Lw}{4m} J_1(m\pi w/L) = -\frac{\sigma(\omega)c^2}{\epsilon_2 \omega \pi} \sum_{n=-\infty}^{\infty} C_n k_{yn}^+ \frac{L}{m-n} \sin \frac{\pi(m-n)w}{L} , \quad (7.54)$$

where $J_1(x)$ is a Bessel function and $\sigma(\omega)$ is the optical conductivity of a graphene ribbon (here assumed to be bulk-like). Equation (7.54) defines the constant χ once the coefficients C_n are known.

For determining the coefficients C_n, we must employ the other boundary condition holding for this system, which states that the $z-$component of the magnetic field is discontinuous across the graphene grating owing to the surface current induced by the electric field. Explicitly, we have

$$B_0^i e^{ik_x x} = \sum_{n=-\infty}^{\infty} C_n e^{ik_{xn}^+ x} - \sum_{n=-\infty}^{\infty} A_n e^{ik_{xn}^- x} - \mu_0 \chi e^{ik_x x} \sqrt{w^2/4 - x^2} . \quad (7.55)$$

Once again, multiplying the previous equation with a basis function and integrating over the unit cell produces

$$B_0^i \delta_{m,0} = C_m - A_m - \mu_0 \chi \frac{w}{4m} J_1(m\pi w/L) . \qquad (7.56)$$

This equation only establishes a relation between the coefficients A_m and C_m. Therefore, another relation is necessary for determining these coefficients. The extra relation comes from the continuity of the $x-$component of the electric field above and below the graphene grid (which we have already used when writing the surface current). Proceeding as in the derivation of Eq. (7.56), we obtain the remaining relation between the coefficients A_m and C_m:

$$A_m = \frac{k_y}{k_{ym}^-} B_0^i \delta_{m,0} - \frac{\epsilon_1}{\epsilon_2} \frac{k_{ym}^+}{k_{ym}^-} C_m . \qquad (7.57)$$

Solving Eqs. (7.56) and (7.57), we obtain the following expression for C_m:

$$C_m = \frac{\epsilon_2 k_{ym}^-}{\epsilon_1 k_{ym}^+ + \epsilon_2 k_{ym}^-} \left[2B_0^i \delta_{m,0} + \chi\mu_0 \frac{w}{4m} J_1(m\pi w/L) \right] . \qquad (7.58)$$

Moreover, by combining Eqs. (7.54) and (7.58), we can now determine both χ and C_m, from which the coefficient A_m directly follows. As usual, the reflectance (for normal incidence) is defined as

$$\mathcal{R} = \left| \frac{A_0}{B_0^i} \right|^2 , \qquad (7.59)$$

where the sum, $\mathcal{R} = \sum_n \frac{|\Re\{k_{yn}^-\}|}{k_y} |A_n/B_0^i|^2$, was carried out over propagating modes only (which correspond to the zero-th mode only for the parameters used here), and where

$$A_0 = B_0^i \left(1 - \frac{2\epsilon_1 k_{y0}^+}{\epsilon_1 k_{y0}^+ + \epsilon_2 k_{y0}^-} \right) - \chi\mu_0 \frac{\pi w^2}{8L} \frac{\epsilon_1 k_{y0}^+}{\epsilon_1 k_{y0}^+ + \epsilon_2 k_{y0}^-} , \qquad (7.60)$$

and

$$\chi = -\frac{B_0^i}{D(\omega)} \frac{\sigma(\omega)c^2}{\epsilon_2\omega} \frac{2\epsilon_2 k_{y0}^- k_{y0}^+}{\epsilon_1 k_{y0}^+ + \epsilon_2 k_{y0}^-} . \qquad (7.61)$$

Here, the quantity $D(\omega)$ is defined as

$$D(\omega) = \frac{w}{4} \sum_{n=-\infty}^{\infty} \frac{1}{n} J_1(n\pi w/L) \left[1 + \frac{\sigma(\omega)}{w\epsilon_0} \frac{k_{yn}^- k_{yn}^+}{\epsilon_1 k_{yn}^+ + \epsilon_2 k_{yn}^-} \right] . \qquad (7.62)$$

It is clear that the plasmonic resonance is controlled by the *poles* of $\chi(\omega)$. Looking carefully at the form of $D(\omega)$, we readily identify that the most important contribution to the resonance comes from the terms with $n = \pm 1$. This leads to the condition:

$$1 + \frac{\sigma(\omega)}{w\epsilon_0} \frac{k_{y\pm 1}^- k_{y\pm 1}^+}{\epsilon_1 k_{y\pm 1}^+ + \epsilon_2 k_{y\pm 1}^-} = 0 \Leftrightarrow \frac{\epsilon_1}{k_{y\pm 1}^-} + \frac{\epsilon_2}{k_{y\pm 1}^+} + \frac{\sigma(\omega)}{w\epsilon_0} = 0 . \qquad (7.63)$$

This last equation is nothing but the implicit expression for the dispersion relation of a graphene surface plasmon-polariton with wavevector $q = 2\pi/L$, as long as $q^2 > \epsilon_j\omega^2/c^2$, with $j = 1, 2$. In the frequency range we are working in (i.e. the THz spectral window) and for ribbons with widths within the order of a micrometer, we can show numerically that the infinite sum can be well approximated by the terms with $n = 0, \pm 1$ alone. Notice that in the case of the calculation carried out in this section both the conductivity and the current are not represented as a Fourier series, since this is not necessary here (owing to our assumption for the current —the edge condition).

In Fig. 7.14 we depict the reflectance (broken curves) as given by Eq. (7.59). It can be observed that the plasmonic resonant frequency experiences a blue-shift as the Fermi energy increases. Such behavior is ascribed to the form of the dispersion relation of graphene surface plasmon-polaritons (GSPs), which disperse as $\omega_{GSP} \propto \sqrt{E_F}$. In addition, one may see that the calculated reflectance curves [out of Eq. (7.63)] for the different E_F's have a maximum that *essentially* coincides —up to a small redshift— with the same quantities computed using the method of Sec. 7.3 (represented by the solid lines in Fig. 7.14). Furthermore, we note that in the latter curves (i.e. computed using the method of Sec. 7.3), it is possible to identify a second little peak at the right of the main resonance. This is due to the excitation in the ribbon of a surface plasmon of shorter wavelength (corresponding to the second diffracted order). Such behavior cannot be captured by the analytical method outlined above, as it assumes that only the plasmon of longer wavelength is excited in the ribbon. This assumption is encoded in the form of Eq. (7.53). Nevertheless, the model captures rather accurately the fundamental plasmonic resonance. The reader should appreciate[13] the advantages of the (approximate) analytical approach developed in this section over the semi-analytical[14] method developed in Sec. 7.3. Having a simple analytical form for the reflectance allows for a straightforward interpretation of the results, as seen in Eq. (7.63). On the other hand, the formalism used in Sec. 7.3 is easily extended to any periodic conductivity pattern, which is not the case of the method used in this section. That is because for a general conductivity pattern the condition to be imposed on the current in the unit cell is not clear, and, in general, it is not known before-hand. However, we might be able to work backwards and fit the current computed numerically, using the method of the previous sections, to a given function $J_x(x)$ and then apply the same procedure employed in this section to determine the reflectance (or, else, the transmittance and/or absorbance). If such a fitting can be accomplished, then the determination of an analytical expression for the reflectance is straightforward.

The arrow in Fig. 7.14 signals the value 0.0051 (baseline) corresponding to the reflectance measure at an interface between two dielectrics with relative permittivities of $\epsilon_1 = 3$ and $\epsilon_2 = 4$ (i.e. with graphene absent), as

---

[13]Namely, in this way one can carry out the computation without the need to solve an eigenproblem with large matrices.

[14]Here, we shall refer to the method of Sec. 7.3 as *semi-analytical* in the sense that at some point one has to solve an eigenproblem numerically using linear-algebra computational packages.

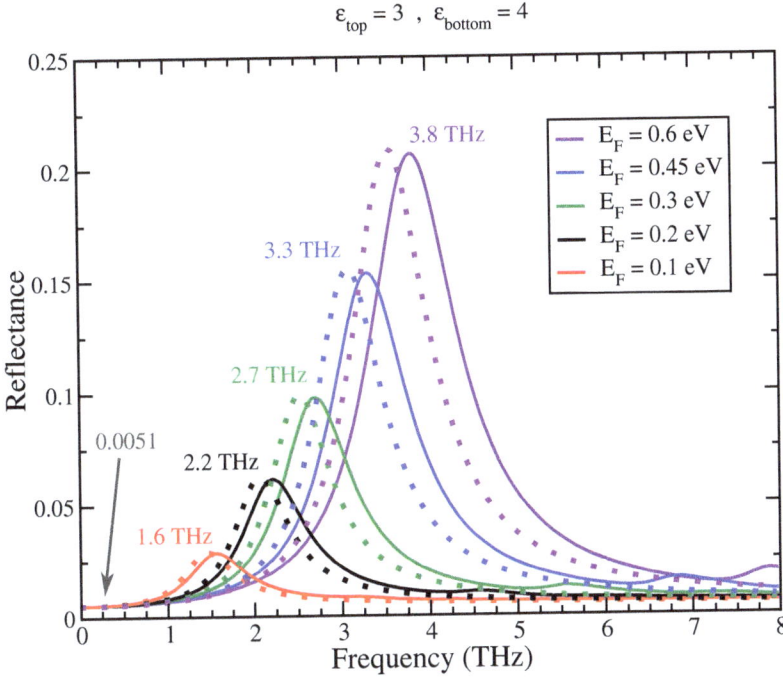

Figure 7.14: Reflectance from a periodic grid of graphene ribbons illuminated by a plane-wave at normal incidence computed using the analytical model developed in this section (broken curves). The parameters are $\Gamma = 2.6$ meV, $L = 8$ μm, $w = L/2 = 4$ μm, $\epsilon_1 = 3$ (region akin to the incoming wave), and $\epsilon_2 = 4$. The maximum of the reflectance —indicated by corresponding $\omega_{GSP}/(2\pi)$ values— disperses as function of E_F (in the same way as ω_{GSP} disperses in proportion to $\sqrt{E_F}$). For the frequency corresponding to the maximum of the blue curve we have $kw \approx 0.27$, which is well within the validity of Eq. (7.53). The broken lines correspond to the results obtained using the approach developed in the present section, whereas the solid lines correspond to the results computed using the method outlined in Sec. 7.3 (with $\Gamma = 3.0$ meV), for the same parameters. [Note that the definition of the axis are different between this and the previous sections; note also that the dielectric constants of the embedding dielectrics are reversed. In the language of Sec. 7.3 (and in the rest of this Chapter for that matter), $R = L$ and $d_g = w$ —cf. Fig. 7.13. These remarks should avoid any confusion.]

given by Eq. (9.60) for $\theta = 0$:

$$\mathcal{R}_{\epsilon_1 \rightarrow \epsilon_2} = \left| \frac{\sqrt{\epsilon_2/\epsilon_1} - 1}{\sqrt{\epsilon_2/\epsilon_1} + 1} \right|^2 \approx 0.0051 . \qquad (7.64)$$

The correspondence between the numerical value in Fig. 7.14 and that given by Eq. (7.64) gives additional confidence on the calculated curves (on top of the agreement between two independent theoretical models). It is interesting that in the presence of a grid the low-frequency reflectance is insensitive to the to the existence of graphene. This can be understood if we note that for low frequencies (large wavelengths) it is not possible to excite plasmons in the system as they would have a very long wavelength and therefore would not fit within a ribbon.

Chapter 8

Plasmons in Graphene Nanostructures and in One-dimensional Channels

The 21st century has undoubtedly brought about the "nano-revolution" that began in the 1990s. It can eventually be argued that this led to an outburst of interest in nanomaterials which may have contributed for the isolation of graphene itself.

So far we have focused our attention in the investigation of plasmon-related effects either in extended graphene or in periodic graphene-based micro/nanostructures. In particular, we have seen that the patterning of an otherwise continuous graphene sheet into a periodic array of graphene ribbons provides an appealing route towards the excitation and control of plasmonic excitations in engineered graphene. In this way, localized plasmons dressed by the grating are produced when the system is illuminated by a laser beam, thanks to the Bloch waves originated via grating coupling.

Looking beyond the two above-mentioned systems, one can also achieve graphene surface plasmons supported by single graphene nanostructures. Such an idea in not necessarily new, since mankind has been doing plasmonics in single or colloidal solutions of metallic nanoparticles for many years (often without knowing it, as in the case of the Lycurgos cup and of stain glasses in cathedrals), in which localized plasmons are used to concentrate the electromagnetic field in tiny regions in the neighborhood of the particles. In parallel, this and related mechanisms may also be applied to plasmonics in graphene-based nanostructures. Typically, these can be fabricated essentially by employing the very same technology and nano-engineering techniques used in the manufacturing of periodic arrays of graphene ribbons or disks. Examples of different geometries in which isolated graphene nanostructures have been contemplated include individual graphene ribbons [Wang and Kinaret (2013); Thongrattanasiri *et al.* (2012b); Christensen *et al.* (2012); Wang *et al.* (2011); Christensen *et al.*

(2015b); Shylau *et al.* (2015)], graphene disks [Wang *et al.* (2012b, 2011); Christensen *et al.* (2014); Wang *et al.* (2016); Fang *et al.* (2013); Yan *et al.* (2012c); Thongrattanasiri *et al.* (2012b)] and rings [Wang (2012); Fang *et al.* (2013); Yan *et al.* (2012c)], and graphene triangles [Wang *et al.* (2015)] or nano-antennas [Liu *et al.* (2014a); Llatser *et al.* (2012)]. All of this became possible thanks to the rapid evolution of our arsenal of "nano-tools" and nanofabrication techniques. Interestingly, plasmons in nanoislands can be induced by adding to them single electrons [Manjavacas *et al.* (2015); Garcia de Abajo and Manjavacas (2015)]. We should also mention that other materials beyond graphene can be tailored in nanostructures, such as nanogranules of hBN [Sun *et al.* (2015)].

Many of the works available in the literature are carried out by discretizing the system and solving Maxwell's equations numerically, irrespective of the structure's geometry, owing to the widespread use of a vast collection of numerical solvers in the nanophotonics community, which supply an abundance of computational approaches, such as boundary element method (BEM), finite element method (FEM), finite-difference time-domain (FDTD), and discrete-dipole approximation (DDA) [Pelton and Bryant (2013)]. Notwithstanding, for typical geometries such as the ones referred in the previous paragraph, it is often possible to construct *semi-analytical* techniques to determine the plasmonic excitations and optoelectronic properties of each individual graphene structure. The prime goal of the current Chapter will be to describe and apply that semi-analytical formalism to the study of plasmons in single nanoribbons and nanodisks (and rings). The main advantage of such an approach is that it provides a deeper insight about the physics, while also providing a more adequate control of the theory and its limitations, both of which can be hindered in some situations when using computational tools (albeit that, if used right, they usually deliver very good results). However, for more intricate geometries, there is little hope for these semi-analytical techniques, and numerical simulations are the norm. Indeed, as we shall see, these semi-analytical methods require the knowledge of the Green's function of the problem, which for intricate geometries is out of reach.

Coming back to the subject of this Chapter, here we shall work in the so-called electrostatic limit or non-retarded regime, which is a good approximation when dealing with plasmonic excitations occurring in structures with dimensions ~ 100 nm, since the corresponding free space resonant wavelength will be orders of magnitude greater than the size of system (this also enable us to assume that the response function is essentially local). In the first case, that of the ribbon geometry, we will consider a waveguide-like plasmon propagating along the ribbon (that is, along the direction parallel to its edges), whereas in the second case, we will study the dispersion of localized plasmons in a circular graphene disk and also in a ring-like configuration.

8.1 Edge Plasmons in a Graphene Nanoribbon

In what concerns isolated graphene-based nanostructures, the ribbon geometry has been extensively studied [Wang and Kinaret (2013); Thongrattanasiri *et al.* (2012b); Christensen *et al.* (2012); Wang *et al.* (2011); Christensen *et al.* (2015b)], and has attracted a great deal of interest in the prospection of its plasmonic properties. In what follows, we shall derive the spectrum of the plasmonic excitations supported by a graphene nanoribbon —represented in Fig. 8.1—, which we can already anticipate it will be made up of several branches associated with different multipolar orders (this shall become clearer later). As shown in the figure, we will consider the general

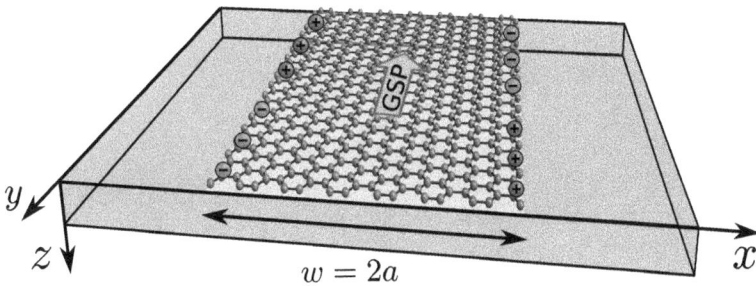

Figure 8.1: Graphene nanoribbon with a dipolar (edge) plasmon represented in a pictorial fashion for eye-guidance. The ribbon is infinite along the y-direction and it is placed on a semi-infinite substrate ($z > 0$) of dielectric constant ϵ_2. The superstrate ($z < 0$) is also semi-infinite, but with a dielectric constant characterized by ϵ_1.

case in which the ribbon sits between two semi-infinite dielectrics, like the case when an individual graphene nanoribbon is deposited onto a dielectric substrate such as silicon dioxide. For describing plasmonic effects in this problem, it will be necessary to find the Green's functions ascribed to Poisson's equation in a 2-layered medium.

8.1.1   *Prelude: Green's functions for a 2-layered medium*

Our first task will be to determine the Green's functions that stems from Poisson's equation in a 2-layered medium, that is, in a system that comprises two semi-infinite dielectrics meeting at an interface (at $z = 0$), as depicted in Fig. 8.2.

8.1.1.1   *Homogeneous medium*

For the moment, let us consider a homogeneous medium with dielectric constant ϵ. In that simple case, the Poisson equation reads

$$\nabla^2 \Phi(\mathbf{r}) = -\frac{\rho(\mathbf{r})}{\epsilon \epsilon_0} \, , \tag{8.1}$$

where $\Phi(\mathbf{r})$ is the potential and $\rho(\mathbf{r})$ stands for the charge density. For a point charge located at $\mathbf{r}'$, the Green's function associated with Eq. (8.1) can be defined through (which basically allows to determine the potential "felt" at position $\mathbf{r}$, when an elementary charge is placed at $\mathbf{r}'$)

$$\nabla^2 G^0(\mathbf{r}, \mathbf{r}') = -\frac{1}{\epsilon \epsilon_0} \delta(\mathbf{r} - \mathbf{r}') \, , \tag{8.2}$$

where the superscript "0" in the Green's function is to remind ourselves that, for the time being, we are treating the homogeneous case (this is also sometimes termed *free-space Green's function*). We also note that Eq. (8.2) implies $G^0(\mathbf{r}, \mathbf{r}') = G^0(\mathbf{r} - \mathbf{r}')$. Performing a Fourier transform in Eq. (8.2), we obtain the result[1]

$$G^0(\mathbf{k}) = \frac{1}{\epsilon \epsilon_0} \frac{1}{k_x^2 + k_y^2 + k_z^2} \, , \tag{8.3}$$

---

[1]Note that:

$$G^0(\mathbf{r} - \mathbf{r}') = \int \frac{d\mathbf{k}}{(2\pi)^3} G^0(\mathbf{k}) e^{i\mathbf{k} \cdot (\mathbf{r} - \mathbf{r}')} \, ,$$

$$\delta(\mathbf{r} - \mathbf{r}') = \int \frac{d\mathbf{k}}{(2\pi)^3} e^{i\mathbf{k} \cdot (\mathbf{r} - \mathbf{r}')} \, .$$

where $|\mathbf{k}| = \sqrt{k_x^2 + k_y^2 + k_z^2}$. Going back to real space, with respect to the z-coordinate only, gives[2]

$$G^0(q; z, z') = \frac{1}{\epsilon \epsilon_0} \int_{-\infty}^{\infty} \frac{dk_z}{2\pi} \frac{e^{ik_z(z-z')}}{k_x^2 + k_y^2 + k_z^2} , \qquad (8.4)$$

$$= \frac{1}{2q\epsilon\epsilon_0} e^{-q|z-z'|} , \qquad (8.5)$$

where $q = \sqrt{k_x^2 + k_y^2}$ is the length of a two-dimensional vector living in the plane orthogonal to $\hat{\mathbf{k}}_z$. The result written in this form is convenient for computing the Green's function for a stratified medium along $\hat{\mathbf{z}}$.

8.1.1.2 *2-layered medium*

Having considered the elementary problem of a homogeneous medium, let us now see how this result is modified in the situation where the space is divided into two semi-infinite regions, occupied by two insulators with different relative permittivities, ϵ_1 and ϵ_2, as in Fig. 8.2.

Figure 8.2: Side view of an inhomo-geneous medium along the z-direction, stratified into two regions (layers) with different dielectric environments. The interface is defined by the $z = 0$ plane.

In the particular case in which an elementary point charge is placed at $\mathbf{r}' = (x', y', z') \in$ medium j, and if the "observation point" is in the same layer, at $\mathbf{r} = (x, y, z) \in$ medium j, the free Green's function corresponding to the homogeneous dielectric case, that is,

$$G_j^0(q; z, z') = \frac{1}{2q\epsilon_j\epsilon_0} e^{-q|z-z'|} , \qquad (8.6)$$

is a particular solution of the Poisson's equation. The reader should note that, by assumption, $z, z' \in$ medium j. Saying differently, the prime use of this free-space-like Green's function will be as a particular solution of

---

[2]This integral is similar to the one computed in Eq. (G.6), appearing in the Appendix G. The interested reader unfamiliar with these type of integrals is advised to check the aforementioned equation and corresponding Appendix, where it is solved using the contour integration technique from basic complex analysis.

the non-homogeneous equation (Poisson's equation after the 2D Fourier transform performed in the x and y variables)

$$\frac{\partial^2}{\partial z^2} G_j^0(q; z, z') - q^2 G_j^0(q; z, z') = -\frac{1}{\epsilon_j \epsilon_0} \delta(z - z') . \tag{8.7}$$

It is well-known from mathematical analysis that one can always build a general solution of a differential equation by adding a particular solution of the non-homogeneous equation [here, Eq. (8.6)] with a general solution of the corresponding homogeneous equation. Thus, assuming that the test charge is in the medium with $j = 1$, we can write

$$G_{11}(q; z, z') = \frac{1}{2q\epsilon_1 \epsilon_0} e^{-q|z-z'|} + A e^{qz} , \tag{8.8}$$

$$G_{21}(q; z, z') = B e^{-qz} , \tag{8.9}$$

where the parameters A and B need to be determined and we have introduced the double-index notation G_{kl} for the Green's functions, where the index k labels the medium where the "observation point" is, and l refers to the medium where the test charge is located. Note that these solutions must be constructed in such a way that they satisfy the conditions

$$\lim_{z \to -\infty} G_{1l}(q; z, z') = 0 \quad \text{and} \quad \lim_{z \to +\infty} G_{2l}(q; z, z') = 0 . \tag{8.10}$$

The parameters A and B may be determined by calling the boundary conditions at the interface [Kang (2015)]; these are

$$G_{11}(q; 0, z') = G_{21}(q; 0, z') , \tag{8.11}$$

$$\epsilon_1 \frac{\partial}{\partial z} G_{11}(q; 0, z') = \epsilon_2 \frac{\partial}{\partial z} G_{21}(q; 0, z') , \tag{8.12}$$

from which, using Eqs. (8.8) and (8.9), one arrives at

$$A = \frac{\epsilon_1 - \epsilon_2}{\epsilon_1 + \epsilon_2} \frac{e^{qz'}}{2q\epsilon_1 \epsilon_0} , \tag{8.13}$$

$$B = \frac{1}{\epsilon_1 + \epsilon_2} \frac{e^{qz'}}{q\epsilon_0} . \tag{8.14}$$

Therefore, for $z' < 0$ (the test charge is in the medium with $j = 1$), the Green's function's (8.8) and (8.9) translate to

$$G_{11}(q; z, z') = \frac{1}{2q\epsilon_1 \epsilon_0} \left[e^{-q|z-z'|} + \frac{\epsilon_1 - \epsilon_2}{\epsilon_1 + \epsilon_2} e^{q(z+z')} \right] ; \ z, z' < 0 , \tag{8.15}$$

$$G_{21}(q; z, z') = \frac{2\epsilon_1}{\epsilon_1 + \epsilon_2} \frac{e^{-q(z-z')}}{2q\epsilon_1 \epsilon_0} ; \ z \geq 0 , \ z' \leq 0 . \tag{8.16}$$

The reader should appreciate the appearance of the reflection and transmission amplitudes [3] in the $G_{11}(q; z, z')$ and $G_{21}(q; z, z')$, respectively [see also Eqs. (O.28) and (O.29)]. Notice that —and perhaps not surprisingly— the result for G_{21} is equivalent to the one obtained after replacing $\epsilon_j \to \frac{\epsilon_1 + \epsilon_2}{2}$ in Eq. (8.6). Similarly, by applying the same procedure to the case where the point charge is in the medium with $j = 2$, i.e. for $z' > 0$, we obtain

$$G_{22}(q; z, z') = \frac{1}{2q\epsilon_2\epsilon_0} \left[e^{-q|z-z'|} + \frac{\epsilon_2 - \epsilon_1}{\epsilon_1 + \epsilon_2} e^{q(z+z')} \right] \; ; \; z, z' > 0 \;, \quad (8.17)$$

$$G_{12}(q; z, z') = \frac{1}{\epsilon_1 + \epsilon_2} \frac{e^{q(z-z')}}{q\epsilon_0} \; ; \; z \le 0 \;, \; z' \ge 0 \;. \quad (8.18)$$

Having worked out the Green's functions for a 2-layered medium, we are now ready to address the problem of finding the spectrum of surface plasmons supported by a single graphene nanoribbon sitting at the interface between dielectrics made up of two distinct materials.

8.1.2 *Plasmonic spectrum of a graphene nanoribbon*

In order to determine the spectrum of the plasmonic modes sustained by a graphene ribbon —see Fig. 8.1— we need to solve Poisson's equation in

---

[3]Using some intuition, we can derive the reflection and transmission amplitudes as follows. In Fourier space, Laplace's equation reads

$$-\frac{\partial \phi(q, z)}{\partial z^2} + q^2 \phi(q, z) = 0 \,,$$

where q is the absolute value of the in-plane momentum. Let us assume we have a charge in medium ϵ_1 (for $z < 0$) located at $z = z_0 < 0$. Then in medium ϵ_2 (for $z > 0$) the potential that satisfies the boundary condition at infinity is $\phi(q, z) = te^{-zq}$, and represents a "transmitted" potential. In the region $z_0 < z < 0$ we have a solution of the form

$$\phi(q, z) = re^{qz} + e^{-qz} \,,$$

where e^{-qz} represents the "impinging" potential at the interface at $z = 0$ and re^{qz} represents a "reflected" potential. The boundary conditions at $z = 0$ are: $\phi(q, 0^-) = \phi(q, 0^+)$ and $\epsilon_1\phi'(q, 0^-) = \epsilon_2\phi'(q, 0^+)$, where the prime stands for derivative in order to z. Solving the boundary conditions for t and r, we obtain

$$t = \frac{2\epsilon_1}{\epsilon_1 + \epsilon_2} \qquad \text{and} \qquad r = \frac{\epsilon_1 - \epsilon_2}{\epsilon_1 + \epsilon_2} \,.$$

Note that $r^2 + t^2 \ne 1$ due to the different value of the speed of light in the two dielectrics, which is proportional to $c_i \propto 1/\sqrt{\epsilon_i}$. Therefore, taking this into account, one has

$$r^2 + \epsilon_2/\epsilon_1 t^2 = 1 \,,$$

an equation written remembering that the flux of energy is proportional to ϵ_i. Please note that this derivation is not accurate, although it reproduces the correct result.

this configuration:

$$\nabla^2 \phi(\mathbf{r}) = -\frac{\rho(\mathbf{r}_{xy}, z)}{\epsilon \epsilon_0} , \qquad (8.19)$$

where we have defined the vector $\mathbf{r}_{xy} = (x, y)$, and the charge-density is naturally limited to the graphene's plane, so that it can be written as

$$\rho(\mathbf{r}_{xy}, z) = -en(x)e^{ik_y y}\delta(z) , \qquad (8.20)$$

where $n(x)$ stands for the density of charge-carriers. Note that in writing Eq. (8.20) we have exploited the system's translational invariance along the y-direction, which has allowed us to make the decomposition $n(\mathbf{r}_{xy}) = n(x)e^{ik_y y}$. Furthermore, although not explicitly written in the previous expressions, the time-dependency of both the potential and the particle density is assumed to be of the form

$$\Phi(\mathbf{r}, t) = \phi(\mathbf{r})e^{-i\omega t} , \qquad (8.21)$$

$$\varrho(\mathbf{r}, t) = \rho(\mathbf{r}_{xy}, z)e^{-i\omega t} , \qquad (8.22)$$

respectively. We further remark that we are working in the electrostatic limit, and we will do so for the remaining of this Chapter. This is justified[4] since we will be describing excitations whose resonant wavelengths will be of order of $10 - 100$ μm in ribbons with dimensions of only a few hundred of nanometers, retardation effects can be safely neglected thereby making the electrostatic limit indeed a very good approximation. Like for the charge-density (8.20), we can also decompose the potential as $\phi(\mathbf{r}) = \phi(x, z)e^{ik_y y}$. Thus, Eq. (8.19) becomes

$$\left[\frac{\partial^2}{\partial x^2} + \frac{\partial^2}{\partial z^2} - k_y^2\right]\phi(x, z) = -\frac{\rho(x, z)}{\epsilon \epsilon_0} , \qquad (8.23)$$

with $\rho(x, z) = -en(x)\delta(z)$, depending only on the x- and $z-$ coordinates. Consequently, the Green's function associated with this last differential equation is defined via

$$\phi(x, z) = -e \int_{-\infty}^{\infty} dx' \int_{-\infty}^{\infty} dz' G(x, z; x', z')n(x')\delta(z') , \qquad (8.24)$$

$$= -e \int_{-a}^{a} dx' G(x, z; x', 0)n(x') , \qquad (8.25)$$

where in the last step we have assume that the density profile of the charge-carriers in the graphene ribbon (whose width is $w = 2a$) has a box-like form, as illustrated in Fig. 8.3. Formally speaking, this is equivalent to

---

[4]The calculations can be done taking retardation into account, as done in the problem discussed in Chapter 6. This additional complication is, however, unnecessary here.

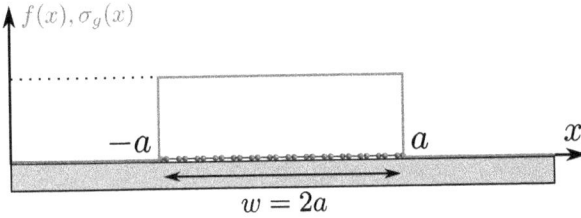

Figure 8.3: Charge-density envelope function/conductivity profile of an individual graphene nanoribbon of width $w = 2a$.

performing $n(x) \to n(x)f(x)$ with $f(x) = \Theta(x + a)\Theta(a - x)$, where $\Theta(x)$ stands for the Heaviside step function, and $f(x)$ is an envelope function that characterizes the profile of the charge-density within the nanoribbon[5]. The next step is then to determine the form of the Green's function figuring in Eq. (8.25). This is where the effort that we made in the previous subsection pays off, by recalling from Eq. (8.9) that

$$G(q; z, z') = \frac{2}{\epsilon_1 + \epsilon_2} \frac{e^{-q|z-z'|}}{2q\epsilon_0} , \qquad (8.26)$$

from which we can now perform a one-dimensional Fourier transform, yielding the Green's function in the same form that appears in expression (8.25), namely

$$G(x, z; x', z') = \int_{-\infty}^{\infty} \frac{dk_x}{2\pi} G(q; z, z') e^{ik_x(x-x')} , \qquad (8.27)$$

$$= \frac{1}{2\pi\epsilon_0} \frac{1}{\epsilon_1 + \epsilon_2} \int_{-\infty}^{\infty} dk_x \frac{e^{-|z-z'|\sqrt{k_x^2+k_y^2}}}{\sqrt{k_x^2 + k_y^2}} e^{ik_x(x-x')} , \qquad (8.28)$$

$$= \frac{1}{\pi\epsilon_0} \frac{1}{\epsilon_1 + \epsilon_2} K_0 \left(k_y \sqrt{(x - x')^2 + (z - z')^2} \right) , \qquad (8.29)$$

where K_0 designates the zero-order modified Bessel function of the second kind. Now, substituting this result back into Eq. (8.25 produces

$$\phi(x, z) = -\frac{e}{\pi\epsilon_0} \frac{1}{\epsilon_1 + \epsilon_2} \int_{-a}^{a} dx' K_0 \left(k_y \sqrt{(x - x')^2 + z^2} \right) n(x') , \qquad (8.30)$$

which hopefully will be solvable once we have the carrier density, $n(x)$. With that end in view, we need to call some more physics in order to arrive at an expression for $n(x)$. We shall tackle this exercise by invoking two

---

[5] Note that one can of course use more realistic forms for the profile function, $f(x)$, at the cost of more labor to perform the integrals. However, based on the results obtained in the problem for the periodic system of ribbons, the sharp rectangular-box profile seems to constitute a good first approximation.

physical laws, the first being charge-conservation, which dictates, through the continuity equation, that

$$\frac{\partial}{\partial t}\varrho(\mathbf{r}, t) + \nabla \cdot \mathbf{J} = 0 ,\tag{8.31}$$

$$\Rightarrow i\omega\varrho(\mathbf{r}, t) = \nabla \cdot \mathbf{J} ,\tag{8.32}$$

and then by using Ohm's law to express the current in terms of the potential,

$$\mathbf{J} = \sigma\mathbf{E} = -\sigma\nabla\Phi(\mathbf{r}, t) .\tag{8.33}$$

Combining these two expressions, yields

$$\nabla \cdot [\sigma\nabla\phi(\mathbf{r})] = -i\omega\rho(\mathbf{r}_{xy}, z) ,\tag{8.34}$$

$$= ie\omega n(\mathbf{r}_{xy})\delta(z) ,\tag{8.35}$$

where the dot product can be expressed as

$$\nabla \cdot [\sigma\nabla\phi(\mathbf{r})] = \frac{\partial\sigma}{\partial x}\frac{\partial\phi}{\partial x} + \sigma\frac{\partial^2\phi}{\partial x^2} + \sigma\frac{\partial^2\phi}{\partial y^2} ,\tag{8.36}$$

$$= \frac{\partial\sigma}{\partial x}\frac{\partial\phi}{\partial x} + \sigma\left[\frac{\partial^2\phi}{\partial x^2} - k_y^2\phi\right] ,\tag{8.37}$$

and therefore, inserting this result into Eq. (8.35), finally gives the sought expression for $n(x)$:

$$n(x) = -\frac{i}{e\omega}\left[\frac{\partial\sigma(x)}{\partial x}\frac{\partial\phi(x, z)}{\partial x} + \sigma(x)\left(\frac{\partial^2\phi(x, z)}{\partial x^2} - k_y^2\phi(x, z)\right)\right]\delta(z) .\tag{8.38}$$

In possession of Eq. (8.38), we can now use this result to rewrite Eq. (8.30) as

$$\phi(x, z) = \frac{i}{\pi\omega\epsilon_0}\frac{1}{\epsilon_1 + \epsilon_2}\int_{-a}^{a} dx' K_0\left(k_y\sqrt{(x - x')^2 + z^2}\right) \times$$

$$\times \left[\frac{\partial\sigma(x')}{\partial x'}\frac{\partial\phi(x', 0)}{\partial x'} + \sigma(x')\left(\frac{\partial^2\phi(x', 0)}{\partial x'^2} - k_y^2\phi(x', 0)\right)\right] ,\tag{8.39}$$

which is a self-consistent integro-differential equation for the potential that must be solved in some approximation, since it cannot be solved exactly in its current form. Note that the determination of the potential in the entire space[6], requires previous knowledge of the potential at the $z = 0$ plane.

---

[6]Recall that $\phi(\mathbf{r}) = \phi(x, z)e^{ik_y y}$.

Thus, setting $z = 0$, we obtain a slightly more friendly expression for the potential along $\hat{\mathbf{x}}$),

$$
\phi(x) = \frac{i\sigma_g}{\pi\omega\epsilon_0} \frac{1}{\epsilon_1 + \epsilon_2} \int_{-a}^{a} dx'\, K_0\left(k_y|x - x'|\right) \times
$$
$$
\times \left[\frac{\partial f(x')}{\partial x'} \frac{\partial \phi(x')}{\partial x'} + f(x') \left(\frac{\partial^2 \phi(x')}{\partial x'^2} - k_y^2 \phi(x') \right) \right], \tag{8.40}
$$

which is valid for $z = 0$, so that the potential in the plane where the graphene ribbon lives can be easily obtained from $\phi(x, y) = \phi(x)e^{ik_y y}$. Also, notice that we have written the spatial dependence of the conductivity in terms of the envelope/profile function, i.e. $\sigma(x) = \sigma_g f(x)$, where $\sigma_g \equiv \sigma_g(\omega)$ is the (bulk) dynamical conductivity of graphene. Moreover, we note that this later integro-differential equation inherits the same difficulties of its predecessor in what concerns its solvability. We shall address this in a moment, but before that, let us work towards a more explicit form for Eq. (8.40). Realizing that,

$$
\frac{\partial f(x)}{\partial x} = \delta(x + a) - \delta(a - x), \tag{8.41}
$$

where we have assumed that $f(x) = \Theta(x+a)\Theta(a-x)$, as portrayed in Fig. 8.3, enables us to express Eq. (8.40) in the following form:

$$
\phi(x) = \frac{i\sigma_g}{\pi\omega\epsilon_0} \frac{1}{\epsilon_1 + \epsilon_2} \left\{ \int_{-a}^{a} dx'\, K_0\left(k_y|x - x'|\right) \left[\frac{\partial^2 \phi(x')}{\partial x'^2} - k_y^2 \phi(x') \right] \right.
$$
$$
\left. + \frac{\partial \phi(x')}{\partial x'}\bigg|_{x'=-a} K_0(k_y|x + a|) - \frac{\partial \phi(x')}{\partial x'}\bigg|_{x'=a} K_0(k_y|x - a|) \right\}, \tag{8.42}
$$

which will be more suitable for what follows. As previously mentioned, this kind of self-consistent integro-differential equation for the potential must be solved numerically in some approximation, since it cannot be solved exactly by analytical means. Hence, we shall attempt to solve Eq. (8.42) by expanding the in-plane potential as [Wang (2012); Beletskii and Bludov (2005)]

$$
\phi(x) = \sum_{m=0}^{\infty} A_m \psi_m(x), \tag{8.43}
$$

where the set of orthonormal basis functions $\psi_m(x)$ are defined in terms of Legendre's polynomials[7], $P_m(x)$, that is [Arfken *et al.* (2012); Beletskii and Bludov (2005)]

$$\psi_m(x) = \sqrt{\frac{2m+1}{2a}} P_m\left(\frac{x}{a}\right) . \tag{8.44}$$

This polynomial expansion technique will allow us to transform the integro-differential equation into a typical eigenvalue equation suitable to be casted in matrix form, which can then be tackled numerically using well-known linear algebra techniques. With that purpose in mind we start by inserting the expansion for the potential (8.43) into Eq. (8.42), which renders

$$\sum_m A_m \psi_m(x) = \frac{i\sigma_g}{\pi\omega\epsilon_0} \frac{1}{\epsilon_1 + \epsilon_2} \times \sum_m A_m$$

$$\times \left\{ \int_{-a}^a dx' K_0\left(k_y|x - x'|\right) \left[\frac{\partial^2 \psi_m(x')}{\partial x'^2} - k_y^2 \psi_m(x')\right] \right.$$

$$+ \left.\frac{\partial \psi_m(x')}{\partial x'}\right|_{x'=-a} K_0(k_y|x+a|)$$

$$\left. - \frac{\partial \psi_m(x')}{\partial x'}\right|_{x'=a} K_0(k_y|x-a|) \right\}, \tag{8.45}$$

Multiplying both sides of the previous equation by $\psi_l(x)$, and then integrating over $x \in [-a, a]$ (thereby exploiting the orthogonality of the basis functions[8]), yields the eigenvalue equation

$$\boxed{(\epsilon_1 + \epsilon_2)\frac{\pi\omega\epsilon_0}{i\sigma_g(\omega)} A_l = \sum_m U_{lm}^{\text{G-ribbon}} A_m} , \tag{8.46}$$

where the matrix elements, $U_{lm}^{\text{G-ribbon}}$, are given by[9]

$$U_{lm}^{\text{G-ribbon}} = \int_{-1}^1 du \int_{-1}^1 dv K_0\left(\beta|u - v|\right) \left[\psi_l(u)\frac{\partial^2 \psi_m(v)}{\partial v^2} - \beta^2 \psi_l(u)\psi_m(v)\right]$$

$$+ \left.\frac{\partial \psi_m(v)}{\partial v}\right|_{v=-1} \int_{-1}^1 du K_0(\beta|u+1|)\psi_l(u)$$

$$- \left.\frac{\partial \psi_m(v)}{\partial v}\right|_{v=1} \int_{-1}^1 du K_0(\beta|u-1|)\psi_l(u) , \tag{8.47}$$

---

[7]These are known to be an appropriate choice as a basis for expansions in many electrostatic problems [Jackson (1998); Arfken *et al.* (2012)].

[8]Explicitly: $\int_{-a}^a \psi_m(x)\psi_l(x)dx = \delta_{m,l}$.

[9]Here [Eq. (8.47)], in writting $\psi_k(u)$ it should be noted that we actually mean $\psi_k(u) = \sqrt{\frac{2k+1}{2a}} P_k(u)$, which is not exactly equivalent to our previous definition (8.44). However, this makes the notation lighter. We thus hope that this remark may eliminate any source of confusion.

where the dimensionless variables $u = x/a$, $v = x'/a$ and $\beta = k_y a$ have been introduced in order to facilitate the numerical implementation.

Finally, at this point the spectrum of plasmons in the graphene nanoribbon stems directly from the diagonalization of the matrix $\mathbf{U}^{\text{G-ribbon}}$, that is, once the eigenvalues of $\mathbf{U}^{\text{G-ribbon}}$ are known, the plasmonic spectrum follows from

$$\boxed{(\epsilon_1 + \epsilon_2)\frac{\pi\omega\epsilon_0}{i\sigma_g(\omega)} = \lambda(\beta)} \,, \qquad (8.48)$$

where $\lambda(\beta)$ is a N-dimensional[10] list containing the calculated eigenvalues of $\mathbf{U}^{\text{G-ribbon}}$, for a given $\beta = k_y a$.

In the limit in which the conductivity of graphene takes a Drude-like form without damping —cf. Eq. (2.72)—, it is feasible to attain a closed-form expression for the dispersion of plasmon modes sustained by the nanoribbon; within that regime, Eq. (8.48) delivers

$$\boxed{\Omega^{\text{ribbon}}_{GSP}(k_\parallel) = \sqrt{\frac{4\alpha E_F}{\epsilon_1 + \epsilon_2}\frac{\hbar c}{\pi}\lambda(\beta)}} \,, \qquad (8.49)$$

where we have written $k_\parallel \equiv k_y$ to highlight that it corresponds to the wavevector parallel to/along the graphene ribbon. On the other hand, if one uses a more generic expression for the conductivity, it will be necessary to do some extra work and solve Eq. (8.48) numerically, which, in any case, should be straightforward.

8.1.2.1 *Plasmonic spectrum of a graphene nanoribbon: results*

The spectra of the plasmonic excitations supported by a 100nm-wide (doped) graphene nanoribbon are shown in Fig. 8.4; the panel on the left follows directly from Eq. (8.49) [where we have assumed a Drude-like conductivity for graphene], whereas the panel on the right stems from the numerical solution of Eq. (8.48) using the local-Kubo conductivity of graphene [cf. Eqs. (2.70) and (2.71)]. In both panels one can observe that the plasmon dispersion in the nanoribbon configuration is dramatically distinct from the usual dispersion found in the case of an infinite (i.e. occupying the entire $z = 0$ plane) graphene-sheet (black-dashed lines in the figure). Broadly speaking, instead of a single band we now have a plasmonic spectrum which comprises multiple well-defined branches, with a

---

[10]Where N is the number of terms that constitute the (truncated) expansion, so that $\mathbf{U}^{\text{G-ribbon}}$ is a $N \times N$ square matrix.

legion of resonant frequencies for the same wavevector k_y. To each of these individual curves (shown in different colors) that together constitute the manifold of modes visible in the figure, there is a corresponding momentum along the x-coordinate such that the plasmon eigenmode resembles a (plasmonic) standing wave along $\hat{\mathbf{x}}$, while travelling parallel to the ribbon's edge. This also makes this geometry potentialy interesting for waveguiding applications, since the ribbon can act as a plasmonic waveguide with multiple acessible guided plasmon modes.

Furthermore, note that for small k_y the plasmonic modes posses little dispersion at threshold energies (with the exception of the fundamental mode), progressively acquiring the same wavevector dependence as in the infinite sheet case as k_y grows. This can be understood as follows: since the dispersion of the plasmon is proportional to $\sqrt{(k_x^n)^2 + k_y^2}$ and $k_x^n \propto \pi n/L$ (standing waves in the transverse direction), with n an integer, then for $k_y < k_x^n$ the plasmon frequency is essentially dispersionless. Note, however, that we can still have a localized surface plasmon oscillating perpendicularly to the boundaries of the ribbon —much like a standing wave— even for $k_y = 0$.

The other aspect that immediately strikes us when looking at Fig. 8.4 is that the two plots differ significantly from each other. Note that all the parameters are the same in both panels with the exception of the approximation in which the conductivity of graphene was computed, as indicated in the figure. In particular, this difference can be ascribed to the inclusion of interband transitions that contribute to the dynamical conductivity of graphene, which are neglected in the first panel. Naturally, this effect becomes increasingly important as we move towards higher plasmon energies.

In addition, these plasmonic excitations in the graphene nanoribbon emerge as a suitable platform for shrinking electromagnetic fields within volumes of a few tens of cubic nanometers. To see this in practice, notice that the wavelength of a free-propagating light wave with a photon energy of 0.5 eV (equivalent to the E_F considered in the figure) is $\lambda_0 \simeq 2480$ nm, which corresponds to a wavevector of $k_0 = 2\pi/\lambda_0 \simeq 0.25 \times 10^{-2}$ nm^{-1}. Then, in the light of Fig. 8.49, we conclude that $k_0 \ll k_y \Leftrightarrow \lambda_0 \gg \lambda_{GSP}^{\text{ribbon}}$, thus implicating the ability of these graphene plasmons to confine the electromagnetic field at nanometer scales. In fact, had we plotted the dispersion of light in the figure, and it would be barely resolved from the figure's vertical axis. This acknowledgement also makes our initial assumption of working in the quasi-static limit entirely justifiable, since retardation ef-

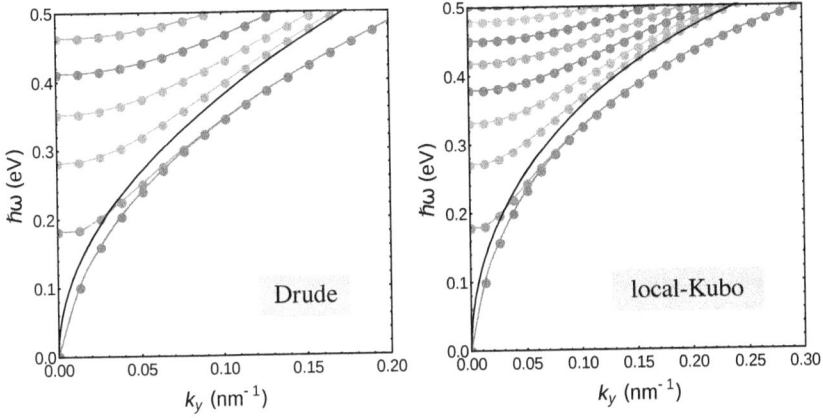

Figure 8.4: Spectra of plasmon excitations in an individual self-standing ($\epsilon_1 = \epsilon_2 = 1$) graphene nanoribbon with $E_F = 0.5$ eV. The ribbon's width is $w = 2a = 100$ nm. Left panel: dispersion of plasmon modes obtained using Drude's conductivity for graphene, Eq. (8.49); right panel: the same as in the left panel but now using the local-Kubo conductivity of graphene and solving Eq. (8.48) numerically. An electronic scattering rate of $\hbar\gamma = 3.7$ meV was also assumed in this latter calculation. Both results were obtained by truncating the eigenvalue problem [cf. Eq. (8.46)] by setting $N = 20$, which provides converged results for the depicted modes.

fects should be inconsequential here. Another relevant remark is to note that the electrostatic limit is scale-invariant, in the sense that an absolute length scale is absent (in this limit, $\lambda_0 \to \infty$). This can be easily seen by studying Eq. 8.48) [or, perhaps even more clearly, Eq. (8.49)], from which we recognize that we can obtain the plasmon's resonant frequency in terms of β; in other words, we only need to solve the eigenvalue problem, Eq. (8.46), once and we can then use this result to determine the plasmon dispersion curves for *any* ribbon width (provided that the electrostatic limit is fulfilled) [Thongrattanasiri *et al.* (2012b)].

In order to shed more light on our previous claim that to each branch of the plasmon spectrum there corresponds a plasmon standing wave along the perpendicular direction (i.e. x-direction), it is useful to determine the charge-density profile as a function of the x-coordinate. This may be done provided that we know the expansion coefficients A_m, which can be readily determined from the eigenvectors that were found when solving the eingensystem defined in Eq. (8.46). From here, it is possible to retrieve the carrier density by recalling, from Eq. (8.38), that

$$\rho(u) \propto \frac{\Im\left[\sigma_g(\omega)\right]}{\omega} \sum_{m=0}^{N-1} A_m \sqrt{\frac{2m+1}{2a}} \left(\beta^2 P_m(u) - \frac{\partial^2 P_m(u)}{\partial u^2}\right) , \quad (8.50)$$

where, as above, $u = x/a$ and $\beta = k_y a$, and such that the pair (k_y, ω) entering in this equation corresponds to a point belonging to a branch of the plasmon dispersion. The smaller panels at the bottom of Fig. 8.5 depict the density of charge-carriers for four such points (each indicated by an encircled "X" in the figure's top panel). Clearly, these can be ascribed to plasmonic excitations that, although propagating along $\hat{\mathbf{y}}$, mimick standing waves of oscillating charge along $\hat{\mathbf{x}}$. The insets in the bottom panels illustrate the (2D) monopolar, dipolar, quadrupolar and octupolar nature of the considered eigenmodes. Not surprisingly, to higher energy modes will correspond concomitant higher multipolar modes. Additionally, note that the relative density of charge-carriers tends to be higher at the edges of the ribbon. For this reason, these type of plasmonic modes are ocasionally referred to as *edge plasmons*.

Finally, we further remark that our semi-analytical results presented in this section —obtained by employing the self-consistent technique outlined above— are consistent with the computational work performed by [Thongrattanasiri *et al.* (2012b)] using the boundary element method (BEM) to solve Maxwell's equations numerically in a discretized space. Apart from the description of plasmon waveguide-like modes in a single graphene nanoribbon, [Thongrattanasiri *et al.* (2012b)] have also studied the hybridization of plasmons in configurations involving pairs of graphene ribbons.

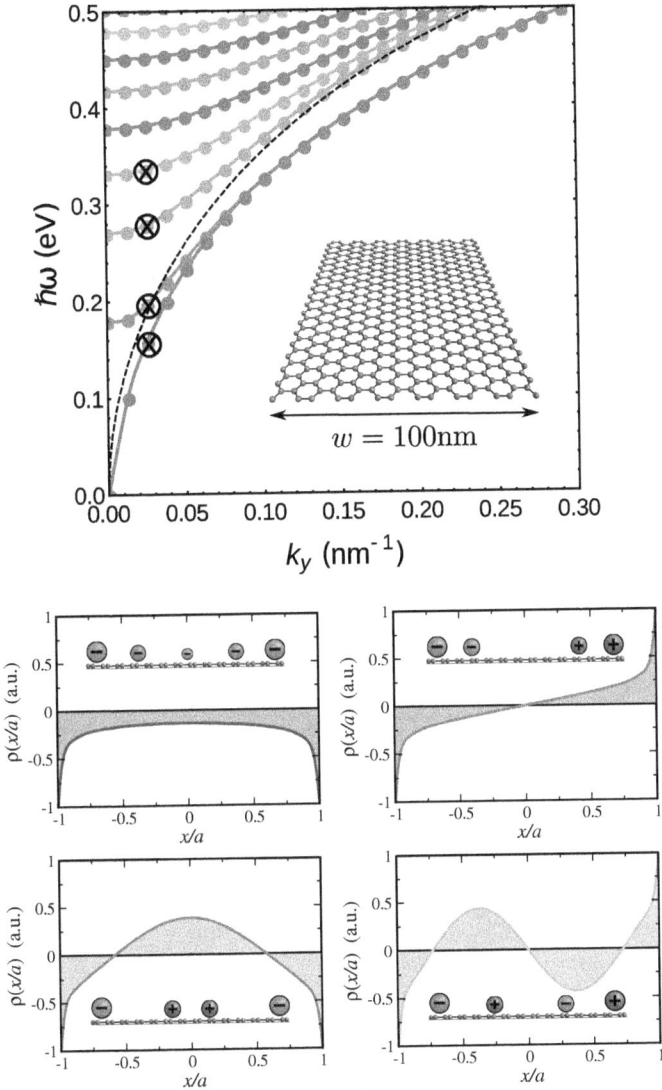

Figure 8.5: Spectra of plasmon excitations in an individual graphene nanorib-bon (top panel) and (normalized) charge-density profiles for the first four modes (four panels at the bottom) for the same $k_y \simeq 0.025$ nm^{-1}. The corresponding energies are $\approx$ 0.158, 0.197, 0.279, 0.336 eV —cf. points indicated by an en-circled "X" in the figure's top panel. The remaining panels below represent the charge-density profiles akin to these selected points. As the figure shows, these correspond to monopolar, dipolar, (linear) quadrupolar and (linear) octupolar modes. Parameters: the same as in the right panel of Fig. 8.4.

8.1.3   *An extension: magneto-plasmons in a graphene ribbon*

Before we move on to the description of plasmonic excitations in another geometries (namely, individual graphene rings and disks), let us present an oriented (but concise) walkthrough showing how one can extend this self-consistent quasi-static model to derive the dispersion of magneto-plasmons in graphene nanoribbons when a static magnetic field is applied perpendicularly to the system. This approach was first applied in the 1980s to the study of lateral surface magnetoplasma waves in semiconductor super-lattices and 2DEGs [Wu *et al.* (1985, 1986b,a)], and more recently also on graphene [Wang *et al.* (2012b,c)].

Recall from our discussion in Sec. 4.4 that in the presence of an uniform dc **B**-field, graphene's conductivity is no longer a scalar, and the linear response is now expressed in terms of a 2-ranked tensor instead —the magneto-optical conductivity tensor—, which in matrix form reads

$$\boldsymbol{\sigma}_g = \begin{pmatrix} \sigma_{xx} & \sigma_{xy} \\ \sigma_{yx} & \sigma_{yy} \end{pmatrix} = \begin{pmatrix} \sigma_{xx} & \sigma_{xy} \\ -\sigma_{xy} & \sigma_{xx} \end{pmatrix} , \tag{8.51}$$

where in the last step we have assumed that the graphene is isotropic. This is obviously not true for a nanostructure like a nanoribbon, but this should be an acceptable approximation for ribbons much smaller that the corresponding Fermi's wavelength.

We stress that the entire formalism developed up until Eq. (8.30) remains valid for the case with the magnetic field. However, in this case the electron number density will no longer be given by Eq. (8.38) due to the presence of the Hall current. Still, the relation that stems for the combination of the continuity equation with Ohm's law, that is,

$$n(x) = -\frac{1}{ie\omega} \nabla_{2D} \cdot \mathbf{J}(\mathbf{r}) , \tag{8.52}$$

is entirely general. The presence of the magnetic field will then be accounted by the 2D current, which follows from

$$J_\alpha = -\sum_\beta \sigma_{\alpha\beta} \frac{\partial \phi}{\partial \beta} , \tag{8.53}$$

where $\alpha, \beta \in \{x, y\}$. The term in the right-hand-side of Eq. (8.52) evaluates to

$$-\nabla_{2D} \cdot \mathbf{J}(\mathbf{r}) = \sigma_{xx} \left(\frac{\partial^2 \phi}{\partial x^2} - k_y^2 \phi \right) + \frac{\partial \sigma_{xx}}{\partial x} \frac{\partial \phi}{\partial x} + ik_y \phi \frac{\partial \sigma_{xy}}{\partial x} , \tag{8.54}$$

and thus, the particle density becomes

$$n(x) = \frac{1}{ie\omega} \left[\sigma_{xx} \left(\frac{\partial^2 \phi}{\partial x^2} - k_y^2 \phi \right) + \frac{\partial \sigma_{xx}}{\partial x} \frac{\partial \phi}{\partial x} + ik_y \phi \frac{\partial \sigma_{xy}}{\partial x} \right], \qquad (8.55)$$

where the novelty is in the last term [compare, for instance, with Eq. (8.38)]. This expression agrees with the one found in [Beletskii and Bludov (2005)] when dealing with magneto-plasmons in arrays of 2DEGs.

From here, the rest of the procedure is completely analogous to the case of zero-field that we have treated above. Hence, this straightforward extension allows the computation of the dispersion of magneto-plasmon modes in graphene nanostructures. These hybrid excitations were already theoretically investigated [Wang *et al.* (2012b)] and experimentally observed in graphene microrings [Yan *et al.* (2012c)].

8.1.4 *Scattering of THz radiation by a graphene micro-ribbon*

In this section we want to compute the scattering of THz radiation by an individual graphene micro-ribbon and show that surface plasmon-polaritons can be excited in graphene using this method. The main result of this section is an analytical expression for the scattering echo[11] of the ribbon. For our system we assume a ribbon lying in the $xz-$plane, upon which a plane wave propagating along $y-$direction impinges. We consider an infinite ribbon along the $z-$direction, characterized by a width w along the $x-$direction (see Fig. 8.6 for details). Notice the different orientation of the axis with relation to the previous subsection.

In contrast to what we have done in the preceding subsections, here we will work with both scalar and vector potentials, as retardation is included in the calculation. To that end, we note that the Lorentz gauge enforces the relation

$$\nabla \cdot \mathbf{A} + \frac{\epsilon}{c^2} \frac{\partial \phi}{\partial t} = 0, \qquad (8.56)$$

where $\mathbf{A}$ and ϕ are the vector and scalar potentials, respectively. In this gauge the potentials obey the following wave equations

$$\nabla^2 \mathbf{A} - \frac{\epsilon}{c^2} \frac{\partial^2 \mathbf{A}}{\partial t^2} = -\mu_0 \mathbf{J}, \qquad (8.57)$$

$$\nabla^2 \phi - \frac{\epsilon}{c^2} \frac{\partial^2 \phi}{\partial t^2} = -\frac{\rho}{\epsilon \epsilon_0}, \qquad (8.58)$$

---

[11]The scattering echo is nothing more than a two-dimensional cross-section.

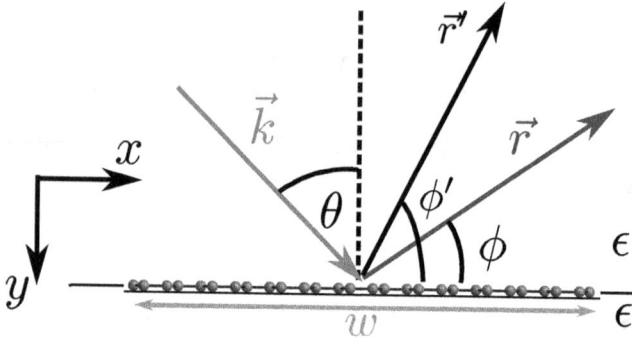

Figure 8.6: Scattering geometry for a plane wave carrying momentum **k**, and impinging on a graphene stripe at an angle of incidence θ. The angle ϕ is the observation angle (and **r** the "observation point"). The graphene ribbon has width w and is embedded in a homogeneous medium with relative permittivity ϵ. The vector **r′** and the angle ϕ' refer to the integration variables.

where **J** and ρ are the volume current and charge densities, respectively. From the charge conservation condition:

$$\nabla \cdot \mathbf{J} + \frac{\partial \rho}{\partial t} = 0 \tag{8.59}$$

we can define a single vector potential to describe both the electric and magnetic fields. At this point recall that

$$\mathbf{E} = -\frac{\partial \mathbf{A}}{\partial t} - \nabla \phi, \tag{8.60}$$

$$\mathbf{B} = \nabla \times \mathbf{A}. \tag{8.61}$$

If we now define a new vector potential $\vec{\pi}$ such that

$$\mathbf{A} = \frac{\epsilon}{c^2} \frac{\partial \vec{\pi}}{\partial t}, \tag{8.62}$$

$$\phi = -\nabla \cdot \vec{\pi}, \tag{8.63}$$

we obtain the following equations for the electric and magnetic fields:

$$\mathbf{E} = \nabla(\nabla \cdot \vec{\pi}) - \frac{\epsilon}{c^2} \frac{\partial^2 \vec{\pi}}{\partial t^2}, \tag{8.64}$$

$$\mathbf{B} = \frac{\epsilon}{c^2} \nabla \times \frac{\partial \vec{\pi}}{\partial t}, \tag{8.65}$$

where the new vector potential $\vec{\pi}$ is termed the *Hertz potential.* Combining the definition of the electric and magnetic fields in terms of the Hertz potential with the wave equation for the former fields, a (single) wave equation for the Hertz potential can be derived, yielding

$$\nabla^2 \vec{\pi} - \frac{\epsilon}{c^2}\frac{\partial^2 \vec{\pi}}{\partial t^2} = -\frac{\mathbf{P}}{\epsilon \epsilon_0}, \qquad (8.66)$$

where $\mathbf{J} = \partial \mathbf{P}/\partial t$. If we now introduce the time-frequency Fourier transform of the fields, Eqs. (8.64) and (8.65) read

$$\mathbf{E} = \nabla(\nabla \cdot \vec{\pi}) + \frac{\epsilon}{c^2}\omega^2 \vec{\pi}, \qquad (8.67)$$

$$\mathbf{B} = -i\omega\frac{\epsilon}{c^2}\nabla \times \vec{\pi}, \qquad (8.68)$$

where we have assumed a time dependence of the form $e^{-i\omega t}$ for the fields. Making the same assumption for the Hertz potential, i.e. Fourier transforming Eq. (8.66), we obtain

$$\nabla^2 \vec{\pi} + \frac{\epsilon}{c^2}\omega^2 \vec{\pi} = -\frac{i}{\omega \epsilon \epsilon_0}\mathbf{J}. \qquad (8.69)$$

In order to solve Eq. (8.69) we introduce the Green's function $G(\mathbf{r}, \mathbf{r}')$ through the differential equation

$$\nabla^2 G(\mathbf{r}, \mathbf{r}') + k^2 G(\mathbf{r}, \mathbf{r}') = -\delta(\mathbf{r} - \mathbf{r}'), \qquad (8.70)$$

with $k^2 = \epsilon \omega^2/c^2$. From the equation for the Green's function it is clear that the Hertz potential is obtained from

$$\vec{\pi} = \frac{i}{\omega \epsilon \epsilon_0}\int_V d\mathbf{r}' \mathbf{J}(\mathbf{r}') G(\mathbf{r}, \mathbf{r}'), \qquad (8.71)$$

where the integral extends over the volume where the current is finite. The Green's function for the Helmholtz equation in three dimensions is well known [Duffy (2015)], allowing us to write a closed-form expression for $\vec{\pi}$:

$$\vec{\pi} = \frac{i}{\omega \epsilon \epsilon_0}\int_V d\mathbf{r}' \mathbf{J}(\mathbf{r}')\frac{e^{ik|\mathbf{r}-\mathbf{r}'|}}{4\pi|\mathbf{r}-\mathbf{r}'|}. \qquad (8.72)$$

Let us now consider an TM-polarized wave impinging on a graphene ribbon of width w, where the magnetic field is given by[12]

$$\mathbf{B}_i = \hat{\mathbf{z}}B_0 e^{i\mathbf{k}\cdot\mathbf{r}}, \qquad (8.73)$$

with the wavevector of the incoming radiation reading

$$\mathbf{k} = \hat{\mathbf{x}}k\sin\theta + \hat{\mathbf{y}}k\cos\theta, \qquad (8.74)$$

---

[12]Recall that, for a TM wave, the magnetic field has only one component (which is perpendicular to the plane of incidence).

and where θ is the angle of incidence. The x-component of the incoming electric field follows from Maxwell's equations, and reads

$$E_{x,i} = -B_0 \frac{c^2}{\epsilon \omega} k_y e^{i\mathbf{k}\cdot\mathbf{r}} . \tag{8.75}$$

Since the problem has translational invariance along the z-direction and the ribbon is sitting in the $y = 0$ plane, the current density reads

$$\mathbf{J}(\mathbf{r}) = \hat{\mathbf{x}} K(x) \delta(y) , \tag{8.76}$$

where $K(x)$ is the surface current-density. Using this current in Eq. (8.72) it follows a simpler form for the x-component of the Hertz potential:

$$\pi_x = -\frac{1}{4\omega\epsilon\epsilon_0} \int_{-w/2}^{w/2} dx' K(x') H_0^{(1)}(k\sqrt{(x-x')^2 + y^2}) , \tag{8.77}$$

where we have introduce the following integral representation of the Hankel function

$$\int_{-\infty}^{\infty} dz' \frac{e^{ik\sqrt{(x-x')^2 + (z-z')^2 + y^2}}}{4\pi\sqrt{(x-x')^2 + (z-z')^2 + y^2}} = \frac{i}{4} H_0^{(1)}(k\sqrt{(x-x')^2 + y^2}) . \tag{8.78}$$

In possession of Eq. (8.77), it is now a simple matter to obtain $\mathbf{E}$ and $\mathbf{B}$ from Eq. (8.77) using Eqs. (8.67) and (8.68).

Let us now proceed with the calculation of the unknown current-density $K(x)$. To determine this quantity we need to use the boundary conditions at the graphene ribbon. The sough-after condition follows directly from Ohm's law, that is, the current $K(x)$ is induced by the field tangential to the graphene stripe (this is termed resistive boundary condition):

$$[\mathbf{E} - (\hat{\mathbf{n}} \cdot \mathbf{E})\hat{\mathbf{n}}] = \hat{\mathbf{x}} \frac{K(x)}{\sigma(\omega)} , \tag{8.79}$$

where $\hat{\mathbf{n}}$ is a unit vector normal to the graphene plane and $\sigma(\omega)$ is the conductivity of graphene in the local approximation. We can compute the scattered electric field $\mathbf{E}^s$ from Eq. (8.67) using Eq. (8.77) and obtain

$$E_x^s(x, y = 0) = E_x^s(x) = -\frac{\mu_0\omega}{4} \int_{-w/2}^{w/2} dx' K(x') H_0^{(1)}(k|x - x'|)$$

$$- \frac{1}{4\omega\epsilon\epsilon_0} \frac{\partial^2}{\partial x^2} \int_{-w/2}^{w/2} dx' K(x') H_0^{(1)}(k|x - x'|) , \tag{8.80}$$

for the x-component of $\mathbf{E}^s$. Considering both the incoming field (8.75) and the scattered field (8.80), the boundary condition (8.79) gives the following integral equation for the current $K(x)$

$$-\frac{B_0}{\mu_0}\cos\theta e^{ik_x x} = \sqrt{\epsilon}\frac{K(x)}{Z_0\sigma(\omega)} + \frac{\sqrt{\epsilon\omega}}{4c}\int_{-w/2}^{w/2}dx' K(x')H_0^{(1)}(k|x-x'|)$$

$$+\frac{c}{4\omega\sqrt{\epsilon}}\frac{\partial^2}{\partial x^2}\int_{-w/2}^{w/2}dx' K(x')H_0^{(1)}(k|x-x'|)\,, \qquad (8.81)$$

where $Z_0 = \sqrt{\mu_0/\epsilon_0}$ is the vacuum impedance and ϵ is the relative permittivity of the medium encapsulating the graphene ribbon. In principle, Eq. (8.81) has to be solved numerically by, say, the method of moments. However, in the limit $kw < 1$ we can invoke the edge condition [Rothwell and Cloud (2009)] and write the current $K(x)$ as[13]

$$K(x) = \chi e^{ik_x x}\sqrt{w^2/4 - x^2}\,, \qquad (8.82)$$

where χ is a constant to be determined using Eq. (8.81). As it is easy to understand, Eq. (8.82) is central in our argument. Indeed, the use of Eq. (8.82) simplifies the problem considerably avoiding time-consuming numerical calculations, such as the use of the FDTD method. Introducing Eq. (8.82) into Eq. (8.81) and integrating over x, it follows an algebraic equation for χ, given by

$$-B_0\cos\theta\frac{2}{\mu_0 k_x}\sin\frac{k_x w}{2} = \chi\frac{\sqrt{\epsilon}}{Z_0\sigma(\omega)}\frac{\pi w}{2k_x}J_1(k_x w/2)$$

$$+\chi\frac{\sqrt{\epsilon\omega}}{4c}\frac{w^3}{8}F(k_x w) + \chi\frac{c}{4\omega\sqrt{\epsilon}}\frac{w}{2}G(k_x w)\,, \qquad (8.83)$$

where $J_1(x)$ is the Bessel function and

$$F(kw) = \int_{-1}^{1}dy\int_{-1}^{1}dy'\sqrt{1-(y')^2}e^{ik_x wy'/2}H_0^{(1)}(kw|y-y'|/2)\,, \qquad (8.84)$$

$$G(kw) = \int_{-1}^{1}dy\frac{\partial^2}{\partial y^2}\int_{-1}^{1}dy'\sqrt{1-(y')^2}e^{ik_x wy'/2}H_0^{(1)}(kw|y-y'|/2)\,. \qquad (8.85)$$

Note that in $G(kw)$ the integration over y is elementary due to the presence of the second derivative of y followed by the use of the relation

$$\frac{dH_n^{(1)}(x)}{dx} = \frac{n}{x}H_n^{(1)}(x) - H_{n+1}^{(1)}(x)\,, \qquad (8.86)$$

---

[13]The edge condition states that the current perpendicular to a sharp edge is proportional to the square root of the distance, ρ, to the edge, that is, $K(\rho) \propto \sqrt{\rho}$. Our assumption is that in the condition $kw < 1$ we can interpolate the current by an expression that includes the edge condition at both edges, $x = \pm w/2$, simultaneously. In the case of a periodic grid of ribbons we have verified numerically that our assumption holds at every ribbon.

j	f_j (THz)	$E_{F,j}$ (eV)	f_j/f_1	$\sqrt{E_{F,j}/E_{F,1}}$
1	1.70	0.1	1	1
2	2.35	0.2	1.38	1.41
3	2.82	0.3	1.66	1.73
4	3.24	0.4	1.91	2.00

Table 8.1: Comparison of the ratio of the frequencies, $f_j = \omega_j/(2\pi)$, where the resonances present their maximum (see Fig. 8.7) with the square root of the ratio of the Fermi energies. The agreement between both quantities is quite good and shows that indeed the incoming radiation is exciting surface plasmons in the graphene stripe [recall that in graphene $f \propto \sqrt{E_F}$ —see Eq. (4.18)].

after the integration. Using Eq. (8.68) we can finally compute the scattered magnetic field, in the far-field, for the case of normal incidence on the graphene ribbon, as

$$B_z^s = -i \sin \phi \frac{\chi \mu_0 w}{8} \sqrt{\frac{2i\pi}{k}} \frac{J_1(kw \cos \phi/2)}{\cos \phi} \frac{e^{ikr}}{\sqrt{r}}, \qquad (8.87)$$

where ϕ is the scattering angle measured relatively to the x−axis (as in Fig. 8.6). Note that B_z^s is a cylindrical wave. The measurable physical quantity is the scattering echo, defined as [Rothwell and Cloud (2009)]

$$\sigma(\phi) = 2\pi r \left| \frac{B_z^s}{B_0} \right|^2, \qquad (8.88)$$

which has dimensions of length. The scattering echo is depicted in Fig. 8.7 as function of frequency, $f = \omega/(2\pi)$, of the incoming radiation for four different values of the Fermi energy. There is a clear dispersion of the resonance of the echo as function of E_F: we observe a blue-shift as E_F increases (red curve for $E_F = 0.1$ eV and blue curve for $E_F = 0.4$). If we plot the position of the resonance as function of the Fermi energy the resulting curve can be fitted by the function $\omega \propto E_F^{1/2}$ in agreement with the expected dispersion of SPP's in graphene. We can make this more explicit if we compute the ratio of f_j/f_1, with $j = 1,2,3,4$ [$f_j = \omega_j/(2\pi)$], corresponding to the Fermi energies of $E_F = 0.1, 0.2, 0.3, 0.4$ eV, respectively. This in done in Table 8.1.

From the previous analysis we conclude that not only it is possible to excite SPP's in a graphene micro-ribbon using a plane wave, but the scattering echo has its maximum at the energy of the SPP. The results of Fig. 8.7 should be compared with those of Fig. 7.4 for the reflectance (dashed lines) of a grid of graphene ribbons. In the bottom panel of Fig. 8.7

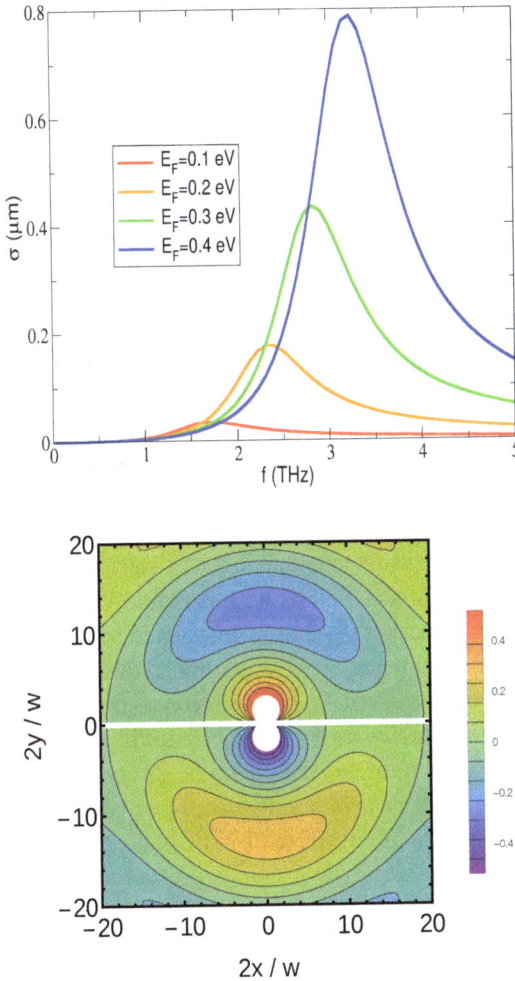

Figure 8.7: Top panel: Scattering echo measured in reflection ($\phi = \pi/2$) of a graphene micro-ribbon illuminated by a plane wave at normal incidence. The parameters are $\Gamma = 4.1$ meV (that is 1 THz), $w = 4$ μm, $\epsilon = 4$. The maximum of the echo disperses as function of E_F. To 5 THz —the maximum frequency considered in the figure— it corresponds a wavelength of 60 μm, thus $2\pi w/\lambda \approx$ 0.4, which is within the regime of validity of Eq. (8.82). Bottom panel: contour plot of the real part of the scattered field B_z^s at resonance for $E_F = 0.4$ eV; the parameters are the same as before. The axes are in units of $w/2$, therefore the ribbon is located in the interval $x = [-1, 1]$. Near the ribbon the field becomes more intense so that no contour lines were drawn.

the real part of B_z^s is depicted; the reader may recognize the field pattern as that produced by an oscillating Hertzian dipole.

The same ideas developed in this section were considered in the problem of a periodic array of graphene micro-ribbons. We can show that the behavior of such system is essentially controlled by the physics of a single ribbon as long as the filling factor, w/R, where R is the period of the array, is smaller than $1/2$. The main difference lies in the form we have chosen to write the fields, which in the periodic case was done in the Bloch representation (recall Chapter 7). Qualitatively the results are the same as those depicted in the top panel of Fig. 8.7.

8.2 Localized Surface Plasmons in a Graphene Ring

In what engineered graphene micro- and nano-structures are concerned, the second most popular configuration (after the ribbon(s) structure) is the circular disk/ring geometry. As for the ribbon case, these two-dimensional disks/rings —typically fabricated from patterning an otherwise homogeneous graphene sheet— were considered in the scientific literature either in a periodic crystal-like structures or as individual circles. In this section, we shall consider yet another geometry, in particular, that of an individual, symmetric graphene nanoring —see Fig. 8.8. Moreover, the treatment of this problem can be considered as an extension of the graphene-disk configuration, since one can arrive at the latter by taking the limit of $R_i \to 0$ in the former[14] (see Fig. 8.8 for the definition of R_i and other relevant parameters). As it will become clear in a moment, we will determine the spectrum of the plasmonic excitations supported by the nanoring using essentially the same technique considered in the previous section for the nanoribbon, but now applied to the ring geometry.

8.2.1 *Spectrum of graphene plasmons in a graphene ring*

Geometry plays a leading role in physics. For example, in the context of condensed matter physics the crystalline structure of solids and molecules largely determines their mechanical, electronic, and chemical properties. Also in plasmonics, the impact delivered by the geometrical properties of the material can affect its plasmonic properties to a great extent. Here,

---

[14]Although the procedure for obtaining the plasmon spectrum in a graphene disk is similar in every way to the one explained in what follows, there are a couple of differences, the most important one being the necessity of expanding the potential using a distinct set of polynomials. The disk geometry will be dealt with later in Sec. 8.3, where these nuances will be detailed.

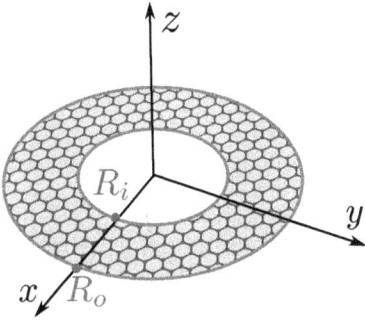

Figure 8.8: Circularly symmetric graphene nanoring embedded in a homogeneous dielectric medium with relative permittivity ϵ. The radius of the inner and outer boundaries are given by R_i and R_o, respectively. The graphene ring is located in the $z = 0$ plane.

we will tailor the theoretical model detailed in Sec. 8.1 to compute the dispersion of graphene plasmons in a graphene ring. As we will see, there will be a number of specificities in this case which were not present in the nanoribbon calculation, all of which can be ultimately be attributed to geometrical aspects.

Since we will be assuming structures much smaller than the light's wavelengths we shall work, as before, in the quasi-static limit. Therefore, we need to solve, as before, Poisson's equation:

$$\nabla^2 \Phi(\mathbf{r}) = -\frac{\rho(\mathbf{r})}{\epsilon_0} \, , \tag{8.89}$$

(assuming $\epsilon = 1$ for the time being) which has already been introduced in Eq. (8.1). The next step is to find the free-space Green's function that solves the homogeneous equation related to Eq. (8.89). This was already done in subsection 8.1.1 in the context of the graphene-ribbon geometry; however, that result is not appropriate for the present geometry, where symmetry dictates a different approach. Before anything else, notice that the system possesses axial symmetry (where the axis of symmetry is along the z-direction), so that one is led to (re)formulate Eq. (8.2) in cylindrical coordinates [Duffy (2015)],

$$\nabla^2 G(r, \theta, z; r', \theta', z') = -\frac{1}{\epsilon_0} \frac{\delta(r - r')\delta(\theta - \theta')\delta(z - z')}{r} \, , \tag{8.90}$$

in which the Laplacian reads

$$\nabla^2 = \frac{\partial^2}{\partial r^2} + \frac{1}{r}\frac{\partial}{\partial r} + \frac{1}{r^2}\frac{\partial^2}{\partial \theta^2} + \frac{\partial^2}{\partial z^2} \, . \tag{8.91}$$

Noting that the delta function with respect to the angular coordinate can be written as[Duffy (2015)]

$$\delta(\theta - \theta') = \frac{1}{2\pi} \sum_{l=-\infty}^{\infty} e^{il(\theta - \theta')} \ , \tag{8.92}$$

and decomposing the Green's function as [Duffy (2015)]

$$G(r, \theta, z; r', \theta', z') = \sum_{l=-\infty}^{\infty} g_l(r, z; r', z') e^{il(\theta - \theta')} \ , \tag{8.93}$$

(owing to the system's symmetry), then Eq. (8.90) transforms to

$$\frac{\partial^2 g_l}{\partial r^2} + \frac{1}{r} \frac{\partial g_l}{\partial r} + \frac{l^2}{r^2} g_l + \frac{\partial^2 g_l}{\partial z^2} = -\frac{1}{2\pi\epsilon_0 r} \delta(r - r') \delta(z - z') \ , \tag{8.94}$$

where we can already anticipate that "l" is an integer that will play the role of angular momentum.

Now, at this point, the typical procedure would be to take the (inverse) Fourier transform of the functions figuring in the above equation, thereby obtaining an expression for the Green's function in terms of its Fourier transform. Yet, it turns out that due to the circular symmetry of the problem, it is much more useful to use the Hankel transform instead, which is basically a one-dimensional transform with a (radial) kernel containing a Bessel function. In particular, the Hankel transform of order ν of a given function $f(r)$, is defined through [Bracewell (1999)]

$$f_\nu(k) = \int_0^\infty f(r) J_\nu(kr) r dr \ , \tag{8.95}$$

while its inverse reads

$$f_\nu(r) = \int_0^\infty f(k) J_\nu(kr) k dk \ . \tag{8.96}$$

In these equations, J_ν stands for the Bessel function of the first kind of order ν [with $\nu \geq -1/2$ in general (we will only be interested in $\nu \in \mathbb{N}_0$)]. Thus, writing the Green's function g_l in terms of its Hankel tranform

$$g_l(r, z; r', z') = \int_0^\infty G_l(k, z; r', z') J_l(kr) k dk \ , \tag{8.97}$$

and using the orthogonality relationship

$$\int_0^\infty J_l(kr') J_l(kr) k dk = \frac{\delta(r - r')}{r} \quad ; \ r, r' > 0 \ , \tag{8.98}$$

allow us to transform Eq. (8.94) into

$$G_l(k,z;r',z') \left[\frac{\partial^2}{\partial r^2} + \frac{1}{r}\frac{\partial}{\partial r} + \frac{l^2}{r^2} \right] J_l(kr) + J_l(kr)\frac{\partial^2}{\partial z^2}G_l(k,z;r',z')$$

$$= -\frac{1}{2\pi\epsilon_0}J_l(kr')J_l(kr)\delta(z-z') \ .$$

$$(8.99)$$

Recognizing that the Bessel differential equation for a given function $w(x)$ reads [Abramowitz and Stegun (1965)]

$$\frac{d^2w}{dx^2} + \frac{1}{x}\frac{dw}{dx} + \left(1 - \frac{l^2}{x^2}\right)w = 0 \ , \qquad (8.100)$$

then (since $w(x) = J_l(x)$ are obviously solutions of the above equation) one may write (making $x \to kr$)

$$\left[\frac{\partial^2}{\partial r^2} + \frac{1}{r}\frac{\partial}{\partial r} + \frac{l^2}{r^2} \right] J_l(kr) = -k^2 J_l(kr) \ , \qquad (8.101)$$

so that Eq. (8.99) conveniently simplifies to

$$\left[\frac{\partial^2}{\partial z^2} - k^2 \right] G_l(k,z;r',z') = -\frac{J_l(kr')}{2\pi\epsilon_0}\delta(z-z') \ . \qquad (8.102)$$

The solution of this differential equation is trivial, that is,

$$G_l(k,z;r',z') = -\frac{J_l(kr')}{4\pi\epsilon_0 k}e^{-k|z-z'|} \ . \qquad (8.103)$$

Recalling that our Green's function, in real space, follows from [cf. Eqs. (8.93) and (8.97)]

$$G(r,\theta,z;r',\theta',z') = \sum_{l=-\infty}^{\infty} g_l(r,z;r',z')e^{il(\theta-\theta')}$$

$$= \sum_{l=-\infty}^{\infty} e^{il(\theta-\theta')} \int_0^{\infty} G_l(k,z;r',z')J_l(kr)k\,dk \ , \quad (8.104)$$

upon which substituting the result (8.103), produces

$$G(r,\theta,z;r',\theta',z') = \frac{1}{4\pi\epsilon_0} \sum_{l=-\infty}^{\infty} e^{il(\theta-\theta')} \int_0^{\infty} dk\, J_l(kr')J_l(kr)e^{-k|z-z'|} \ ,$$

$$(8.105)$$

which is the soughted free-space Green's function; hence, this will enable us to find the potential that is solution of Poisson's equation in cylindrical coordinates, according to

$$\Phi(r,\theta,z) = \int_{R_i}^{R_o} r'dr' \int_0^{2\pi} d\theta' \int_{-\infty}^{\infty} dz'G(r,\theta,z;r',\theta',z')\rho(r',\theta',z')$$

$$= -e\int_{R_i}^{R_o} r'dr' \int_0^{2\pi} d\theta'G(r,\theta,z;r',\theta',0)n(r',\theta') \ , \qquad (8.106)$$

where we have used the relation $\rho(r',\theta',z') = -en(r',\theta')\delta(z')$ and explicitly assumed that the carrier-density is non-zero only in the interval $r' \in [R_i, R_o]$, that is, within the boundaries of the (doped) graphene ring.

Since the graphene ring is circularly symmetric, we can separate the angular part of the potential from the remaining spatial variables as (a time dependence of $e^{-i\omega t}$ is assumed throughout)

$$\Phi(r,\theta,z) = \sum_{l=-\infty}^{\infty} \phi_l(r,z)e^{il\theta}. \tag{8.107}$$

At the same time, the electron number density takes the form

$$n(r,\theta) = \sum_{l=-\infty}^{\infty} n_l(r)e^{il\theta}. \tag{8.108}$$

The next step involves the substitution of Eqs. (8.105), (8.107) and (8.108) into Eq. (8.106), which results in the following expression:

$$\sum_p \phi_p(r,z)e^{ip\theta} + \frac{e}{4\pi\epsilon_0}\int_{R_i}^{R_o} r'dr' \int_0^{2\pi} d\theta' \sum_{q,l} e^{il(\theta-\theta')}e^{iq\theta'}$$
$$\times \int_0^\infty dk J_l(kr')J_l(kr)n_q(r')e^{-k|z|} = 0. \tag{8.109}$$

On top of that, we further project the previous equation onto the subspace defined by the basis $e^{-in\theta}$, which results in[15]

$$2\pi\sum_p \phi_p(r,z)\delta_{pn} + \frac{e}{4\pi\epsilon_0}\int_{R_i}^{R_o} r'dr' \sum_{q,l} 4\pi^2\delta_{ln}\delta_{lq}$$
$$\times \int_0^\infty dk J_l(kr')J_l(kr)n_q(r')e^{-k|z|} = 0. \tag{8.110}$$

Finally, the Kronecker deltas effectively "kill" the sums (thus, only the "global" index n survives), granting us with an expression for the potential, i.e.

$$\phi_l(r,z) = -\frac{e}{2\epsilon_0}\int_{R_i}^{R_o} r'dr' \int_0^\infty dk J_l(kr')J_l(kr)n_l(r')e^{-k|z|}, \tag{8.111}$$

where we have renamed the index, $n \to l$, as the the letter "l" is usually assigned to angular momentum. In order to carry on with our computation, it is now necessary to determine the form of $n_l(r)$. To that end, we shall adopt the same line of thought as in the nanoribbon case that we treated in the Sec. 8.1, namely, by calling the continuity equation and Ohm's law.

---
[15]Notice that $\int_0^{2\pi} d\varphi\, e^{i\varphi(i-j)} = 2\pi\delta_{ij}$.

In this way, we can write the particle-density in terms of the potential [cf. Eq.(8.35)]:

$$n(r,\theta) = \frac{1}{ie\omega}\nabla_{2D} \cdot [\sigma(r)\nabla_{2D}\Phi(r,\theta,0)] \; , \qquad (8.112)$$

where $\nabla_{2D} \equiv \left(\frac{\partial}{\partial_r}, r^{-1}\frac{\partial}{\partial_\theta}\right)$ in cylindrical coordinates and the conductivity depends on distance r to the origin (more on this in what follows). The term on the ringht-hand-side of this equation can be readily evaluated, yielding (now for each l-th component)

$$n_l(r) = \frac{1}{ie\omega}\left[\sigma(r)\left(\frac{\partial^2\phi_l(r,0)}{\partial r^2} + \frac{1}{r}\frac{\partial\phi_l(r,0)}{\partial r} - \frac{l^2}{r^2}\phi_l(r,0)\right) + \frac{\partial\sigma(r)}{\partial r}\frac{\partial\phi_l(r,0)}{\partial r}\right] \; . $$
$$ (8.113) $$

Again, similarly to what we have done in Sec. 8.1, we shall assume that the radial dependence of the conductivity is accounted by a suitable envelope function $f(r)$, so that $\sigma(r) = \sigma_g f(r)$, where σ_g refers to the dynamical conductivity of bulk graphene. Henceforth, we take $f(r) = \Theta(r - R_i)\Theta(R_o - r)$ as portrayed in Fig. 8.9. Therefore, the radial derivative of $\sigma(r)$ entering in Eq. (8.113) becomes

$$\frac{\partial\sigma(r)}{\partial r} = \sigma_g(\omega)\left[\delta(r - R_i) - \delta(r - R_o)\right] \; , \qquad (8.114)$$

allowing us to write the potential, using the two previous equations in

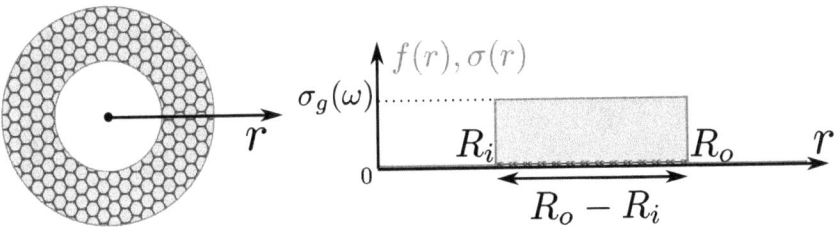

Figure 8.9: Conductivity's radial profile and envelope function, $f(r)$, for an individual graphene nanoring.

Eq. (8.111), as (at $z = 0$)

$$\phi_l(u) = \frac{\sigma_g(\omega)}{2i\omega\epsilon_0 R_o} \left\{ \int_s^1 du' \mathcal{K}_l(u, u') \left[\frac{l^2}{u'} \phi_l(u') - u' \frac{\partial^2 \phi_l(u')}{\partial u'^2} - \frac{\partial \phi_l(u')}{\partial u'} \right] \right. $$
$$\left. + \frac{\partial \phi_l(u')}{\partial u'} \Big|_{u'=1} \mathcal{K}_l(u, 1) - s \frac{\partial \phi_l(u')}{\partial u'} \Big|_{u'=s} \mathcal{K}_l(u, s) \right\},$$

(8.115)

where we have changed to a dimensionless variable according to $u = r/R_o$, and the integral kernel, $\mathcal{K}_l$, reads[16]

$$\mathcal{K}_l(u, u') = \int_0^\infty d\beta J_l(\beta u') J_l(\beta u) ,$$

(8.116)

with $\beta = kR_o$. Here, we have also introduced the scaling parameter $s = R_i/R_o$ relating the inner and outer radius of the graphene ring. Notice that $0 < s < 1$, and that to thicker rings correspond larger s ratios.

By inspection of Eq. (8.115), one notices that the expression for the electric potential depends on the potential itself. For this reason, that equation is known as a self-consistent integro-differential equation, being the analogue of Eq. (8.42), but now for the ring geometry. On these grounds, we shall employ the same strategy that we have used in Sec. 8.1 to solve Eq. (8.115) numerically, since an analytical solution cannot be carried out. Following [Wang (2012)], we introduce the variable $y = \frac{2u-(1+s)}{1-s}$ which makes the integration interval $u \in [s, 1]$ transform to $y \in [-1, 1]$. This transformation will demonstrate its utility shortly. In pursuit of the solutions of the above equation, we write the potential using the orthogonal polynomials expansion technique:

$$\phi_l(u) = \sum_{m=0}^\infty A_m \psi_m(y) ,$$

(8.117)

---

[16]It should be noted that this integral kernel as an analytical solution in terms of other special functions, namely

$$\mathcal{K}_l(u, v) = \frac{1}{\sqrt{\pi}} u^{-l-1} v^l \, \Gamma\left(l + \frac{1}{2}\right) {}_2\tilde{F}_1\left(\frac{1}{2}, l + \frac{1}{2}; l + 1; \frac{v^2}{u^2}\right)$$

for $u > v$; for $u < v$ we only need to exchange $u \leftrightarrow v$ in the previous equation. Here, $\Gamma(z)$ stands for the Gamma function and ${}_2\tilde{F}_1(a, b; c, z)$ is a regularized hypergeometric function. It is assumed that both variables are between 0 and 1 and that l is a non-negative integer.

where the ψ_m's are proportional to the Legendre's polynomials, P_m,

$$\psi_m(y) = \sqrt{\frac{2m+1}{2}} P_m(y) , \qquad (8.118)$$

thereby making them orthonormal (which also justifies the introduction of the variable y)

$$\int_{-1}^{1} \psi_i(y)\psi_j(y)dy = \delta_{ij} . \qquad (8.119)$$

Thus, expressing the potential figuring in Eq. (8.115 as a polynomial expansion (8.117), and then multiplying the resulting expression by $\psi_k(y)$ and integrating over the interval spanned by $y \in [-1, 1]$, one obtains

$$\boxed{\frac{2i\omega\epsilon_0 R_o}{\sigma_g(\omega)} A_k = \sum_{m=0}^{\infty} U_{km}^{\text{G-ring}} A_m ,} \qquad (8.120)$$

where the property 8.119 has been used, and the (rather gigantic) matrix elements $U_{km}^{\text{G-ring}}$ are defined as

$$U_{km}^{\text{G-ring}} = \int_{-1}^{1} dy \int_{-1}^{1} dy' \, \mathcal{K}_l(y, y')\psi_k(y) \left[\frac{l^2(1-s)}{y'(1-s)+1+s}\psi_m(y') \right.$$
$$- \frac{y'(1-s)+1+s}{1-s} \frac{\partial^2 \psi_m(y')}{\partial y'^2} - \frac{\partial \psi_m(y')}{\partial y'} \right]$$
$$+ \frac{2}{1-s} \left. \frac{\partial \psi_m(y')}{\partial y'} \right|_{y'=1} \int_{-1}^{1} dy \, \psi_k(y)\mathcal{K}_l(y, 1)$$
$$- \frac{2s}{1-s} \left. \frac{\partial \psi_m(y')}{\partial y'} \right|_{y'=-1} \int_{-1}^{1} dy \, \psi_k(y)\mathcal{K}_l(y, -1) . \qquad (8.121)$$

By applying this procedure we have transformed the integro-differential equation (8.115) into a regular eigenvalue equation (8.120), from which the eigenvalues (and corresponding eigenvectors) can be readily obtained using numerical means. From here, the plasmonic spectrum of a graphene nanoring (for a fixed s) follows in conformity with

$$\boxed{\frac{2i\omega\bar{\epsilon}\epsilon_0 R_o}{\sigma_g(\omega)} = \lambda_l ,} \qquad (8.122)$$

where the eigenvalue depends on the angular momentum l (and for each l there will be N corresponding eigenfrequencies, where $N - 1$ refers to largest integer that truncates the expansion for the potential), and we have also restored $\epsilon_0 \to \bar{\epsilon}\epsilon_0 = (\epsilon_1 + \epsilon_2)\epsilon_0/2$ to account for an arbitrary dielectric environment.

At this point it is important to highlight that in the quasi-static limit the value of the eigenvalues λ_l do not depend on the specific size of the graphene ring (e.g. R_o and/or R_i); in fact, the physical dimension of the system only appears in Eq. (8.122) as a proportional constant, whose function is only to adjudicate the scaling of the resonant frequencies with the system's dimension.

Furthermore, we note that depending on the framework in which the conductivity is calculated, Eq. (8.122) is, in general, a transcendental equation, and thus it is only solvable numerically (although such a computation is acessible enough). Nonetheless, if we choose to use Drude's expression with negligible damping for the conductivity of graphene, i.e. setting $\sigma_g(\omega) = \sigma_D(\omega)|_{\gamma \to 0}$ [cf. Eq. (2.72)], it becomes possible to derive from the result (8.122) an analytical expression for the spectrum of plasmons in a 2D graphene ring, reading

$$\boxed{\Omega_l = \Omega_{R_o} \sqrt{\frac{\lambda_l}{2\pi}}} \ , \tag{8.123}$$

where $\Omega_{R_o} = \sqrt{\frac{2\alpha}{\epsilon} E_F \hbar c \left(\frac{2\pi}{R_o}\right)}$ is energy of a plasmon carrying an in-plane momentum of $q = 2\pi/R_o$ in a homogeneous graphene sheet. In this latter equation, the electrostatic scaling law for the plasmon's dispersion is self-evident.

8.2.1.1 *Plasmonic spectrum of a graphene ring: results*

The four lowest eigenfrequencies coming out of Eq. (8.123), relative to the plasmonic excitations sustained by an individual graphene ring, are plotted in Fig. 8.10 for a fixed $s = 0.5$. Like in the case of the graphene ribbon, the spectrum of plasmons in the graphene ring is also comprised by a set of modes with threshold energies at $l = 0$ (excluding the mode with lowest eigenfrequency). Here, the angular momentum l plays a role analogous to the parallel momentum k_y in the ribbon case. However, the present case is distinct, since these excitations correspond to non-propagating plasmons which are bounded to the ring —*localized surface plasmons*— whereas for the ribbon geometry, plasmons with finite k_y travel through the ribbon along the direction parallel to its edges.

Among the set of modes visible in the figure, the dipolar mode, that is, the mode with $l = 1$, is arguably the most interesting, since it is the only one that strongly couples to radiation modes, and therefore it is also the only mode that can be optically excited using plane waves.

Having discussed the spectrum of localized plasmons in a graphene ring within the Drude regime, let us now investigate how this result compares to the one obtained numerically from the general expression (8.122) using the (local) conductivity of graphene within the framework of the Kubo formular, i.e. as given by Eqs. (2.70) and (2.71). Under these assumptions, the resulting dispersion relation of graphene plasmons supported by a 2D graphene ring, is represented in Fig. 8.11 (squares; solid lines). In the making of the figure, we have considered a graphene ring with an outer radius of $R_o = 100$ nm and ratio $s = 0.5$, and graphene is assumed to be doped so that $E_F = 0.5$ eV. Additionally, we have also included the plasmonic spectrum within Drude's approximation for the conductivity of graphene, which was already illustrated in Fig. 8.10, but now it was (re)calculated taking into account the specific parameters that we have just mentioned in the

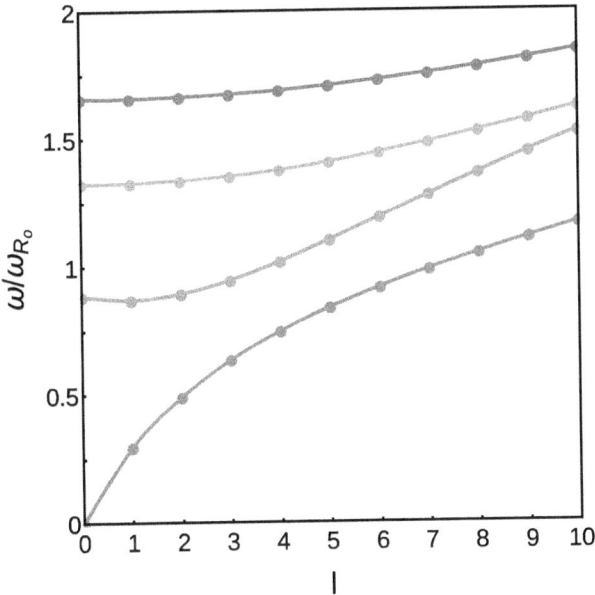

Figure 8.10: Spectrum of plasmonic excitations in an individual self-standing ($\bar{\epsilon} = 1$) graphene ring (four lowest eigenfrequencies) as a function of the angular momentum l, obtained via Eq. (8.123), for a fixed $s = 0.5 \Rightarrow R_o = 2R_i$. Notice that this remains valid for any R_o, obeying the (electrostatic) scaling law $\omega_l = \omega_{R_o}\sqrt{\lambda_l/(2\pi)}$, where $\omega_{R_o} = \Omega_{R_o}/\hbar$. We have used $N = 10$ modes for which the first four eigenmodes achieve satisfactory convergence (for higher-frequency modes, one would only need to use a larger N).

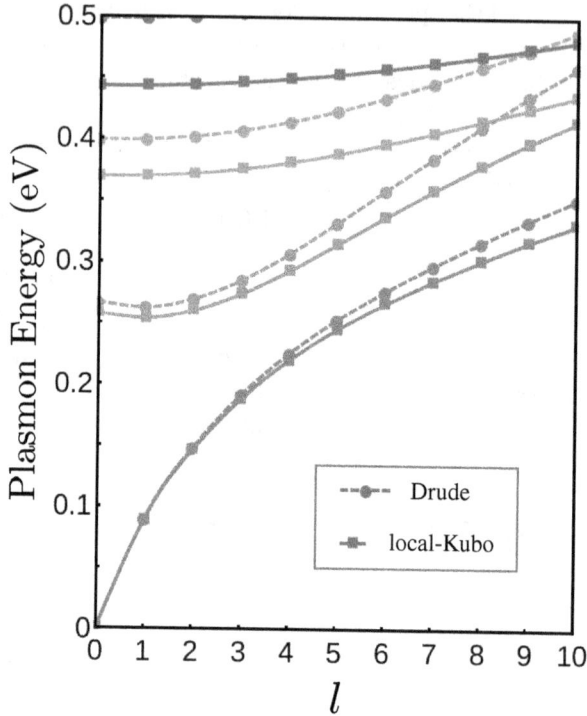

Figure 8.11: Dispersion of localized surface plasmons in a 2D graphene nanoring, with the first four eigenfrequencies akin to each l are shown. The points identified by a circle (and connected by a dashed line) correspond to the outcome of Eq. (8.123), that is, using Drude's approximation for the conductivity of graphene, whereas the points identified by the squares (and interconnected by a solid line) result from the numerical solution of Eq. (8.122) using the local-Kubo expression for the conductivity. Parameters: $R_o = 100$ nm, $s = 0.5$, $E_F = 0.5$ eV, $\bar{\epsilon} = 1$ and $N = 10$ for the matrix truncation. We note in passing that the mode $l = 0$ and $\omega = 0$ is not really a plasmonic mode, in the sense that the electronic density is homogeneous. For this reason, sometimes this point is not depicted in the literature.

previous sentence. Looking at Fig. 8.11 —and labeling each curve based on the approximation in which the conductivity is derived—, it is clear that, although both results exhibit the same qualitative behavior, quantitatively the curves deviate from each other, with the Drude's result systematically overestimating the plasmon energy. This effect becomes more prominent as we move up towards higher energies, where the interband contribution for the conductivity becomes non-zero. Since Drude's theory only accounts for intraband transitions, this must be the source of the observed differences. The plasmon energies akin to the first four dipolar ($l = 1$) eigenmodes in the

$\hbar\omega_l(n)$	local-Kubo (eV)	Drude (eV)
$\hbar\omega_1(1)$	0.089	0.089
$\hbar\omega_1(2)$	0.254	0.262
$\hbar\omega_1(3)$	0.370	0.399
$\hbar\omega_1(4)$	0.443	0.499

Table 8.2: Plasmon energies relative to the first four dipolar eigenmodes in a 2D graphene nanoring. The modes are labelled by the integer index n, in which $n = 1$ refers to the fundamental mode. These energies correspond to the points with $l = 1$ in Fig. 8.11.

doped graphene nanoring are shown in Table 8.2. From the table it is apparent that the differences between "Drude's energies" and their "Kubo's" counterparts increase as the energy goes up, being as large as ~ 55 meV for the fourth mode.

After studying the plasmonic spectrum, it is interesting to uncover how the density of charge-carriers is rearranged and oscillates in response to these excitations. This can be achieved by calling Eq. (8.113), from which we can write

$$\rho_{l=1}(u) \propto \frac{\Im m\{\sigma_g[\omega_1(n)]\}}{\omega_1(n)R_o^2} \sum_{m=0}^{N-1} A_m \sqrt{\frac{2m+1}{2}} \left(\frac{\partial^2 P_m[y(u)]}{\partial u^2} \right.$$
$$\left. + \frac{\partial P_m[y(u)]}{\partial u} - \frac{1}{u^2} P_m[y(u)] \right) \qquad (8.124)$$

which is valid for small damping, and where we have used the notation introduced in Table 8.2. Recall that the A_m's are determined from the solution of the eigenvalue problem (namely, from the eigenvectors) [cf. Eq. (8.120)]. The spatial distributions of the density of charges are plotted in Fig. 8.12 for the two lowest eigenfrequencies of angular momentum $l = 1$, as derived from the above expression. In resemblance with what we have observed in the case of a 2D graphene nanoribbon, here we also realize that the profile of the density of carriers is (approximately) *symmetric* for the fundamental dipolar mode, and (approximately) *antisymmetric* with opposite signs at the edges for the second lowest dipolar plasmon mode. The two above-mentioned modes can also be labeled as bonding (for the symmetric) and antibonding mode (for the antisymmetric) [Prodan *et al.* (2003)], and arise from the hybridization between charges at antipodal points. Notice, however, that here the symmetry point is not exactly equidistant from the inner and outer edges of the ring, and the absolute value of the charge-density is also distinct at each edge, i.e. $|\rho(r = R_i)| \neq |\rho(r = R_o)|$, both of which contrast to what we have verified for the ribbon geometry (cf. Fig. 8.5). In addition, it happens that the distance of this symmetry point with respect to the ring's edges, along with the difference in the modulus of ρ at the boundaries, are both strongly dependent on the ratio s [Wang (2012)], which essentially controls the width of the ring. Moreover, notice that the spatial profile of the charge-density distributions reverses at the other extremity of the nanoring, that is to say, after half a round-trip around the perimeter of the ring, or, else, after travelling the length of a circular arc

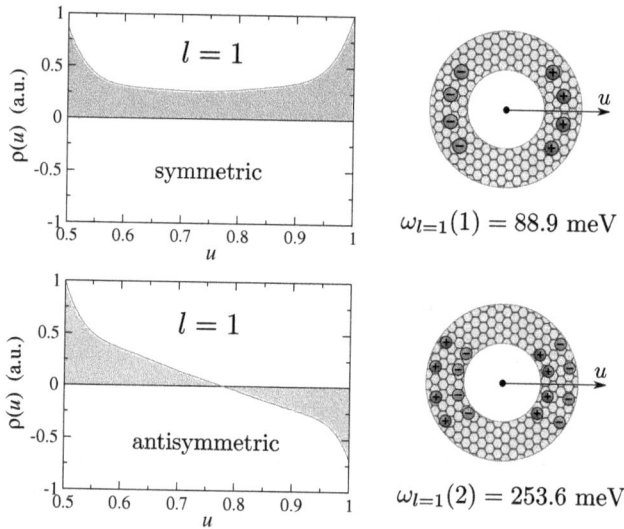

Charge-density spatial distributions for the first two dipolar eigenfrequencies.

Figure 8.12: Radial charge-density (normalized) distributions akin to the first two dipolar plasmon modes in a graphene nanoring (the dependence on the azimuthal angle θ is of the form $e^{il\theta}$). We have dubbed the fundamental mode as "symmetric" and the second dipolar eigenmode as "antisymmetric" (note that this is only approximately), owing to their spatial distributions of charge-carriers. In this figure, $\hbar = 1$. The parameters used in this computation are the same as in Fig. 8.11. Note that the figures for the charge density in the left panels refer to the case where the polar angle is zero.

subtended by an angle of π. Indeed, for an arbitrary value of angular momentum l, this occurs after a rotation by an angle of π/l, with nodal lines (in which $\rho \to 0$) at $\pi/(2l)$ (as measured from a maximum or minimum of ρ).

This analysis endows us with suficient intuition and should be enough to plot how the electrons in the 2D ring oscillate in response to a resonant excitation, for any eigenmode corresponding to any angular momentum l.

So far, we have analyzed how the plasmon's resonant frequencies change as a function of the angular momentum, for a fixed s parameter. Now, we shall study the opposite case, in which we fix the angular momentum and vary the quotient between the inner and outer radius of the ring, s. The out-

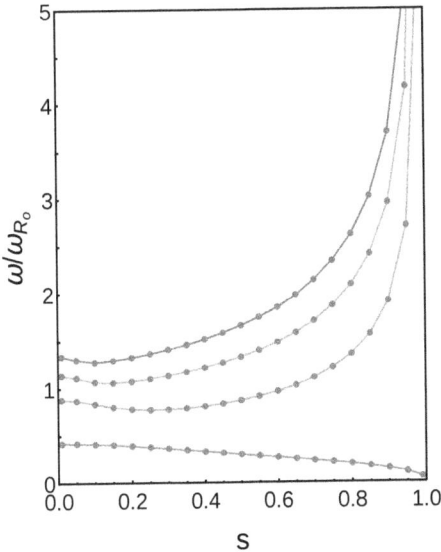

Figure 8.13: Plasmon eigenfrequencies akin to the dipolar mode ($l = 1$) in graphene rings with different s ratios. Only the first four eigenmodes are shown. The remaining parameters are the same as in Fig. 8.10.

come of such operation is displayed in Fig. 8.13, where we have picked the dipolar mode ($l = 1$) as it is the most relevant for optical experiments, being the only that is optically active. The figure shows a number of features, the most striking being the frequency-scaling with increasing s, for which the lowest frequency mode exhibits a different behavior with respect to the other modes. This eigenmode approaches a static frequency as the ratio s increases, whereas the higher-frequency modes seem to grow exponentially in frequency at larger s values. This behavior is natural, since these latter eigenmodes possess nodes within the ring; as a consequence, charges with opposite signs become increasingly squeezed into smaller and smaller areas as the ring's width diminishes (that is, as s increases), and thus the corresponding eigenfrequencies must increase. A note of caution is in order: for s-values close to either extreme of the interval $0 < s < 1$, the model can become inaccurate both owing to non-local effects and to quantum finite-size correction to the conductivity, including the dependence on the type of edge termination. Nevertheless, we expect that the model outlined in this section to be robust for typical s ratios. This claim is supported by benchmarking our quasi-analytical results with full-wave electromagnetic simulations [Wang (2012)], to which our results compare extremely well.

Let us close this section by noting that these hybrid plasmon resonances can be tuned by varying the width of the 2D graphene rings, even when the outer radius R_o is fixed. This supplementary "degree of freedoom", along with the possibility of controlling the Fermi level using a gate, can be put into use to uncover graphene plasmons with extreme tunability. From the experimental side, THz plasmonic excitations in arrays of graphene microrings have been observed by [Yan *et al.* (2012c)], which may constitute a tantalizing pathway towards plasmonic crystals using this kind of geometry.

8.3   Plasmonic Excitations in a Graphene Nanodisk

In classical metal-based plasmonics, 3D spherical nanoparticles have been dominating the ever-growing list of preferred nano-sized plasmonic structures. This is because nanoparticles with spherical symmetry are easy to manufacture. In what regards graphene plasmonics, the closest relative of such a structure is the 2D graphene disk. Like its 3D cousin, a doped 2D graphene nanodisk can also sustain multiple localized plasmon resonances. In traditional metallic nanoparticles, these are fixed by the particle's dimensions; however, in graphene disks, although they are largely determined by the disk radius, they can be tuned in real-time by changing graphene's Fermi level using a gate voltage. This type of active control cannot be achieved in an assemble of spherical nanoparticles.

The existence and tunability of localized plasmons in engineered arrays of graphene disks has been demonstrated in many experiments, in which the dipolar plasmon resonances where excited using plane waves [Yan *et al.* (2012c,a); Fang *et al.* (2014, 2013); Zhu *et al.* (2014)]. The aim of the current section is to derive the spectrum of plasmons in a graphene nanodisk —see Fig. 8.14. We shall do so using the quasi-analytical model that we

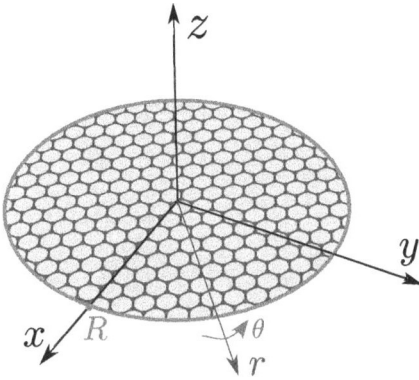

Figure 8.14: Circularly symmetric graphene nanodisk, of radius R embedded in a homogeneous dielectric medium. The graphene disk is located in the $z = 0$ plane.

outlined in the two previous sections for the ribbon and ring geometries. Since both the disk and ring structures exhibit cylindrical symmetry, it should come as no surprise that the derivation of the plasmon spectrum in a graphene disk will bear a strong resemblance with the procedure adopted in Sec. 8.2 for the graphene ring. For this reason, and in order to avoid repetition and duplicate results, we will only revisit briefly the model detailed in the aforementioned section and tailor it to the disk geometry. In reality, the disk can be thought as being the limiting case of a ring when $R_i \to 0$; while this is admittedly true, one cannot simply set $s = 0$ in the matrix elements $U_{km}^{\text{G-ring}}$ —cf. Eq. (8.121)— and then solve the eigenvalue problem (8.122) in that limit. That is because Legendre's polynomials are no longer an appropriate choice of basis for expanding the electric potential, since now the conductivity must be continuous at the origin (which would coincide with $r = R_i = 0$ in that case). Yet, prior to that step, the equations for the disk can be easily reformulated from the corresponding ring's expressions. In particular, it is apparent from Eq. (8.115) [for the ring] that the integro-differential equation for the in-plane (radial) potential akin to an individual graphene disk should be

$$
\phi_l(u) = \frac{\sigma_g(\omega)}{2i\omega\epsilon_0 R} \left\{ \int_0^1 du' \mathcal{K}_l(u,u') \left[\frac{l^2}{u'}\phi_l(u') - u'\frac{\partial^2 \phi_l(u')}{\partial u'^2} - \frac{\partial \phi_l(u')}{\partial u'} \right] \right.
$$
$$
\left. + \left. \frac{\partial \phi_l(u')}{\partial u'}\right|_{u'=1} \mathcal{K}_l(u,1) \right\},
$$

(8.125)

owing to the fact that in this case $\frac{\partial\sigma(r)}{\partial r} = -\sigma_g(\omega)\delta(r - R)$, so that the last term in Eq. (8.115) vanishes and the integral's lower limit changes to zero. Here, R designates the disk's radius.

Tantamount to what we did for the other geometries, we now seek an approximate solution for Eq. (8.125) by expanding the potential using a basis of orthogonal polynomials. For a 2D disk, the most appropriate expansion for the potential is of the form [Fetter (1986); Wang *et al.* (2011)]

$$
\phi_l(u) = \sum_{m=0}^{\infty} A_m \sqrt{2(2m + 1 + l)}\, \varphi_m^l(u) ,
$$

(8.126)

where the functions $\varphi_m^l(u)$ are characterized in terms of Jacobi's polynomials [Abramowitz and Stegun (1965)], $J_m^{(l,0)}$, that is

$$
\varphi_m^l(u) = u^l\, J_m^{(l,0)}(1 - 2u^2) ,
$$

(8.127)

which obey the orthogonality relation expressed by

$$
\int_0^1 du\, u^{2l+1} J_k^{(l,0)}(1 - 2u^2)\, J_m^{(l,0)}(1 - 2u^2) = \frac{1}{2}\frac{\delta_{km}}{2m + 1 + l} .
$$

(8.128)

In possession of Eqs. (8.126)–(8.128), the integro-differential equation (8.125) transforms into the eigenvalue matrix equation defined by

$$
\frac{2i\omega\bar{\epsilon}\epsilon_0 R}{\sigma_g(\omega)} A_k = \sum_{m=0}^{\infty} U_{km}^{\text{G-disk}} A_m ,
$$

(8.129)

with

$$
U_{km}^{\text{G-disk}} = C_{km} \left\{ \int_0^1 du \int_0^1 du'\, \mathcal{K}_l(u,u') u^{l+1} J_k^{(l,0)}(1 - 2u^2) \times \right.
$$
$$
\times \left[l^2 u'^{l-1} J_m^{(l,0)}(1 - 2u'^2) - u'\frac{\partial^2 \varphi_m^l(u')}{\partial u'^2} - \frac{\partial \varphi_m^l(u')}{\partial u'} \right]
$$
$$
\left. + \left.\frac{\partial \varphi_m^l(u')}{\partial u'}\right|_{u'=1} \int_0^1 du\, u^{l+1} J_k^{(l,0)}(1 - 2u^2)\mathcal{K}_l(u,1) \right\} ,
$$

(8.130)

where we have defined $C_{km} = 2\sqrt{(2k+1+l)(2m+1+l)}$ and performed the substitution $\epsilon_0 \to \bar{\epsilon} = (\epsilon_1 + \epsilon_2)/2$ to adequate the expression for the case of two different cladding dielectrics. The dispersion of plasmons in a 2D graphene disk then follows from the determined set of eigenvalues λ_l (for each angular momentum, l):

$$\lambda_l = \frac{2i\omega\bar{\epsilon}\epsilon_0 R}{\sigma_g(\omega)} \qquad \text{(general)} \qquad (8.131)$$

$$\Omega_l = \Omega_R \sqrt{\frac{\lambda_l}{2\pi}} \qquad \text{(Drude)} \qquad (8.132)$$

respectively for a graphene conductivity obtained in a given (local) approximation and in the Drude limit, and where the energy scale defined by $\Omega_R = \sqrt{\frac{2\alpha}{\bar{\epsilon}} E_F \hbar c \left(\frac{2\pi}{R}\right)}$ has been introduced.

The spectrum of plasmonic excitations supported by an individual 2D graphene disk is shown in Fig. 8.15 according to results coming out of Eqs. (8.131) and (8.132). In parallel with what we have obtained for the previous structures, the spectrum contains multiple branches akin to different multipole plasmon resonances. Among these, the dipolar plasmon resonances can be excited using plane waves, in which the charges oscillate in response to the driving electromagnetic field[17]. It is interesting to note that for the first few values of angular momentum, l, the lowest eigenfrequencies seem to follow the gray curve, which behaves as $\omega(l) = \omega_R \sqrt{l/2\pi}$ and represents the dispersion of a plasmon in a homogeneous (infinite) graphene sheet with in-plane momentum $q = l/R$. Notice the similarity of these with quantized circular orbits. Focusing on the first dipolar ($l = 1$) mode, it has a resonant frequency *approximately* equivalent to a plasmon with $q = R^{-1}$ (since $l = qR = 1$) propagating along an infinite graphene layer. Notwithstanding, this behavior is only approximate and we have found that this dipolar resonance occurs at $\omega \simeq 1.05\omega_R/\sqrt{2\pi}$, in conformity with the values reported in the literature [Wang *et al.* (2012b)].

We further stress that using this model one can determine the plasmon frequencies for graphene disks of *any* radius (within the range of validity of the quasi-static limit), due to the lack (in electrostatics) of another length-scale other than R, the latter fixing an overall energy scale only (if

---

[17]Modes with different azimuthal symmetry, i.e. $l \neq 1$, can still be optically excited, for example, by placing a metallic particle or a dipole in the vicinity of the disk [Silveiro and de Abajo (2014)].

retardation effects are allowed, another length scale given by c/ω, where c is the speed of light, emerges).

In Fig. 8.16 the plasmon energy of the lowest dipolar mode as a function of the disk radius for several doping concentrations (described by different Fermi energies). These calculations were carried out using Eq. (8.131) and Kubo's formula for the conductivity of graphene, and correspond to the colored dashed curves in the figure; the solid gray lines represent the scaling found by [Thongrattanasiri *et al.* (2012a)]. The figure shows a remarkable agreement between our results and the ones reported in the literature, thereby validating the employed model for describing localized plasmonic excitations in graphene nanodisks.

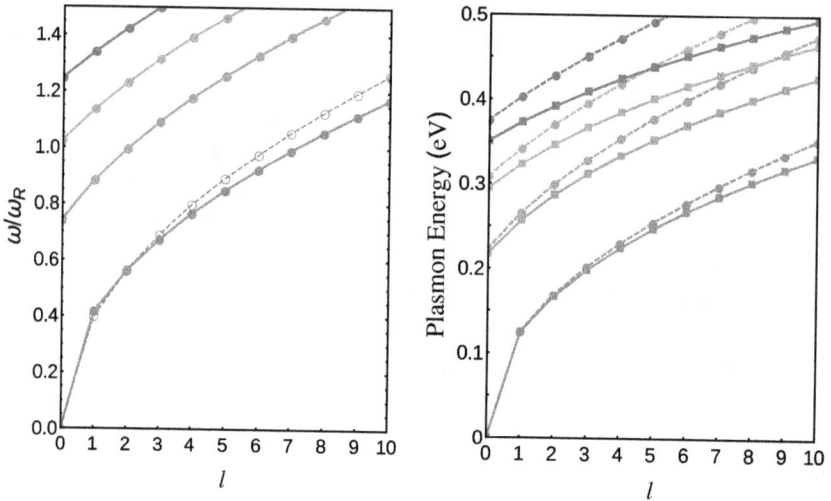

(a) Plasmonic resonances (normalized with respect to ω_R) of a graphene disk for different values of l, as given by Eq. (8.132). The gray circumferences correspond to a "bulk"-equivalent plasmon with momentum $q = l/R$, that is, represented by $\omega = \omega_R \sqrt{l/2\pi}$.

(b) Energies of the manifold of localized plasmon excitations in a graphene nanodisk of radius $R = 100$ nm and $E_F = 0.5$ eV, obtained via Eq. (8.131) using the Drude (circles joined by dashed lines) and local-Kubo conductivity (squares joined by solid lines).

Figure 8.15: Plasmonic spectrum of an individual graphene nanodisk as a function of the angular momentum, l. Only the four lowest eigenfrequencies (for each l) are shown. We have used $N = 12$ modes for the matrix truncation and assumed $\epsilon_1 = \epsilon_2 = 1$.

Having uncovered the plamon spectrum, let us now study how the plasmon-induced charge density oscillates under resonant excitation. As for the previously considered geometries, here we will also focus on the dipolar modes with the two lowest eigenfrequencies. The spatial (radial) charge-density distribution ascribed to those modes are depicted in Fig. 8.17. For the lowest order (dipolar) mode, the figure clearly demonstrates that the induced charge-density piles up near the egdes of the disks, with charges of opposite signs inhabiting antipodal edges. Also, we note that azimuthal symmetry, for these $l = 1$ modes, dictates an angular dependence of the form $e^{i\theta}$. On the other hand, although the plasmon-induced density for second order dipolar mode is also higher at the disk's edges, it has yet another regions, within the disk, where the carrier-density is higher, so that is looks essentially as a (linear) quadrupolar plasmon mode along the radial direction. Naturally, for higher-order modes the induced density of charges will resemble higher (linear) multipolar orders confined to the "potential well" that is the 2D graphene disk.

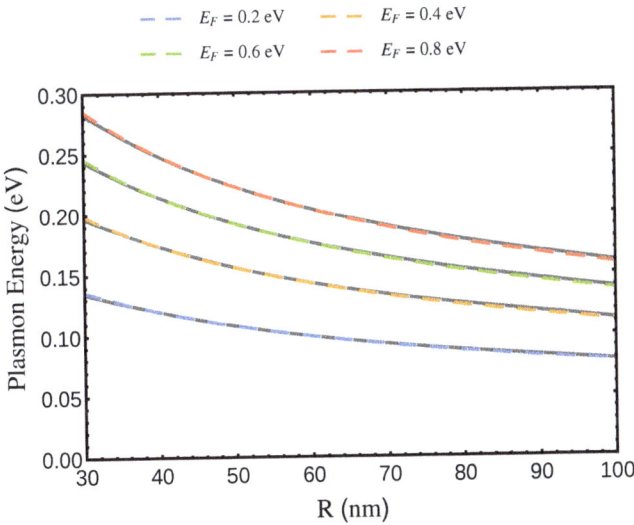

Figure 8.16: Dependence of the first dipolar resonance akin to localized plasmons in 2D graphene disks as a function of the disk's radius, for different doping levels. The colored dashed curves were retrieved from Eq. (8.131) using Kubo's formula for the conductivity of graphene with negligible damping, and the solid gray lines correspond to the scaling reported by [Thongrattanasiri et al. (2012a)].

To conclude the discussion of the plasmonic spectrum in individual graphene nanostructures, it is imperative to remark that the eigenvalue problem represented by Eqs. (8.46), (8.120) and (8.129) —respectively, for the ribbon, ring, and disk— is purely dependent on the structure's geometry; in particular, it does not depend neither on the system's response [i.e. $\sigma_g(\omega)$ does not influences the values of the computed eigenvalues/eigenvectors, and it only mediates the frequencies at which excitations do occur], nor on its dimensions. This can also be seen by realizing that Eqs. (8.46), (8.120) and (8.129) are completely analogous, only differing in the matrices $\mathbf{U}$ in which the geometrical form the system is accounted for.

Finally, a rather obvious extension of this self-consistent quasi-static model is to allow the vertical stacking of these graphene nanostructures separated by a dielectric layers. Cutting-edge experiments involving such

Charge-density spatial distributions for the first two dipolar eigenfrequencies

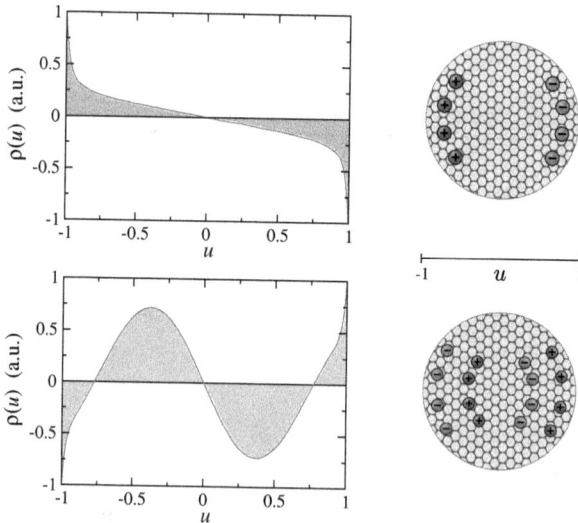

Figure 8.17: Normalized spatial charge-density distributions akin to the first two dipolar plasmon modes in a graphene nanodisk. Negative values of $u = r/R$ are to be understood as points after a π-rotation. We have only plotted the radial dependence since the dependence on the azimuthal angle θ is of the form $e^{il\theta}$. The parameters used in the making of this figure are the same as in Fig. 8.15 (under the Kubo's formula framework for graphene's conductivity).

vertical graphene/insulator stack have been pursued by [Yan *et al.* (2012a)]. This proof-of-concept experimental effort have demonstrated that these photonic-crystal-like plasmonic structures are viable platforms for tunable plasmonic devices. In fact, in that experiment an electromagnetic radiation shield, a tunable notch filter, and a tunable THz linear polarizer have been fabricated.

8.4 Scattering of Graphene Plasmons by a Conductivity Step

Incidentally, the determination of the electrostatic potential for the ribbon problem also allows to describe the scattering of a surface plasmon-polariton by a conductivity step. This problem has been considered by [Rejaei and Khavasi (2015)], using a slightly different approach. Let us assume a graphene sheet that has an inhomogeneous doping: n_L to the left of $x = 0$ and n_R to the right of the same point (eventually, such inhomogeneous doping can be induced by a split gate). As a consequence, the system will have two different conductivities, σ_L and σ_R, respectively at the left and at the right of $x = 0$; this forms a conductivity step. The geometry of the problem is depicted in Fig. 8.18: a surface plasmon-polariton impinges on the conductivity step located at $x = 0$ and can be either reflected, with wave number k_L, or transmitted, with wave number k_R. When $n_R = 0$ the reflection probability is equal to 1, as there are no electrons for $x > 0$ to support a surface plasmon-polariton. If we assume that some external

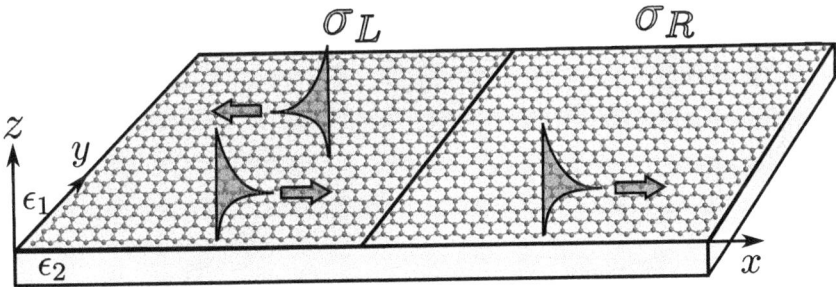

Figure 8.18: Pictorial representation of the scattering of graphene plasmons at sharp discontinuity in the conductivity. The incident plasmon is assumed to be travelling from the left. The parameters σ_L and σ_R indicate the conductivity of graphene at the left and right of the discontinuity, respectively.

source of field induces a surface plasmon-polariton propagating from left to right, the total current in the system is that associated with the propagating SPP plus a reflected and transmitted SPP at the conductivity step. From Eq. (8.30), and in the limit $a \to \infty$, we can describe the induced electrostatic potential created by external sources in the graphene as

$$\phi(x) = \frac{1}{2\pi\epsilon_0\epsilon} \int_{-\infty}^{\infty} dx' K_0(k_y|x - x'|)\rho(x'), \qquad (8.133)$$

where $\epsilon = (\epsilon_1 + \epsilon_2)/2$. It follows that the electric field of the surface plasmon-polaritons is given by

$$E_x(x) = -\frac{d\phi(x)}{dx}. \qquad (8.134)$$

Since we are interested in head-on collisions of the SPP in the conductivity step, we take the limit $k_y \to 0$. This implies that

$$K_0(k_y|x - x'|) \to -\gamma - \ln(k_y) - \ln(|x - x'|/2), \qquad (8.135)$$

where γ is the Euler's gamma-constant. From the previous result, it follows that

$$\frac{d}{dx}[K_0(k_y|x - x'|)] \to -\frac{1}{x - x'}. \qquad (8.136)$$

Therefore the electrostatic field due to the induced SPP's is given by

$$E_x(x) = \frac{1}{2\pi i\omega\epsilon_0\epsilon} \int_{-\infty}^{\infty} \frac{dx'}{x - x'} \frac{dJ_x(x')}{dx'}, \qquad (8.137)$$

where the continuity equation in the frequency domain

$$\rho(x) = \frac{1}{i\omega} \frac{dJ_x(x)}{dx} \qquad (8.138)$$

has been used. Equation (8.137) represents the field created by an inhomogeneous surface charge density $dJ_x(x)/dx$. Thus, the total current induced in the system is given by

$$\frac{J_x(x)}{\sigma(x)} = E_x^{ex}(x) + \frac{1}{2\pi i\omega\epsilon_0\epsilon} \int_{-\infty}^{\infty} \frac{dx'}{x - x'} \frac{dJ_x(x')}{dx'}, \qquad (8.139)$$

which is an integro-differential equation for $J_x(x)$, and $E_x^{ex}(x)$ describes the propagating surface plasmon-polariton impinging on the conductivity step. That is, Eq. (8.139) represents the current due to the external field (first term) and the current due to the field created by a surface charge density $dJ_x(x)/dx$. Let us first analyze the case of a graphene sheet with a homogeneous conductivity σ_L. Introducing the Fourier representation of the current

$$J_x(x) = \int_{-\infty}^{\infty} dq e^{iqx} J_x(q),$$
(8.140)

Eq. (8.139) can be written as

$$J_x(q) = \frac{2i\omega}{k_L - |q|} E_x^{ex}(q),$$
(8.141)

where

$$k_L = \frac{2i\omega\epsilon_0\epsilon}{\sigma_L}$$
(8.142)

is the plasmon wave number [cf. Eq. (4.15) when $c \to \infty$]. Thus, it follows from Eq. (8.141) that for a field describing a free propagating plasmon, that is $E_x(q) = E_0\delta(q-k_L)$, the current has a pole at $q = k_L$, which signals that a SPPs is a normal mode of the system. We note that in the derivation of Eq. (8.141) the following integral has been used:

$$\int_{-\infty}^{\infty} \frac{dx'}{x - x'} e^{iqx'} = -i\pi \operatorname{sgn}(q) e^{iqx}.$$
(8.143)

In what follows, we want to consider the problem of the conductivity step. To obtain, in Fourier space, the equation describing this problem we have to Fourier transform Eq. (8.139), taking into account that $\sigma(x) = \sigma_L\theta(-x) + \sigma_R\theta(x)$, where $\theta(x)$ is the Heaviside step function.

8.4.1   *The mathematical problem*

We now introduce the mathematical problem we need to solve in order compute the scattering coefficients for the surface plasmon-polariton impinging in the conductivity step. Therefore, the step-like form of $\sigma(x)$ naturally introduces the one-side Fourier transforms

$$J_L(q) = \int_{-\infty}^{0} \frac{dx}{2\pi} e^{-ixq} J(x),$$
(8.144)

and

$$J_R(q) = -\int_{0}^{\infty} \frac{dx}{2\pi} e^{-ixq} J(x),$$
(8.145)

thus $J(q) = J_L(q) - J_R(q)$, where $J(q)$ is the Fourier transform of $J(x)$. Letting $q = q' + iq''$, we see that $J_L(q)$ is analytic in the upper half of the complex plane, whereas $J_R(q)$ is analytic in the lower half of the same plane[18].

Fourier transforming Eq. (8.139) and using again Eq. (8.143) we obtain after some algebra

$$J_L(q) = \nu(q)J_R(q) + s(q),$$ (8.146)

where

$$\nu(q) = \frac{k_R - |q|}{k_L - |q|},$$ (8.147)

with $k_R = 2i\omega\epsilon_0\epsilon/\sigma_R$, and the source function reading

$$s(q) = \frac{k_L\sigma_L}{k_L - |q|} \int_{-\infty}^{\infty} \frac{dx}{2\pi} e^{-ixq} E^{ex}(x) = \frac{k_L\sigma_L}{k_L - |q|} E^{ex}(q),$$ (8.148)

which represents the Fourier transform of the impinging current on the conductivity step. The function $\nu(q)$ has zeros at $q = \pm k_R$ and poles at $q = \pm k_L$, where k_R and k_L are real positive numbers. Equation (8.146) defines a (non-normal) Riemann-Hilbert problem [Lu (1993); Ablowitz and Fokas (1997)] along the real axis, with $-\infty < q < \infty$. Using a more familiar notation, we rewrite Eq. (8.146) as

$$\Phi^+(q) = g(q)\Phi^-(q) + f(q),$$ (8.149)

where $\Phi^+(q) = J_L(q)$, $\Phi^-(q) = J_R(q)$, $g(q) = \nu(q)$, and $f(q) = s(q)$.

8.4.2   *Solution of the Riemann-Hilbert problem*

For solving the Riemann-Hilbert problem (8.149) we first consider the homogeneous problem [Lu (1993); Ablowitz and Fokas (1997)]:

$$X^+(q) = g(q)X^-(q).$$ (8.150)

It is known that its solution is given by [Lu (1993); Ablowitz and Fokas (1997)]

$$X^\pm(q) = e^{\Gamma^\pm(q)},$$ (8.151)

with

$$\Gamma^\pm(q) = \pm\frac{1}{2}\ln[g(q)] + \frac{1}{2\pi i}\mathcal{P}\int_{-\infty}^{\infty}\frac{\ln[g(q')]}{q' - q}dq',$$ (8.152)

where $\mathcal{P}$ stands for the principal value.

---

[18]The reader should recall at this point the definitions of retarded and advanced Green's functions in quantum field theory as well as their analytical properties.

Let us now consider the inhomogeneous Riemann-Hilbert problem defined by Eq. (8.149). We write it as

$$\Phi^+(q) = g(q)\Phi^-(q) + f(q) = \frac{X^+(q)}{X^-(q)}\Phi^-(q) + f(q). \tag{8.153}$$

Rearranging, we obtain

$$\frac{\Phi^+(q)}{X^+(q)} - \frac{\Phi^-(q)}{X^-(q)} = \frac{f(q)}{X^+(q)}, \tag{8.154}$$

which defines a jump-problem of the form[19]

$$\Psi^+(q) - \Psi^-(q) = \frac{f(q)}{X^+(q)}. \tag{8.155}$$

whose particular solution is given by the Plemelj formulae[Lu (1993)]:

$$\Psi^+(q) = \left(1 - \frac{\theta}{2\pi}\right)\frac{f(q)}{X^+(q)} + \frac{1}{2\pi i}\int_{-\infty}^{\infty}\frac{f(q')}{X^+(q')}\frac{dq'}{q'-q}, \tag{8.156}$$

and

$$\Psi^-(q) = -\frac{\theta}{2\pi}\frac{f(q)}{X^+(q)} + \frac{1}{2\pi i}\int_{-\infty}^{\infty}\frac{f(q')}{X^+(q')}\frac{dq'}{q'-q}, \tag{8.157}$$

where θ is defined as

$$\theta = [\arg g(q)]_L = [\ln g(q)]_L \tag{8.158}$$

and L is the arc coinciding with the real axis with $-\infty < q < \infty$. Finally, we have

$$\Phi^+(q) = X^+(q)\Psi^+(q) = \left(1 - \frac{\theta}{2\pi}\right)f(q) + \frac{X^+(q)}{2\pi i}\int_{-\infty}^{\infty}\frac{f(q')}{X^+(q')}\frac{dq'}{q'-q}, \tag{8.159}$$

and

$$\Phi^-(q) = X^-(q)\Psi^-(q) = -\frac{\theta}{2\pi}\frac{f(q)X^-(q)}{X^+(q)} + \frac{X^-(q)}{2\pi i}\int_{-\infty}^{\infty}\frac{f(q')}{X^+(q')}\frac{dq'}{q'-q}. \tag{8.160}$$

Reverting to the original notation, and noting that $\theta = \pi$, we have

$$J_L(q) = \frac{s(q)}{2} + \frac{X^+(q)}{2\pi i}\int_{-\infty}^{\infty}\frac{s(q')}{X^+(q')}\frac{dq'}{q'-q}, \tag{8.161}$$

and

$$J_R(q) = -\frac{s(q)}{2\nu(q)} + \frac{X^-(q)}{2\pi i}\int_{-\infty}^{\infty}\frac{s(q')}{X^+(q')}\frac{dq'}{q'-q}. \tag{8.162}$$

Equations (8.161) and (8.162) are the formal solutions of the Riemann-Hilbert problem.

---

[19]We recognize in Eq. (8.155), and in the context of many-body physics, the difference between retarded and advanced functions, which equals the spectral function.

8.4.3   *Further details*

To proceed further, we need to compute the integral in the $\Gamma^{\pm}(q)$ function. First we note that the real function $\nu(q)$ is negative in the intervals $q \in\ ]-k_L, -k_R]$ and $q \in [k_R, k_L[$. From here on we assume $k_L > k_R$, without loss of generality. Thus, the function

$$\ln[\nu(q)] = \ln \frac{k_R - |q|}{k_L - |q|}, \tag{8.163}$$

has an imaginary part equal to $i\pi$ in those two intervals. Using this result we write [from Eq. (8.152)]

$$\frac{1}{2\pi i} \mathcal{P} \int_{-\infty}^{\infty} \frac{dq'}{q' - q} \ln \frac{k_R - |q'|}{k_L - |q'|} = \frac{1}{2} \mathcal{P} \int_{-k_L}^{-k_R} \frac{dq'}{q' - q} + \frac{1}{2} \mathcal{P} \int_{k_R}^{k_L} \frac{dq'}{q' - q}$$

$$+ \frac{1}{2\pi i} \mathcal{P} \int_{-\infty}^{\infty} \ln \left| \frac{k_R - |q'|}{k_L - |q'|} \right| \frac{dq'}{q' - q}. \tag{8.164}$$

The first two integrals in the previous equation are elementary, reading

$$\mathcal{P} \int_{-k_L}^{-k_R} \frac{dq'}{q' - q} = \ln \left| \frac{k_R + q}{k_L + q} \right|, \tag{8.165}$$

$$\mathcal{P} \int_{k_R}^{k_L} \frac{dq'}{q' - q} = \ln \left| \frac{k_L - q}{k_R - q} \right|. \tag{8.166}$$

The last integral in Eq. (8.164) can be transformed as

$$\frac{1}{2\pi i} \mathcal{P} \int_{-\infty}^{\infty} \ln \left| \frac{k_R - |q'|}{k_L - |q'|} \right| \frac{dq'}{q' - q} = \frac{1}{2\pi i} \mathcal{P} \int_{-\infty}^{0} \ln \left| \frac{k_R + q'}{k_L + q'} \right| \frac{dq'}{q' - q}$$

$$+ \frac{1}{2\pi i} \mathcal{P} \int_{0}^{\infty} \ln \left| \frac{k_L - q'}{k_L - q'} \right| \frac{dq'}{q' - q}$$

$$= -i \frac{q}{\pi} \mathcal{P} \int_{0}^{\infty} \ln \left| \frac{k_L - q'}{k_L - q'} \right| \frac{dq'}{(q')^2 - q^2}. \tag{8.167}$$

Denoting the last integral by,

$$\beta(q) = -\frac{q}{\pi} \mathcal{P} \int_{0}^{\infty} \ln \left| \frac{k_R - q'}{k_L - q'} \right| \frac{dq'}{(q')^2 - q^2} \tag{8.168}$$

we can show that the integration path can be rotated to the imaginary axis reading

$$\beta(q) = \frac{q}{\pi} \int_{0}^{\infty} dy \left[\arctan \frac{y}{k_R} - \arctan \frac{y}{k_L} \right] \frac{1}{y^2 + q^2}, \tag{8.169}$$

which is convenient for numerical calculations. Finally, Eq. (8.164) can be written as

$$\frac{1}{2\pi i}\mathcal{P}\int_{-\infty}^{\infty}\frac{dq'}{q'-q}\ln\frac{k_R-|q'|}{k_L-|q'|}=\frac{1}{2}\ln\left|\frac{k_R+q}{k_L+q}\right|+\frac{1}{2}\ln\left|\frac{k_L-q}{k_R-q}\right|+i\beta(q)\,.$$

(8.170)

Using the previous result, the $\Gamma^\pm(q)$ function reads

$$\Gamma^\pm(q)=\pm\frac{1}{2}\ln\frac{k_R-|q|}{k_L-|q|}+\frac{1}{2}\ln\left|\frac{k_R+q}{k_L+q}\right|+\frac{1}{2}\ln\left|\frac{k_L-q}{k_R-q}\right|+i\beta(q)\,,\quad(8.171)$$

and the function $X^+(q)$ is given by

$$X^+(q)=e^{i\beta(q)}e^{i\tilde\pi/2}\sqrt{\left|\frac{k_R-|q|}{k_L-|q|}\right|}\sqrt{\left|\frac{k_R+q}{k_L+q}\right|}\sqrt{\left|\frac{k_L-q}{k_R-q}\right|}\,,\qquad(8.172)$$

where

$$\tilde\pi=\pi\quad\text{if}\quad q\in]-k_L,-k_R]\quad\text{or}\quad q\in[k_R,k_L[\,,\qquad(8.173)$$

and $\tilde\pi=0$ otherwise. We note that

$$X^+(k_L)=\sqrt{\frac{k_R+k_L}{2k_L}}e^{i\beta(k_L)}e^{i\tilde\pi/2}\,,\qquad(8.174)$$

a result used ahead.

8.4.4 *Calculation of the reflection amplitude*

From the solution of Eq. (8.139) for a continuous sheet of graphene with conductivity σ_L we found that

$$J_x^0(q)=\frac{k_L\sigma_L}{k_L-|q|}E^{ex}(q)\,.\qquad(8.175)$$

Let us now assume that the external field propagating on left part of the graphene sheet and far away from $x=0$, created by some external source, induces a propagating current of the form $J^{ex}(x)=J_0e^{ik_Lx}$. In this case, the Fourier transform reads

$$J^{ex}(q)=J_x^0(q)=J_0\delta(q-k_L)\,.\qquad(8.176)$$

This result makes the evaluation of the integrals in Eqs. (8.161) and (8.162) elementary, since

$$s(q)=J_x^0(q)=J_0\delta(q-k_L)\,,\qquad(8.177)$$

as follows from the interpretation of Eq. (8.148). The Fourier transform of the current, $J_L(q)$, thus reads

$$
\begin{aligned}
J_L(q) &= \frac{s(q)}{2} + \frac{X^+(q)}{2\pi i} \frac{J_0}{X^+(k_L)} \frac{1}{k_L - q}, \\
&= \frac{s(q)}{2} + J_0 \frac{X^+(q) - X^+(k_L) + X^+(k_L)}{2\pi i X^+(k_L)(k_L - q)}, \\
&= \frac{s(q)}{2} + \frac{1}{2\pi i} \frac{J_0}{k_L - q} + J_0 \frac{X^+(q) - X^+(k_L)}{2\pi i X^+(k_L)(k_L - q)}.
\end{aligned} \tag{8.178}
$$

Transforming $J_L(q)$ to real space, the current at the left of the conductivity step reads

$$
\begin{aligned}
J_L(x) &= \int_{-\infty}^{\infty} dq J_L(q) e^{iqx} = \frac{1}{2} \int_{-\infty}^{\infty} dq e^{ikq} J_0 \delta(q - k_L) + \frac{1}{2\pi i} \int_{-\infty}^{\infty} dq \frac{J_0 e^{ikq}}{k_L - q} \\
&\quad + \frac{J_0}{2\pi i} \int_{-\infty}^{\infty} dq e^{ikq} \frac{X^+(q) - X^+(k_L)}{X^+(k_L)(k_L - q)}, \\
&= J_0 e^{ik_L x} + \frac{J_0}{2\pi i} \int_{-\infty}^{\infty} dq e^{ikq} \frac{X^+(q) - X^+(k_L)}{X^+(k_L)(k_L - q)}.
\end{aligned} \tag{8.179}
$$

We note that the function

$$
\frac{X^+(q) - X^+(k_L)}{X^+(k_L)(k_L - q)}
$$

has a pole at $q = -k_L$ with residue, $R(-k_L)$, given by

$$
R(-k_L) = \lim_{q \to -k_L} (q + k_L) \frac{X^+(q) - X^+(k_L)}{X^+(k_L)(k_L - q)} = e^{-i2\beta(k_L)} \frac{k_L - k_R}{k_L + k_R}, \tag{8.180}
$$

from which follows that

$$
J_L(x) = J_0 e^{ik_L x} + e^{-i2\beta(k_L)} \frac{k_L - k_R}{k_L + k_R} J_0 e^{-ik_L x}. \tag{8.181}
$$

The previous expression allows to obtain the reflection amplitude as[20]

$$
r = e^{-i2\beta(k_L)} \frac{k_L - k_R}{k_L + k_R} \Rightarrow \mathcal{R} = |r|^2 = \frac{(k_L - k_R)^2}{(k_L + k_R)^2}, \tag{8.182}
$$

which implies that the transmission probability is given by

$$
\mathcal{T} = \frac{4 k_L k_R}{(k_L + k_R)^2}. \tag{8.183}
$$

We also note that in the limit $k_R \to 0$ we have total reflection. This makes sense, since there are no charges at the right of $x = 0$ to form the

---

[20]It has been suggested that this result may be obtained by imposing the continuity of both the electric current and the electromagnetic flux at $x = 0$ and $z = 0$ (this comment is due to Yuliy Bludov —private communication).

SPP. In this case, the phase $-2\beta(k_L) = -\pi/4$, differing from the value $-\pi$ characterizing the collision of an electron on a hard-hall. In Fig. 8.19 we depict the probabilities of reflection and transmission as function of the Fermi energy, $E_{F,R}$, of the right graphene sheet $(x > 0)$, for three different frequencies, keeping constant the $E_{F,L}$=0.4 eV. As expected, when $E_{F,L} = E_{F,R}$ the transmission is equal to 1. In Fig. 8.20 we depict an intensity plot of the transmission (left panel) and reflectance (right panel) as function of frequency and Fermi energy of the right graphene sheet, $E_{F,R}$. It is clear that both the transmittance and the reflectance are essentially independent of the frequency. This is corroborated by the results of Fig. 8.19.

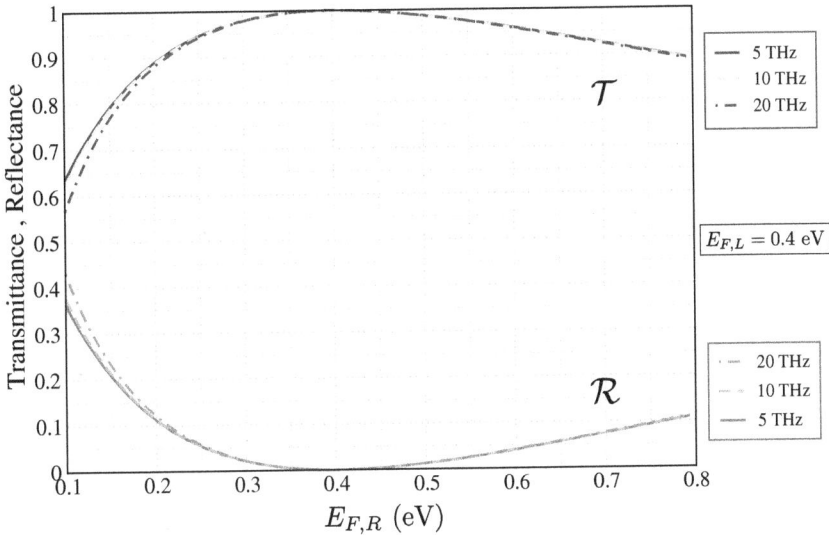

Figure 8.19: Transmittance and reflectance of GSPs scattered by a sharp discontinuity in graphene's conductivity, for different plasmon frequencies, as a function of the Fermi energy akin to the graphene on the right, $E_{F,R}$. The Fermi level corresponding to the patch of graphene on the left is fixed at $E_{F,L} = 0.4$ eV.

8.4.5   An alternative derivation of Eq. (8.137)

Here we provide an alternative derivation of Eq. (8.137). As before we assume a finite stripe of width $2a$, oriented along the $y-$axis; in the end we take the limit $a \to \infty$. Since both the current and the charge are confined

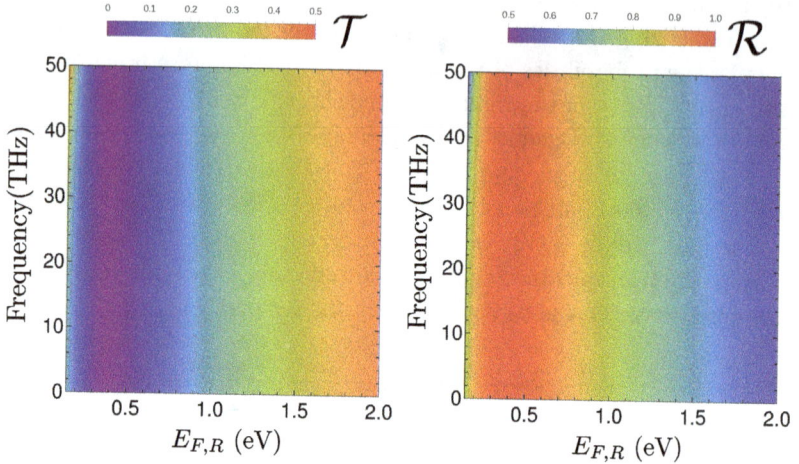

Figure 8.20: Dependence of the scattering probabilities —i.e. transmittance ($\mathcal{T}$) and reflectance ($\mathcal{R}$)— on the frequency and conductivity-mismatch for graphene plasmons encountering a sharp discontinuity in the conductivity. The semi-infinite region at the left has a fixed doping concentration corresponding to $E_{F,L} = 0.4$ eV.

to the $xy-$plane, we write the current and the charge density as

$$j_x(\mathbf{r}_\|, z, t) = J_s(\mathbf{r}_\|, t)\delta(z) \,, \tag{8.184}$$

$$\rho(\mathbf{r}_\|, z, t) = \sigma_s(\mathbf{r}_\|, t)\delta(z) \,, \tag{8.185}$$

where $J_s(\mathbf{r}_\|, t)$ and $\sigma_s(\mathbf{r}_\|, t)$ are the surface current and surface charge densities, respectively, and $\mathbf{r}_\| = (x, y)$. In Fourier space we have the additional relation between the charge and the current surface densities

$$i\omega\sigma_s(x, \omega) = \frac{dJ_s(x, \omega)}{dx} \,, \tag{8.186}$$

which originates from the continuity equation. Translational invariance along the $y-$direction makes the charge and the current surface densities independent of y, except for a possible phase $e^{ik_y y}$, which we take equal to one ($k_y = 0$). In the Lorentz gauge (v is the speed of light in a medium characterized by ϵ and μ),

$$\nabla \cdot \mathbf{A}(\mathbf{r}_\|, z, t) + \frac{1}{v^2}\frac{\partial\phi(\mathbf{r}_\|, z, t)}{\partial t} = 0 \,, \tag{8.187}$$

the scalar and vector potentials, $\phi(\mathbf{r}_\|, z, t)$ and $\mathbf{A}(\mathbf{r}_\|, z, t)$, are given by

$$\mathbf{A}(\mathbf{r}_\|, z, t) = \mu_0 \mu \int \frac{d^3 k}{(2\pi)^2} \frac{d\omega}{2\pi} G(\mathbf{k}, \omega) \mathbf{j}(\mathbf{k}, \omega) e^{i(\mathbf{k}\cdot\mathbf{r} - \omega t)}, \quad (8.188)$$

$$\phi(\mathbf{r}_\|, z, t) = \frac{1}{\epsilon_0 \epsilon} \int \frac{d^3 k}{(2\pi)^2} \frac{d\omega}{2\pi} G(\mathbf{k}, \omega) \rho(\mathbf{k}, \omega) e^{i(\mathbf{k}\cdot\mathbf{r} - \omega t)}, \quad (8.189)$$

where $G(\mathbf{k}, \omega) = 1/[k^2 - (\omega + i0^+)^2/v^2]$ is the free-space Green's function in momentum space, obtained from the Fourier transform of

$$[\nabla^2 + \omega^2/v^2] G(\mathbf{r}, \omega) = -\delta(\mathbf{r}), \quad (8.190)$$

and $\mu_0 \mu$ and $\epsilon_0 \epsilon$ are the medium permeability and permittivity, respectively, and $v = \sqrt{1/(\mu_0 \mu \epsilon_0 \epsilon)}$. The current density $\mathbf{j}(\mathbf{k}, \omega)$ and the charge density $\rho(\mathbf{k}, \omega)$ are defined in terms of the Fourier transform of the surface current and the surface charge as

$$\mathbf{j}(\mathbf{k}, \omega) = J_s(k_x, \omega) 2\pi \delta(k_y) \hat{\mathbf{u}}_x, \quad (8.191)$$

$$\rho(\mathbf{k}, \omega) = \sigma_s(k_x, \omega) 2\pi \delta(k_y). \quad (8.192)$$

Let us find the explicit form of the equation obeyed by the electric field in the stripe. The equation for the scalar potential reads

$$\phi(x, z, \omega) = \frac{1}{\epsilon_0 \epsilon} \int_{-\infty}^{\infty} \frac{dk_x}{2\pi} \frac{dk_y}{2\pi} G(k_x, 0, k_z, \omega) \sigma_s(k_x, \omega) e^{i(k_x x + k_z z)}. \quad (8.193)$$

Introducing

$$\sigma_s(k_x, \omega) = \int_{-\infty}^{\infty} dx' \sigma_s(x', \omega) e^{-i k_x x'}, \quad (8.194)$$

the potential $\phi(x, z, \omega)$ obeys

$$\phi(x, z, \omega) = \frac{1}{\epsilon_0 \epsilon} \int_{-\infty}^{\infty} dx' \sigma_s(x', \omega) \int_{-\infty}^{\infty} \frac{dk_x}{2\pi} \frac{dk_y}{2\pi} G(k_x, 0, k_z, \omega) e^{i[k_x(x-x') + k_z z]}. \quad (8.195)$$

The integration over k_z is done in the complex plane (see, for example, Appendix G). We thus obtain

$$G(x - x', 0, z, \omega) = \int_{-\infty}^{\infty} \frac{dk_x}{2\pi} \frac{dk_y}{2\pi} \frac{e^{i[k_x(x-x') + k_z z]}}{k_x^2 + k_z^2 - (\omega + i0^+)^2/v^2}, \quad (8.196)$$

$$= \int_{-\infty}^{\infty} \frac{dk_x}{2\pi} e^{i k_x(x-x')} \frac{e^{-\sqrt{k_x^2 - \omega^2/v^2}|z|}}{2\sqrt{k_x^2 - \omega^2/v^2}}. \quad (8.197)$$

Since we only need the potential in the graphene sheet we take $z = 0$. Thus, the Green's function becomes

$$G(x - x', 0, 0, \omega) = \int_0^\infty \frac{dk_x}{2\pi} \frac{\cos[k_x(x - x')]}{\sqrt{k_x^2 - \omega^2/v^2}},$$

$$= -\frac{i}{4}[J_0(|x - x'|\omega/v) + iY_0(|x - x'|\omega/v)],$$

$$\equiv -\frac{i}{4}H_0^{(1)}(|x - x'|\omega/v) \tag{8.198}$$

where $J_0(x)$ and $Y_0(x)$ are the usual Bessel functions of order zero, and $H_0^{(1)}(x)$ is the Hankel function of the first kind and order zero. In the electrostatic limit $v \to \infty$, following that

$$G(x - x', 0, z, \omega) \to -\frac{i}{4} - \frac{1}{2\pi}\left[\gamma + \ln\frac{\omega|x - x'|}{2v}\right]. \tag{8.199}$$

Since the electric field in-plane is given by[21]

$$E_x(x, \omega) = -\frac{d\phi(x, 0, \omega)}{dx}, \tag{8.200}$$

it follows from Eqs. (8.195) and (8.199) that

$$E_x(x, \omega) = \frac{1}{2\pi\epsilon_0\epsilon}\int_{-\infty}^\infty dx' \frac{\sigma_s(x', \omega)}{x - x'},$$

$$= \frac{1}{2\pi i\omega\epsilon_0\epsilon}\int_{-\infty}^\infty \frac{dx'}{x - x'}\frac{dJ_s(x', \omega)}{dx'}, \tag{8.201}$$

a result in agreement with Eq. (8.137), when the limit $a \to \infty$ is taken [note that for a finite ribbon $\sigma_s(x, \omega)$ is constraint to the width of the ribbon and so is $J_s(x, \omega)$]. The differences between the two derivations are the following: the derivation of Eq. (8.137) assumed the electrostatic limit right from the start whereas in Eq. (8.201) the electrostatic limit was taken only in the end of the calculation. On the other hand, the derivation of Eq. (8.201) assumed that the momentum along the stripe was zero, whereas in the derivation of Eq. (8.137) it was assumed that k_y was finite at the start of the calculation.

8.5 Localized Surface Plasmons in a Graphene Sheet with a Gaussian Groove

When light is shone on a defect, either in graphene or in a metallic surface, and if the defect has the right dimensions, the excitation of surface plasmon-polaritons becomes possible. On the other hand, the presence of the defect

---

[21]In the electrostatic limit the contribution coming from the vector potential is negligible.

induces changes in the spectrum of surface plasmon-polaritons. Indeed, for a single defect the momentum is no longer a well-defined quantity and therefore the spectrum cannot be parametrized by it. In what follows, we shall consider both the case of graphene and that of a dielectric-metal interface, which is a particular case of the former. Contrary to the cases discussed in the previous sections, in this problem graphene is taken as a continuous, pristine sheet. The (one-dimensional) defect under consideration here is a micro-groove; this constitutes our "microstructure", in the sense that graphene is microstructured by the groove. In addition, we shall consider the case of even grooves (or bumps) only. This system gives rise to the so-called channel plasmons.

To address this problem we follow the methodology introduced by Maradudin in the context of conventional 3D metals [Maradudin (1986)] (see also [Rahman and Maradudin (1980)]).

Let us assume a continuous graphene sheet lying in the xy−plane. We then consider that the graphene sheet has been corrugated to form a 1D-groove —cf. Fig. 8.21— with a surface profile given by $z = \zeta(\mathbf{r}_\parallel)$, where $\mathbf{r}_{xy} = \mathbf{r}_\parallel = (x, y)$ is the in-plane (xy) vector; any position in the graphene membrane can thus be defined by $\mathbf{r} = \mathbf{r}_\parallel + z\hat{\mathbf{z}}$. The geometry of the problem in depicted in Fig. 8.21, together with some of the relevant vectors needed to define the curvature of the graphene sheet. The goal is to determine how the spectrum of plasmon-polaritons in graphene is modified by the presence of the groove. The same question will be answered for a metallic interface, as a basis for comparison. The two situations will be different, since in the electrostatic limit the surface plasmons at a metallic interface are essentially non-dispersive, whereas graphene surface plasmons disperse as $\sqrt{q}$, with q being the wavevector (see Chapters 3 and 4).

8.5.1 *Few useful definitions*

Let us start by introducing few useful definitions. Defining the graphene surface by $f(\mathbf{r}) = z - \zeta(\mathbf{r}_\parallel)$, a unit vector normal to this surface is given by

$$\hat{\mathbf{n}} = \left[1 + \left(\frac{\partial \zeta}{\partial x}\right)^2\right]^{-1/2} \left(-\frac{\partial \zeta}{\partial x}, 0, 1\right), \tag{8.202}$$

noting that, for a one-dimensional groove like the one illustrated in Fig. 8.21, we have $\zeta(\mathbf{r}_\parallel) \equiv \zeta(x)$, so that $\partial \zeta / \partial y = 0$; this is the case we consider

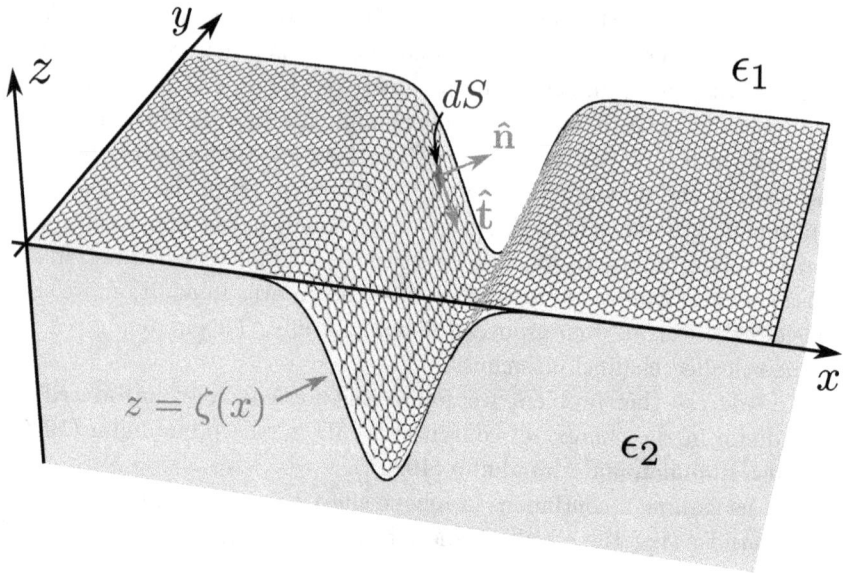

Figure 8.21: Illustration of the geometry of the problem, depicting a topographic 1D-groove in the dielectric substrate hosting the graphene sheet. The surface profile of this 1D-deformation is characterized by $\zeta(x)$. The unit vector $\hat{\mathbf{n}}$ is normal to the surface modulation whereas the unit vector $\hat{\mathbf{t}}$ is tangent to it. In the numerical calculations ahead we will consider a smooth gaussian 1D-groove.

in this section. The normal derivative to the surface can be defined as

$$\frac{\partial}{\partial n} = \hat{\mathbf{n}} \cdot \nabla = \left[1 + \left(\frac{\partial \zeta}{\partial x} \right)^2 \right]^{-1/2} \left(-\frac{\partial \zeta}{\partial x} \frac{\partial}{\partial x} + \frac{\partial}{\partial z} \right) , \qquad (8.203)$$

and the differential element of surface area, dS, as

$$dS = \frac{d\mathbf{r}_{\|}}{\hat{\mathbf{n}} \cdot \hat{\mathbf{z}}} , \qquad (8.204)$$

where $d\mathbf{r}_{\|}$ stands for the "in-plane" (i.e. at the graphene sheet) elementary area. Ahead we shall consider a one-dimentional Gaussian groove of the form[22]

$$z = \zeta(x) = A e^{-x^2/R^2} , \qquad (8.205)$$

where A and R are constants characterizing, repectively, the amplitude and width of the surface defect; other shapes, such as a Lorentzian, can also be

---

[22]We note that we can define our 1D surface defect either as a bump, with $A > 0$, or as a groove, for which we have $A < 0$.

considered. Both the Gaussian and the Lorentzian grooves allow to make much analytical progress, as some integrals can be computed exactly upon making a perturbative expansion in the parameter $A/R \ll 1$ [Maradudin (1986)]. However, in order to attain the non-perturbative regime where $A/R \sim 1$, we will need to compute the integrals numerically (see below).

8.5.2 *Formulation of the problem*

We shall consider the problem of determining the spectrum of surface plasmons in the so-called electrostatic limit, that is, the limit in which the speed of light is considered to be infinite and, therefore, no retardation effects are included (as such, this regime is sometimes referred to as the non-retarded limit). Let us assume that there are no charges in the space surrounding the graphene sheet, except on the graphene itself. The (surface) charge-density ascribed to graphene shall be denoted by ρ_{2D}.

Since we will be working in the quasi-static limit, we seek solutions of Laplace's equation for the potential above and below the graphene sheet, i.e. $z \gtrless \zeta(x)$:

$$
\begin{cases}
\nabla^2 \phi^>(\mathbf{r};\omega) = 0 & \text{if } z > \zeta(x) \\
\nabla^2 \phi^<(\mathbf{r};\omega) = 0 & \text{if } z < \zeta(x)
\end{cases} ,
\tag{8.206}
$$

where the time-dependence (or Fourier decomposition) is assumed to be of the form

$$
\phi^\gtrless(\mathbf{r};\omega) = \phi^\gtrless(\mathbf{r})e^{-i\omega t} ,
\tag{8.207}
$$

hereafter. The solutions of Eq. (8.206) satisfying the boundary conditions at infinity, that is $\lim_{z \to \pm\infty} \phi^\gtrless(\mathbf{r}) = 0$, can be written as (akin to a plane-wave expansion)

$$
\phi^>(\mathbf{r}) = \int \frac{dk_x}{2\pi} A(k_x)e^{i(k_x x + k_y y)}e^{-k_\| z} ; \quad z > \zeta(x) ,
\tag{8.208}
$$

$$
\phi^<(\mathbf{r}) = \int \frac{dk_x}{2\pi} B(k_x)e^{i(k_x x + k_y y)}e^{k_\| z} ; \quad z < \zeta(x) ,
\tag{8.209}
$$

where $\mathbf{k}_\| = (k_x, k_y)$, $k_\| = |\mathbf{k}_\|| = \sqrt{k_x^2 + k_y^2}$, and we have used the fact that the system is translational invariant along the y-direction, so that k_y is well-defined along this direction:

$$
\phi^\gtrless(\mathbf{r}) = \phi^\gtrless(x, z)e^{ik_y y} .
\tag{8.210}
$$

The electric potential is subjected to the following set of boundary conditions (see Chapter 2):

$$\phi^>(\mathbf{r})|_{z=\zeta(x)} = \phi^<(\mathbf{r})|_{z=\zeta(x)} \,, \tag{8.211}$$

$$\epsilon_1 \frac{\partial \phi^>(\mathbf{r})}{\partial n}\Big|_{z=\zeta(x)} - \epsilon_2 \frac{\partial \phi^<(\mathbf{r})}{\partial n}\Big|_{z=\zeta(x)} = -\frac{\rho_{2D}}{\epsilon_0} \,, \tag{8.212}$$

that is, the continuity of the potential and the discontinuity of the normal component of the electric displacement field. Here, ϵ_0 is the vacuum permittivity, ϵ_1 is the relative permittivity above the graphene sheet and ϵ_2 the same quantity below it. In principle, by imposing the boundary conditions for the potential, the Fourier amplitudes $A(k_x)$ and $B(k_x)$ can be determined. However, using Green's theorem it is possible to write the boundary conditions as function of the potential in the region $z > \zeta(x)$ alone. The advantage of this approach is that the eigenvalue problem (see below) can be formulated in terms of one coefficient alone, say $A(k_x)$, and therefore is more amenable for a numerical solution.

8.5.3 Green's theorem, Green's functions, and the eigen-value problem

Green's theorem, for two scalar functions u and v, states that

$$\int_V d\mathbf{r}(u\nabla^2 v - v\nabla^2 u) = \int_\Sigma dS \left(u \frac{\partial v}{\partial n} - v \frac{\partial u}{\partial n} \right) \,, \tag{8.213}$$

where V is a volume limited by the surface Σ. For our problem the volume is that enclosed by a hemispherical cap of infinite radius in the lower half-space (this part will give a null contribution since $\lim_{z\to-\infty} \phi^< = 0$) and by the surface defined by $z = \zeta(x)$. Let us now consider that u is the potential $\phi^<(\mathbf{r})$ and v is the Green's function defined as the solution of

$$\nabla^2 G(\mathbf{r};\mathbf{r}') = -\delta(\mathbf{r} - \mathbf{r}') \,, \tag{8.214}$$

that is, the free-space Green's function. The solution of the previous equation that vanishes at infinity is well known (see Sec. 8.1), reading

$$G(\mathbf{r};\mathbf{r}') = \int \frac{d\mathbf{q}_\parallel}{(2\pi)^2} \frac{e^{-q_\parallel|z-z'|}}{2q_\parallel} e^{i\mathbf{q}_\parallel \cdot (\mathbf{r}_\parallel - \mathbf{r}'_\parallel)} \,, \tag{8.215}$$

where $\mathbf{q}_\parallel = (q_x, q_y)$ and $q_\parallel = |\mathbf{q}_\parallel| = \sqrt{q_x^2 + q_y^2}$ (and similarly for $\mathbf{r}_\parallel$). Since the Green's function satisfies Eq. (8.214) and the potential $\phi^<(\mathbf{r})$ satisfies Laplace's equation, we can write Green's theorem as

$$-\theta_V(\mathbf{r})\phi^<(\mathbf{r}) = \int_S dS' \left[\phi^<(\mathbf{r}') \frac{\partial G(\mathbf{r};\mathbf{r}')}{\partial n'} - G(\mathbf{r};\mathbf{r}') \frac{\partial \phi^<(\mathbf{r}')}{\partial n'} \right] \,, \tag{8.216}$$

where the symmetry property of the Green's function, $G(\mathbf{r}; \mathbf{r}') = G(\mathbf{r}'; \mathbf{r})$, has been used and we have defined

$$\theta_V(\mathbf{r}) = \begin{cases} 1 & , \ \mathbf{r} \in V \\ 0 & , \ \text{otherwise} \end{cases} . \tag{8.217}$$

Considering $z > \zeta(x)$ (and thus $\theta_V(\mathbf{r}) = 0$), with the help of the boundary conditions (8.211) and (8.212) we get $[z > \zeta(x)]$

$$0 = \int_S dS' \left[\phi^>(\mathbf{r}') \frac{\partial G(\mathbf{r}; \mathbf{r}')}{\partial n'} - G(\mathbf{r}; \mathbf{r}') \left(\frac{\epsilon_1}{\epsilon_2} \frac{\partial \phi^>(\mathbf{r}')}{\partial n'} + \frac{\rho_{2D}}{\epsilon_2 \epsilon_0} \right) \right]_{z'=\zeta(x')} . \tag{8.218}$$

where the index S in the integral represents the surface defined by $z = \zeta(x)$; note that the integration over the hemispherical cap is zero due to the form of the potential and of the Green's function. Noting that (the reader may want to revisit Sec. 8.5.1 to check the definitions introduced there)

$$\int dS' \frac{\partial G(\mathbf{r}; \mathbf{r}')}{\partial n'} = \int d\mathbf{r}'_{\parallel} G(\mathbf{r}; \mathbf{r}') \left(q_{\parallel} + i q_x \frac{\partial \zeta(x')}{\partial x'} \right) , \tag{8.219}$$

and

$$\int dS' \frac{\epsilon_1}{\epsilon_2} \frac{\phi^>(\mathbf{r}')}{\partial n'} = \int d\mathbf{r}'_{\parallel} \phi^>(\mathbf{r}') \frac{\epsilon_1}{\epsilon_2} \left(-k_{\parallel} - i k_x \frac{\partial \zeta(x')}{\partial x'} \right) , \tag{8.220}$$

then, Eq. (8.218) can be casted in the form

$$0 = \int d\mathbf{r}'_{\parallel} \left[\phi^>(\mathbf{r}') G(\mathbf{r}; \mathbf{r}') \left(q_{\parallel} + \frac{\epsilon_1}{\epsilon_2} k_{\parallel} + i \left[q_x + \frac{\epsilon_1}{\epsilon_2} k_x \right] \frac{\partial \zeta(x')}{\partial x'} \right) \right]_{z'=\zeta(x')}$$

$$- \int d\mathbf{r}'_{\parallel} \sqrt{1 + \left(\frac{\partial \zeta(x')}{\partial x'} \right)^2} \left[G(\mathbf{r}; \mathbf{r}') \frac{\rho_{2D}}{\epsilon_2 \epsilon_0} \right]_{z'=\zeta(x')} . \tag{8.221}$$

Later on, the equations will take a simpler form (although that may not be evident at this stage) if we restrict ourselves to the particular case where $k_y = 0$, for which we have $k_{\parallel} \equiv |k_x|$. We note that we shall assume this case, i.e. that with $k_y = 0$, henceforth. In such scenario, the previous equation reduces to (note the replacement $k_{\parallel} \to |k_x|$)

$$0 = \int d\mathbf{r}'_{\parallel} \left[\phi^>(\mathbf{r}') G(\mathbf{r}; \mathbf{r}') \left(q_{\parallel} + \frac{\epsilon_1}{\epsilon_2} |k_x| + i \left[q_x + \frac{\epsilon_1}{\epsilon_2} k_x \right] \frac{\partial \zeta(x')}{\partial x'} \right) \right]_{z'=\zeta(x')}$$

$$- \int d\mathbf{r}'_{\parallel} \sqrt{1 + \left(\frac{\partial \zeta(x')}{\partial x'} \right)^2} \left[G(\mathbf{r}; \mathbf{r}') \frac{\rho_{2D}}{\epsilon_2 \epsilon_0} \right]_{z'=\zeta(x')} . \tag{8.222}$$

At this stage of the calculation, our progress is halted until we find an expression for ρ_{2D}.

As such, the next step in the derivation towards an eigenvalue problem is to write the surface charge-density ρ_{2D} in terms of potential. To that end we use the continuity equation, which reads (in frequency space)

$$\frac{\partial \rho_{2D}}{\partial t} + \nabla \cdot \mathbf{J}_{2D} = 0 \quad \Rightarrow \quad \nabla \cdot \mathbf{J}_{2D} = i\omega \rho_{2D} \, , \qquad (8.223)$$

where $\mathbf{J}_{2D}$ is the surface current-density in graphene. We note that saying that the surface charge-density depends on time is a simple form of introducing dynamics in this problem. The other source of dynamics is the use of the optical conductivity of graphene or, in the case of the metal, the use of its frequency dependent dielectric function. The surface current-density relates to the conductivity via Ohm's law[23]

$$\mathbf{J}_{2D} = \sigma_g \mathbf{E}(\mathbf{r})|_{z=\zeta(x)} = -\sigma_g \left[\hat{\mathbf{t}} \cdot \nabla \phi^{>}(\mathbf{r})|_{z=\zeta(x)} \right] \hat{\mathbf{t}} \, , \qquad (8.224)$$

where σ_g is the optical 2D conductivity of graphene and the unit-vector $\hat{\mathbf{t}}$ tangent to graphene's surface is defined through

$$\hat{\mathbf{t}} = \left[1 + \left(\frac{\partial \zeta}{\partial x} \right)^2 \right]^{-1/2} \left(1, 0, \frac{\partial \zeta}{\partial x} \right) \, . \qquad (8.225)$$

Taking the divergence of Eq. (8.224) and combining the result with Eq. (8.223) yields

$$\rho_{2D} = \frac{i\sigma_g}{\omega} \left[\frac{\partial^2}{\partial x^2} + 2 \frac{\partial \zeta}{\partial x} \frac{\partial^2}{\partial x \partial z} + \frac{\partial^2 \zeta}{\partial x^2} \frac{\partial}{\partial z} \right] \phi^{>}(\mathbf{r})|_{z=\zeta(x)} + \mathcal{O}\left(\zeta^2 \right) \, , \quad (8.226)$$

up to first order in ζ; recall that here we are interested in the $k_y = 0$ case only —trapped modes in the groove. The $k_y \neq 0$ case would describe localized modes guided along the groove with finite momentum along $\hat{\mathbf{y}}$. This expansion allow us to make analytical progress in the calculations, but it is not essential, since we can workout the full expression for the divergence of the current and use numerical quadrature to evaluate the integrals. The first term in Eq. (8.226) represents the field in flat graphene, whereas the second and third terms are associated with the groove's slope and curvature, respectively. Therefore, the last two terms probe the structure of the groove.

---

[23]In general, we should also include a term of the form $-\sigma_g \left[\hat{\mathbf{y}} \cdot \nabla \phi^{>}(\mathbf{r}) \right] \hat{\mathbf{y}}$. However, since we are interested in the simple case of $k_y = 0$, this term gives no contribution.

Having introduced the dynamics in the problem via Eq. (8.226), we can use it to rewrite expression (8.222) as (again, up to first order)

$$0 = \int d\mathbf{r}'_\| \left[\phi^>(\mathbf{r}')G(\mathbf{r};\mathbf{r}') \left(q_\| + \frac{\epsilon_1}{\epsilon_2}|k_x| + i \left[q_x + \frac{\epsilon_1}{\epsilon_2}k_x \right] \frac{\partial \zeta}{\partial x'} \right) \right]_{z'=\zeta(x')}$$

$$+ \frac{i\sigma_g}{\omega\epsilon_2\epsilon_0} \int d\mathbf{r}'_\| \left[G(\mathbf{r};\mathbf{r}')\phi^>(\mathbf{r}') \left(k_x^2 + 2ik_x|k_x|\frac{\partial\zeta}{\partial x'} + |k_x|\frac{\partial^2\zeta}{\partial x'^2} \right) \right]_{z'=\zeta(x')},$$
$$(8.227)$$

so that $\phi^>(\mathbf{r}')G(\mathbf{r};\mathbf{r}')$ is effectively factorized, thereby being a common term with respect to the others. This makes the calculation more compliable. Now, substituting both the potential [cf. Eq. (8.208) for $k_y = 0$]

$$\phi^>(\mathbf{r}) = \int \frac{dk_x}{2\pi} A(k_x)e^{ik_x x}e^{-|k_x|z} \; ; \; z > \zeta(x) , \qquad (8.228)$$

and the Green's function (8.215) explicitly, produces ($z > z'$)

$$0 = \int \frac{dq_x}{2\pi} \int \frac{dx'}{2\pi} \int dk_x A(k_x)\frac{e^{-|q_x|z}}{2|q_x|}e^{iq_x x}e^{ix'(k_x - q_x)}$$

$$\times e^{(|q_x|-|k_x|)\zeta(x')} \left[|q_x| + \frac{\epsilon_1}{\epsilon_2}|k_x| + i \left(q_x + \frac{\epsilon_1}{\epsilon_2}k_x \right)\frac{\partial\zeta}{\partial x'} \right.$$

$$\left. + f_\sigma(\omega) \left(k_x^2 + 2ik_x|k_x|\frac{\partial\zeta}{\partial x'} + |k_x|\frac{\partial^2\zeta}{\partial x'^2} \right) \right], \qquad (8.229)$$

where we have defined the function $f_\sigma(\omega) = \frac{i\sigma_g(\omega)}{\omega\epsilon_2\epsilon_0}$, and q_y vanishes from integrating the delta function $\delta(q_y)$ (owing to the fact that we have assumed $k_y = 0$).

Now, we introduce the following representations for the terms involving the surface profile function [Maradudin (1986); Farias and Maradudin (1983)]:

$$e^{\alpha\zeta(x')} = 1 + \alpha \int \frac{dQ}{2\pi} J(\alpha; Q)e^{iQx'} , \qquad (8.230)$$

$$\frac{\partial\zeta}{\partial x'}e^{\alpha\zeta(x')} = \int \frac{dQ}{2\pi} iQ J(\alpha; Q)e^{iQx'} , \qquad (8.231)$$

$$\frac{\partial^2\zeta}{\partial x'^2}e^{\alpha\zeta(x')} = -\int \frac{dQ}{2\pi} Q^2 J(\alpha; Q)e^{iQx'} . \qquad (8.232)$$

to write out the corresponding terms in Eq. (8.229). Doing so and then projecting with e^{-iqx} in the resulting expression (followed by an integration over x), yields, after some lenghty calculations, an integral equation for the potential's Fourier amplitude, namely

$$\left[\epsilon_1 + \epsilon_2 + \frac{i\sigma_g}{\omega\epsilon_0}|q| \right] A(q) = \int \frac{dk_x}{2\pi} A(k_x) \, J(|q| - |k_x|; q - k_x) \, |k_x|$$

$$\times \left[(\epsilon_2 - \epsilon_1)\left[1 - \mathrm{sgn}(q)\,\mathrm{sgn}(k_x)\right] + \frac{i\sigma_g}{\omega\epsilon_0}\left(|q| - |k_x|\right) \right] . \quad (8.233)$$

Moreover, it is interesting to note that for flat graphene, i.e. $\zeta = 0$, we have $J(|q| - |k_x|; q - k_x) = 0$, and thus, in that scenario, we recover the non-retarded condition for the spectrum of graphene surface plasmons:

$$\epsilon_1 + \epsilon_2 + \frac{i\sigma_g}{\omega\epsilon_0}|q| = 0, \qquad (8.234)$$

as the reader may check by comparing this result with Eq. (4.13) and taking the electrostatic limit $(q \gg \omega/c)$ in that expression.

Naturally, the equivalent result for a dielectric-metal interface [i.e. with graphene absent and $\epsilon_2 \to \epsilon_2(\omega)$] can be obtained by taking the limit $\sigma_g \to 0$ in Eq. (8.233), which gives

$$\frac{\epsilon_2(\omega) + \epsilon_1}{\epsilon_2(\omega) - \epsilon_1} A(q) = \int \frac{dk_x}{2\pi} A(k_x) \, J(|q| - |k_x|; q - k_x) \, |k_x|[1 - \text{sgn}(q)\,\text{sgn}(k_x)], \qquad (8.235)$$

a result that agrees with the one found in [Maradudin (1986)]. As we show below, the integral equation (8.233) can be transformed into a matrix eigenvalue problem. Assuming that the profile function $\zeta(x)$ is even with respect to the x-coordinate, that is, $\zeta(x) = \zeta(-x)$, the solutions of Laplace's equation can be classified as either even or odd, $\phi^>(-x, z) = \pm\phi^>(x, z)$. From the definition (8.228), this implies that $A(-p) = \pm A(p)$, which allows us rewrite Eq. (8.233) involving only positive values of q and k_x. Taking into to account both types of solutions, that procedure leads to

$$\left[\epsilon_1 + \epsilon_2 + \frac{i\sigma_g}{\omega\epsilon_0} q\right] A(q) =$$
$$\frac{i\sigma_g}{\omega\epsilon_0} \int_0^\infty \frac{dk_x}{2\pi} A(k_x) \, k_x(q - k_x) \, J(q - k_x; q - k_x)$$
$$\pm \int_0^\infty \frac{dk_x}{2\pi} A(k_x) \, J(q - k_x; q + k_x) \, k_x \left[2(\epsilon_2 - \epsilon_1) + \frac{i\sigma_g}{\omega\epsilon_0}(q - k_x)\right], \qquad (8.236)$$

with $k_x, q \geq 0$ and where $\pm$ refers to the even/odd eigenvalue problem. Again, for the dielectric-metal interface, this result simplifies considerably. However, for finite $\sigma_g(\omega)$ the eigenvalue problem has some degree of complexity, since the frequency, ω, appears both at the left- and righ-hand sides of the integral equation (8.236). Because of this, we shall consider here only the specific case in which $\epsilon \equiv \epsilon_1 = \epsilon_2$, that is, that of a graphene sheet (with a 1D-groove) encapsulated in a symmetric dielectric environment[24] (e.g. free-standing graphene or graphene encapsulated in hBN). This removes the aforementioned complication, allowing us to write

$$\frac{2i\omega\epsilon\epsilon_0}{\sigma_g(\omega)} A(q) = qA(q) - \int_0^\infty \frac{dk_x}{2\pi} A(k_x)\, k_x(q-k_x)\left[J(q-k_x; q-k_x) \right.$$
$$\left. \pm J(q-k_x; q+k_x) \right] . \tag{8.237}$$

In possession of Eq. (8.237), we now show how to employ a Gauss-Laguerre numerical quadrature to transform the above integral equation into a standard eigenvalue problem. Equation (8.237) contains two quadratures: one in the variable k_x and another one in the variable x due to the function $J(\alpha; Q)$ [see Eq. 8.239]. The integral in $J(\alpha; Q)$ can be computed either numerically or analytically for some special profiles, such as the Gaussian one we are considering. However, as already mentioned, the latter approach can only be applied for small values of the ratio A/R. Therefore, we shall proceed in computing the $J(\alpha; Q)$ numerically.

8.5.4   *Spectrum of surface plasmon-polaritons in the presence of a Gaussian groove*

The spectrum of surface plasmon-polaritons —in the particular case that we have assumed above— is obtained from the solution of the integral equation (8.237). To proceed further we have to specify the profile of the groove. We choose a Gaussian groove defined by [recall Eq. (8.205)]

$$\zeta(x) = Ae^{-x^2/R^2} , \tag{8.238}$$

which is also symmetric with respect to the x-coordinate. Having an explicit form for $\zeta(x)$, we can now proceed in the computation of Eq. (8.237). In particular, from the definition of the functions $J(\alpha; Q)$ [Maradudin (1986); Farias and Maradudin (1983)],

$$J(\alpha; Q) = \int dx\, e^{-iQx} \frac{e^{\alpha\zeta(x)} - 1}{\alpha} , \tag{8.239}$$

which for an even profile reduces to

$$J(\alpha; Q) = 2 \int_0^\infty dx\, \cos(Qx) \frac{e^{\alpha\zeta(x)} - 1}{\alpha} . \tag{8.240}$$

---

[24]In spite of that, Eq. (8.236) could still be tacked numerically by transforming that integral equation into a homogeneous matrix equation, after which one imposes that the determinant of the resulting matrix must be zero, a condition that yields the eigenfrequencies.

Introducing the following change of variables (transformation to dimensionless variables)

$$u = qR, \qquad v = k_x R, \qquad x = x'R, \qquad (8.241)$$

the integral equation (8.237) translates to

$$\frac{2i\omega\epsilon\epsilon_0 R}{\sigma_g(\omega)} a(u) = ua(u) - \frac{1}{2\pi} \int_0^\infty dv \, a(v)v(u-v)[\mathcal{Q}_1(u,v,A/R)$$
$$\pm \mathcal{Q}_2(u,v,A/R)], \qquad (8.242)$$

where the Kernels $\mathcal{Q}_1(u,v,A/R)$ and $\mathcal{Q}_2(u,v,A/R)$ read

$$\mathcal{Q}_1(u,v,A/R) = 2\int_0^\infty dx' \cos[(v-u)x'] \frac{e^{(u-v)\zeta(x')} - 1}{u-v} \qquad (8.243)$$

and

$$\mathcal{Q}_2(u,v,A/R) = 2\int_0^\infty dx' \cos[(u+v)x'] \frac{e^{(u-v)\zeta(x')} - 1}{u-v}. \qquad (8.244)$$

The integral equation (8.242) is easily tractable numerically as it can be transformed into an eigenvalue problem. To that end we discretize the integration over the variable v into a finite sum. This procedure has to be judiciously made to account for the improper nature of the integral. Fortunately, our integration kernel $\mathcal{Q}_j(u,v,A/R)$ behaves as $\sim e^{-v}$, for large v (compared to u) and, therefore, the integral of the kernel can be tackled using the Gauss-Laguerre quadrature scheme. In this approach an integral of the form

$$I = \int_0^\infty dy f(y)e^{-y}, \qquad (8.245)$$

is approximated by

$$\int_0^\infty dy f(y)e^{-y} \cong \sum_{j=1}^N w_j f(y_j), \qquad (8.246)$$

where $\{w_j\}$ are the weights and y_j is the j-th zero of the Laguerre polynomial of order N, i.e. satisfying the equation $L_N(y_j) = 0$. The weights are obtained via

$$\{w_j\} = \frac{\{y_j\}}{[(N+1)L_{N+1}(\{y_j\})]^2}. \qquad (8.247)$$

Finally, applying this technique to Eq. (8.242), we obtain

$$\frac{2i\omega\epsilon\epsilon_0 R}{\sigma_g(\omega)} a(u_i) \cong u_i \, a(u_i) - \sum_{j=1}^N w_j \, e^{v_j} \mathcal{Q}(u_i, v_j, A/R) \, a(v_j), \qquad (8.248)$$

which is an usual eigenvalue problem that can be solved by any linear algebra numerical package. More explicitly, one has

$$\frac{2i\omega\epsilon\epsilon_0 R}{\sigma_g(\omega)}a(u_i) = \sum_{j=1}^{N} \mathcal{M}_{ij}\, a(v_j)\,, \qquad (8.249)$$

where the matrix elements $\mathcal{M}_{ij}$ stem from

$$\mathcal{M}_{ij} = u_j\delta_{ij} - w_j\, e^{v_j}\, \mathcal{Q}(u_i, v_j, A/R)\,,$$

where the matrix $\mathcal{Q}(u_i, v_j, A/R)$ is defined as

$$\mathcal{Q}(u_i, v_j, A/R) = \frac{1}{2\pi}v_i(u_j - v_i)[\mathcal{Q}_1(u_i, v_j, A/R) \pm \mathcal{Q}_2(u_i, v_j, A/R)]\,. \qquad (8.250)$$

Once the eigenvalues λ_i of the eigenvalue problem (8.249) are known, the spectrum of surface plasmon-polaritons, $\omega = \omega(\lambda_i)$, can be fetched from the numerical solution of

$$\lambda_i = \frac{2i\omega\epsilon\epsilon_0 R}{\sigma_g(\omega)}\,. \qquad (8.251)$$

Note the similarity of the last result with Eq. (8.48). This is no coincidence, since both methods of solution are, in the end, quite similar, except that in this case we are expanding the potential in plane waves. In the case in which $\sigma_g(\omega)$ is given by graphene's Drude conductivity with negligible broadening $\gamma = 0$, the previous equation can be solved analytically in order to obtain $\omega = \omega(\lambda_i)$, which avoids an extra numerical step. We note that is not difficult to generalize the previous paragraphs taking into account the non-local nature of the conductivity of graphene; the equations will be slightly different though.

In the problem of the bump at the interface between a metal and a dielectric the electrostatic surface plasmon mode is split into two modes —one even and one odd. The frequency of the even mode softens whereas that of the odd mode stiffs, as can be seen in Fig. 8.22 (left panel) for the first[25] two modes. The softening and the hardening the surface plasmon's spectrum can be interpreted as follows: the groove/bump, to a first approximation, breaks up the system into two disjoint semi-infinite metal-dielectric interfaces, each supporting SPP's of the costumary type observed in a flat metallic interface. To a second approximation the surface plasmons of the two semi-infinite systems interact with each other, leading to

[25]We will be order the eigenvalues in decreasing order, i.e. the largest ones being the first ones (cf. Tab. 8.3).

anti-crossing of the energy levels or, if one prefers, to bonding (even) and
anti-bonding (odd) states, if we establish an analogy with the energy levels
in molecules. We note that the presence of the groove (or bump for the
same matter) renders the concept of a mode with a well defined wavevector
along $\hat{x}$ meaningless. In fact, these modes correspond to localized electrostatic plasmons in the groove. For graphene, in the right panel of Fig.
8.22 the dependence of the mode λ_7 as function of $|A/R|$ (for a groove) is
depicted (top panel). There is a clear dependence of the value of λ_7 on the
ratio $|A/R|$. Interestingly enough, and contrary to the case of a metal with
a Lorentzian bump, the odd and even modes are affected essentially in the
same way for graphene deposited onto a Gaussian groove, with no clear dis-

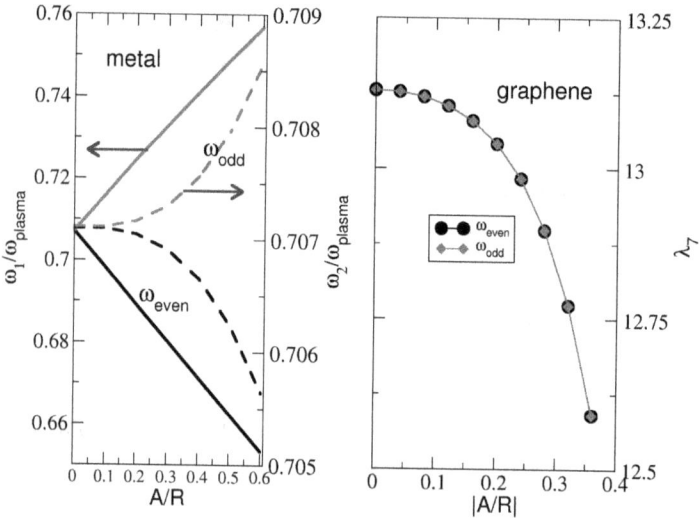

Figure 8.22: Dispersion relation of SPP's in the electrostatic limit for a metallic
Lorentzian bump (left panel), and for corrugated graphene (squares and circles in
the right top and bottom panels), with a Gaussian profile. For the metallic case
(left panel) the two first modes, ω_1 (solid lines) and ω_2 (dashed lines) are depicted,
which branch up with the parameter A/R in even and odd modes (the frequency
is given in units of the plasma frequency). For graphene, two of the eigenvalues,
λ_7 (top) and λ_{14} (bottom), are depicted. Although there is a clear dependence
of the frequency modes on $|A/R|$, both the even and the odd modes present the
same dependence for the case λ_7, which corresponds to a larger eigenvalue than
for the mode λ_{14} (check Table 8.3). The total number of modes considered is
$N = 15$.

tinction between the two cases. On the other hand, for small eigenvalues, as is the case of the mode λ_{14} (Fig. 8.22 —right bottom panel) the even and odd modes are affected differently upon varying the parameter $|A/R|$, with the even mode dispersing more with $|A/R|$ than the odd mode.

In Fig. 8.23 we depict the profile of the electrostatic potential at the surface of graphene, that is $\phi[x, \zeta(x)]$. The modes λ_7, λ_8, λ_9, and λ_{10} are represented in that figure. The larger the number of the mode the larger is the period of the electrostatic potential, which translates into a smaller degree of confinement of the mode around the groove. Also, smaller values of λ_i corresponds to more intense fields located near the apex of the groove. This is depicted in the bottom panel of Fig. 8.23 for the mode λ_7, which is the one, from the four represented in the top panel of this figure, that presents the largest degree of confinement and more intense local field. In Fig. 8.24 the density plot of the field $E_z(x, z)$ is depicted for the modes λ_8 and λ_{10}. From a glance at this figure we immediately see the aforementioned behavior of the electric field. If we had kept a finite value of k_y we could have plot the dispersion of the modes along the groove for each of the localized modes, a situation describing guided modes in an

i	u_i	$\lambda_{i,+}$	$\lambda_{i,-}$
1	48.03	48.15	48.15
2	38.53	37.71	37.71
3	31.41	30.45	30.45
4	25.62	28.25	28.25
5	20.78	21.78	21.78
6	16.65	16.13	16.13
7	13.13	10.98	10.98
8	10.12	9.507	9.507
9	7.566	8.284	8.284
10	5.425	5.681	5.681
11	3.668	3.868	3.870
12	2.270	2.362	2.374
13	1.216	1.250	1.248
14	0.4927	0.5061	0.4954
15	0.09331	0.09481	0.09332

Table 8.3: Eigenvalues $\lambda_{i,+}$ (even modes) and $\lambda_{i,-}$ (odd modes) for $N = 15$ and $A/R = -0.5$. The effect of the groove is present in the deviation of the eigenvalues λ_i from u_i.

indentation covered with graphene, that is, channel plasmons.

In table 8.3 we present the numerical eigenvalues λ_i for both even and odd modes for a particular value of A/R. It is clear from the table that the effect of the groove is largest for the smallest eigenvalues. In Fig. 8.22 we plot the even and odd modes $\lambda_{7,+}$ and $\lambda_{7,-}$, respectively. It is clear from both this figure and table 8.3 that these modes are affected in the same way and by the same amount by the presence of the groove. Although, as already mentioned, the concept of momentum along the $x-$direction is rendered meaningless by the presence of the groove, we can still argue that period of the electric potential (or electric field) is controlled by the value of λ_i. For larger λ_i there is no difference between the momentum u_i and the eigenvalue λ_i. In this condition we can argue that the system is characterized by a wavelength $\Lambda_i = 2\pi/u_i$. Since u_i is large, Λ_i is small and the system is sensitive to the presence of the groove, leading to well localized modes. As λ_i decreases Λ_i increases and the presence groove is essentially invisible to the modes, which leads to poor localization of the modes and to small values of the local electric field. Indeed, we see that for the mode λ_{14} in Fig. 8.22 the modification of the eigenvalue upon varying $|A/R|$ is, at most, of 0.1. On the contrary, for the mode λ_7 the variation is much larger. All this is also clear from Figs. 8.22 and 8.23.

We conclude this Chapter showing how to adapt the formalism to a periodic groove, that is, a grating. To that end we recall that:

$$\phi^>(\mathbf{r};\omega) = \int \frac{d^2 \mathbf{k}_\parallel}{(2\pi)^2} A(\mathbf{k}_\parallel;\omega) e^{i\mathbf{k}_\parallel \cdot \mathbf{r}_\parallel} e^{-k_\parallel z}. \qquad (8.252)$$

In a periodic landscape the potential must obey the Bloch condition $\phi^>(\mathbf{r}+\mathbf{a};\omega) = e^{i\mathbf{q}\cdot\mathbf{a}}\phi^>(\mathbf{r};\omega)$, where $\mathbf{a}$ is an in-plane vector encoding the period of the grating (for a one-dimensional grating we have $\mathbf{a} = a\hat{\mathbf{x}}$) and $\mathbf{q}$ is the Bloch wavevector. If we write the amplitudes $A(\mathbf{k}_\parallel;\omega)$ as

$$A(\mathbf{k}_\parallel;\omega) = \sum_{\mathbf{G}} \tilde{A}_{\mathbf{G}}(\mathbf{q};\omega)\delta(\mathbf{G} + \mathbf{q} - \mathbf{k}_\parallel), \qquad (8.253)$$

and introduce it in Eq. (8.252) we obtain

$$\phi^>(\mathbf{r};\omega) = \sum_{\mathbf{G}} \tilde{A}_{\mathbf{G}}(\mathbf{q};\omega) e^{i(\mathbf{G}+\mathbf{q})\cdot\mathbf{r}_\parallel} e^{-|\mathbf{G}+\mathbf{q}|z}$$

$$= e^{i\mathbf{q}\cdot\mathbf{r}_\parallel} \sum_{\mathbf{G}} \tilde{A}_{\mathbf{G}}(\mathbf{q};\omega) e^{i\mathbf{G}\cdot\mathbf{r}_\parallel} e^{-|\mathbf{G}+\mathbf{q}|z}, \qquad (8.254)$$

where $\mathbf{G}$ are the reciprocal lattice vectors of the grating and $\mathbf{G}\cdot\mathbf{a} = 2\pi m$, with m an integer. In this latter form $\phi^>(\mathbf{r};\omega)$ obeys the Bloch condition, as required. We can now proceed along the same lines of this Section and derive an eigenvalue problem where the eigenvectors are composed by the amplitudes $\tilde{A}_{\mathbf{G}}(\mathbf{q};\omega)$. This derivation sets the stage for the next chapter, which deals with the excitation of SPP's in gratings, including retardation effects.

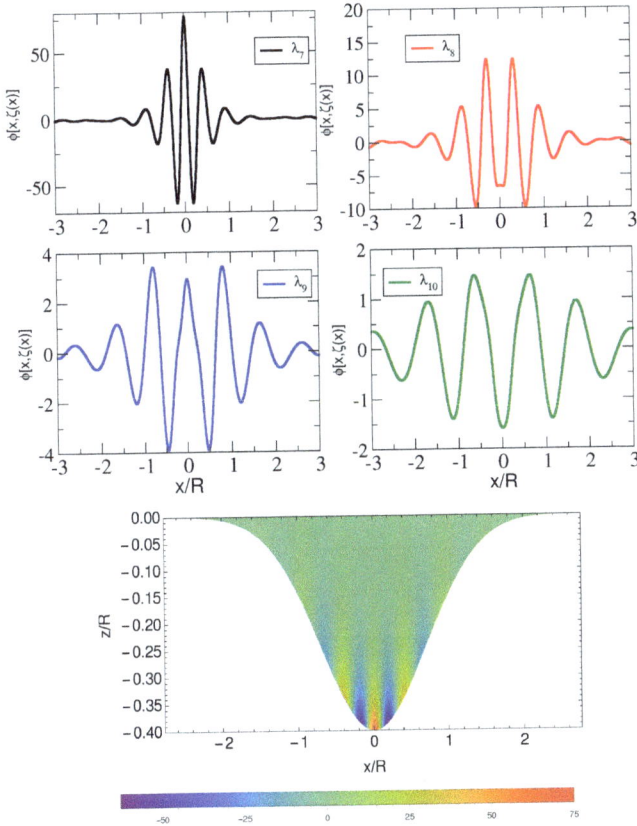

Figure 8.23: Profile of the potential at the surface of graphene, that is, $\phi[x, \zeta(x)]$, for four different modes, in the case of a Gaussian groove (arbitrary units are used). Clearly, the four modes have different degrees of confinement, with the mode labeled λ_7 the more confined of the four. In the bottom panel, an intensity plot of the potential, for the mode λ_7, as function of x/R and y/R, is depicted. There is a clear large potential value at the tip of the groove, which translates into a large electric field. Note the different scales in the $x-$ and $z-$directions, which differ by about an order of magnitude. We have use $N = 15$ and $A/R = -0.4$.

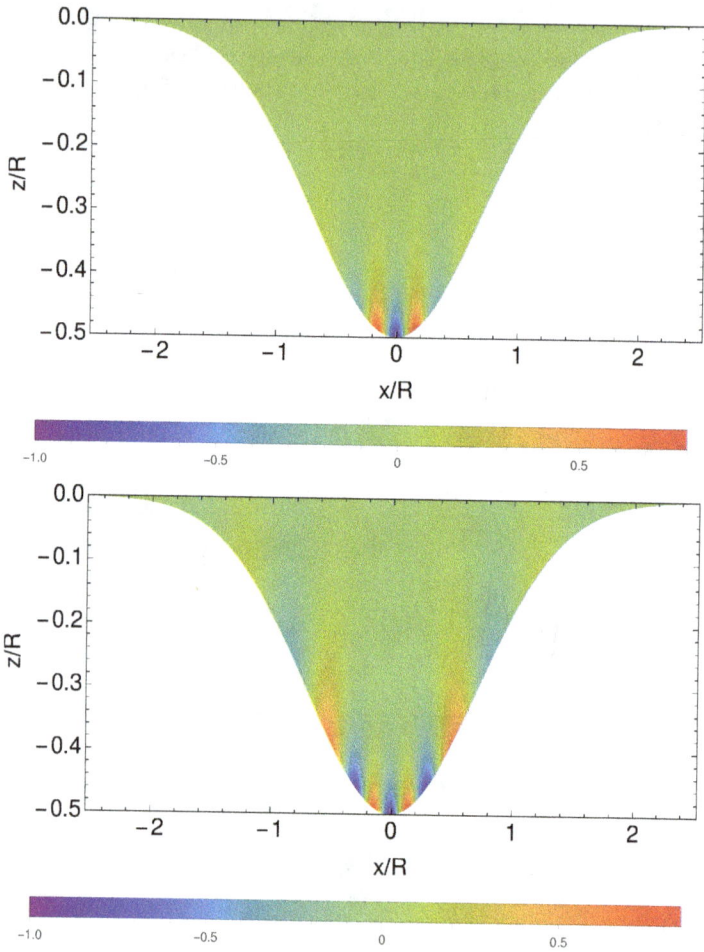

Figure 8.24: Density plot of the electric field E_z for λ_7 (top) and λ_9 (bottom) corresponding to the potential profiles depicted in Fig. 8.23. We have use $N = 15$ and $A/R = -0.5$. Note the different scales in the vertical and horizontal axes. Clearly, the λ_7 mode is more confined than the λ_9 one. The field is normalized to its value at the apex of the groove.

Chapter 9

Excitation of Surface Plasmon-Polaritons Using Dielectric Gratings

The study of gratings and rough surfaces [Maradudin (2007a)] as methods of exciting surface plasmon-polaritons was quite popular in the eighties in the context of noble-metals plasmonics [Farias and Maradudin (1983)]. The interest in this topic has been kept alive due to the problem of light scattering by rough surfaces at the nanoscale [Maradudin (2007b)]. With the birth of graphene plasmonics came a renewed interest in the problem of grids and gratings based on graphene. In Chapter 7 we have discussed the grid problem, which was the first method used to induce plasmons in graphene-based periodic structures. As expected, the excitation of graphene plasmons by gratings soon followed (see Chapter 1). The advantage of gratings over grids is that the quality of the graphene sheet is not affected by the lithographic process, since graphene keeps its integrity. On the other hand, the fabrication of gratings with high dielectric contrast may present some difficulties. This is specially true for square-wave gratings. If the grating has some profile, as the one discussed in this Chapter, a graphene sheet may follow that profile thus making a natural graphene-based grating.

As discussed briefly Chapter 5 a method for inducing surface plasmon-polaritons in graphene is depositing it on a dielectric grating [Gao *et al.* (2013)]. In that chapter the example of a square wave grating was given. In this chapter we discuss the excitation of surface plasmon-polaritons by a sine-grating, following a method due to Rayleigh [Voronovich (2007)] (for a different method see Appendix J). Our approach is mainly analytical, which is a consequence of the simplicity of the profile used in the calculations. Interesting enough these periodic structures give rise to plasmonic bands [Bludov *et al.* (2012a)] which is the fundamental reason why surface plasmon-polaritons can be excited via grating coupling. The excitation of SPP's belonging to different dispersion branches depends, however, on symmetry considerations [Yeung *et al.* (2015)]. Periodic graphene-antidot arrays are another way of inducing SPP's in graphene [Nikitin *et al.* (2012b)].

In this problem we discuss the induction of SPP's by a grating. However the method can also work in reverse. For example, using SNOM coupling we can induce a propagating SPP along a flat graphene portion of a sheet. This SPP will propagate along the sheet until it reaches a region where a grating has been paterned. Upon hitting this region the SPP radiates into space and this radiation can be collected for some purpose.

9.1 Some Basic Definitions and Results

We want to solve the scattering problem of light impinging on a diffraction grating, as represented in Fig. 9.1. It is assumed that both in D_+ and D_- the dielectric functions $\epsilon_{1/2}$ are homogeneous (they can be frequency dependent however, as in the example considered in Sec. 5.1.1). The regions D_+ and D_- define the upper and lower half-planes relatively to the curve $y = a(x)$. For a TM wave we have $\mathbf{B} = (0, 0, B_z)$ and $\mathbf{E} = (E_x, E_y, 0)$. Maxwell's equations for the TM wave are

$$\partial_x E_y^{(p)} - \partial_y E_x^{(p)} = i\omega B_z^{(p)}, \tag{9.1}$$

$$E_x^{(p)} = i\frac{v_p^2}{\omega}\frac{\partial B_z^{(p)}}{\partial y}, \tag{9.2}$$

$$E_y^{(p)} = -i\frac{v_p^2}{\omega}\frac{\partial B_z^{(p)}}{\partial x}, \tag{9.3}$$

where the superscript p defines de region above or below the grating and $v_p = c/\sqrt{\epsilon_p^r}$ is the speed of light in the medium p[1]. In a region of constant dielectric function the wave equation has the form (assuming $TM-$polarization)

$$\left(\frac{\partial^2}{\partial x^2} + \frac{\partial^2}{\partial y^2} + \omega^2 \mu \epsilon_p\right) B_z^{(p)} = 0, \tag{9.4}$$

where $\mu = \mu_0 \mu^r$ is the magnetic permeability of the grating's material, μ_0 is the vacuum permeability, and μ^r is the relative permeability. In the regions D_+ and D_-, and in the limit $y \to \pm\infty$, the scattered and transmitted fields are given by the Rayleigh expansions (Bloch representation of the field)

$$B_z^{(1)} = \sum_{m=-\infty}^{\infty} A_{1;m} e^{i(x\alpha_n + y\beta_m^{(1)})}, \tag{9.5}$$

---

[1]In this chapter $\epsilon_p = \epsilon_0 \epsilon_p^r$, where ϵ_p^r is the relative permitivity of medium p.

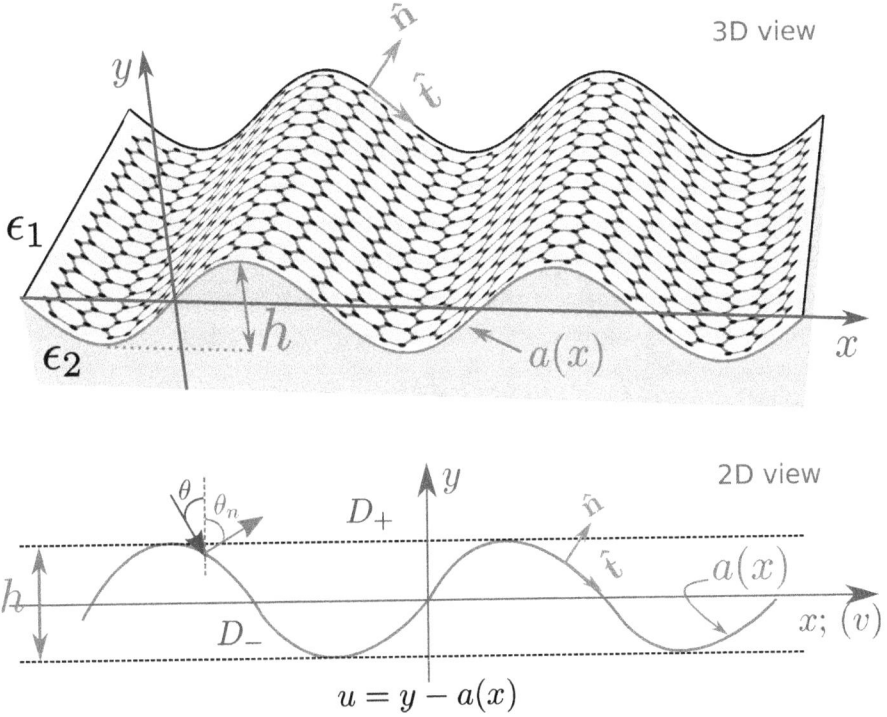

Figure 9.1: Grating geometry and different regions. The grating is covered with graphene. The curve defining the grating is $y = a(x)$. Above the grating we have region D_+ with dielectric permittivity ϵ_1, whereas below the curve $y = a(x)$ we have region D_- with dielectric permittivity ϵ_2.

and

$$B_z^{(2)} = \sum_{m=-\infty}^{\infty} A_{2;m} e^{i(x\alpha_n - y\beta_m^{(2)})} . \tag{9.6}$$

The incoming field is given by

$$B_{\text{inc};z} = e^{i(xk\sin\theta - yk\cos\theta)} , \tag{9.7}$$

where $k = \omega/v_1 = k_0\sqrt{\epsilon_1^r\mu_1^r}$, where k_0 is the wave number in vacuum, and ϵ_1^r and μ_1^r are the relative dielectric permittivity and relative magnetic permeability, respectively. The constants α_n and $\beta_{p;m}$ are given by

$$\alpha_n = k\sin\theta - n2\pi/D = k\sin\theta - nG , \tag{9.8}$$

$$\beta_n^{(p)} = \sqrt{k_p^2 - \alpha_n^2} , \tag{9.9}$$

where $k_p = \omega/v_p$. The form of $\beta_n^{(p)}$ follows from the solution of the wave equation and we must have $\Re(\beta_n^{(p)})$, $\Im(\beta_n^{(p)}) > 0$, for respecting the propagating and evanescent nature of the different harmonics. The number α_n/k is interpreted as the scattering angle of the Fourier mode n, that is,

$$e^{ix\alpha_n} = e^{ikx\sin\theta_n}, \tag{9.10}$$

leading to

$$\theta_n = \arcsin(\alpha_n/k). \tag{9.11}$$

Whenever the argument of the inverse sine function, in the previous equation, is smaller than one, we have propagating diffracted orders. In fact, $\alpha_n^2 + [\beta_n^{(1)}]^2 = k^2$.

The solution of the problem in the regions D_+ and D_-, that is, above and below the curve $y = a(x)$, respectively, can be transformed into one of a homogeneous boundary condition by transforming from the x, y reference frame to the $u, a(u)$ one. We introduce the change of coordinates

$$y = a(x) + u. \tag{9.12}$$

Clearly, the interface is represented by the homogeneous boundary $u = 0$.

9.2 Tangent and Normal Vectors, and Boundary Conditions

To define the boundary condition in a curved interface we need to find the tangent and normal vectors to that interface. A vector tangent to the curve $y = a(x)$ at point x is given by

$$\mathbf{t} = \hat{\mathbf{x}} + \dot{a}\hat{\mathbf{y}}, \tag{9.13}$$

where $\hat{\mathbf{x}}$ and $\hat{\mathbf{y}}$ are unit vectors, and $\dot{a} = da(x)/dx$. Thus, the unit vector tangent to that curve is

$$\boxed{\hat{\mathbf{t}} = \frac{1}{\sqrt{1 + \dot{a}^2}} \left(\hat{\mathbf{x}} + \dot{a}\hat{\mathbf{y}}\right).} \tag{9.14}$$

The normal vector to the same curve reads

$$\boxed{\hat{\mathbf{n}} = \frac{1}{\sqrt{1 + \dot{a}^2}} \left(-\dot{a}\hat{\mathbf{x}} + \hat{\mathbf{y}}\right).} \tag{9.15}$$

Clearly $\hat{\mathbf{n}} \cdot \hat{\mathbf{t}} = 0$. Since the current is tangent to the graphene sheet we must have $\mathbf{J} = \sigma \mathbf{E}_t = \sigma E_t \hat{\mathbf{t}}$, where σ is the optical conductivity and E_t is given by

$$E_t = \mathbf{E} \cdot \hat{\mathbf{t}} = \frac{1}{\sqrt{1 + \dot{a}^2}} \left(E_x + \dot{a}E_y\right) \tag{9.16}$$

One of the boundary conditions requires the calculation of vector product $\mathbf{J} \times \hat{\mathbf{n}}$:

$$\mathbf{J} \times \hat{\mathbf{n}} = \frac{\sigma}{\sqrt{1 + \dot{a}^2}} \left(E_x + \dot{a} E_y \right) \hat{\mathbf{t}} \times \hat{\mathbf{n}} = \frac{\sigma}{\sqrt{1 + \dot{a}^2}} \left(E_x + \dot{a} E_y \right) \hat{\mathbf{z}} . \tag{9.17}$$

Thus, the boundary condition $B_z^+ \hat{\mathbf{z}} - B_z^- \hat{\mathbf{z}} = \mu \sigma \mathbf{J} \times \hat{\mathbf{n}}$ reads

$$B_z^+ - B_z^- = \frac{\mu \sigma}{\sqrt{1 + \dot{a}^2}} \left(E_x + \dot{a} E_y \right) . \tag{9.18}$$

The second boundary condition is $\mathbf{E}_t^+ = \mathbf{E}_t^-$, which can be written as

$$E_x^+ + \dot{a} E_y^+ = E_x^- + \dot{a} E_y^- . \tag{9.19}$$

9.3 A Trivial Example: Recovering Previous Results

When the surface is flat we have $\dot{a} = 0$. Then, the boundary conditions reduce to

$$B_z^+ - B_z^- = \mu \sigma E_x , \tag{9.20}$$
$$E_x^+ = E_x^- . \tag{9.21}$$

The second boundary condition can be written as

$$\frac{1}{\epsilon_1} \partial_y B_z^+ = \frac{1}{\epsilon_2} \partial_y B_z^- . \tag{9.22}$$

Since the B^+ and B^- behave, generically, as $e^{-q_1 y}$ and $e^{q_2 y}$, respectively, we have

$$-\frac{q_1}{\epsilon_1} B_z^+ = \frac{q_2}{\epsilon_2} B_z^- \Leftrightarrow B_z^- = -\frac{q_1 \epsilon_2}{q_2 \epsilon_1} B_z^+ . \tag{9.23}$$

Then, the first boundary condition reads

$$B_z^+ + \frac{q_1 \epsilon_2}{q_2 \epsilon_1} B_z^+ = \mu \sigma \frac{i}{\omega \mu \epsilon_1} \partial_y B_z^+ = -i \mu \sigma \frac{q_1}{\omega \mu \epsilon_1} B_z^+ , \tag{9.24}$$

which gives the spectrum of the surface plasmon-polaritons in graphene, since we obtain

$$\boxed{\frac{\epsilon_1}{q_1} + \frac{\epsilon_2}{q_2} + i \frac{\sigma}{\omega} = 0 ,} \tag{9.25}$$

in agreement with the results of Chap. 4.

9.4   The Fields in D_+ and D_-

The fields in region D_+ are (the harmonic convention $e^{-i\omega t}$ is used):

$$B_z^+ = e^{i(\alpha_0 x - \beta_0^{(1)} y)} + \sum_n e^{i(\alpha_n x + \beta_n^{(1)} y)} A_n \,, \tag{9.26}$$

$$E_x^+ = \frac{1}{\omega\mu\epsilon_1}\left(\beta_0^{(1)} e^{i(\alpha_0 x - \beta_0^{(1)} y)} - \sum_n \beta_n^{(1)} e^{i(\alpha_n x + \beta_n^{(1)} y)} A_n\right) , \tag{9.27}$$

and

$$E_y^+ = \frac{1}{\omega\mu\epsilon_1}\left(\alpha_0 e^{i(\alpha_0 x - \beta_0^{(1)} y)} + \sum_n \alpha_n e^{i(\alpha_n x + \beta_n^{(1)} y)} A_n\right) . \tag{9.28}$$

The fields in region D_- are

$$B_z^- = \sum_n e^{i(\alpha_n x - \beta_n^{(2)} y)} C_n \,, \tag{9.29}$$

$$E_x^- = \frac{1}{\omega\mu\epsilon_2}\sum_n \beta_n^{(2)} e^{i(\alpha_n x - \beta_n^{(2)} y)} C_n \,, \tag{9.30}$$

and

$$E_y^- = \frac{1}{\omega\mu\epsilon_2}\sum_n \alpha_n e^{i(\alpha_n x - \beta_n^{(2)} y)} C_n \,. \tag{9.31}$$

Recall that the constants α_n and $\beta_n^{(p)}$ are given by

$$\alpha_n = k\sin\theta - n2\pi/D = k\sin\theta - nG \,, \tag{9.32}$$

$$\beta_n^{(p)} = \sqrt{k_p^2 - \alpha_n^2}, \quad k_p > \alpha_n \,, \tag{9.33}$$

$$\beta_n^{(p)} = i\sqrt{\alpha_n^2 - k_p^2}, \quad k_p < \alpha_n \,, \tag{9.34}$$

where $k_p = \omega/v_p$ and $p = 1, 2$; also we have $k_1 \equiv k$. In what follows, we need the Fourier transforms

$$\dot{a} = \sum_p \dot{a}_p e^{iGpx} \,, \tag{9.35}$$

$$\frac{1}{\sqrt{1+\dot{a}^2}} = \sum_p K_p e^{iGpx} \,, \tag{9.36}$$

$$\frac{\dot{a}}{\sqrt{1+\dot{a}^2}} = \sum_p R_p e^{iGpx} \,, \tag{9.37}$$

where $\dot{a}_p$, K_p, and R_p have the generic definition

$$C_p = \frac{1}{D} \int_0^D e^{-iGpx} c(x)\,dx\,. \tag{9.38}$$

When $\dot{a} \ll 1$ we have $R_p = \dot{a}_p$ and $K_p = \delta_{p,0}$, which is the case for shallow gratings. If we now introduce the change of variables $y = u + a(x)$, the boundary conditions are applied at $u = 0$. Then we replace y by $u + a(x)$ in the fields and put $u = 0$ when we consider the boundary conditions. We also define

$$e^{i\beta_n^{(1)} a(x)} = \sum_j L_{j;n}^{(1)} e^{iGjx}\,, \tag{9.39}$$

$$e^{-i\beta_0^{(1)} a(x)} = \sum_j L_{j;0}^{(1)} e^{iGjx}\,, \tag{9.40}$$

$$e^{-i\beta_n^{(2)} a(x)} = \sum_j L_{j;n}^{(2)} e^{iGjx}\,, \tag{9.41}$$

where

$$L_{m;n}^{(\kappa)} = \frac{1}{D} \int_0^D e^{-iGmx} e^{(-1)^{\kappa+1} i\beta_n^{(\kappa)} a(x)}\,dx\,, \tag{9.42}$$

with $\kappa = 1, 2$ and

$$L_{p;0}^{(1)} = \frac{1}{D} \int_0^D e^{-iGpx} e^{-i\beta_0^{(1)} a(x)}\,dx\,. \tag{9.43}$$

After a lengthy calculation, the boundary condition (9.18) reads

$$L_{k;0}^{(1)} + \sum_n L_{n+k;n}^{(1)} A_n - \sum_n L_{n+k;n}^{(2)} C_n = \frac{\sigma}{\omega\epsilon_2}\left(\sum_{n,j} K_{k-j}\beta_n^{(2)} L_{n+j;n}^{(2)} C_n\right.$$

$$\left. + \sum_{n,j} R_{k-j}\alpha_n L_{n+j;n}^{(2)} C_n\right)\,, \tag{9.44}$$

and the boundary condition (9.19) reads

$$\frac{\epsilon_2}{\epsilon_1}\left(\beta_0 L_{k;0}^{(1)} + \alpha_0 \sum_j \dot{a}_{k-j} L_{j;0}^{(1)} - \sum_n \beta_n^{(1)} L_{n+k;n}^{(1)} A_n\right.$$

$$\left. + \sum_{n,j} \dot{a}_{k-j}\alpha_n L_{n+j;n}^{(1)} A_n\right) = \sum_n \beta_n^{(2)} L_{n+k;n}^{(2)} C_n + \sum_{n,j} \dot{a}_{k-j}\alpha_n L_{n+j;n}^{(2)} C_n\,. \tag{9.45}$$

We can use Eqs. (9.44) and (9.45) to compute the intensity of the reflect orders.

9.4.1 *Reflectance and transmittance efficiencies*

In the end we want to compute the amount of reflected, transmitted, and absorbed radiation. To that end we have to defined the different electromagnetic fluxes. The pointing vector $\mathbf{S} = \mathbf{E} \times \mathbf{B}^*/\mu$ is given by

$$\mathbf{S} = \hat{\mathbf{x}} E_y B_z^* - \hat{\mathbf{y}} E_x B_z^* . \tag{9.46}$$

From Maxwell's equations we have

$$E_x = i\frac{v^2}{\omega}\partial_y B_z . \tag{9.47}$$

Thus for a mode m we have

$$S_{\text{inc}} = -\frac{v_1^2}{\omega}\beta_0^{(1)} , \tag{9.48}$$

$$S_{\text{refl}} = \frac{v_1^2}{\omega}\beta_n^{(1)}|A_m|^2 , \tag{9.49}$$

$$S_{\text{tran}} = -\frac{v_2^2}{\omega}\beta_n^{(2)}|C_m|^2 \tag{9.50}$$

for the incident, reflected, and transmitted pointing vector, respectively. Thus the reflectance and transmittance efficiencies are

$$\mathcal{R}_n = \frac{\beta_n^{(1)}}{\beta_0^{(1)}}|A_m|^2 , \tag{9.51}$$

$$\mathcal{T}_n = \frac{v_2^2\beta_n^{(2)}}{v_1^2\beta_0^{(1)}}|C_m|^2 = \frac{\epsilon_1\beta_n^{(2)}}{\epsilon_2\beta_0^{(1)}}|C_m|^2 , \tag{9.52}$$

respectively. Naturally, the absorbance is defined as $\mathcal{A} = 1 - \sum_n(\mathcal{R}_n + \mathcal{T}_n)$, where n extends over the propagating orders only.

9.4.2 *Particular limits for the transmittance and the reflectance*

As a rather simple limit, we want to obtain from Eqs. (9.44) and (9.45) the transmittance and the reflectance of a flat graphene sheet in vacuum, and the same quantities for an interface composed of two flat dielectrics. For these two cases we have $a(x) = 0$ which implies: $\dot{a}(x) = 0$, $R_p = 0$, $K_p = \delta_{p,0}$, and $L_p = \delta_{p,0}$. In this case, Eqs. (9.44) and (9.45) reduce to

$$\delta_{k,0} + A_{-k} - C_{-k} = \frac{\sigma}{\omega\epsilon_2}\beta^{(2)}C_{-k} , \tag{9.53}$$

$$\beta^{(1)}\delta_{k,0} - \beta_{-k}^{(1)}A_{-k} = \frac{\epsilon_1}{\epsilon_2}\beta_{-k}^{(2)}C_{-k} . \tag{9.54}$$

The last two equations imply that only A_0 and C_0 are non-zero. Then,

$$1 + A_0 - C_0 = \frac{\sigma}{\omega \epsilon_2} \beta_0^{(2)} C_0 , \qquad (9.55)$$

$$1 - A_0 = \frac{\epsilon_1 \beta_0^{(2)}}{\epsilon_2 \beta_0^{(1)}} C_0 , \qquad (9.56)$$

from which we obtain

$$2 = \left(1 + \frac{\epsilon_1 \beta_0^{(2)}}{\epsilon_2 \beta_0^{(1)}} + \frac{\sigma}{\omega \epsilon_2} \beta_0^{(2)} \right) C_0 . \qquad (9.57)$$

In the particular limit $\epsilon_1 = \epsilon_2$, C_0 reduces to

$$C_0 = \frac{2}{2 + \sigma \cos \theta / (\epsilon_1 v_1)} , \qquad (9.58)$$

which is the transmittance amplitude through graphene at an incident angle θ. The transmittance follows from $T_0 = |C_0|^2$. Another particular limit is $\sigma = 0$. In this case we have

$$C_0 = \frac{2 \epsilon_2 \beta_0^{(1)}}{\epsilon_2 \beta_0^{(1)} + \epsilon_1 \beta_0^{(2)}} , \qquad (9.59)$$

from which follows

$$\begin{aligned} A_0 = -1 + C_0 &= \frac{\epsilon_2 \beta_0^{(1)} - \epsilon_1 \beta_0^{(2)}}{\epsilon_2 \beta_0^{(1)} + \epsilon_1 \beta_0^{(2)}} \\ &= \frac{\sqrt{\epsilon_2/\epsilon_1} \cos \theta - \sqrt{1 - \epsilon_1 \sin^2 \theta / \epsilon_2}}{\sqrt{\epsilon_2/\epsilon_1} \cos \theta + \sqrt{1 - \epsilon_1 \sin^2 \theta / \epsilon_2}} , \end{aligned} \qquad (9.60)$$

which reproduces the well known result from elementary optics for the reflectance amplitude at the interface between two dielectrics.

9.5 A Non-Trivial Example: A Grating with a Sine-Profile

Let us now consider a non-trivial example, where the expressions for R_p, K_p, and L_p are not expressed in terms of simple numbers, as in the previous example. We assume a profile of the form

$$a(x) = h \sin(2\pi x / D) , \qquad (9.61)$$

from which follows

$$\dot{a} = 2\pi h / D \cos(2\pi x / D) . \qquad (9.62)$$

In the regime $2\pi h / D \ll 1$, we have $R_p \approx \dot{a}_p$ and $K_p \approx \delta_{p,0}$, with

$$\dot{a}_p = \frac{\pi h}{D} (\delta_{m,1} + \delta_{m,-1}) . \qquad (9.63)$$

In this case, Eqs. (9.44) and (9.45) simplify to

$$\sum_{n=-N}^{N} f_a(n+k;n)A_n + \sum_{n=-N}^{N} f_c(n+k;n)C_n = L_{k;0}^{(1)}, \qquad (9.64)$$

$$\sum_{n=-N}^{N} g_a(n+k;n)A_n + \sum_{n=-N}^{N} g_c(n+k;n)C_n = L_{k;0}^{(1)}$$

$$+ \frac{\alpha_0}{\beta_0}\frac{\pi h}{D}(L_{k-1;0}^{(1)} + L_{k+1;0}^{(1)}), \qquad (9.65)$$

where the functions $f_a(n+k;n)$, $f_c(n+k;n)$, $g_a(n+k;n)$, and $g_c(n+k;n)$ read

$$f_a(n+k;n) = -L_{n+k;n}^{(1)}, \qquad (9.66)$$

$$f_c(n+k;n) = \left(1 + \frac{\sigma}{w\epsilon_2 c\hbar}\hbar c\beta_n^{(2)}\right)L_{n+k;n}^{(2)}$$

$$+ \frac{\pi h}{D}\frac{\sigma}{w\epsilon_2 c\hbar}\hbar c\alpha_n(L_{n+k-1;n}^{(2)} + L_{n+k+1;n}^{(2)}), \qquad (9.67)$$

$$g_a(n+k;n) = \frac{\beta_n^{(1)}}{\beta_0^{(1)}}L_{n+k;n}^{(1)} - \frac{\pi h}{D}\frac{\alpha_n}{\beta_0}(L_{n+k-1;n}^{(1)} + L_{n+k+1;n}^{(1)}), \qquad (9.68)$$

$$g_c(n+k;n) = \frac{\epsilon_1}{\epsilon_2}\frac{\beta_n^{(2)}}{\beta_0^{(1)}}L_{n+k;n}^{(2)} + \frac{\pi h}{D}\frac{\epsilon_1}{\epsilon_2}\frac{\alpha_n}{\beta_0^{(1)}}(L_{n+k-1;n}^{(2)} + L_{n+k+1;n}^{(2)}).$$

$$(9.69)$$

We also have the following definitions

$$L_{m;n}^{(1)} = J_m(\beta_n^{(1)}h), \qquad (9.70)$$

$$L_{m;0}^{(1)} = J_m(-\beta_0^{(1)}h), \qquad (9.71)$$

$$L_{m;n}^{(2)} = J_m(-\beta_n^{(2)}h), \qquad (9.72)$$

where $J_m(x)$ is the regular Bessel function at the origin. It is convenient for numerical purposes to write

$$\hbar c\beta_0^{(1)} = \sqrt{\epsilon_1^r}\hbar w \cos\theta, \qquad (9.73)$$

$$\hbar c\alpha_n = \sqrt{\epsilon_1^r}\hbar w \sin\theta - 2\pi nhc/D, \qquad (9.74)$$

$$\hbar c\beta_n^{(p)} = \sqrt{\epsilon_p^r(\hbar w)^2 - (\sqrt{\epsilon_1^r}\hbar w \sin\theta - 2\pi nhc/D)^2}, \qquad (9.75)$$

and

$$\frac{\sigma}{w\epsilon_2}\beta_n^{(2)} = \alpha\frac{4}{\epsilon_2^r}\frac{E_F}{\hbar w}\frac{\hbar c\beta_n^{(2)}}{\hbar\gamma - i\hbar w}, \qquad (9.76)$$

where $\epsilon_{1/2}^r$ are the relative dielectric permittivities, E_F is the Fermi energy, and α is the fine structure constant. In Fig. 9.2 we represent the efficien-

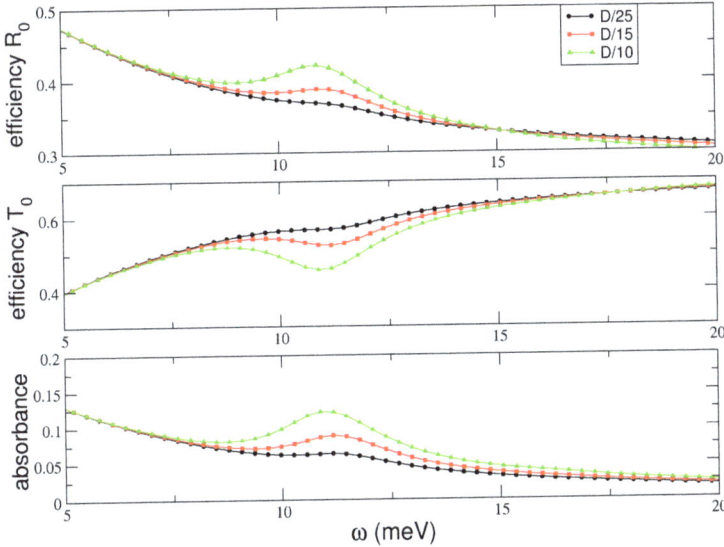

Figure 9.2: Dependence of the reflectance, transmittance efficiencies, and absorbance as function of frequency for different depth of the grooves for a sine-grating. The parameters are $D = 10$ μm, $\hbar\gamma = 2.6$ meV, $E_F = 0.45$ eV, $\epsilon_1 = 1$, $\epsilon_2 = 11$, and $h = D/25$ (circles), $h = D/15$ (squares), and $h = D/10$ (triangles).

cies $\mathcal{R}_0$ and $\mathcal{T}_0$, top and central panels, respectively, and the absorbance, $\mathcal{A} = 1 - \mathcal{R}_0 - \mathcal{T}_0$, as function of the energy of the incoming photon, for different values of the ratio h/D. When $h/D \ll 1$ we recover the properties of a flat graphene sheet. In Fig. 9.3 we compare results from the Rayleigh procedure, described in this chapter, with those produced by the extinction theorem (dashed curves) [Ferreira and Peres (2012)]. We see that the agreement is excellent. In the figure, results for several angles of incidence θ are given for $h/D = 0.1$. The resonance seen in Figs. 9.2 and 9.3 around 10 meV is due to the excitation of a surface-plasmon-polariton of energy

$$\hbar\omega \approx \sqrt{\frac{4\alpha}{\epsilon_1^r + \epsilon_2^r} E_F c\hbar \frac{2\pi}{D}} = 11 \text{ meV}. \tag{9.77}$$

It should now be clear that the technique described above is suitable for the excitation of SPP's. A different approach to the same problem was considered in [Slipchenko *et al.* (2013b)]. We note that we have not included the effect of the grating profile, which also acts as a gate, in the

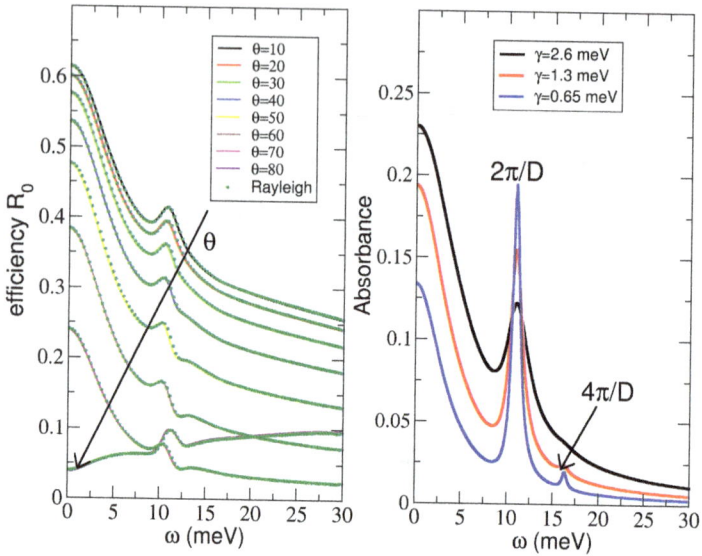

Figure 9.3: Reflectance and absorbance of graphene on a silicon sine-grating. Left: Dependence of the reflectance on the angle of incidence. We compare the Rayleigh method (solid circles) with the results from the extinction theorem [Ferreira and Peres (2012)] (solid lines); the parameters are $D = 10$ μm, $\hbar\gamma = 2.6$ meV, $E_F = 0.45$ eV, $\epsilon_1 = 1$, $\epsilon_2 = 11$, $h = D/10$. The arrow indicates the direction of growth of the incident angle θ for the several curves. Right: Absorbance for the same parameters as before but different γ values. The prominent peak see in the absorbance curves is the fingerprint of the excitation of an SPP.

conductivity of the system. This effect will induce an additional periodicity into the problem [Ochiai (2015)] which can be tackled by methods similar to those explained above by [Peres *et al.* (2012)]. An interesting variation of the problem we have considered here was proposed in [Ferreira and Peres (2012)], where a periodic meta-material grating was covered with graphene. In this work it was show that it is possible to obtain 100% of light absorption (see also [Thongrattanasiri *et al.* (2012a)]) in graphene due to the excitation of surface plasmon-polaritons. The tuning parameter was (maybe as expected) the amplitude h of the grating profile. It was also found (and this maybe less obvious) that the dependence of the absorption on h is non-monotonous. Interesting enough, the corrugation needed to achieve total absorption is not very large: $h \sim 10 - 100$ nm, and decreases as the Fermi energy increases. These findings are summarized in Fig. 9.4. It should be appreciate that central to the mechanism of total absorption

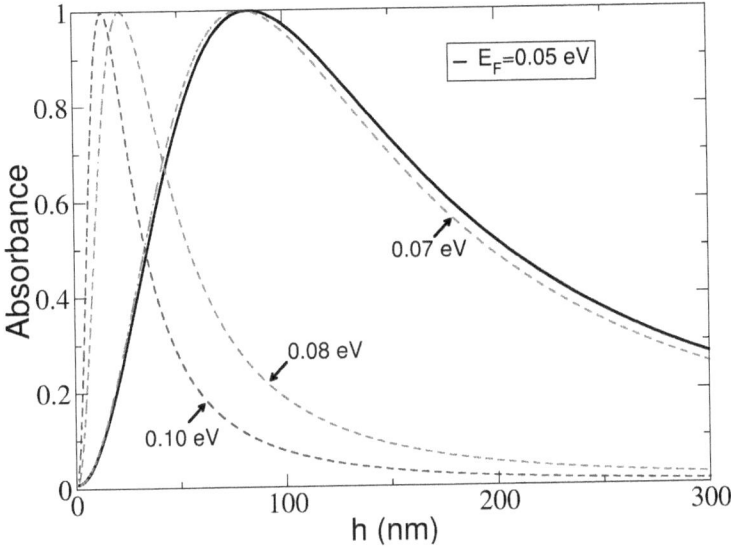

Figure 9.4: Dependence of the absorption at resonance (that is, at the frequency ω_{spp} of the surface plasmon-polariton) on the corrugation height for different graphene Fermi energies, E_F. The fast change in the absorbance as E_F varies from 0.07 to 0.08 eV results from a suppression of interband transitions in graphene as $2E_F \gtrsim \hbar\omega_{spp}$. In this calculation, both the inter- and intra-band conductivity have been included.

is the negative value of the dielectric function of the meta-material in the spectral region of interest (see also Fig. 5.2, where a similar effect has been explored). Indeed, we can imagine that by changing $\epsilon(\omega)$, for example, by considering a meta-material with an appropriate plasma frequency, the resonant frequencies ω_{spp} can be controlled within a broad range of values, thus tailoring the hybrid structure for a desired spectral region.

An interesting variation of the above problem is that of a square-profile grating [Peres *et al.* (2013)] (see also [Sheng *et al.* (1982)]). This problem is rather pathological due to the infinite derivative of the profile at the transition from the tooth to the trenches. However, the problem can still be solved exactly by considering a Kronig-Penney model in the grating zone. This method was applied in Sec. 5.1.1 to the solution of graphene on a Gallium-Arsenide substrate with a square grating in between, and the technical details can be found in [Peres *et al.* (2013)]. As expected, this configuration is quite efficient in exciting surface plasmon-polaritons on graphene. The efficiency of absorption is a function of the depth of the trenches, as in the case of Fig. 9.4 albeit with a different behavior.

Chapter 10

Excitation of Plasmons by an Emitting Dipole

The problem of the emission of an excited dye (or quantum dot) over a graphene sheet has been studied both theoretically [Gómez-Santos and Stauber (2011); Koppens *et al.* (2011)] and experimentally [Kasry *et al.* (2012); Gaudreau *et al.* (2013)]. The case of Förster resonant energy transfer (FRET) between two dyes in the presence of graphene has also been considered [Biehs and Agarwal (2013)]. In this chapter we show that a dye can, in principle, de-excite via the excitation of plasmons in graphene. This possibility has its fingerprint in a resonance in the effective dielectric function of the system, or in the dispersion of the life-time of the excited state of the dye with the distance to the graphene sheet. We first formulate the problem in the electrostatic approximation and use Fermi's golden rule, following [Swathi and Sebastian (2009)]. We show that this approach corresponds to a decaying pathway associated with the induction of particle-hole excitations in graphene. We end the chapter with an exact calculation of the total decaying rate of the dye in the presence of graphene, where the fingerprint of plasmon excitation becomes manifest. We show that the decaying rate of an emitter in the presence of graphene differs substantially from the free space decaying rate, a manifestation of the Purcell effect. This effect has been demonstrated in a variety of physical systems, such as radiating molecules near a metallic-dielectric interface, Rydberg atoms in an optical cavity, and nuclear transitions in small metallic particles, the latter example being the basis of the work of Purcell [Purcell (1995)]. A decaying emitter can couple to surface plasmons when the distance to the metallic interface (which can be doped graphene) is smaller than the emission wavelength (but much larger than typical atomic distances), the reason being that the dipole field contains large in-plane wavevectors, associated with evanescent modes, that do not propagate in the far field but can couple to the SPP's at the surface of the conductor [Toropov and Shubina (2015)]. For distances much smaller than the emission wavelength, the emitter couples to the conductor by inducing either inter-band or intra-band absorption. For neutral graphene only the inter-band transitions pathway exists. This latter case is discussed in the appendix section M.2.

10.1 Statement of the Problem and a Bit of Electrostatics

Let us consider a point-like dipole embedded, at a distance $z > 0$, in a dielectric medium of permittivity ϵ_1 above a doped graphene sheet ($E_F > 0$). This dipole represents a quantum emitter (such as a dye). The sheet rests on a dielectric of permittivity ϵ_2 (located at $z < 0$). We want to derive the explicit boundary conditions obeyed by the field (or the potential for that matter) at the graphene plane ($z = 0$). We define **s** as the in-plane ($z = 0$) position vector, as shown in Fig. 10.1. We want to compute the rate at

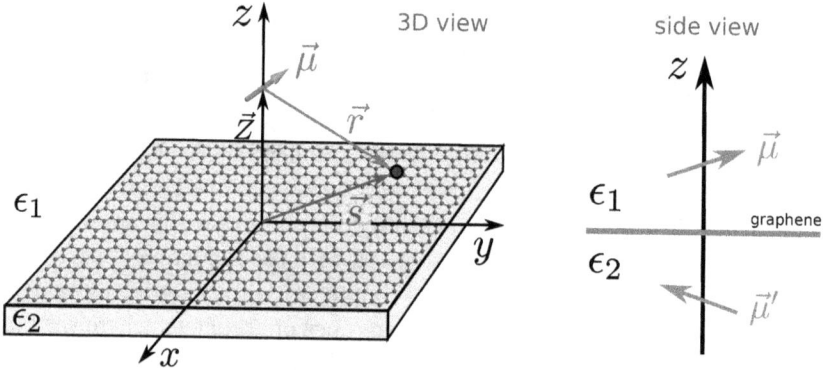

Figure 10.1: An emitting dipole (e.g. a dye), characterized by a dipole moment μ, over a graphene sheet. Above the graphene the dielectric constant is ϵ_1 and below it is ϵ_2. The vector **r** can be written as $\mathbf{r} = \mathbf{s} - z\hat{\mathbf{u}}_\mathbf{z}$, and **s** an in-plane vector. At the right we have a side view of the geometry under consideration together with the image dipole in medium 2.

which the dye decays to the ground state via an energy transfer to electronic particle-hole excitations in graphene, that is, we want to compute the non-radiative decaying rate of the dye. We further assume that graphene is doped with electrons and the energy between the excited state and the ground state of the dye, $\hbar\omega$, is smaller than $2E_F$. In this case, there are no electronic inter-band excitations in graphene; all excitation processes are intra-band particle-hole excitations. The rate of excitation of intra-band electronic processes is given by Fermi's golden rule (see Sec. 2.3.1 for the definitions of the different quantities)

$$\frac{1}{\tau} = g_s g_v \frac{2\pi}{\hbar} \sum_{\mathbf{k}} \sum_{\mathbf{q}} |\langle \psi_{\mathbf{k+q},+} | V | \psi_{\mathbf{k},+} \rangle|^2 F(\mathbf{k},\mathbf{q}) \delta(E_{\mathbf{k+q},+} - E_{\mathbf{k},+} - \hbar\omega),$$

(10.1)

where g_s and g_v are the spin and valley degeneracies, respectively, $F(\mathbf{k},\mathbf{q})$ is a combination of Fermi distribution functions (see Appendix M):

$$F(\mathbf{k},\mathbf{q}) = f(v_F \hbar k)[1 - f(v_F \hbar |\mathbf{k} + \mathbf{q}|)],$$

(10.2)

taking into account the empty and occupied electronic states, and V represents the dipole potential. The next step is to compute the potential felt by the electrons due to the dipole at position $\mathbf{z}$. This is not as simple as it may sound because of the existence of two different dielectrics. Indeed, we have to compute the image potential created by the real dipole. Let us do this in two steps. First we consider the case where graphene is absent and later we introduce graphene into the problem. In the first case, the boundary conditions are the continuity of the potential and of the component perpendicular to the interface of the electric displacement vector $\mathbf{D}$ at $z = 0$. Following the method of images, we assume that there is a image dipole at $z < 0$ with dipole moment $\boldsymbol{\mu}'$ (see Fig. 10.2). The original dipole is located at $z = h$ and the image dipole at $z = -h$. The potential of the

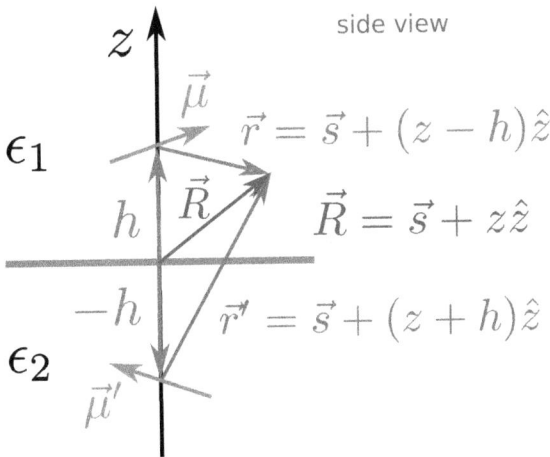

Figure 10.2: Position vectors $\mathbf{r}$ and $\mathbf{r}'$ of the point $\mathbf{R} = (x, y, z)$ relative to the dipoles. The vector $\mathbf{s}$ has coordinates (x, y) and $\hat{\mathbf{z}} = \hat{\mathbf{u}}_{\mathbf{z}}$.

original dipole reads

$$V = \frac{1}{4\pi\epsilon_0\epsilon_1} \frac{\boldsymbol{\mu} \cdot \mathbf{r}}{r^3}, \qquad (10.3)$$

and that of the image dipole reads

$$V_i = \frac{1}{4\pi\epsilon_0\epsilon_1} \frac{\boldsymbol{\mu}' \cdot \mathbf{r}}{r^3}, \qquad (10.4)$$

where $\mathbf{r} = (x, y, z - h)$ and $\mathbf{r}' = (x, y, z + h)$ (see Fig. 10.2). At $z > 0$ the total potential reads $V_1 = V + AV_i$ and at $z < 0$ we have $V_2 = BV$, where A and B are to be determined from the boundary conditions. We assume that $\boldsymbol{\mu}' = (\mu_z, -\mu_x, -\mu_y)$ and check whether this choice allows the determination of A and B uniquely. The $z-$component of the electric field is determined from the derivative of the potential in order to z. The condition $V_1 = V_2$ at $z = 0$ implies $A + B = 1$ and the condition $\epsilon_1 E_{z,1} = \epsilon_2 E_{z,2}$ implies $\epsilon_1(1 + A) = \epsilon_2 B$. Solving these two equations we obtain

$$A = \frac{\epsilon_2 - \epsilon_1}{\epsilon_2 + \epsilon_1}, \qquad (10.5)$$

$$B = \frac{2\epsilon_1}{\epsilon_2 + \epsilon_1}. \qquad (10.6)$$

Then, the potential felt by the electrons in graphene would be

$$V_2 = BV = \frac{2}{\epsilon_2 + \epsilon_1} \frac{1}{4\pi\epsilon_0} \frac{\boldsymbol{\mu} \cdot \mathbf{r}}{r^3}, \qquad (10.7)$$

with $z = 0$. This is true for neutral graphene. However, the electrons in doped graphene change the second boundary condition. Indeed, the correct boundary condition is given by Eq. (K.1). Furthermore, due to the presence of the dipole, the electronic charge is not homogeneous, becoming a function of the vector $\mathbf{s}$. As shown in Appendix K, and ignoring diffusion, the correct boundary condition is

$$\epsilon_1 E_z^1(z = 0, \mathbf{q}) - \epsilon_2 E_z^2(z = 0, \mathbf{q}) = \sigma_D(\omega)\mathbf{q} \cdot \mathbf{E}(z = 0, \mathbf{q})/\omega, \qquad (10.8)$$

where, due to the dot product of $\mathbf{E}(\mathbf{q})$ with $\mathbf{q}$, only the component of the field parallel to the graphene plane enters in the right-hand-side of the boundary condition. Note that the boundary condition (10.8) is written in a mixed representation, where we have kept the spatial dependence along the $z-$direction but have Fourier transformed the x- and y-coordinates of the fields. We can relate the field in-plane, $\mathbf{E}_\parallel(\mathbf{q})$, to the potential in the following way: $\mathbf{E}_\parallel(z, \mathbf{s}) = -\nabla_\mathbf{s}\phi(z, \mathbf{s})$, where $\phi(z, \mathbf{s})$ is the electrostatic potential. Introducing the Fourier representation of the field and of the potential we obtain $\mathbf{E}_\parallel(\mathbf{q}) = -i\mathbf{q}\phi(z, \mathbf{q})$, which leads to

$$\epsilon_0\epsilon_1 \frac{\partial}{\partial z}\phi(0^+, \mathbf{q}) - \epsilon_0\epsilon_2 \frac{\partial}{\partial z}\phi(0^-, \mathbf{q}) = i\sigma_D(\omega)q^2\phi(0, \mathbf{q})/\omega \equiv \lambda\phi(0, \mathbf{q}), \qquad (10.9)$$

where $\lambda = i\sigma_D(\omega)q^2/\omega$. Thus, we need to determine the Fourier transform of the potential (see Appendix L). For a dipole $\boldsymbol{\mu}_a$, at $z = h$, and for $z < h$ we obtain

$$V_a = \frac{\boldsymbol{\mu}_a}{2\epsilon_0\epsilon_1} \cdot \left(-i\frac{\mathbf{q}}{q} - \hat{u}_z\right) e^{-q(h-z)}. \tag{10.10}$$

For a dipole $\boldsymbol{\mu}_b$, at $z = -h$, and for $z > -h$ we obtain

$$V_b = \frac{\boldsymbol{\mu}_b}{2\epsilon_0\epsilon_2} \cdot \left(-i\frac{\mathbf{q}}{q} + \hat{u}_z\right) e^{-q(h+z)}. \tag{10.11}$$

To determine the potential (in Fourier space) felt by the dipole we again employ the method of images. In the region $0 < z < h$ the total potential is a sum of the real dipole plus the image dipole, that is

$$V_+(z) = V_a + V_b, \tag{10.12}$$

and for $z < 0$, the potential is due to a third dipole located at $z = h$, that is,

$$V_-(z) = \frac{\boldsymbol{\mu}_c}{2\epsilon_0\epsilon_1} \cdot \left(-i\frac{\mathbf{q}}{q} - \hat{u}_z\right) e^{-q(h-z)}. \tag{10.13}$$

It follows from the boundary conditions that $V_+(0) = V_-(0)$ and

$$\epsilon_0\epsilon_1\frac{\partial}{\partial z}V_+(0^+) - \epsilon_0\epsilon_2\frac{\partial}{\partial z}V_-(0^-) = \lambda V_-(0). \tag{10.14}$$

We choose to write $\boldsymbol{\mu}_a = (\mu_x^a, \mu_y^a, \mu_z^a)$, $\boldsymbol{\mu}_b = (-\mu_x^b, -\mu_y^b, \mu_z^b)$, and $\boldsymbol{\mu}_c = (\mu_x^c, \mu_y^c, \mu_z^c)$. Introducing these in the boundary conditions we obtain

$$\frac{\mu_z^b}{\epsilon_2} + \frac{\mu_z^c}{\epsilon_1} = \frac{\mu_z^a}{\epsilon_1}, \tag{10.15}$$

and

$$-\frac{\epsilon_1}{\epsilon_2}\mu_z^b + \left(\frac{\epsilon_2}{\epsilon_1} + \frac{\lambda}{q\epsilon_0\epsilon_1}\right)\mu_z^c = \mu_z^a. \tag{10.16}$$

Solving the last two equations for μ_z^b and μ_z^c we obtain

$$\mu_z^b = \mu_z^a\frac{\epsilon_2}{\epsilon_1}\frac{\epsilon_2 - \epsilon_1 + \lambda/(\epsilon_0 q)}{\epsilon_2 + \epsilon_1 + \lambda/(\epsilon_0 q)} \tag{10.17}$$

$$\mu_z^c = \mu_z^a\frac{2\epsilon_1}{\epsilon_2 + \epsilon_1 + \lambda/(\epsilon_0 q)}, \tag{10.18}$$

which is the generalization of the results (10.5) and (10.6). Identical equations hold for the other components of the dipole moments, $\mu_{x/y}^b$ and $\mu_{x/y}^c$. Finally, the potential (in Fourier space) felt by the electrons in graphene is

$$V_-(0) = \frac{\boldsymbol{\mu}_a \cdot (-i\mathbf{q}/q - \hat{u}_z)}{\epsilon_0[\epsilon_2 + \epsilon_1 + \lambda/(\epsilon_0 q)]} e^{-qh}. \tag{10.19}$$

We are now in position of computing the matrix element of the potential entering in Fermi's golden rule. We can show (see Appendix L) that the matrix element is given by[1]

$$\langle \psi_{\mathbf{k+q},+} | V | \psi_{\mathbf{k},+} \rangle = -e \frac{1}{2A} [1 + e^{i(\theta_{\mathbf{k}} - \theta_{\mathbf{k+q}})}] V_{-}(0) \,. \tag{10.20}$$

It is clear from the above expression that

$$|\langle \psi_{\mathbf{k+q},+} | V | \psi_{\mathbf{k},+} \rangle|^2 \propto \frac{1}{|\epsilon_2 + \epsilon_1 + \lambda/(\epsilon_0 q)|^2}$$

$$= \frac{1}{[\epsilon_1 + \epsilon_2 - q\sigma_D''/(\omega\epsilon_0)]^2 + q^2\sigma_D'^2/(\omega\epsilon_0)^2} \,, \tag{10.21}$$

where the conductivity has been separated in its real, σ_D', and imaginary, σ_D'', parts. The previous denominator has a resonance when the frequency obeys the condition

$$\epsilon_1 + \epsilon_2 - q\sigma_D''/(\omega\epsilon_0) = 0 \,, \tag{10.22}$$

which is nothing but the equation for the spectrum of graphene surface plasmon-polaritons in the electrostatic limit [see Eq. (4.16)]. Thus, at least in principle, the decay of a quantum emitter (a dye, a quantum dot, etc.) may excite SPP's in graphene.

A note on the results (10.17) and (10.18) is in order. Let us consider, for simplicity, the result for (10.18) in vacuum. In this case we obtain

$$\mu_z^c = \mu_z^a \frac{2}{2 + \lambda/(\epsilon_0 q)} \,. \tag{10.23}$$

We now show that the factor multiplying μ_z^a is nothing but the transmission amplitude of a plane wave through graphene in vacuum. In fact, that proportionally factor can be written as

$$t_p = \frac{2}{2 + i\sigma_D q/(\epsilon_0 \omega)} \,. \tag{10.24}$$

On the other hand, from the results of Appendix N, the transmission amplitude for a p-polarized wave reads

$$t_p = \frac{2}{2 + \pi\alpha f(\omega)\cos\theta} \,. \tag{10.25}$$

---

[1] In Eq. (10.20) we are using the spinors representing the electronic motion in graphene in the form:

$$\chi_\alpha = \frac{1}{\sqrt{2}} \begin{pmatrix} 1 \\ \alpha e^{i\theta_{\mathbf{q}}} \end{pmatrix},$$

and not as written in Eq. (2.49). The two representations are equivalent as they differ from an overall multiplicative phase. What representation to use is a matter of convenience.

Since $\sigma_D = \sigma_0 f(\omega)$ and $\cos\theta = k_z/k = c\sqrt{k^2 - q^2}/\omega$, the two results coincide when $qc/\omega \gg 1$, that is, in the electrostatic limit (in this case we have $\cos\theta \to icq/\omega$). This result brings about an interesting connection between the boundary conditions in electrostatics at the interface of two dielectrics and the transmittance and the reflectance amplitudes of the same interface. It should therefore come as no surprise to learn that the electromagnetic fields of a dipole at an interface can be expressed in terms of these amplitudes (more on this in Sec. 10.3).

10.2 Calculation of the Non-Radiative Transition Rate: Particle-Hole Excitations Pathway

We now compute the non-radiative transition-rate of a quantum emitter via the particle-hole excitations pathway. We will see that within the model of the previous section the calculation of $1/\tau$ cannot account for the presence of the SPP's. That is because we are assuming that the only possible excitation induced in graphene is the creation of particle-hole pairs. From Appendix M we have [see Eq. (M.15)]

$$\frac{1}{\tau} = 2\frac{2\pi}{\hbar}\frac{e^2}{\epsilon_0^2}\frac{2\pi\mu_z^2 + \pi\mu_\parallel^2}{(2\pi)^4\hbar v_F}\int_\Omega^{2k_F+\Omega}\frac{dq}{|\epsilon(\mathbf{q},\omega)|^2}\frac{q^3 e^{-2qz}}{\sqrt{q^2 - \Omega^2}}G(q). \quad (10.26)$$

From Fig. M.1 we see that the integral extends only over the regions 1A and 2A of Fig. 2.5. However, plasmons, as long-lived excitations, exist in region 1B (see Fig. 4.5). Nevertheless, the presence of the plasmons can still be felt rather indirectly. The idea is simple: the tail of the spectral function associated with the presence of the plasmons "leaks" from region 1B into region 1A, this is possible because of damping effects. This leaking, however, is appreciable only at frequencies and momenta where the dispersion is close to the line $\omega = v_F q$ (see Fig. 4.5) and therefore has no significant impact on the results computed in this section (the impact is higher if the dielectric constant of the substrate is large; see the right panel of Fig. 4.5). In technical terms, we say that the SPPs are, in this case, excited off-shell. Thus the calculation presented in this section refers to the decay pathway associated with Ohmic losses.

As noted in Appendix M, the integral over q cannot be performed analytically. Therefore we present below the result of a numerical calculation of $1/\tau$. For simplicity, we assume that graphene is in vacuum and compare the non-radiative decaying rate with the radiative decaying rate of an isolated dipole, given by [Novotny and Hecht (2014); Zettili (2009)]

$$\frac{1}{\tau_{\text{vac}}} = \frac{\omega^3 |\boldsymbol{\mu}|^2}{3\pi\epsilon_0 \hbar c^3} \,. \tag{10.27}$$

We then define the ratio

$$\Phi(z,\omega) = \frac{\tau_{\text{vac}}}{\tau}, \tag{10.28}$$

which allows to compare the two decaying rates. Next, we introduce the change of variables $q/\Omega = x$. This allows to write the decaying rate as

$$\frac{1}{\tau} = 2 \frac{2\pi}{\hbar} \frac{e^2 \omega^3}{\epsilon_0^2} \frac{2\pi\mu_z^2 + \pi\mu_\parallel^2}{(2\pi)^4 \hbar v_F^4} \int_1^{2k_F/\Omega+1} \frac{dx}{|\epsilon(x,\omega)|^2} \frac{x^3 e^{-2x\Omega z}}{\sqrt{x^2-1}} G(x\Omega), \tag{10.29}$$

and allows to rewrite $\Phi(z,\omega)$ as

$$\Phi(z,\omega) = 6\alpha_g \frac{\mu_z^2 + \mu_\parallel^2/2}{|\boldsymbol{\mu}|^2} \frac{c^3}{v_F^3} J(z,\omega), \tag{10.30}$$

where $\alpha_g = e^2/(4\pi\epsilon_0 \hbar v_F) \approx 2.3$ and

$$J(z,\omega) = \int_1^{2E_F/(\hbar\omega)+1} \frac{dx}{|\epsilon(x,\omega)|^2} \frac{x^3 e^{-2x\Omega z}}{\sqrt{x^2-1}} G(x\Omega), \tag{10.31}$$

where

$$\frac{1}{|\epsilon(x,\omega)|^2} = \frac{1}{[2 - x\sigma_D''/(v_F\epsilon_0)]^2 + x^2\sigma_D'^2/(v_F\epsilon_0)^2} \,. \tag{10.32}$$

For σ_D we use the value of the longitudinal non-homogeneous conductivity $\sigma_{xx}(\omega, q)$ derived in Appendix C. In Fig. 10.3 we plot both $\Phi(z,\omega)$, for three different values of z, and $1/|\epsilon(\mathbf{q},\omega)|^2$, for two different values of x, corresponding to the extremes of the integration interval in $\Phi(z,\omega)$. Whenever $\Phi(z,\omega)$ grows the non-radiative decaying rate is enhanced. We see an enhancement of $\Phi(z,\omega)$ at low frequencies. The effect of increasing z is to suppress the effect of the non-radiative energy transfer, as can be seen in the left panel of Fig. 10.3, by comparing the different curves. This is due to the exponential in the integral, which depends both on z and ω. In the right panel of Fig. 10.3 we see that $1/|\epsilon(\mathbf{q},\omega)|^2$ has no resonance in the integration interval.

We should note that, by construction, our model cannot be extended to frequencies higher that $2E_F$, since in this case, inter-band transitions start to play a role, an effect we have not included in the model (its inclusion is, however, straightforward and does not change the conclusions).

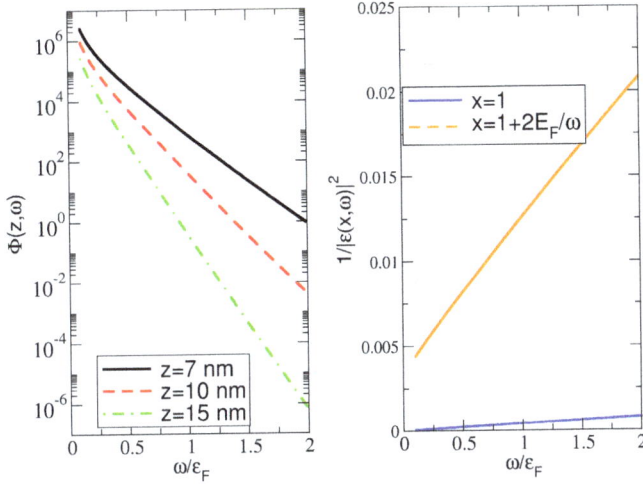

Figure 10.3: Non-radiative decaying rate of a dipole, as measured by the function $\Phi(z,\omega)$ (left panel), for different values of z, $z = 7, 10, 15$ nm. Right panel: Inverse of the absolute value of the dielectric function for different values of x, $x = 1$ and $x = 1 + 2E_F/(\hbar\omega)$. The parameters are $E_F = 0.3$ eV and $\hbar\gamma = 4$ meV.

10.3 Calculation of the Total Transition Rate: Full Electromagnetic Calculation

As we have seen, the decaying rate of a quantum emitter is influenced by the environment. It is possible to compute this effect using a full electromagnetic theory. What we need to know is the full electromagnetic field in the presence of the environment. This can be determined using dyadic Green's functions [Novotny and Hecht (2014)]. Here we consider a doped graphene sheet in vacuum with a quantum emitter located at distance z above graphene. To determine the fields, we need to know the reflectance amplitudes of a plane wave through graphene. These coefficients are derived in Appendix N and read

$$r_p = \frac{\pi\alpha f_L(\omega,q)s_z}{2 + \pi\alpha f_L(\omega,q)s_z}, \tag{10.33}$$

$$r_s = -\frac{\pi\alpha f_T(\omega,q)}{2s_z + \pi\alpha f_T(\omega,q)}, \tag{10.34}$$

for $p-$ and $s-$polarizations, respectively, and $s_z = \sqrt{1 - s^2}$. Taking into account intra-band transitions only, the functions $f_T(\omega,q)$ and $f_L(\omega,q)$ read

(see Appendix C)

$$f_L(\omega, q) = \frac{8E_F}{\pi} \frac{1}{\hbar\gamma - i\hbar\omega} \frac{1}{\beta^2(\omega, q)} \left(-1 + \frac{1}{\sqrt{1 - \beta^2(\omega, q)}} \right), \quad (10.35)$$

$$f_T(\omega, q) = \frac{8E_F}{\pi} \frac{1}{\hbar\gamma - i\hbar\omega} \frac{1}{\beta^2(\omega, q)} \left(-1 - \sqrt{1 - \beta^2(\omega, q)} \right), \quad (10.36)$$

where $\beta(\omega, q)$ is defined as

$$\beta(\omega, q) = \frac{i\hbar v_F q}{\Gamma - i\hbar\omega}, \quad (10.37)$$

and $\Gamma = \hbar\gamma$. It can be shown [Novotny and Hecht (2014)] that the rate of decay of the dipole, when compared with the decaying rate in vacuum is given by (see Appendix P)

$$
\begin{aligned}
\frac{1/\tau_g}{1/\tau_{\text{vac}}} = \frac{\tau_{\text{vac}}}{\tau_g} &= 1 + \frac{\mu_x^2 + \mu_y^2}{\mu^2} \frac{3}{4} \int_0^\infty ds \Re \left[\frac{s}{s_z}(r_s - s_z^2 r_p) e^{2i\omega z s_z/c} \right] \\
&+ \frac{\mu_z^2}{\mu^2} \frac{3}{2} \int_0^\infty ds \Re \left[\frac{s^3}{s_z} r_p e^{2i\omega z s_z/c} \right]
\end{aligned}
\quad (10.38)
$$

where $1/\tau_g$ is the decaying rate of the quantum emitter in the presence of graphene; the ratio τ_{vac}/τ_g is called the Purcell factor[2]. For $s > 1$ we have $\sqrt{1 - s^2} = i\sqrt{s^2 - 1}$. In general, decaying of a quantum emitter can occur through several channels, but in our case only three are active: radiative decay, particle-hole excitations in graphene, and excitation of surface plasmon-polaritons propagating at the surface of graphene. From Eq. (10.38) we see the existence of two regimes. In the first the emission is controlled by propagating waves, corresponding to $0 < s < 1$; in interval $1 < s < \infty$ the emission is by evanescent waves, which can excite surface plasmon-polaritons. The result of the calculation of the ratio τ_{vac}/τ_g is given in Fig. 10.4.

As in the case of Fig. 10.3, an increase of the distance z leads to a reduction of the decaying rate (and therefore to an increase of τ_g). This leads to a similar behaviour of τ_{vac}/τ_g and τ_{vac}/τ, as can be seen by comparing the left panels of Figs. 10.3 and 10.4. However, in Fig. 10.3 the mechanism of excitation of surface plasmon-polaritons is not present (although the dielectric function does have a pole at the plasmon dispersion the momentum is off-shell).

---

[2]The Purcell effect refers to the change of the decaying rate of an emitter due to the proximity of a system that can support both losses (as either metal or graphene) and/or surface modes (as, for example, a metal, graphene, and hBN).

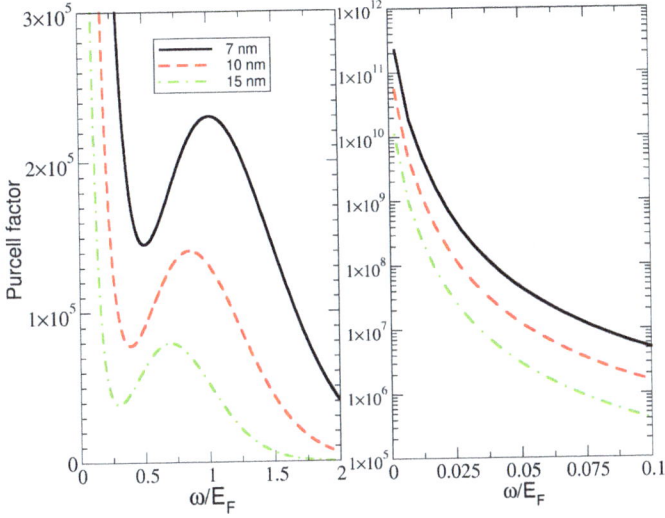

Figure 10.4: Radiative and non-radiative decaying rate of a dipole, as measured by the ratio τ_{vac}/τ_g (Purcell factor), for three different values of $z = 7, 10, 15$ nm. The parameters are $E_F = 0.3$ eV, $\hbar\gamma = 4$ meV, and $\mu_x = \mu_y = \mu_z = 1$. The right panel is a zoom-in (in log-scale) of the low frequency behavior of the data in the left panel. The broad bump in the data in the left panel is due to the excitation of surface-plasmon-polaritons, as argued in Table 10.1.

We can also show numerically that the largest contribution to the integral τ_{vac}/τ_g comes from evanescent waves in the TM-channel (p–polarization). In fact, we can approximate the ratio τ_{vac}/τ_g by

$$
\frac{\tau_{\text{vac}}}{\tau_g} \approx 1 + \frac{\mu_x^2 + \mu_y^2}{\mu^2}\frac{3}{4}\int_1^\infty ds\,\Im m[r_p]\frac{s(s^2-1)}{\sqrt{s^2-1}}e^{-2\omega z\sqrt{s^2-1}/c}
$$

$$
+ \frac{\mu_z^2}{\mu^2}\frac{3}{2}\int_1^\infty ds\,\Im m[r_p]\frac{s^3}{\sqrt{s^2-1}}e^{-2\omega z\sqrt{s^2-1}/c},
$$

$$
\approx 1 + \frac{3}{2}\frac{\mu_z^2 + \mu_\parallel^2/2}{\mu^2}\int_1^\infty ds\,\Im m[r_p]\frac{s^3}{\sqrt{s^2-1}}e^{-2\omega z\sqrt{s^2-1}/c} \quad (10.39)
$$

where $\Im m[r_p]$, in the limit $\beta(\omega, q) \to 0$, reads

$$
\Im m[r_p] = \frac{2\alpha E_F \hbar\gamma\sqrt{s^2-1}}{(\hbar\gamma)^2 + (\hbar\omega - 2E_F\alpha\sqrt{s^2-1})^2}, \quad (10.40)
$$

and has a resonance at the surface plasmon-polariton energy. Indeed, we see that when $s = q/k \gg 1$ (where q is the in-plane momentum) we have a resonance at the energy

$$
\hbar\omega - 2E_F\alpha q/k = 0 \Leftrightarrow \hbar\omega = 2E_F\alpha q/(\omega/c) \Leftrightarrow \hbar\omega = \sqrt{2E_F\alpha\hbar cq}, \quad (10.41)
$$

which agrees with the result of Eq. (4.16). We note that the result (10.39) is rather similar to the that derived from the Fermi's golden rule (10.31) for the non-radiative rate of decay.

On the other hand it is, clear from Fig. 10.4 that the peak of the broad bump is dispersing as function of z. We can show that this dispersion is due to the excitation of surface plasmons. We assume that the distance z introduces a momentum of the order of $q_{SPP} \propto 1/z$. In this case, the ratio of energy of two maxima has to coincide with the ratio of the square root of q_{SPP}. This analysis is done in Table 10.1 and confirms the above statement.

z (nm)	ω_z/E_F	ω_7/ω_z	$\sqrt{z/7}$
$z = 7$	$\omega_7/E_F = 1.01$	$\omega_7/\omega_7 = 1$	$\sqrt{7/7} = 1.00$
$z = 10$	$\omega_{10}/E_F = 0.86$	$\omega_7/\omega_{10} = 1.17$	$\sqrt{10/7} = 1.20$
$z = 15$	$\omega_{15}/E_F = 0.71$	$\omega_7/\omega_{15} = 1.42$	$\sqrt{15/7} = 1.46$

Table 10.1: Analysis of the dispersion, as function of z, of the bump seen in the curves of Fig. 10.4. We note that the $\sqrt{z_i/7}$, with $z_i = 7, 10, 15$ was computed assuming that the electrostatic limit for ω_{SPP} is valid.

10.4   Purcell Effect in Hyperbolic Materials: Few Results

In this section we note that tunable strong light-matter interaction can be obtained by depositing graphene on a hyperbolic material, as is the case of hexagonal Boron Nitride (hBN) [see also Sec. 4.5]. The surface phonon-polaritons of hBN [Dai *et al.* (2014)] can couple to the surface plasmon-polaritons on graphene giving rise to tunable surface phonon-polaritons which in turn can interact to a quantum emitter thus altering its decaying rate [Kumar *et al.* (2015)]. However, although the presence of graphene can lead to tunable phonon-polaritons, hBN alone can have a strong effect on the Purcell factor of a quantum emitter. Following [Kumar *et al.* (2015)], we have computed the Purcell factor for a quantum emitter in the presence of a 50 nm thick hBN crystal, considering that only the μ_z component of the dipole moment is finite. The results are presented in Fig. 10.5 and are identical to those reported in [Kumar *et al.* (2015)]. In this figure we see a dramatic enhancement of the Purcell factor, by several orders of magnitude (the effect is more pronounced for smaller distances to the hBN crystal), when the quantum emitter is located in the proximity of hBN, in the spectral range where the system presents type II hyperbolic behavior.

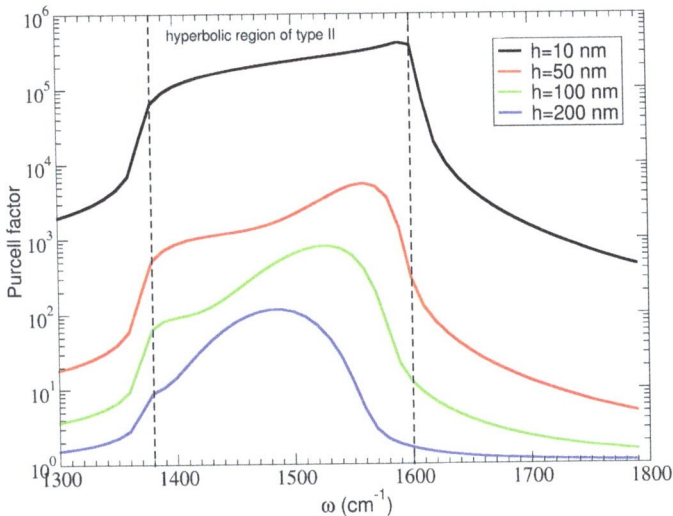

Figure 10.5: Purcell factor of a quantum emitter in the presence of finite thickness (50 nm) hBN crystal. We took $\mu_x = \mu_y = 0$ and $\mu_z = 1$. The spectral range scanned in between the vertical dashed lines is that where the system possesses type II hyperbolic behavior. The dielectric function used for hBN is defined in 4.5 and the crystal is standing on air.

As discussed in Sec. 4.5, in this spectral range the system supports surface phonon-polaritons, which can be excited by the proximity of a quantum emitter. That is, the emitter can decay via the excitation of surface waves in the hBN waveguide. We then see that this type of polaritons is an alternative (to plasmons) for controlling light-matter interactions at the nanoscale.

Chapter 11

Concluding Remarks

The field of plasmonics is growing, as it can be seen from the data-plot in a recent editorial in *Nature Photonics* [Editorial (2012)]. In that article it is shown that the number of publications in plasmonics has been growing steadily over the last twenty years. Broadly speaking, we can divide the field of plasmonics according to the nature of the surface-plasmon resonances. Whenever these resonances occur in localized nano-particles we talk about localized surface plasmon-polaritons [Willets and Duyne (2007); Pelton and Bryant (2013); Klimov (2014)]. When they occur at metal/dielectric surfaces one usually denotes them as (propagating) surface plasmon-polaritons [Sarid and Challener (2010)]. In both cases, these plasmonic resonances have wavelengths below the diffraction limit [Armstrong (2012)] and exhibit intense electric fields. For propagating surface plasmon-polaritons we are generally interested in the case of propagation along a thin metal film. Since 2004 the thinnest conductive film available has been a graphene sheet. It is natural to inquire whether graphene can support propagating surface plasmon-polaritons [Low and Avouris (2014)] (which can be excited by different methods; see Chapter 5 and [Bludov *et al.* (2013); Zhang *et al.* (2014)]). Indeed, the combination of graphene and plasmonics opened a new avenue to this exciting field. This recent branch of nanophotonics is nowadays known as graphene (nano)plasmonics. For example, graphene plasmons can be used to engineer nanoscale infrared light sources [Manjavacas *et al.* (2014)], and intrinsic plasmons on graphene have already been used for enhancing photodetection [Kim *et al.* (2014); Koppens *et al.* (2014)].

Graphene, however, is not the only two-dimensional material available these days [Castro Neto and Novoselov (2011); Novoselov and Castro Neto (2012)]. Many others, as dichalcogenides (MX_2), hexagonal boron nitride (hBN), phosphorene [Low *et al.* (2014b)], silicene, some oxides, and graphene multilayers [Low *et al.* (2014a); Wu *et al.* (2014b)], are waiting in line to be explored from the point of view of their electronic properties and light-matter interactions [Kim *et al.* (2014); Koppens *et al.* (2014)]. Indeed, a new class of materials, dubbed "van der Waals heterostructures" [Geim and Grigorieva (2013)], were conjectured to unleash a number of both new fundamental physics and new applications unforeseen until recently. On one hand, these new heterostructures [Kretinin *et al.* (2014)] can be made of two-dimensional materials (or, for that matter, of few layers of those materials); on the other hand, can be combined with more traditional materials, like noble metals [Yamaguchi *et al.* (2014); Chaves *et al.* (2014); Moriya *et al.* (2015)]. This opens a multitude of possibilities for light-matter interactions where plasmonics will certainly play a key role. Since graphene is intrinsically a surface, the coupling of surface plasmons to surface polar phonons of the underlying substrate, for instance hBN, is an interesting prospect, leading to hybrid surface phonon-plasmon polaritons [Luxmoore *et al.* (2014); Brar *et al.* (2014)], which appear as multiple dispersion branches. Furthermore, it has been shown that plasmons in graphene encapsulated in hBN possess very low damping [Woessner *et al.* (2015)]. Thus, this type of van der Waals heterostructure emerges as a plasmonic medium with high uniformity. As the field matures, low losses in plasmons propagation can be translated into long propagation lengths, which might open new routes for photonic/plasmonic-based communications and data transfer. On another topic, graphene-dichalcogenides van der Waals heterostructures, coupled to noble metals plasmonics, seem to work better as a chemical sensor than the same system in the absence of the heterostructure [Zeng *et al.* (2015)] (see also [Kravets *et al.* (2014)]); in this case, the plasmonic medium is the noble metal and not graphene, the latter acting as an intermediate for enhancing light-matter interactions. Also worth noting is the interaction of surface plasmons with grain boundaries in CVD-grown graphene. It has been shown that the reflection of plasmons at grain boundaries in graphene presents an anomalous scattering phase (different from π) [Fei *et al.* (2013); Nikitin *et al.* (2014)] (see also [Sánchez-Gil and Maradudin (1999); Polanco *et al.* (2013)] for calculations of SPPs scattering amplitudes in traditional plasmonics).

Strongly related to the field of surface plasmon-polaritons is the emerging field of surface phonon-polaritons in van der Waals heterostructures, hBN and hBN/graphene structures being the most studied. Hexagonal Boron Nitride is a hyperbolic material, that is, the in-plane and the out-of-plane dielectric constants have different signs [Jacob (2014); Dai *et al.* (2014); Caldwell *et al.* (2014); Kumar *et al.* (2015)]. As a consequence, hBN presents two Reststrahlen bands where free radiation cannot propagate. Within these bands the system behaves like a conventional metal (as concerns radiation propagation). The existence of these Reststrahlen bands allows the propagation of surface phonon-polaritons in hBN. When graphene is deposited on hBN, an additional coupling between the surface phonon-polaritons in hBN and the surface plasmon-polaritons in graphene becomes possible, giving rise to tunable plasmon-phonon-polaritons [Dai *et al.* (2015)]. These results are opening a new field in van der Waals heterostructures.

Additionally, with the simple stacking of layered materials, a new degree of freedom enters in the game: the relative rotation of the lattices associated with different layers [Mishchenko *et al.* (2014)]. This rotation, say between a graphene layer on a boron nitride buffer layer, creates an external potential that splits the graphene bands, leading to the emergence of further plasmonic branches (dubbed spectrum reconstruction). The new avenue opened by this additional degree of freedom has not yet been fully explored [Tomadin *et al.* (2014)].

Another interesting line of research is that of plasmonics in graphene nano-structures [Christensen *et al.* (2014); Cox and de Abajo (2014); Nene *et al.* (2014); Wang *et al.* (2015); Yeung *et al.* (2014); Zhu *et al.* (2014)] or even in nanoparticles coated with graphene [Christensen *et al.* (2015a)] which may also exhibit non-linear optical response [Peres *et al.* (2014); Cox and de Abajo (2015)]. In fact, the non-linear optical response of graphene has recently been used to excite surface plasmons in this material [Constant *et al.* (2015)]. Graphene's plasmonic response can also be put to good use for non-linear optics [Gullans *et al.* (2013)], and graphene itself is able to exhibit non-linear GSPs, as predicted by a recent work [Gorbach (2013)], and highly confined TE waves [Bludov *et al.* (2014)]. On a related topic, we can consider the combination of metallic nanoparticles, acting as a plasmonic material, with graphene. It has recently been shown that gold nano-disks can interact strongly with graphene, generating locally hot carriers which can be probed by femtosecond pump-probe measurements [Gilbertson *et al.* (2015)].

In this book we have only covered some aspects related to graphene plasmonics [Politano and Chiarello (2014); Low and Avouris (2014); Grigorenko *et al.* (2012); Xiao *et al.* (2016)]. We started by reviewing the basic properties of SPPs sustained at dielectric-metal interfaces, setting the stage for the discussion of SPPs in doped graphene. We then derived the dispersion relation of SPP in graphene, both in single-layer graphene and in double-layer graphene. We have seen that even for these simple geometries, the spectrum needs to be obtained by numerical means. From the corresponding solutions, it was clear that GSP-excitation is not feasible by simply striking light directly on graphene, since the dispersion curve always lies to the right of the light line. Therefore, in Chapter 5 we reviewed the most popular methods to couple electromagnetic radiation to GSPs (see also [Torre *et al.* (2015)] for a full-electrical plasmon detection method). Another interesting idea has been put forward in [Liu *et al.* (2015a)]. The excitation methods described in Chapter 5 cannot distinguish left from right propagation of plasmons. However, by putting graphene on a magneto-optical substrate, its dispersion relation becomes asymmetric in the forward and backward directions, thus allowing unidirectional excitation [Liu *et al.* (2015a)].

Motivated by a recent experiment reporting the excitation of GSPs using a metallic antenna [Alonso-González *et al.* (2014)], in Chapter 6 we have conducted a theoretical study of a similar system consisting of an infinitely long and highly conducting metal stripe placed on top of a graphene layer.

In Chapter 7 we developed a general theoretical framework for describing the excitation of GSPs in systems where the graphene's conductivity is periodically modulated. In particular, we have applied our analytical model to the case of a periodic array of graphene ribbons, since this setup has been extensively studied experimentally. We then confronted our theoretical results against the experimental data [Ju *et al.* (2011)], having observed a good agreement between theory and experiment (see also [Velizhanin (2015)]). In a recent experiment [Sorger *et al.* (2015)], GSPs' excitation in graphene stripes grown in silicon carbide has been achieved, and the same model that was used in Chapter 7 was applied to the description of the experimental data. In Chapter 8 the case of a single graphene ribbon was considered in the electrostatic limit. The approach used was then applied to graphene disks and rings. The case of a groove was also discussed. If the groove is periodically repeated we achieve a diffraction grating, a problem that can be analyzed semi-analytically. In Chapter 9 an extension of Chapter 7 was presented, which consisted of an approach to deal with graphene on periodically corrugated surfaces, that is diffraction gratings. These can be made of either a dielectric or a metamaterial, as in the case discussed at the end of that Chapter. It is worth mentioning that graphene resting on a grating can be a source of THz radiation [Zhan *et al.* (2014)], when a beam of electrons moves over its surface, a problem that we have analysed at the end of Chapter 5 . Finally, in Chapter 10 we discussed the excitation of SPP's by an emitting dipole. We have shown that the decaying rate of a dipole is intimately tied to the excitation of SPP's on graphene, a manifestation of the Purcell effect.

Appendix A

Derivation of the Susceptibility of Graphene (*written with André Chaves*)

In this Appendix we compute the electronic susceptibility of graphene. This quantity is defined by Eq. (2.75), which we now write with a slightly different, more convenient, notation[1]:

$$\chi(\mathbf{q}, \omega) = 4 \sum_{\alpha, \alpha' = \pm} \int \frac{d^2\mathbf{k}}{4\pi^2} \frac{n_F(\alpha E_\mathbf{k}) - n_F(\alpha' E_{\mathbf{k+q}})}{\omega + \alpha E_\mathbf{k} - \alpha' E_{\mathbf{k+q}} + i\varepsilon} F_{\alpha\alpha'}(\mathbf{q}, \mathbf{k}), \qquad (A.1)$$

where $\varepsilon = 0^+$,

$$F_{\alpha\alpha'}(\mathbf{q}, \mathbf{k}) = \frac{1}{2}\left(1 + \alpha\alpha' \frac{k^2 + \mathbf{k} \cdot \mathbf{q}}{k|\mathbf{k+q}|}\right), \qquad (A.2)$$

and

$$n_F(E) = \left[e^{\beta(E - E_F)} + 1\right]^{-1}, \qquad (A.3)$$

where $\beta = 1/(k_B T)$, with T the temperature and k_B the Boltzmann constant. From here on we treat the case of zero temperature only. First, we write the susceptibility as $\chi(\mathbf{q}, \omega) = \sum_{\alpha\alpha'} \chi^{\alpha\alpha'}(\mathbf{q}, \omega)$. Next, we separate the susceptibility into *doped*, $\chi_D(\mathbf{q}, \omega)$, and *undoped*, $\chi_U(\mathbf{q}, \omega)$, parts:

$$\chi(\mathbf{q}, \omega) = \chi_D(\mathbf{q}, \omega) + \chi_U(\mathbf{q}, \omega). \qquad (A.4)$$

By *undoped* part we mean the expression of $\chi(\mathbf{q}, \omega)$ that appears in the limit $E_F = 0$, which is also present for $E_F \neq 0$. The *doped* part is given by the remaining terms.

A.1 Undoped Susceptibility

The term χ_U is the susceptibility when $E_F = 0$ and is given by:

$$\chi_U(\mathbf{q}, \omega) = \chi_U^{+-}(\mathbf{q}, \omega) + \chi_U^{-+}(\mathbf{q}, \omega). \qquad (A.5)$$

It is convenient to determine both the imaginary and real parts separately.

---

[1] In what follows we choose $v_F = \hbar = 1$, which simplifies the calculations.

A.1.1   *The imaginary part of the undoped susceptibility*

The imaginary part of χ_U^{+-} is null, because the argument of the δ-function is always positive, and the imaginary part of χ_U^{-+} is given by:

$$\Im m \chi_U(\mathbf{q}, \omega) = -\pi \int \frac{kdkd\theta}{\pi^2} \delta(\omega - k - |\mathbf{k} + \mathbf{q}|) F_{-+}(\mathbf{q}, \mathbf{k}), \qquad (A.6)$$

where we have used the Sokhotski-Plemelj formula. First we perform the integral over θ and we obtain:

$$\Im m \chi_U(\mathbf{q}, \omega) = -2 \frac{1}{2\pi} \int_{\frac{q+\omega}{2}}^{\omega} kdk \frac{2|\omega - k|}{\sqrt{4k^2 q^2 - (\omega^2 - 2k\omega - q^2)^2}}$$
$$\times \left(1 - \frac{2k^2 + \omega^2 - 2k\omega - q^2}{2k(\omega - k)} \right), \qquad q > \omega \qquad (A.7)$$

$$\Im m \chi_U(\mathbf{q}, \omega) = -2 \frac{1}{2\pi} \int_{\frac{\omega-q}{2}}^{\frac{q+\omega}{2}} kdk \frac{2|\omega - k|}{\sqrt{4k^2 q^2 - (\omega^2 - 2k\omega - q^2)^2}}$$
$$\times \left(1 - \frac{2k^2 + \omega^2 - 2k\omega - q^2}{2k(\omega - k)} \right), \qquad q < \omega \qquad (A.8)$$

where we have also used $\omega - k = |\mathbf{k} + \mathbf{q}|$. A factor of 2 appears because of there are two solutions of the Dirac δ's in the trigonometric plane. The first integral is zero because necessarily we have $\omega < (q + \omega)/2$. The above integrals can be simplified to

$$\Im m \chi_U(\mathbf{q}, \omega) = 0, \qquad q > \omega$$
$$\Im m \chi_U(\mathbf{q}, \omega) = \frac{-1}{\pi} \int_{\frac{\omega-q}{2}}^{\frac{q+\omega}{2}} dk \frac{q^2 - \omega^2 + 4k\omega - 4k^2}{\sqrt{(\omega^2 - q^2)(q^2 - \omega^2 + 4k\omega - 4k^2)}}, q < \omega \quad (A.9)$$

or still

$$\Im m \chi_U(\mathbf{q}, \omega) = 0, \qquad q > \omega$$
$$\Im m \chi_U(\mathbf{q}, \omega) = \frac{-1}{\pi \sqrt{\omega^2 - q^2}} \int_{\frac{\omega-q}{2}}^{\frac{q+\omega}{2}} dk \sqrt{q^2 - \omega^2 + 4k\omega - 4k^2}, q < \omega \quad (A.10)$$

and finally

$$\Im m \chi_U(\mathbf{q}, \omega) = 0, \qquad q > \omega$$
$$\Im m \chi_U(\mathbf{q}, \omega) = -\frac{q^2}{4\sqrt{\omega^2 - q^2}}, \qquad q < \omega \qquad (A.11)$$

Let us now compute the real part of χ_U.

A.1.2 The real part of the undoped susceptibility

To obtain the real part of the susceptibility we use the Kramers-Kronig relation:

$$\Re\chi(\mathbf{q}, \omega) = \frac{2}{\pi} \mathcal{P} \int_0^\infty d\omega' \frac{\omega' \Im\mathfrak{m}\chi(\mathbf{q}, \omega')}{\omega'^2 - \omega^2}. \tag{A.12}$$

Introducing equation (A.11) in the Kramers-Kronig relation we have:

$$\Re\chi_U(\mathbf{q}, \omega) = -\frac{q^2}{2\pi} \mathcal{P} \int_q^\infty d\omega' \frac{\omega'}{\omega'^2 - \omega^2} \frac{1}{\sqrt{\omega'^2 - q^2}}. \tag{A.13}$$

Making the change of variables $x = \omega'^2 - \omega^2$ with $dx = 2\omega' d\omega'$ we write $\Re\chi_U(\mathbf{q}, \omega)$ as

$$\Re\chi_U(\mathbf{q}, \omega) = -\frac{q^2}{4\pi} \mathcal{P} \int_{q^2-\omega^2}^\infty dx \frac{1}{x} \frac{1}{\sqrt{x + \omega^2 - q^2}}. \tag{A.14}$$

If $q > \omega$ we make the substitution $x = (q^2 - \omega^2)y$ and it follows

$$\Re\chi_U(\mathbf{q}, \omega) = -\frac{q^2}{4\pi\sqrt{q^2 - \omega^2}} \mathcal{P} \int_1^\infty dy \frac{1}{y} \frac{1}{\sqrt{y - 1}}, \tag{A.15}$$

or else,

$$\Re\chi_U(\mathbf{q}, \omega) = -\frac{q^2}{4\sqrt{q^2 - \omega^2}}. \quad q > \omega \tag{A.16}$$

For $q < \omega$ we have

$$\Re\chi_U(\mathbf{q}, \omega) = 0. \tag{A.17}$$

We note that we can combine the real and imaginary part of $\chi_U(\mathbf{q}, \omega)$ in a single expression as:

$$\chi_U(\mathbf{q}, \omega) = -\frac{q^2 \operatorname{sign}[q - \omega]}{4\sqrt{q^2 - \omega^2}}. \tag{A.18}$$

where we are using the convention $\sqrt{-x} = i\sqrt{x}$ for $x > 0$.

A.2 Doped Part: Fermi Sea Contributions

A.2.1 Intraband term

First we workout the intraband term

$$\chi_D^{++} = 4 \int \frac{d^2\mathbf{k}}{4\pi^2} \frac{n_F(E_\mathbf{k}) - n_F(E_{\mathbf{k}+\mathbf{q}})}{\omega + E_\mathbf{k} - E_{\mathbf{k}+\mathbf{q}} + i\varepsilon} F_{++}(\mathbf{q}, \mathbf{k}). \tag{A.19}$$

Splitting the above integral into two new integrals

$$\chi_D^{++} = \chi_D^{\text{intra1}} + \chi_D^{\text{intra2}} = 4 \int \frac{d^2\mathbf{k}}{4\pi^2} \frac{n_F(E_\mathbf{k})}{\omega + E_\mathbf{k} - E_{\mathbf{k+q}} + i\varepsilon} F_{++}(\mathbf{q}, \mathbf{k})$$

$$+ 4 \int \frac{d^2\mathbf{k}}{4\pi^2} \frac{-n_F(E_{\mathbf{k+q}})}{\omega + E_\mathbf{k} - E_{\mathbf{k+q}} + i\varepsilon} F_{++}(\mathbf{q}, \mathbf{k}), \tag{A.20}$$

and making in the second integral the substitution $\mathbf{k} \to \mathbf{k} - \mathbf{q}$ we obtain

$$\chi_D^{\text{intra2}} = 4 \int \frac{d^2\mathbf{k}}{4\pi^2} \frac{-n_F(E_\mathbf{k})}{\omega + E_{\mathbf{k-q}} - E_\mathbf{k} + i\varepsilon} F_{++}(\mathbf{q}, \mathbf{k} - \mathbf{q}). \tag{A.21}$$

Noting that

$$F_{\alpha\alpha'}(\mathbf{q}, \mathbf{k} - \mathbf{q}) = \frac{1}{2}\left(1 + \alpha\alpha' \frac{|\mathbf{k} - \mathbf{q}|^2 + (\mathbf{k} - \mathbf{q})\cdot\mathbf{q}}{|\mathbf{k} - \mathbf{q}|k}\right)$$

$$= \frac{1}{2}\left(1 + \alpha\alpha' \frac{k^2 - \mathbf{k}\cdot\mathbf{q}}{k|\mathbf{k} - \mathbf{q}|}\right) = F_{\alpha\alpha'}(\mathbf{q}, -\mathbf{k}), \tag{A.22}$$

and making the substitution $k \to -k$ in (A.21) and using $E_\mathbf{k} = E_{-\mathbf{k}}$ we arrive at:

$$\chi_D^{\text{intra2}} = 4 \int \frac{d^2\mathbf{k}}{4\pi^2} \frac{-n_F(E_\mathbf{k})}{\omega + E_{\mathbf{k+q}} - E_\mathbf{k} + i\varepsilon} F_{++}(\mathbf{q}, \mathbf{k}) \tag{A.23}$$

$$\chi_D^{\text{intra1}} = 4 \int \frac{d^2\mathbf{k}}{4\pi^2} \frac{n_F(E_\mathbf{k})}{\omega + E_\mathbf{k} - E_{\mathbf{k+q}} + i\varepsilon} F_{++}(\mathbf{q}, \mathbf{k}) \tag{A.24}$$

A.2.2 Interband term

The interband part is:

$$\chi_D^{+-}(\mathbf{q}, \omega) = 4 \int \frac{d^2\mathbf{k}}{4\pi^2} \frac{n_F(E_\mathbf{k})}{\omega + E_\mathbf{k} + E_{\mathbf{k+q}} + i\varepsilon} F_{+-}(\mathbf{q}, \mathbf{k}), \tag{A.25}$$

and

$$\chi_D^{-+}(\mathbf{q}, \omega) = 4 \int \frac{d^2\mathbf{k}}{4\pi^2} \frac{-n_F(E_{\mathbf{k+q}})}{\omega - E_\mathbf{k} - E_{\mathbf{k+q}} + i\varepsilon} F_{-+}(\mathbf{q}, \mathbf{k}). \tag{A.26}$$

Using the same manipulations of the intraband part, the above integral transforms into

$$\chi_D^{-+}(\mathbf{q}, \omega) = -4 \int \frac{d^2\mathbf{k}}{4\pi^2} \frac{n_F(E_\mathbf{k})}{\omega - E_{\mathbf{k+q}} - E_\mathbf{k} + i\varepsilon} F_{+-}(\mathbf{q}, \mathbf{k}). \tag{A.27}$$

We now sum A.25 and A.24 to obtain χ_D^1, and A.27 and A.23 to obtain χ_D^2:

$$\chi_D^1 = \frac{1}{\pi^2} \int_0^{2\pi} d\theta \int_0^{k_F} dk \frac{2k^2 + k\omega + kq\cos\theta}{\omega^2 + 2\omega k - q^2 - 2kq\cos\theta + i\varepsilon}, \tag{A.28}$$

$$\chi_D^2 = -\frac{1}{\pi^2} \int_0^{2\pi} d\theta \int_0^{k_F} dk \frac{k\omega - 2k^2 - kq\cos\theta}{\omega^2 - 2\omega k - q^2 - 2kq\cos\theta + i\varepsilon\mathrm{sign}(\omega - k)}. \tag{A.29}$$

Using again the Sokhotski-Plemelj formula we can calculate the imaginary part as:

$$\Im\chi_D^1 = -\frac{1}{\pi} \int_0^{2\pi} d\theta \int_0^{k_F} dk \left[2k^2 + k\omega + kq\cos\theta\right] \\ \times \delta(\omega^2 + 2\omega k - q^2 - 2kq\cos\theta), \tag{A.30}$$

$$\Im\chi_D^2 = \frac{1}{\pi} \int_0^{2\pi} d\theta \int_0^{k_F} dk \left[k\omega - 2k^2 - kq\cos\theta\right] \\ \times \delta(\omega^2 - 2\omega k - q^2 - 2kq\cos\theta)\mathrm{sign}(\omega - k). \tag{A.31}$$

The δ−functions can be integrated by performing the angular integral first, resulting in:

$$\Im\chi_D^1 = -\frac{2}{2\pi} \int_0^{k_F} dk \frac{4k^2 + 2k\omega + A_+}{\sqrt{B^2 - A_+^2}} G_1(k), \tag{A.32}$$

and

$$\Im\chi_D^2 = \frac{2}{2\pi} \int_0^{k_F} dk \frac{2k\omega - 2k^2 - A_-}{\sqrt{B^2 - A_-^2}} G_2(k)\mathrm{sign}(\omega - k), \tag{A.33}$$

where we have used

$$\cos\theta = \frac{A_\pm}{B}, \tag{A.34}$$

with

$$A_\pm = \omega^2 \pm 2k\omega - q^2, \quad B = 2kq, \tag{A.35}$$

and

$$G_i(k) = \begin{cases} 1, & \text{if } k \text{ solves the } \delta \text{ function} \\ 0, & \text{otherwise} \end{cases}. \tag{A.36}$$

Finally, after some algebra we can write:

$$\Im\chi_D^1 = \frac{2}{2\pi} \int_0^{k_F} dk \frac{q^2 - (2k + \omega)^2}{\sqrt{(\omega^2 - q^2)(q^2 - (2k + \omega)^2)}} G_1(k), \tag{A.37}$$

and

$$\Im\chi_D^2 = \frac{1}{\pi} \int_0^{k_F} dk \frac{q^2 - (\omega - 2k)^2}{\sqrt{(\omega^2 - q^2)(q^2 - (2k - \omega)^2)}} G_2(k)\mathrm{sign}(\omega - k). \tag{A.38}$$

A.2.3 Imaginary part of χ_D^1

For $q > \omega > q - 2k_F$:

$$\Im m \chi_D^1 = -\frac{1}{\pi\sqrt{q^2 - \omega^2}} \int_{\frac{q-\omega}{2}}^{k_F} dk \sqrt{(2k+\omega)^2 - q^2},$$ (A.39)

and

$$\Im m \chi_D^1 = -\frac{q^2}{4\pi\sqrt{q^2 - \omega^2}}\left[-\operatorname{arccosh}\left(\frac{2k_F+\omega}{q}\right) + \frac{2k_F+\omega}{q}\sqrt{\left(\frac{\omega+2k_F}{q}\right)^2 - 1}\right]$$ (A.40)

otherwise the result is null.

A.2.4 Imaginary part of χ_D^2

First for $\omega > q$:

$$\Im m \chi_D^2 = \frac{1}{\pi\sqrt{\omega^2 - q^2}} \int_{\frac{\omega-q}{2}}^{k_F} dk \sqrt{q^2 - (\omega - 2k)^2}\theta(\omega + q - 2k)\operatorname{sign}(\omega - k).$$ (A.41)

If $2k_F > \omega + q$ then $\operatorname{sign}(\omega - k) = +1$ and therefore

$$\Im m \chi_D^2 = \frac{q^2}{4\sqrt{\omega^2 - q^2}}.$$ (A.42)

On the other hand, if $\omega - q < 2k_F$ and $\omega + q > 2k_F$, it follows

$$\Im m \chi_D^2 = \frac{q^2}{4\pi\sqrt{\omega^2 - q^2}}\left[\frac{\pi}{2} + \arcsin\left(\frac{2k_F - \omega}{q}\right) + \frac{2k_F - \omega}{q}\sqrt{1 - \left(\frac{\omega - 2k_F}{q}\right)^2}\right].$$ (A.43)

After some trigonometric manipulations we can write

$$\Im m \chi_D^2 = \frac{q^2}{4\pi\sqrt{\omega^2 - q^2}}\left[\pi - \arccos\left(\frac{2k_F - \omega}{q}\right) + \frac{2k_F - \omega}{q}\sqrt{1 - \left(\frac{\omega - 2k_F}{q}\right)^2}\right].$$ (A.44)

For $q > \omega$ and $2k_F > \omega + q$ we obtain

$$\Im m \chi_D^2 = \frac{1}{\pi\sqrt{q^2 - \omega^2}} \int_{\frac{\omega+q}{2}}^{k_F} dk \sqrt{(\omega - 2k)^2 - q^2},$$ (A.45)

which after integration, gives

$$\Im m \chi_D^2 = \frac{q^2}{4\pi\sqrt{q^2 - \omega^2}}\left[\frac{2k_F - \omega}{q}\sqrt{\left(\frac{2k_F - \omega}{q}\right)^2 - 1} - \operatorname{arccosh}\left(\frac{2k_F - \omega}{q}\right)\right].$$ (A.46)

In all other cases $\Im m \chi_D^2$ is zero.

A.2.5 Real Part of χ_D^1 and χ_D^2

The real part of χ_D^1 and χ_D^2 is given by the Cauchy principal value of the expressions (A.28) and (A.29), that is

$$\Re\chi_D^1 = \frac{1}{\pi^2}\mathcal{P}\int_0^{2\pi}d\theta\int_0^{k_F}dk\frac{2k^2 + k\omega + kq\cos\theta}{\omega^2 + 2\omega k - q^2 - 2kq\cos\theta}, \tag{A.47}$$

$$\Re\chi_D^2 = -\frac{1}{\pi^2}\mathcal{P}\int_0^{2\pi}d\theta\int_0^{k_F}dk\frac{k\omega - 2k^2 - kq\cos\theta}{\omega^2 - 2\omega k - q^2 - 2kq\cos\theta}. \tag{A.48}$$

Equations A.47 and A.48 can be written as

$$\Re\chi_D^1 = \frac{1}{2\pi^2}\mathcal{P}\int_0^{2\pi}d\theta\int_0^{k_F}dk\left[\frac{\omega^2 - q^2 + 4k^2 + 4k\omega}{\omega^2 + 2\omega k - q^2 - 2kq\cos\theta} - 1\right], \tag{A.49}$$

$$\Re\chi_D^2 = -\frac{1}{2\pi^2}\mathcal{P}\int_0^{2\pi}d\theta\int_0^{k_F}dk\left[\frac{q^2 - \omega^2 + 4\omega k - 4k^2}{\omega^2 - 2\omega k - q^2 - 2kq\cos\theta} + 1\right]. \tag{A.50}$$

and

$$\Re\chi_D^1 = \frac{1}{2\pi^2}\left[-2\pi k_F + \mathcal{P}\int_0^{2\pi}d\theta\int_0^{k_F}dk\frac{(2k+\omega)^2 - q^2}{\omega^2 + 2\omega k - q^2 - 2kq\cos\theta}\right], \tag{A.51}$$

$$\Re\chi_D^2 = -\frac{1}{2\pi^2}\left[2\pi k_F + \mathcal{P}\int_0^{2\pi}d\theta\int_0^{k_F}dk\frac{q^2 - (\omega - 2k)^2}{\omega^2 - 2\omega k - q^2 - 2kq\cos\theta}\right]. \tag{A.52}$$

Performing the integration over θ with the help of the results of Sec. A.4 we obtain:

$$\Re\chi_D^1 = \frac{1}{2\pi^2}\left[-2\pi k_F + 2\pi\theta(\omega - q)\int_0^{k_F}dk\,\frac{(2k+\omega)^2 - q^2}{\sqrt{(\omega^2 + 2\omega k - q^2)^2 - 4k^2q^2}} + \right.$$
$$\left. +2\pi\theta(q - \omega)\int_0^{k_2}dk\,\frac{(2k+\omega)^2 - q^2}{\sqrt{(\omega^2 + 2\omega k - q^2)^2 - 4k^2q^2}}\mathrm{sign}(2k + \omega - q)\right], \tag{A.53}$$

$$\Re\chi_D^2 = \frac{1}{2\pi^2}\left[-2\pi k_F - 2\pi\theta(\omega - q)\left(\int_0^{k_1}dk + \int_{k_2}^{k_F}dk\right)\right.$$
$$\times\frac{q^2 - (\omega - 2k)^2}{\sqrt{(\omega^2 - 2\omega k - q^2)^2 - 4k^2q^2}} - 2\pi\theta(q - \omega)\int_0^{k_2}dk$$
$$\left.\times\frac{q^2 - (\omega - 2k)^2}{\sqrt{(\omega^2 - 2\omega k - q^2)^2 - 4k^2q^2}}\mathrm{sign}(2k - \omega - q)\right] \tag{A.54}$$

where $k_1 = \min(|\omega - q|/2, k_F)$, $k_2 = \min[(\omega + q)/2, k_F]$. We can recast the above integrals to a simpler form:

$$\Re\chi_D^1 = \frac{1}{2\pi^2}\left[-2\pi k_F + \frac{2\pi\theta(\omega - q)}{\sqrt{\omega^2 - q^2}} \int_0^{k_F} dk\, \sqrt{(2k + \omega)^2 - q^2} + \right.$$
$$\left. -\frac{2\pi\theta(q - \omega)}{\sqrt{q^2 - \omega^2}} \int_0^{k_1} dk\, \sqrt{q^2 - (2k + \omega)^2}\,\mathrm{sign}(2k + \omega - q) \right], \quad \text{(A.55)}$$

$$\Re\chi_D^2 = \frac{1}{2\pi^2}\left[-2\pi k_F + \frac{2\pi\theta(\omega - q)}{\sqrt{\omega^2 - q^2}} \int_0^{k_1} + \int_{k_2}^{k_F} dk\sqrt{(2k - \omega)^2 - q^2} \right.$$
$$\left. -\frac{2\pi\theta(q - \omega)}{\sqrt{q^2 - \omega^2}} \int_0^{k_2} dk\, \sqrt{q^2 - (2k - \omega)^2}\,\mathrm{sign}(2k - \omega - q) \right].$$
$$\text{(A.56)}$$

Finally, integrating over the wave number k we arrive at the final expressions for the real part of the susceptibility:

$$\Re\chi_D^1 = -\frac{k_F}{\pi} + \frac{q^2}{4\pi\sqrt{\omega^2 - q^2}}\left[F\left(\frac{2k_F + \omega}{q}\right) - F\left(\frac{\omega}{q}\right) \right]\theta(\omega - q) +$$
$$+\frac{q^2}{4\pi\sqrt{q^2 - \omega^2}}\left[F_2\left(\frac{2k_1 + \omega}{q}\right) - F_2\left(\frac{\omega}{q}\right) \right]\theta(q - \omega).$$
$$\text{(A.57)}$$

For $\Re\chi_D^2$ we have different branches. If $2k_F < \omega - q$ we have:

$$\Re\chi_D^2 = -\frac{k_F}{\pi} + \frac{q^2}{4\pi\sqrt{\omega^2 - q^2}}\left[F\left(\frac{\omega}{q}\right) - F\left(\frac{\omega - 2k_F}{q}\right) \right]. \quad \text{(A.58)}$$

For $\omega - q < 2k_F < \omega + q$ the result reads:

$$\Re\chi_D^2 = -\frac{k_F}{\pi} + \frac{q^2}{4\pi\sqrt{\omega^2 - q^2}}F\left(\frac{\omega}{q}\right). \quad \text{(A.59)}$$

For $2k_F > \omega + q$ we have:

$$\Re\chi_D^2 = -\frac{k_F}{\pi} + \frac{q^2}{4\pi\sqrt{\omega^2 - q^2}}\left[F\left(\frac{\omega}{q}\right) - F\left(\frac{2k_F - \omega}{q}\right) \right], \quad \text{(A.60)}$$

and, finally, if $q < \omega$ we obtain

$$\Re\chi_D^2 = -\frac{k_F}{\pi} + \frac{q^2}{4\pi\sqrt{q^2 - \omega^2}}\left[F_2\left(\frac{2k_2 - \omega}{q}\right) - F_2\left(-\frac{\omega}{q}\right) \right], \quad \text{(A.61)}$$

where

$$F(x) = x\sqrt{x^2 - 1} - \mathrm{arccosh}(x). \quad \text{(A.62)}$$

$$F_2(x) = x\sqrt{1 - x^2} + \arcsin(x). \tag{A.63}$$

we can still rewrite $F_2(x)$ as:

$$F_2(x) = x\sqrt{1 - x^2} + \frac{\pi}{2} - \arccos(x) = \frac{\pi}{2} + C(x) \tag{A.64}$$

where

$$C(x) = x\sqrt{1 - x^2} - \arccos(x) \tag{A.65}$$

Summing all the contributions we finally have:

- region 1B: $\omega > q$ and $\omega + q < 2k_F$

$$\Re\chi_D = -\frac{2k_F}{\pi} + \frac{q^2}{4\pi\sqrt{\omega^2 - q^2}} \left[F\left(\frac{2k_F + \omega}{q}\right) - F\left(\frac{2k_F - \omega}{q}\right) \right],$$
$$\tag{A.66}$$

- region 2B: $\omega > q$, $\omega + q > 2k_F$ and $\omega - q < 2k_F$

$$\Re\chi_D = -\frac{2k_F}{\pi} + \frac{q^2}{4\pi\sqrt{\omega^2 - q^2}} F\left(\frac{2k_F + \omega}{q}\right), \tag{A.67}$$

- region 3B: $\omega - q > 2k_F$

$$\Re\chi_D = -\frac{2k_F}{\pi} + \frac{q^2}{4\pi\sqrt{\omega^2 - q^2}} \left[F\left(\frac{2k_F + \omega}{q}\right) - F\left(\frac{\omega - 2k_F}{q}\right) \right],$$
$$\tag{A.68}$$

- region 1A: $q > \omega$ and $\omega + q < 2k_F$:

$$\Re\chi_D = -\frac{2k_F}{\pi} + \frac{q^2}{4\sqrt{q^2 - \omega^2}}, \tag{A.69}$$

- region 2A: $q - 2k_F < \omega$ $q > \omega$ and $\omega + q > 2k_F$,

$$\Re\chi_D = -\frac{2k_F}{\pi} + \frac{q^2}{4\pi\sqrt{q^2 - \omega^2}} \left[\pi - C\left(\frac{\omega - 2k_F}{q}\right) \right], \tag{A.70}$$

- region 3A: $q - 2k_F > \omega$

$$\Re\chi_D = -\frac{2k_F}{\pi} + \frac{q^2}{4\pi\sqrt{q^2 - \omega^2}}$$
$$\times \left[\pi + C\left(\frac{2k_F + \omega}{q}\right) - C\left(\frac{\omega - 2k_F}{q}\right) \right], \tag{A.71}$$

where we have used that $C(-x) = -C(x) - \pi$.

A.3 Summary for the Real and Imaginary Parts of Graphene Susceptibility

We now provide a summary of the real and imaginary parts of the graphene susceptibility. Since these quantities are defined by branches, Fig. A.1 summarizes the regions in the $\omega - q$ plane where the different branches are defined.

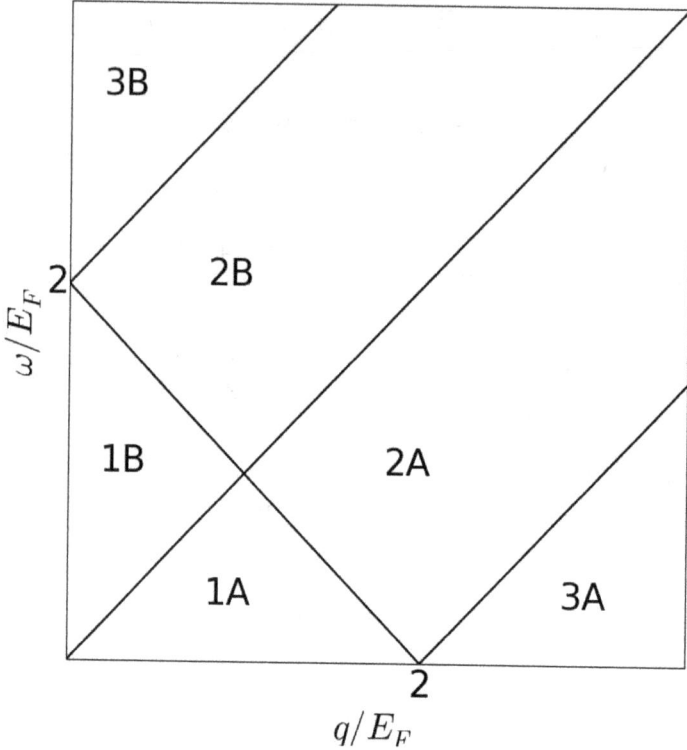

Figure A.1: Regions for the susceptibility of graphene used in the summary of the real and imaginary parts of that quantity.

A.3.1　*Real part of the susceptibility*

- region 1B: $\omega > q$ and $\omega + q < 2k_F$

$$\Re\chi = -\frac{2k_F}{\pi} + \frac{q^2}{4\pi\sqrt{\omega^2 - q^2}} \left[F\left(\frac{2k_F + \omega}{q}\right) - F\left(\frac{2k_F - \omega}{q}\right) \right],$$

$$(\text{A.72})$$

- region 2B: $\omega > q$, $\omega + q > 2k_F$ and $\omega - q < 2k_F$

$$\Re\chi = -\frac{2k_F}{\pi} + \frac{q^2}{4\pi\sqrt{\omega^2 - q^2}} F\left(\frac{2k_F + \omega}{q}\right),$$

$$(\text{A.73})$$

- region 3B: $\omega - q > 2k_F$

$$\Re\chi = -\frac{2k_F}{\pi} + \frac{q^2}{4\pi\sqrt{\omega^2 - q^2}} \left[F\left(\frac{2k_F + \omega}{q}\right) - F\left(\frac{\omega - 2k_F}{q}\right) \right],$$
(A.74)

- region 1A: $q > \omega$ and $\omega + q < 2k_F$:

$$\Re\chi = -\frac{2k_F}{\pi},$$
(A.75)

- region 2A: $q - 2k_F < \omega$ $q > \omega$ and $\omega + q > 2k_F$,

$$\Re\chi = -\frac{2k_F}{\pi} - \frac{q^2}{4\pi\sqrt{q^2 - \omega^2}} C\left(\frac{\omega - 2k_F}{q}\right),$$
(A.76)

- region 3A: $q - 2k_F > \omega$

$$\Re\chi = -\frac{2k_F}{\pi} + \frac{q^2}{4\pi\sqrt{q^2 - \omega^2}} \left[C\left(\frac{2k_F + \omega}{q}\right) - C\left(\frac{\omega - 2k_F}{q}\right) \right],$$
(A.77)

A.3.2 *Imaginary part of the susceptibility*

- region 1B:

$$\Im\chi = 0,$$
(A.78)

- region 2B:

$$\Im\chi = \frac{q^2}{4\pi\sqrt{\omega^2 - q^2}} C\left(\frac{2k_F - \omega}{q}\right),$$
(A.79)

- region 3B:

$$\Im\chi = -\frac{q^2}{4\sqrt{\omega^2 - q^2}},$$
(A.80)

- region 1A:

$$\Im\chi = \frac{q^2}{4\pi\sqrt{q^2 - \omega^2}} \left[F\left(\frac{2k_F - \omega}{q}\right) - F\left(\frac{2k_F + \omega}{q}\right) \right], \quad (A.81)$$

- region 2A:

$$\Im\chi = -\frac{q^2}{4\pi\sqrt{q^2 - \omega^2}} F\left(\frac{2k_F + \omega}{q}\right),$$
(A.82)

- region 3A:

$$\Im\chi = 0,$$
(A.83)

A.4   Some Useful Integrals

We now list some useful integrals that we have used in the calculations of this appendix. The integral

$$I = \mathcal{P} \int_0^{2\pi} d\theta \frac{1}{A - B\cos\theta} = 2 \int_0^{\pi} d\theta \frac{1}{A - B\cos\theta}, \qquad (A.84)$$

can be performed making $t = \tan\theta/2$. It follows:

$$2dt = \sec^2\theta/2 d\theta = (1 + \tan^2\theta/2)d\theta = (1 + t^2)d\theta, \qquad (A.85)$$

and

$$I = 4\mathcal{P} \int_0^{\infty} \frac{dt}{1 + t^2} \frac{1}{A - B\cos\theta}, \qquad (A.86)$$

but $\cos\theta = 2\cos^2\theta/2 - 1 = 2/(\sec^2\theta/2) - 1 = 2/(t^2 + 1) - 1 = (1 - t^2)/(1 + t^2)$, then:

$$I = 4\mathcal{P} \int_0^{\infty} dt \frac{1}{A(1 + t^2) - B(1 - t^2)} = 4\mathcal{P} \int_0^{\infty} dt \frac{1}{A - B + (A + B)t^2}. \qquad (A.87)$$

Making $y = (A - B)/(A + B)$ we have:

$$I = 4 \frac{1}{A + B} \mathcal{P} \int_0^{\infty} dt \frac{1}{y + t^2} \qquad (A.88)$$

If $y > 0$, we make the substitution $t = \sqrt{y}x$ and it follows

$$I = 4\mathrm{sign}(A + B)\sqrt{\frac{A - B}{A + B}} \int_0^{\infty} dx \frac{1}{1 + x^2} = 2\pi\sqrt{\frac{A - B}{A + B}}\mathrm{sign}(A + B). \qquad (A.89)$$

If $y < 0$, we make the substitution $t = \sqrt{-y}x$ and it follows

$$I = 4\sqrt{\frac{B - A}{A + B}} \mathcal{P} \int_0^{\infty} dx \frac{1}{1 - x^2} = 0. \qquad (A.90)$$

The same results can be derived using complex-plane integration.

Appendix B

Derivation of the Intra- and Inter-band Conductivity of Graphene

In this Appendix we derive the expressions for the intra- and interband optical conductivity of graphene, reproducing the results of Eqs. (2.67) and (2.68). We want to compute the response of the system to a time-dependent electric field of the form $E_x(t) = E_0 \cos(\omega t)$. The full Hamiltonian reads

$$H = v_F \boldsymbol{\sigma} \cdot \mathbf{p} + e x E_x(t) \equiv H_0 + e x E_x(t) \,. \qquad (B.1)$$

The eigenstates of H_0 are of the form

$$\psi_{\lambda,\mathbf{k}}(\mathbf{r}) = \frac{1}{\sqrt{2A}} \begin{pmatrix} e^{-i\theta_\mathbf{k}/2} \\ \lambda e^{i\theta_\mathbf{k}/2} \end{pmatrix} e^{i\mathbf{k}\cdot\mathbf{r}} \,, \qquad (B.2)$$

where $\lambda = c(+1), v(-1)$ refers to the band index and A is the area of the system. The current operator, in second quantization, reads

$$j_x = -e g_v g_s v_F \sum_{\mathbf{k},\lambda,\lambda'} a^\dagger_{\mathbf{k},\lambda} a_{\mathbf{k},\lambda'} \langle \mathbf{k}\lambda | \sigma_x | \mathbf{k}\lambda' \rangle \,, \qquad (B.3)$$

which was derived from the first quantization form of the current operator, $j_x = -e v_F \sigma_x$, and the matrix element $\langle \mathbf{k}\lambda | \sigma_x | \mathbf{k}\lambda' \rangle$ is defined as

$$\langle \mathbf{k}\lambda | \sigma_x | \mathbf{k}\lambda' \rangle = \int d^2\mathbf{r} \, \psi^\dagger_{\lambda,\mathbf{k}}(\mathbf{r}) \sigma_x \psi_{\lambda',\mathbf{k}}(\mathbf{r}) \,. \qquad (B.4)$$

The coefficient g_s and g_v stand for the spin and valley degeneracies, respectively. The matrix element $\langle \mathbf{k}\lambda | \sigma_x | \mathbf{k}\lambda' \rangle$ reads

$$\langle \mathbf{k}\lambda | \sigma_x | \mathbf{k}\lambda \rangle = \lambda \cos\theta \,, \qquad (B.5)$$

$$\langle \mathbf{k}c | \sigma_x | \mathbf{k}v \rangle = -i \sin\theta \,, \qquad (B.6)$$

$$\langle \mathbf{k}v | \sigma_x | \mathbf{k}c \rangle = i \sin\theta \,. \qquad (B.7)$$

We can separate the current into two contributions, the intraband one:

$$j_x^{\text{intra}} = -e g_v g_s v_F \sum_{\mathbf{k},\lambda} a^\dagger_{\mathbf{k},\lambda} a_{\mathbf{k},\lambda} \langle \mathbf{k}\lambda | \sigma_x | \mathbf{k}\lambda \rangle \,, \qquad (B.8)$$

and the interband contribution

$$j_x^{\text{inter}} = -eg_v g_s v_F \sum_{\mathbf{k}, \lambda \neq \lambda'} a_{\mathbf{k},\lambda}^\dagger a_{\mathbf{k},\lambda'} \langle \mathbf{k}\lambda | \sigma_x | \mathbf{k}\lambda' \rangle. \tag{B.9}$$

To compute the current we need the average value of the operator $\hat{P}_{\lambda,\lambda'}(\mathbf{k}, \mathbf{k}') = a_{\mathbf{k},\lambda}^\dagger a_{\mathbf{k}',\lambda'}$ taken using the exact eigenstates of the Hamiltonian H; this will be done using the equation of motion method.

In second quantization, the Hamiltonian H reads

$$H = \sum_{\mathbf{k},\lambda} \epsilon_{\mathbf{k},\lambda} a_{\mathbf{k},\lambda}^\dagger a_{\mathbf{k},\lambda} + eE_x(t) \sum_{\mathbf{k}_1,\lambda_1} \sum_{\mathbf{k}_2,\lambda_2} a_{\mathbf{k}_1,\lambda_1}^\dagger a_{\mathbf{k}_2,\lambda_2} \langle \mathbf{k}_1\lambda_1 | x | \mathbf{k}_2\lambda_2 \rangle,$$
$$\tag{B.10}$$

where the calculation of $\langle \mathbf{k}_1\lambda_1 | x | \mathbf{k}_2\lambda_2 \rangle$ is delicate [Blount (1962); Aversa and Sipe (1995); Gu *et al.* (2013)]:

$$\int d^2\mathbf{r}\, \psi_{\lambda_1,\mathbf{k}_1}^\dagger(\mathbf{r}) x \psi_{\lambda_2,\mathbf{k}_2}(\mathbf{r}) = \frac{1}{i}\frac{(2\pi)^2}{A}\delta_{\lambda_1,\lambda_2}\frac{\partial}{\partial k_{2,x}}\delta(\mathbf{k}_1 - \mathbf{k}_2)$$
$$- \frac{\sin\theta}{2k_1}(1 - \delta_{\lambda_1,\lambda_2})\frac{(2\pi)^2}{A}\delta(\mathbf{k}_1 - \mathbf{k}_2)$$
$$\equiv F_{\lambda_1\lambda_2}(\mathbf{k}_1, \mathbf{k}_2). \tag{B.11}$$

Clearly, the first term on the right-hand-side of the previous equation accounts for intraband transitions whereas the second term accounts for interband ones. The choice of gauge implied in Eq. (B.2) allows a simplified expression for (B.11), which is gauge dependent. A physical quantity, as the electric current, is, however, gauge independent.

The equation of motion for the operator $\hat{P}_{\lambda,\lambda'}(\mathbf{k}, \mathbf{k}')$ is defined in the usual way as

$$i\hbar\frac{d}{dt}\hat{P}_{\lambda_1,\lambda_2}(\mathbf{k}_1, \mathbf{k}_2) = [H, \hat{P}_{\lambda_1,\lambda_2}(\mathbf{k}_1, \mathbf{k}_2)]. \tag{B.12}$$

Explicitly we have

$$i\hbar\frac{d}{dt}\hat{P}_{\lambda_1,\lambda_2}(\mathbf{k}_1, \mathbf{k}_2) = (E_{\mathbf{k}_1,\lambda_1} - E_{\mathbf{k}_2,\lambda_2})\hat{P}_{\lambda_1,\lambda_2}(\mathbf{k}_1, \mathbf{k}_2)$$
$$+ eE_x(t)\sum_{\mathbf{q},\alpha} F_{\lambda_2\alpha}(\mathbf{k}_2, \mathbf{q})\hat{P}_{\lambda_1,\alpha}(\mathbf{k}_1, \mathbf{q})$$
$$- eE_x(t)\sum_{\mathbf{q},\alpha} F_{\alpha\lambda_1}(\mathbf{q}, \mathbf{k}_1)\hat{P}_{\alpha\lambda_2}(\mathbf{q}, \mathbf{k}_2). \tag{B.13}$$

We now define the thermal average of $\hat{P}_{\lambda_1,\lambda_2}(\mathbf{k}_1, \mathbf{k}_2)$ by $P_{\lambda_1,\lambda_2}(\mathbf{k}_1, \mathbf{k}_2)$, which, to first order in the electric field, has been taken relatively to the non-perturbed density matrix of the Hamiltonian H_0. We then have

$$i\hbar \frac{d}{dt} P_{cc}(\mathbf{k}_1, \mathbf{k}_2) = (E_{\mathbf{k}_1,c} - E_{\mathbf{k}_2,c}) P_{cc}(\mathbf{k}_1, \mathbf{k}_2)$$

$$+ eE_x(t) \sum_{\mathbf{q}} F_{cc}(\mathbf{k}_2, \mathbf{q}) P_{cc}^0(\mathbf{k}_1, \mathbf{q})$$

$$- eE_x(t) \sum_{\mathbf{q}} F_{cc}(\mathbf{q}, \mathbf{k}_1) P_{cc}^0(\mathbf{q}, \mathbf{k}_2), \qquad (B.14)$$

with $P_{cc}^0(\mathbf{q}, \mathbf{k}_2) = \delta_{\mathbf{q},\mathbf{k}_2} N_{cc}(\mathbf{q})$, where $N_{cc}(\mathbf{q})$ is the Fermi distribution function of the electrons in the $c-$band. Furthermore, from Eq. (B.11) we have

$$F_{cc}(\mathbf{k}_2, \mathbf{q}) = \frac{1}{i} \frac{(2\pi)^2}{A} \frac{\partial}{\partial q_x} \delta(\mathbf{k}_2 - \mathbf{q}), \qquad (B.15)$$

which allows to write Eq. (B.14) as

$$i\hbar \frac{d}{dt} P_{cc}(\mathbf{k}_1, \mathbf{k}_2) = (E_{\mathbf{k}_1,c} - E_{\mathbf{k}_2,c}) P_{cc}(\mathbf{k}_1, \mathbf{k}_2)$$

$$+ eE_x(t) \sum_{\mathbf{q}} \frac{1}{i} \frac{(2\pi)^4}{A^2} \frac{\partial}{\partial q_x} [\delta(\mathbf{k}_2 - \mathbf{q})\delta(\mathbf{k}_1 - \mathbf{q})] N_{cc}(\mathbf{q}),$$

$$(B.16)$$

and integrating the derivative of the $\delta-$functions by parts we find

$$i\hbar \frac{d}{dt} P_{cc}(\mathbf{k}_1, \mathbf{k}_2) = (E_{\mathbf{k}_1,c} - E_{\mathbf{k}_2,c}) P_{cc}(\mathbf{k}_1, \mathbf{k}_2)$$

$$+ ieE_x(t)\delta_{\mathbf{k}_1,\mathbf{k}_2} \frac{\partial}{\partial k_{1,x}} N_{cc}(\mathbf{k}_1). \qquad (B.17)$$

A similar equation holds for $P_{vv}(\mathbf{k}_1, \mathbf{k}_2)$ upon the replacement of c by v in the previous result. The first order differential equation (B.17) can easily be integrated, introducing a relaxation term for the average of $P_{cc}(\mathbf{k}_1, \mathbf{k}_2)$. Upon integration we obtain

$$P_{cc}(\mathbf{k}_1, \mathbf{k}_2) = \frac{eE_0}{2\hbar} \delta_{\mathbf{k}_1,\mathbf{k}_2} \frac{\partial}{\partial k_{1,x}} N_{cc}(\mathbf{k}_1) G(t), \qquad (B.18)$$

where $G(t)$ reads

$$G(t) = \frac{e^{i\omega t}}{\gamma + i\omega} + \frac{e^{-i\omega t}}{\gamma - i\omega}, \qquad (B.19)$$

and γ is the relaxation rate. In addition, we have the following result

$$\frac{\partial}{\partial k_{1,x}} N_{cc}(\mathbf{k}_1) = v_F \hbar \cos\theta \frac{\partial N_{cc}(E_{\mathbf{k}_1,c})}{\partial E_{\mathbf{k}_1,c}}. \qquad (B.20)$$

We are now in position of computing the linear intraband current. This reads

$$j_x^{\text{intra}} = -\sigma_0 \frac{4}{\pi} \int_0^\infty \epsilon d\epsilon \left[\frac{\partial N_{cc}(\epsilon)}{\partial \epsilon} - \frac{\partial N_{vv}(\epsilon)}{\partial \epsilon} \right] G(t) \frac{E_0}{2}, \qquad \text{(B.21)}$$

where $\sigma_0 = \pi e^2/(2h)$ and $N_{vv}(\epsilon) = N_{cc}(-\epsilon)$. Performing the integration in Eq. (B.21) we obtain the intraband conductivity, given by Eq. (2.67).

For computing the interband optical conductivity we proceed along the same lines, but now we have to compute the equation of motion for the average $P_{cv}(\mathbf{k}_1, \mathbf{k}_2)$. From Eq. (B.13) we find

$$\begin{aligned} i\hbar \frac{d}{dt} P_{cv}(\mathbf{k}_1, \mathbf{k}_2) &= (E_{\mathbf{k}_2,v} - E_{\mathbf{k}_1,c}) P_{cv}(\mathbf{k}_1, \mathbf{k}_2) \\ &+ eE_x(t) \sum_{\mathbf{q}} F_{vc}(\mathbf{k}_2, \mathbf{q}) P_{cc}^0(\mathbf{k}_1, \mathbf{q}) \\ &- eE_x(t) \sum_{\mathbf{q}} F_{vc}(\mathbf{q}, \mathbf{k}_1) P_{vv}^0(\mathbf{q}, \mathbf{k}_2). \end{aligned} \qquad \text{(B.22)}$$

For the interband transition the matrix element $F_{vc}(\mathbf{k}_1, \mathbf{k}_2)$ reads

$$F_{vc}(\mathbf{k}_1, \mathbf{k}_2) = -\frac{\sin\theta}{2k_1} \frac{(2\pi)^2}{A} \delta(\mathbf{k}_1 - \mathbf{k}_2). \qquad \text{(B.23)}$$

With the latter result, the equation of motion for $P_{cv}(\mathbf{k}, \mathbf{k}) = P_{cv}(\mathbf{k})$ reads

$$\frac{d}{dt} P_{cv}(\mathbf{k}) = -(\gamma - i\omega_{cv}) P_{cv}(\mathbf{k}) - \frac{ieE_x(t)}{2\hbar} F_{vc}(\mathbf{k}) [N_{cc}(\mathbf{k}) - N_{vv}(\mathbf{k})], \qquad \text{(B.24)}$$

where a relaxation term proportional to γ was introduced and $\omega_{cv} = (E_{\mathbf{k},c} - E_{\mathbf{k},v})/\hbar$. Integrating the equation of motion we obtain

$$P_{cv}(\mathbf{k}) = -\frac{ieE_0}{2\hbar} F_{vc}(\mathbf{k}) [N_{cc}(\mathbf{k}) - N_{vv}(\mathbf{k})] W_{cv}(t, \omega, \omega_{cv}), \qquad \text{(B.25)}$$

with $W_{cv}(t, \omega, \omega_{cv})$ given by

$$W_{cv}(t, \omega, \omega_{cv}) = \frac{e^{i\omega t}}{\gamma + i(\omega - \omega_{cv})} + \frac{e^{-i\omega t}}{\gamma - i(\omega + \omega_{cv})}. \qquad \text{(B.26)}$$

To compute the interband current we still need $P_{vc}(\mathbf{k})$, which is related to $P_{cv}(\mathbf{k})$ by $P_{vc}(\mathbf{k}) = P_{cv}^*(\mathbf{k})$. We have all we need to compute Eq. (B.9):

$$j_x^{\text{inter}} = ieg_s g_v v_F \sum_{\mathbf{k}} \sin\theta \left(\frac{ieE_0}{2\hbar} \frac{\sin\theta}{2k} [N_{cc}(\mathbf{k}) - N_{vv}(\mathbf{k})] W_{cv}(t) - \text{H.c.} \right). \qquad \text{(B.27)}$$

Taking the limit of zero temperature in Eq. (B.27) we obtain

$$j_x^{\text{inter}} = \sigma_0 \frac{E_0}{\pi} \int_{E_F}^\infty d\epsilon \left[W_{cv}(t, \hbar\omega, \hbar\omega_{cv}) + W_{cv}^*(t, \hbar\omega, \hbar\omega_{cv}) \right], \qquad \text{(B.28)}$$

where E_F is the Fermi energy and $\epsilon = \omega_{cv}\hbar/2$. The integral in Eq. (B.28) is easy to perform. If we further take the limit $\gamma \to 0$ we obtain the result

$$j_x^{\text{inter}} = \sigma_0 \left[\left(\Theta(\omega\hbar - 2E_F) + \frac{i}{\pi} \ln \frac{|\hbar\omega + 2E_F|}{|\hbar\omega - 2E_F|} \right) e^{i\omega t} \frac{E_0}{2} + \text{H.c.} \right], \quad \text{(B.29)}$$

where $\Theta(x)$ is the Heaviside step-function.

Appendix C

Inhomogeneous Drude Conductivity

The intraband conductivity of an electron gas follows from the solution of Boltzmann equation, which has the form [Ziman (1972)]

$$- \mathbf{v_k} \cdot \nabla_r f_\mathbf{k} - e/\hbar(\mathbf{E} + \mathbf{v_k} \times \mathbf{H}) \cdot \nabla_\mathbf{k} f_\mathbf{k} = - \left(\frac{\partial f_\mathbf{k}}{\partial t} \right)_{\text{coll.}} + \frac{\partial f_\mathbf{k}}{\partial t} , \quad \text{(C.1)}$$

where $\mathbf{v_k}$ is the group velocity, $\mathbf{E}$ is the electric field, $\mathbf{H}$ is the magnetic field, $e < 0$ is the electron charge, and $f_\mathbf{k} = f(\epsilon_\mathbf{k})$ is the distribution function, with $\epsilon_\mathbf{k}$ the electronic dispersion relation. The solution of the Boltzmann equation in its general form is difficult and one needs therefore to rely upon some approximation. The first step in the usual approximation scheme is to write the distribution as $f(\epsilon_\mathbf{k}) = f^0(\epsilon_\mathbf{k}) + g(\epsilon_\mathbf{k})$ where $f^0(\epsilon_\mathbf{k})$ is the steady state distribution function and $g(\epsilon_\mathbf{k})$ is assumed to be small. Inserting this ansatz in Eq. (C.1) and keeping only terms that are linear in the external fields one obtains the linearized Boltzmann equation which reads

$$- \frac{\partial f^0(\epsilon_\mathbf{k})}{\partial \epsilon_\mathbf{k}} \mathbf{v_k} \cdot \left[\left(-\frac{\epsilon_\mathbf{k} - E_F}{T} \right) \nabla_r T + e \left(\mathbf{E} - \frac{1}{e} \nabla_r E_F \right) \right] =$$

$$- \frac{\partial f_\mathbf{k}}{\partial t} \bigg|_{scatt.} + \mathbf{v_k} \cdot \nabla_r g_\mathbf{k} + \frac{e}{\hbar} (\mathbf{v_k} \times \mathbf{H}) \cdot \nabla_\mathbf{k} g_\mathbf{k} + \frac{\partial f_\mathbf{k}}{\partial t} . \quad \text{(C.2)}$$

The second approximation has to do with the form of the scattering term. The simplest approach is to introduce a relaxation time into the formalism. This is done by considering the approximation

$$\left(\frac{\partial f_\mathbf{k}}{\partial t} \right)_{\text{coll.}} \rightarrow - \frac{g_\mathbf{k}}{\tau_\mathbf{k}} , \quad \text{(C.3)}$$

where $\tau_\mathbf{k}$ is the relaxation time, assumed to be momentum dependent.

Our aim is to obtain the response of the electronic system to an external electric field of the form

$$\mathbf{E} = \mathbf{E}_0 e^{i(\mathbf{K} \cdot \mathbf{r} - \omega t)} . \quad \text{(C.4)}$$

The Boltzmann equation has, for this problem, the form

$$-\frac{\partial f^0(\epsilon_\mathbf{k})}{\partial \epsilon_\mathbf{k}}e v_\mathbf{k} \cdot \mathbf{E} = \frac{g_\mathbf{k}}{\tau_\mathbf{k}} + \mathbf{v_k} \cdot \nabla_r g_\mathbf{k} + \frac{\partial g_\mathbf{k}}{\partial t}. \tag{C.5}$$

The solution of the linearized Boltzmann's equation (C.5) is well known [Ziman (1972)], reading

$$g_\mathbf{k} = -\frac{\partial f^0(\epsilon_\mathbf{k})}{\partial \epsilon_\mathbf{k}} \Phi_\mathbf{K}(\omega, \mathbf{k}) e^{i(\mathbf{K}\cdot\mathbf{r} - \omega t)}, \tag{C.6}$$

where

$$\Phi_\mathbf{K}(\omega, \mathbf{k}) = \frac{e\tau \mathbf{v_k} \cdot \mathbf{E}}{1 - i\omega\tau + i\tau \mathbf{K} \cdot \mathbf{v_k}}, \tag{C.7}$$

where $\mathbf{E}$ is the electric field, $\mathbf{v_k}$ is the electron velocity, $\mathbf{K} = q\mathbf{u}_\parallel + q_z \mathbf{u}_z$ is the radiation wavevector and q is the in plane (coincident with the graphene plane) wave number, and we have assumed a momentum independent relaxation time, $\tau_\mathbf{k} = \tau$. We further assume that $\mathbf{q} = q\mathbf{u}_\parallel$ is along the x-direction.

C.1   Conductivity Along the Direction of Propagation in Plane (longitudinal)

In this case we have $v_\mathbf{k} \cdot \mathbf{E} = v_F \cos\theta E$ and $\mathbf{K} \cdot \mathbf{v_k} = q v_\mathbf{k} \cos\theta$. The current is, counting both spin and valley degeneracies, given by

$$\mathbf{J} = \frac{1}{\pi^2} \int d^2 k e v_\mathbf{k} \Phi_\mathbf{K}(\omega, \mathbf{k}) \left[-\frac{\partial f(\epsilon_\mathbf{k})}{\partial \epsilon_\mathbf{k}} \right]. \tag{C.8}$$

Then

$$J_x = E \frac{2e^2}{\pi^2 \hbar^2} \frac{\tau \epsilon_F}{1 - i\omega\tau} \int_0^\pi \frac{\cos^2\theta}{1 + s\cos\theta}, \tag{C.9}$$

where

$$s = \frac{i\tau v_F q}{1 - i\omega\tau}. \tag{C.10}$$

Making the transformation $x = \cos\theta$, we end up with

$$J_x = E \frac{2e^2}{\pi^2 \hbar} \frac{\epsilon_F}{\hbar/\tau - i\hbar\omega} \int_{-1}^1 \frac{x^2}{\sqrt{1 - x^2}} \frac{1}{1 + sx}. \tag{C.11}$$

Defining $\sigma_0 = \pi e^2/(2h)$ and performing the integral we obtain

$$\boxed{\sigma_{xx}(\omega, q) \equiv \sigma_L(\omega, q) = \sigma_0 \frac{8}{\pi} \frac{\epsilon_F}{\hbar/\tau - i\hbar\omega} \frac{1}{s^2} \left(-1 + \frac{1}{\sqrt{1 - s^2}} \right).} \tag{C.12}$$

For the calculation of J_y we have to perform the integral

$$\int_0^{2\pi} \frac{\sin\theta\cos\theta}{1 + s\cos\theta}, \tag{C.13}$$

which evaluates to zero. Then $\sigma_{yx} = 0$.

C.2 Conductivity Perpendicular to the Direction of Propagation in Plane (transverse)

In this case we have $v_{\mathbf{k}} \cdot \mathbf{E} = v_F \sin\theta E$ and $\mathbf{K} \cdot \mathbf{v_k} = q v_{\mathbf{k}} \cos\theta$. Then we have

$$J_y = E \frac{2e^2}{\pi^2 \hbar^2} \frac{\tau \epsilon_F}{1 - i\omega\tau} \int_0^\pi \frac{\sin^2\theta}{1 + s\cos\theta}, \qquad (C.14)$$

Making the same transformation as before we obtain

$$J_y = E \frac{2e^2}{\pi^2 \hbar} \frac{\epsilon_F}{\hbar/\tau - i\hbar\omega} \int_{-1}^1 \frac{1-x^2}{\sqrt{1-x^2}} \frac{1}{1+sx}. \qquad (C.15)$$

Performing the integral we obtain

$$\boxed{\sigma_{yy}(\omega,q) \equiv \sigma_T(\omega,q) = \sigma_0 \frac{8}{\pi} \frac{\epsilon_F}{\hbar/\tau - i\hbar\omega} \frac{1}{s^2}\left(1 - \sqrt{1-s^2}\right),} \qquad (C.16)$$

which is different from σ_{xx} as long as s is finite.

Appendix D

Derivation of the Expression Relating the Longitudinal Conductivity with the Polarizability (written with Bruno Amorim)

Here we derive and expression relating the longitudinal conductivity with the polarizability an electron gas. It is assumed that the Hamiltonian is linear in the vector potential, as in the case of the low effective model for the electronic properties of graphene (Dirac Hamiltonian). The final result, however, is general. The (charge) current-current correlation function is defined as follows:

$$\Pi^{\mu\nu}(\mathbf{x}, t; \mathbf{x}', t') = -\frac{i}{\hbar}\Theta(t - t') \langle [J^{\mu}(\mathbf{x}, t), J^{\nu}(\mathbf{x}', t')] \rangle , \qquad (D.1)$$

where

$$J^{\mu}(\mathbf{x}, t) = (\rho(\mathbf{x}, t), \mathbf{J}(\mathbf{x}, t)) , \qquad (D.2)$$

is the 4-current. On the other hand, charge conservation implies through the continuity equation, that

$$\partial_t \rho + \nabla \cdot \mathbf{J} = 0 , \qquad (D.3)$$

and therefore the current-current correlation function must obey

$$\partial_t \Pi^{0\nu}(\mathbf{x}, t; \mathbf{x}', t') + \partial_i \Pi^{i\nu}(\mathbf{x}, t; \mathbf{x}', t') = 0 . \qquad (D.4)$$

By separating the time and spatial components, we arrive to

$$\partial_t \Pi^{00}(\mathbf{x}, t; \mathbf{x}', t') + \partial_j \Pi^{j0}(\mathbf{x}, t; \mathbf{x}', t') = 0 , \qquad (D.5)$$

$$\partial_t \Pi^{0i}(\mathbf{x}, t; \mathbf{x}', t') + \partial_j \Pi^{ji}(\mathbf{x}, t; \mathbf{x}', t') = 0 , \qquad (D.6)$$

where the latin indexes represent spatial coordinates (whereas the greek indices present so far correspond to space-time variables). For a translational invariant system one can perform a Fourier transform and obtain

$$-i\omega\Pi^{00}(\mathbf{q}, \omega) + iq_j\Pi^{j0}(\mathbf{q}, \omega) = 0 , \qquad (D.7)$$

$$-i\omega\Pi^{0i}(\mathbf{q}, \omega) + iq_j\Pi^{ji}(\mathbf{q}, \omega) = 0 . \qquad (D.8)$$

From the last equation we can write

$$\Pi^{0i}(\mathbf{q}, \omega) = \frac{q_j}{\omega} \Pi^{ji}(\mathbf{q}, \omega) , \qquad (D.9)$$

after which we insert this into Eq. (D.7), and find that

$$\frac{q_i q_j}{\omega^2} \Pi^{ij}(\mathbf{q}, \omega) = \Pi^{00}(\mathbf{q}, \omega) . \qquad (D.10)$$

Moreover, for and isotropic system (in the absence of a magnetic field), the most general form of $\Pi^{ij}(\mathbf{q}, \omega)$ is[2] [Chaikin and Lubensky (2000); Dressel and Grüner (2002); Poole Jr. (2004)]

$$\Pi^{ij}(\mathbf{q}, \omega) = \Pi^L(q, \omega) \frac{q^i q^j}{q^2} + \Pi^T(q, \omega) \left(\delta^{ij} - \frac{q^i q^j}{q^2} \right) , \qquad (D.11)$$

as a second-order-rank tensor can only be built from the identity, δ_{ij}, and the a bilinear term $q_i q_j$ if no other vector but the wavevector exists. Multiplying the previous equation by $q_i q_j / \omega^2$ and contracting the indices, we obtain

$$\frac{q_i q_j}{\omega^2} \Pi^{ij}(\mathbf{q}, \omega) = \frac{q^2}{\omega^2} \Pi^L(q, \omega) . \qquad (D.12)$$

From Eq. (D.12), it follows that the relation between the density-density correlation function and the longitudinal component of the current-current correlation function reads:

$$e^2 \chi_{\rho\rho}(q, \omega) \equiv \Pi^{00}(q, \omega) = \frac{q^2}{\omega^2} \Pi^L(q, \omega) . \qquad (D.13)$$

For a system in which the Hamiltonian is linear in momentum (diamagnetic term absent), the current-current correlation function is related to the conductivity by

$$\sigma^{ij}(\mathbf{q}, \omega) = \frac{i}{\omega + i\eta} \Pi^{ij}(\mathbf{q}, \omega) , \qquad (D.14)$$

where $\eta = 0^+$. In particular, focusing in the longitudinal component, we have

$$\sigma^L(q, \omega) = \frac{i}{\omega + i\eta} \Pi^L(q, \omega) . \qquad (D.15)$$

---

[2]The longitudinal projector comes from $\mathbf{E}^L = \mathbf{q}(\mathbf{q} \cdot \mathbf{E})/q^2$ or $E_i^L = q_i q_j E_j / q^2 = P_{ij}^L E_j$, with P_{ij}^L being the longitudinal projector.

The transverse projector comes from $\mathbf{E}^T = -\mathbf{q} \times (\mathbf{q} \times \mathbf{E}) = \mathbf{E} - \mathbf{q}(\mathbf{q} \cdot \mathbf{E})/q^2$ or $E_i^T = P_{ij}^T E_j$ with $P_{ij}^T = \delta_{ij} - q_i q_j / q^2$ being the transverse projector.

The longitudinal and transverse conductivity follows from Ohm's law: $J_i = \sigma_{ij} E_j$. The longitudinal current is obtained from $J_l^L = P_{li}^L J_i = P_{li}^L \sigma_{ij} E_j = P_{li}^L \sigma_{ij} (P_{jk}^L E_k + P_{jk}^T E_k)$ which implies $\sigma_{kl}^L = P_{li}^L \sigma_{ij} P_{jk}^L$. In the same manner, $\sigma_{lk}^T = P_{li}^T \sigma_{ij} P_{jk}^T$ [Wooten (1972)].

Thus, by combining Eqs. (D.13) and (D.15), one obtains the desired relation connecting the density-density correlation function with the longitudinal conductivity,

$$\boxed{\sigma^L(q,\omega) = ie^2 \frac{\omega}{q^2} \chi_{\rho\rho}(q,\omega)} \,, \tag{D.16}$$

or, in the language of Eq. (2.84),

$$\boxed{\sigma(q,\omega) = ie^2 \frac{\omega}{q^2} P(q,\omega)} \,. \tag{D.17}$$

The result of Eq. (D.17) is entirely general for any electron gas in the absence of a magnetic field.

Appendix E

Derivation of the Polarization of Graphene in the Relaxation-Time Approximation

In this Appendix we provide a derivation of Eq. (2.85) in the context of graphene. We assume the existence of scattering centres which introduce a relaxation time, τ, for the electronic momentum $\mathbf{k}$. In addition, we consider the response of the electron gas to an external potential $\phi(\mathbf{r}, t)$ (the potential should be considered the result of a self-consistent solution of Poisson's equation). As a response to the presence of the potential the chemical potential becomes a (smooth) function of the position $\mathbf{r}$ and time t. We then write the chemical potential as $\mu = \epsilon_F + \delta\mu(\mathbf{r}, t)$, where $\delta\mu(\mathbf{r}, t)$ is the fluctuation relatively to the homogeneous chemical potential ϵ_F. The polarization is defined by

$$\delta n(\mathbf{q}, \omega) = P_\gamma(\mathbf{q}, \omega)\phi(\mathbf{q}, \omega), \tag{E.1}$$

where $\delta n(\mathbf{q}, \omega)$ is the Fourier component of the fluctuation of the electronic density, $\phi(\mathbf{q}, \omega)$ is the Fourier component of the potential, and $P_\gamma(\mathbf{q}, \omega)$ is the polarization. The goal of this Appendix is to determine $P_\gamma(\mathbf{q}, \omega)$ in the given conditions.

E.1 Hamiltonian and Particle and Current Densities

We write the Hamiltonian in second quantization as

$$\hat{H} = \sum_{\mathbf{k}, \lambda}(E_{k\lambda} - \epsilon_F)c^\dagger_{\mathbf{k}, \lambda}c_{\mathbf{k}, \lambda} + \hat{H}_I + \hat{H}_{\delta\mu}, \tag{E.2}$$

where $c^\dagger_{\mathbf{k}, \lambda}$ is the creation operator of an electrons in band λ ($\lambda = \pm$) with momentum $\mathbf{k}$, $E_{k\lambda} = v_F \hbar k$ is the electronic dispersion (see Sec. 2.3.1), and H_I is the external potential in the form

$$\hat{H}_I = \sum_{\mathbf{k}, \mathbf{q}} \sum_{\alpha, \beta} \phi(\mathbf{q}, t)\gamma_{\alpha, \beta}(\mathbf{k} + \mathbf{q}, \mathbf{k})c^\dagger_{\mathbf{k}+\mathbf{q}, \alpha}c_{\mathbf{k}, \beta}, \tag{E.3}$$

where

$$\gamma_{\alpha,\beta}(\mathbf{k} + \mathbf{q}, \mathbf{k}) = \chi_\alpha^\dagger(\mathbf{k} + \mathbf{q})\chi_\beta(\mathbf{k}), \tag{E.4}$$

and the spinor $\chi_\lambda = \chi_\lambda(\mathbf{k})$ is defined in Sec. 2.3.1. The presence of the external potential makes the chemical potential inhomogeneous $\mu = \epsilon_F \rightarrow \mu = \epsilon_F + \delta\mu(\mathbf{r}, t)$. Then, the electrons may scatter off $\delta\mu(\mathbf{r}, t)$, which leads to the Hamiltonian term

$$\hat{H}_{\delta\mu} = -\sum_{\mathbf{k},\mathbf{q}} \sum_{\alpha,\beta} \delta\mu(\mathbf{q}, t)\gamma_{\alpha,\beta}(\mathbf{k} + \mathbf{q}, \mathbf{k})c_{\mathbf{k}+\mathbf{q},\alpha}^\dagger c_{\mathbf{k},\beta}. \tag{E.5}$$

In what follows we also need the particle density and current density operators. The density operator is written as[3]

$$\hat{n} = \sum_{\mathbf{k},\mathbf{q}} \sum_{\alpha,\beta} \gamma_{\alpha,\beta}(\mathbf{k} + \mathbf{q}, \mathbf{k})c_{\mathbf{k}+\mathbf{q},\alpha}^\dagger c_{\mathbf{k},\beta} = \sum_{\mathbf{q}} \hat{n}_{\mathbf{q}} \tag{E.6}$$

and the current density reads[4]

$$\hat{\mathbf{J}} = v_F \sum_{\mathbf{k},\mathbf{q}} \sum_{\alpha,\beta} \vec{\gamma}_{\alpha,\beta}(\mathbf{k} + \mathbf{q}, \mathbf{k})c_{\mathbf{k}+\mathbf{q},\alpha}^\dagger c_{\mathbf{k},\beta} = \sum_{\mathbf{q}} \hat{\mathbf{J}}_{\mathbf{q}}, \tag{E.7}$$

where

$$\vec{\gamma}_{\alpha,\beta}(\mathbf{k} + \mathbf{q}, \mathbf{k}) = \chi_\alpha^\dagger(\mathbf{k} + \mathbf{q})\sigma\chi_\beta(\mathbf{k}), \tag{E.8}$$

and σ is defined as $\sigma = (\sigma_x, \sigma_y)$, with σ_i the i Pauli's matrix $(i = x, y)$.

E.2   Local Equilibrium Density Matrix

We shall describe the response of the system using the density matrix formalism, which allows to determine the relation between the fluctuations in particle density and the external potential. Due to the momentum relaxation (characterized by τ) the density matrix will attain a local equilibrium after some "long" time. The local equilibrium density matrix can be computed as follows. The density matrix operator can be written as

$$\frac{1}{e^{\beta(\hat{H}_0+\hat{V})} + 1} = \frac{1}{2\pi i} \oint dE \frac{f(E)}{E - \hat{H}_0 - \hat{V}}, \tag{E.9}$$

with $\hat{V} = \hat{H}_{\delta\mu}$,

$$f(E) = 1/(e^{\beta E} + 1), \tag{E.10}$$

---

[3]In first quantization the density operator is simply given by $\delta(\mathbf{r})$

[4]In first quantization the density current operator is defined as $\delta(\mathbf{r})v_F\sigma$, where v_F is the Fermi velocity of the electrons in graphene.

and

$$\hat{H}_0 = \sum_{\mathbf{k},\lambda}(E_{k\lambda} - \epsilon_F)c_{\mathbf{k},\lambda}^\dagger c_{\mathbf{k},\lambda} \, . \qquad (\text{E.11})$$

The operator $(E - \hat{H}_0 - \hat{V})^{-1}$ is to be understood as giving the result $(E - E_n)^{-1}$ when acting on an eigenstate $|\Psi_n\rangle$ of $\hat{H} = \hat{H}_0 + \hat{V}$; the contour in the integral encloses all the eigenvalues of $\hat{H}$ (on the real axis), and extends from $-\infty$ to $+\infty$ below the real axis and from $+\infty$ to $-\infty$ above the real axis. Equation (E.9) refers to a snapshot of the system at "long" times, since we are not discussing its time evolution at this point. Since

$$\frac{1}{E - \hat{H}_0 - \hat{V}} \approx \frac{1}{E - \hat{H}_0} + \frac{1}{E - \hat{H}_0}\hat{V}\frac{1}{E - \hat{H}_0} \, , \qquad (\text{E.12})$$

the matrix element of the density matrix operator reads

$$\begin{aligned}
\langle k\alpha|\hat{\rho}_{loc.}|\mathbf{k} + \mathbf{q}\beta\rangle &\equiv \langle k\alpha|[e^{\beta(\hat{H}_0 + \hat{V})} + 1]^{-1}|\mathbf{k} + \mathbf{q}\beta\rangle \\
&= \frac{1}{2\pi i}\oint dE\frac{f(E)\delta_{q,0}\delta_{\alpha,\beta}}{E - \tilde{E}_{k\alpha}} \\
&\quad + \frac{1}{2\pi i}\oint dE\frac{f(E)\langle k\alpha|\hat{V}|\mathbf{k} + \mathbf{q}\beta\rangle}{(E - \tilde{E}_{k\alpha})(E - \tilde{E}_{\mathbf{k}+\mathbf{q}\beta})} \, , \qquad (\text{E.13})
\end{aligned}$$

where $\tilde{E}_{k\alpha} = E_{k\alpha} - \epsilon_F$ and $H_0|k\alpha\rangle = \tilde{E}_{k\alpha}|k\alpha\rangle$. The matrix element $\langle k\alpha|\hat{V}|\mathbf{k} + \mathbf{q}\beta\rangle$ is given by

$$\langle k\alpha|\hat{V}|\mathbf{k} + \mathbf{q}\beta\rangle = -\gamma_{\alpha,\beta}(\mathbf{k},\mathbf{k} + \mathbf{q})\delta\mu(\mathbf{q},t) \, , \qquad (\text{E.14})$$

where $\delta\mu(\mathbf{q},t)$ is the Fourier transform of $\delta\mu(\mathbf{r},t)$. Finally, we have $(\mathbf{q} \neq 0)$

$$\langle k\alpha|\hat{\rho}_{loc.}|\mathbf{k} + \mathbf{q}\beta\rangle \approx -\delta\mu(\mathbf{q},t)\gamma_{\alpha\beta}(\mathbf{k},\mathbf{k} + \mathbf{q})\frac{f(\tilde{E}_{k\alpha}) - f(\tilde{E}_{\mathbf{k}+\mathbf{q}\beta})}{E_{\mathbf{k},\alpha} - E_{\mathbf{k}+\mathbf{q},\beta}} \, , \qquad (\text{E.15})$$

which is the density matrix characterizing the local equilibrium to first order in $\delta\mu(\mathbf{r},t)$ (the quantity $\delta\mu(\mathbf{r},t)$ has still to be found, a task to be accomplished ahead). The reader may have noticed that the perturbation $\hat{H}_I$ has not been included explicitly in the derivation. That is because the local equilibrium density matrix is only an explicit function of density. Alternatively, we can assume that the potential $\phi(\mathbf{r},t)$ has been absorbed in the variation of the chemical potential $\delta\mu(\mathbf{r},t)$ [Kragler and Thomas (1980)].

E.3 A Useful Identity

The Heisenberg equation of motion for the operator $\hat{n}_{\mathbf{q}}$ reads

$$\frac{\partial \hat{n}_{\mathbf{q}}}{\partial t} = \frac{i}{\hbar}[\hat{H}, \hat{n}_{\mathbf{q}}] = \frac{i}{\hbar}[\hat{H}_0, \hat{n}_{\mathbf{q}}]. \tag{E.16}$$

Evaluating the commutator, we obtain

$$\frac{\partial \hat{n}_{\mathbf{q}}}{\partial t} = \frac{i}{\hbar}\sum_{\mathbf{k}}\sum_{\alpha,\beta}(E_{\mathbf{k}+\mathbf{q}\beta} - E_{\mathbf{k}\alpha})\gamma_{\alpha,\beta}(\mathbf{k}+\mathbf{q},\mathbf{k})c_{\mathbf{k}+\mathbf{q},\alpha}^{\dagger}c_{\mathbf{k},\beta}. \tag{E.17}$$

We now recall the continuity equation, which in momentum space reads

$$\frac{\partial \hat{n}_{\mathbf{q}}}{\partial t} = -i\mathbf{q} \cdot \hat{\mathbf{J}}_{\mathbf{q}}$$
$$= -iv_F\sum_{\mathbf{k}}\sum_{\alpha,\beta}\mathbf{q} \cdot \vec{\gamma}_{\alpha,\beta}(\mathbf{k}+\mathbf{q},\mathbf{k})c_{\mathbf{k}+\mathbf{q},\alpha}^{\dagger}c_{\mathbf{k},\beta}, \tag{E.18}$$

from where it follows, by comparing with Eq. (E.17), the identity

$$-v_F\hbar\mathbf{q} \cdot \vec{\gamma}_{\alpha,\beta}(\mathbf{k}+\mathbf{q},\mathbf{k}) = (E_{\mathbf{k}+\mathbf{q}\beta} - E_{\mathbf{k}\alpha})\gamma_{\alpha,\beta}(\mathbf{k}+\mathbf{q},\mathbf{k}), \tag{E.19}$$

a result that will be used ahead.

E.4 Equation of Motion for the Density Matrix

The equation of motion for the density matrix reads

$$\frac{\partial \hat{\rho}}{\partial t} = -\frac{i}{\hbar}[H, \hat{\rho}] + \frac{\partial \hat{\rho}}{\partial t}\bigg|_{coll.}, \tag{E.20}$$

where the second term on the right-hand-side accounts for the relaxation of the density matrix to a density matrix characterizing the local equilibrium of the electronic system (see Sec. E.2). Given the complexity of computing the collision term, we introduce the relaxation-time approximation by making the replacement

$$\frac{\partial \hat{\rho}}{\partial t}\bigg|_{coll.} \rightarrow -\frac{1}{\tau}(\hat{\rho}_I - \hat{\rho}_{loc.}), \tag{E.21}$$

where we write $\hat{\rho} \approx \rho_0 + \hat{\rho}_I$ and $\hat{\rho}_I$ contains the effect of the potential $\hat{H}_I$ (to first order). The form of the collision term accounts for deviations relatively to the density matrix characterizing the local equilibrium. In this approximation, the linearized (in $\hat{H}_I$) equation of motion for the density matrix reads

$$i\hbar\frac{\partial \hat{\rho}_I}{\partial t} = [\hat{H}_0, \hat{\rho}_I] + [\hat{H}_I, \hat{\rho}_0] - \frac{i\hbar}{\tau}(\hat{\rho}_I - \hat{\rho}_{loc.}). \tag{E.22}$$

From the previous equation it follows

$$i\hbar\frac{\partial\rho_I^{\alpha\beta}(\mathbf{k},\mathbf{k}+\mathbf{q})}{\partial t} = (E_{\mathbf{k}\alpha} - E_{\mathbf{k}+\mathbf{q}\beta})\rho_I^{\alpha\beta}(\mathbf{k},\mathbf{k}+\mathbf{q})$$
$$+ [f(\tilde{E}_{\mathbf{k}+\mathbf{q}\beta}) - f(\tilde{E}_{\mathbf{k}\alpha})]\langle\mathbf{k}\alpha|\hat{H}_I|\mathbf{k}+\mathbf{q}\beta\rangle$$
$$- \frac{i\hbar}{\tau}\rho_I^{\alpha\beta}(\mathbf{k},\mathbf{k}+\mathbf{q})$$
$$+ \frac{i\hbar}{\tau}\langle\mathbf{k}\alpha|\hat{\rho}_{loc.}|\mathbf{k}+\mathbf{q}\beta\rangle\,, \tag{E.23}$$

where

$$\rho_I^{\alpha\beta}(\mathbf{k},\mathbf{k}+\mathbf{q}) = \langle\mathbf{k}\alpha|\hat{\rho}_I|\mathbf{k}+\mathbf{q}\beta\rangle\,. \tag{E.24}$$

The matrix element $\langle\mathbf{k}\alpha|\hat{H}_I|\mathbf{k}+\mathbf{q}\beta\rangle$ reads

$$\langle\mathbf{k}\alpha|\hat{H}_I|\mathbf{k}+\mathbf{q}\beta\rangle = \phi(\mathbf{q},t)\gamma_{\alpha,\beta}(\mathbf{k},\mathbf{k}+\mathbf{q})\,. \tag{E.25}$$

If we take the limit $\tau \to \infty$, Eq. (E.23) reads

$$\hbar\omega\rho_I^{\alpha\beta}(\mathbf{k},\mathbf{k}+\mathbf{q}) = (E_{\mathbf{k}\alpha} - E_{\mathbf{k}+\mathbf{q}\beta})\rho_I^{\alpha\beta}(\mathbf{k}+\mathbf{q},\mathbf{q})$$
$$+ \phi(\mathbf{q},\omega)\gamma_{\alpha,\beta}(\mathbf{k},\mathbf{k}+\mathbf{q})[f(\tilde{E}_{\mathbf{k}+\mathbf{q}\beta}) - f(\tilde{E}_{\mathbf{k}\alpha})]\,, \tag{E.26}$$

from where it follows

$$\rho_I^{\alpha\beta}(\mathbf{k}+\mathbf{q},\mathbf{q}) = \phi(\mathbf{q},\omega)\gamma_{\alpha,\beta}(\mathbf{k},\mathbf{k}+\mathbf{q})\frac{f(\tilde{E}_{\mathbf{k}+\mathbf{q}\beta}) - f(\tilde{E}_{\mathbf{k}\alpha})}{\hbar\omega + E_{\mathbf{k}+\mathbf{q}\beta} - E_{\mathbf{k}\alpha}}\,. \tag{E.27}$$

Upon multiplication by $\gamma_{\alpha,\beta}(\mathbf{k}+\mathbf{q},\mathbf{k})$ and summation over $\mathbf{k}$, α, and β the last equation reduces to

$$n_I(\mathbf{q},\omega) = \phi(\mathbf{q},\omega)P(\mathbf{q},\omega)\,, \tag{E.28}$$

where

$$P(\mathbf{q},\omega) = \sum_{\mathbf{k},\alpha,\beta} \gamma_{\alpha,\beta}(\mathbf{k}+\mathbf{q},\mathbf{k})\gamma_{\alpha,\beta}(\mathbf{k},\mathbf{k}+\mathbf{q})\frac{f(\tilde{E}_{\mathbf{k}+\mathbf{q}\beta}) - f(\tilde{E}_{\mathbf{k}\alpha})}{\hbar\omega + E_{\mathbf{k}+\mathbf{q}\beta} - E_{\mathbf{k}\alpha}}\,, \tag{E.29}$$

and

$$n_I(\mathbf{q},\omega) = \sum_{\mathbf{k},\alpha,\beta} \gamma_{\alpha,\beta}(\mathbf{k}+\mathbf{q},\mathbf{k})\rho_I^{\alpha\beta}(\mathbf{k},\mathbf{k}+\mathbf{q})\,. \tag{E.30}$$

The quantity $P(\mathbf{q},\omega)$ is the electronic polarization of graphene.

E.5   Polarization in the Relaxation-Time Approximation

The continuity equation can be written as $\omega n_I(\mathbf{q}, \omega) = \mathbf{q} \cdot \mathbf{J_q}$ or as

$$\sum_{\mathbf{k}, \alpha, \beta} \omega \gamma_{\alpha, \beta}(\mathbf{k} + \mathbf{q}, \mathbf{k}) \rho_I^{\alpha\beta}(\mathbf{k}, \mathbf{k} + \mathbf{q}) = \sum_{\mathbf{k}, \alpha, \beta} v_F \mathbf{q} \cdot \vec{\gamma}_{\alpha, \beta}(\mathbf{k} + \mathbf{q}, \mathbf{k}) \rho_I^{\alpha\beta}(\mathbf{k}, \mathbf{k} + \mathbf{q}).$$

(E.31)

We note that the continuity equation involves the quantity $n_I(\mathbf{q}, \omega)$ only; the equilibrium term is not time dependent. In frequency space, Eq. (E.23) reads [upon multiplication by $\gamma_{\alpha, \beta}(\mathbf{k} + \mathbf{q}, \mathbf{k})$]

$$\omega \hbar \gamma_{\alpha, \beta}(\mathbf{k} + \mathbf{q}, \mathbf{k}) \rho_I^{\alpha\beta}(\mathbf{k}, \mathbf{k} + \mathbf{q}) = \gamma_{\alpha, \beta}(\mathbf{k} + \mathbf{q}, \mathbf{k})(E_{\mathbf{k}\alpha} - E_{\mathbf{k}+\mathbf{q}\beta}) \rho_I^{\alpha\beta}$$
$$\times \, (\mathbf{k}, \mathbf{k} + \mathbf{q}) + \phi(\mathbf{q}, \omega)[f(\tilde{E}_{\mathbf{k}+\mathbf{q}\beta}) - f(\tilde{E}_{\mathbf{k}\alpha})] \gamma_{\alpha, \beta}(\mathbf{k} + \mathbf{q}, \mathbf{k}) \gamma_{\alpha, \beta}(\mathbf{k}, \mathbf{k} + \mathbf{q})$$
$$- \, \frac{i\hbar}{\tau} \gamma_{\alpha, \beta}(\mathbf{k} + \mathbf{q}, \mathbf{k}) \rho_I^{\alpha\beta}(\mathbf{k}, \mathbf{k} + \mathbf{q})$$
$$- \, \frac{i\hbar}{\tau} \delta\mu(\mathbf{q}, \omega) \gamma_{\alpha, \beta}(\mathbf{k} + \mathbf{q}, \mathbf{k}) \gamma_{\alpha, \beta}(\mathbf{k}, \mathbf{k} + \mathbf{q}) \frac{f(\tilde{E}_{\mathbf{k}\alpha}) - f(\tilde{E}_{\mathbf{k}+\mathbf{q}\beta})}{E_{\mathbf{k}\alpha} - E_{\mathbf{k}+\mathbf{q}\beta}}. \quad (E.32)$$

Summing Eq. (E.32) over $\mathbf{k}$, α, and β, and using the identity (E.19) and the continuity equation (E.31) it follows

$$\delta\mu(\mathbf{q}, \omega) = -n_I(\mathbf{q}, \omega)/P(\mathbf{q}, 0), \quad (E.33)$$

where the result

$$\sum_{\mathbf{k}, \alpha, \beta} [f(\tilde{E}_{\mathbf{k}+\mathbf{q}\beta}) - f(\tilde{E}_{\mathbf{k}\alpha})] \gamma_{\alpha, \beta}(\mathbf{k} + \mathbf{q}, \mathbf{k}) \gamma_{\alpha, \beta}(\mathbf{k}, \mathbf{k} + \mathbf{q}) = 0 \quad (E.34)$$

has been used. Having determined $\delta\mu(\mathbf{q}, \omega)$ we replace it back in Eq. (E.32) and solve it for $n_I(\mathbf{q}, \omega)$:

$$n_I(\mathbf{q}, \omega) = \phi(\mathbf{q}, \omega) P(\mathbf{q}, \omega + i/\tau)$$
$$+ \, \frac{i\hbar}{\tau} \frac{n_I(\mathbf{q}, \omega)}{P(\mathbf{q}, 0)} \frac{1}{\omega\hbar + i\hbar/\tau} [P(\mathbf{q}, 0) - P(\mathbf{q}, \omega + i/\tau)], \quad (E.35)$$

a result that can be further simplified to:

$$n_I(\mathbf{q}, \omega) = \phi(\mathbf{q}, \omega) \frac{(1 - i\omega\tau) P(\mathbf{q}, 0) P(\mathbf{q}, \omega + i/\tau)}{P(\mathbf{q}, \omega + i/\tau) - i\omega\tau P(\mathbf{q}, 0) P}, \quad (E.36)$$

from where it follows $P_\gamma(\mathbf{q}, \omega)$

$$\boxed{P_\gamma(\mathbf{q}, \omega) = \frac{n_I(\mathbf{q}, \omega)}{\phi(\mathbf{q}, \omega)} = \frac{(1 - i\omega\tau) P(\mathbf{q}, 0) P(\mathbf{q}, \omega + i/\tau)}{P(\mathbf{q}, \omega + i/\tau) - i\omega\tau P(\mathbf{q}, 0)}}, \quad (E.37)$$

a result that agrees with Eq. (2.85). In the above derivation the identity

$$\frac{1}{(E_{\mathbf{k}\alpha} - E_{\mathbf{k}+\mathbf{q}\beta})(\omega\hbar + i\hbar/\tau + E_{\mathbf{k}+\mathbf{q}\beta} - E_{\mathbf{k}\alpha})} =$$
$$\frac{1}{\hbar\omega + i\hbar/\tau} \left[\frac{1}{E_{\mathbf{k}\alpha} - E_{\mathbf{k}+\mathbf{q}\beta}} + \frac{1}{\omega\hbar + i\hbar/\tau + E_{\mathbf{k}+\mathbf{q}\beta} - E_{\mathbf{k}\alpha}} \right] \quad (E.38)$$

proved useful.

Appendix F

RPA for Double-Layer Graphene

In this Appendix we derive the RPA dielectric function of a graphene double-layer, with layers separated by a distance d from one another. We shall conduct the derivation twice, using two different approaches: the equation of motion method (sometimes also called "self-consistent method") and also using diagrammatic techniques.

F.1 Equation of Motion Method

The Coulomb potential between the electrons within one layer is, in Fourier space[5], given by

$$V_1 = V_2 = \frac{e^2}{2\epsilon_0 q}, \tag{F.1}$$

whereas the inter-layer Coulomb potential reads

$$V_{12} = V_{21} = \frac{e^2 e^{-qd}}{2\epsilon_0 q}. \tag{F.2}$$

We want to describe the response of the electron gas to an external potential $U(q)$. The Hamiltonian of the system reads

$$H = H_0 + H_I + H_U, \tag{F.3}$$

where H_0 is the independent electrons Hamiltonian of graphene, H_I is the Coulomb interaction between electrons, and H_U is the hamiltonian describing the interaction of the electrons with the external potential $U(q)$. In second quantization, we have:

$$H_0 = v_F \hbar \sum_{\mathbf{k},\lambda} (\phi_{\mathbf{k}} a_{\mathbf{k}\lambda}^\dagger b_{\mathbf{k}\lambda} + \phi_{\mathbf{k}}^* b_{\mathbf{k}\lambda}^\dagger a_{\mathbf{k}\lambda}), \tag{F.4}$$

---

[5]See Appendix G.

333

where $\phi_{\mathbf{k}} = -k_x + ik_y$, $a^{\dagger}_{\mathbf{k}\lambda}$ creates an electron is sub-lattice A, in state $\mathbf{k}$ (the spin index is included in $\mathbf{k}$), and in the layer λ ($\lambda = 1, 2$). The Hamiltonian H_0 can be diagonalized introducing a new set of operators $c^{\dagger}_{\mathbf{k}\alpha\lambda}$, defined as

$$a^{\dagger}_{\mathbf{k}\lambda} = \frac{1}{\sqrt{2}} \sum_{\alpha=\pm} c^{\dagger}_{\mathbf{k}\alpha\lambda}, \tag{F.5}$$

$$b^{\dagger}_{\mathbf{k}\lambda} = \frac{1}{\sqrt{2}} \sum_{\alpha=\pm} \alpha e^{-i\theta_{\mathbf{k}}} c^{\dagger}_{\mathbf{k}\alpha\lambda}, \tag{F.6}$$

where, as before, $\mathbf{k}$ includes the spin σ, $\theta_{\mathbf{k}} = \arctan(q_y/q_x)$, and α is the band index. In the new basis, H_0 reads

$$H_0 = \sum_{\mathbf{k},\alpha,\lambda} E_{\mathbf{k}\alpha} c^{\dagger}_{\mathbf{k}\alpha\lambda} c_{\mathbf{k}\alpha\lambda}, \tag{F.7}$$

where $E_{\mathbf{k}\alpha} = \alpha v_F \hbar k$. The electron-electron interaction part of the Hamiltonian, H_I, reads

$$H_I = \frac{1}{2} \sum_{\mathbf{q},\lambda} V_{\lambda}(q) n_{\lambda}(-\mathbf{q}) n_{\lambda}(\mathbf{q}) + \sum_{\mathbf{q},\lambda} V_{12}(q) n_1(-\mathbf{q}) n_2(\mathbf{q}), \tag{F.8}$$

where the density operator reads

$$n_{\lambda}(\mathbf{q}) = \sum_{\mathbf{k}} (a^{\dagger}_{\mathbf{k}+\mathbf{q}\lambda} a_{\mathbf{k}\lambda} + b^{\dagger}_{\mathbf{k}+\mathbf{q}\lambda} b_{\mathbf{k}\lambda})$$

$$= \sum_{\mathbf{k}} \sum_{\alpha_1,\alpha_2} \gamma^{\alpha_1\alpha_2}_{\mathbf{k}+\mathbf{q},\mathbf{k}} c^{\dagger}_{\mathbf{k}+\mathbf{q}\alpha_1\lambda} c_{\mathbf{k}\alpha_2\lambda}, \tag{F.9}$$

where

$$\gamma^{\alpha_1\alpha_2}_{\mathbf{k}+\mathbf{q},\mathbf{k}} = \frac{1}{2}[1 + \alpha_1\alpha_2 e^{i(\theta_{\mathbf{k}} - \theta_{\mathbf{k}+\mathbf{q}})}]. \tag{F.10}$$

Finally, the interaction of the electrons with the external potential $U(q)$ reads

$$H_U = \sum_{\mathbf{q},\lambda} U(q) n_{\lambda}(\mathbf{q}). \tag{F.11}$$

The next step in the derivation is to solve the equation of motion for the density operator $n_{\lambda}(\mathbf{q})$:

$$\frac{dn_{\lambda}(\mathbf{q})}{dt} = \frac{i}{\hbar}[H, n_{\lambda}(\mathbf{q})]. \tag{F.12}$$

To that end, a number of commutators have to be computed and the following relations

$$[AB, C] = A[B, C] + [A, C]B, \tag{F.13}$$

$$[AB, C] = A\{B, C\} - \{A, C\}B, \tag{F.14}$$

prove useful. The commutator with the kinetic energy reads

$$[H_0, n_\lambda(\mathbf{q})] = \sum_{\mathbf{k},\alpha_1,\alpha_2} (E_{\mathbf{k}+\mathbf{q}\alpha_1} - E_{\mathbf{k}\alpha_2})\gamma_{\mathbf{k}+\mathbf{q},\mathbf{k}}^{\alpha_1\alpha_2} c_{\mathbf{k}+\mathbf{q}\alpha_1\lambda}^\dagger c_{\mathbf{k}\alpha_2\lambda}. \qquad \text{(F.15)}$$

The contribution coming from the interaction with the external potential has the form

$$[H_U, n_\lambda(\mathbf{q})] = \sum_{\mathbf{q}'} U(q') \sum_{\mathbf{k},\alpha_1,\alpha_2} \left(\gamma_{\mathbf{k}+\mathbf{q}+\mathbf{q}',\mathbf{k}+\mathbf{q}}^{\alpha_1\alpha_2} \sum_{\alpha_4} \gamma_{\mathbf{k}+\mathbf{q},\mathbf{k}}^{\alpha_2\alpha_4} c_{\mathbf{k}+\mathbf{q}+\mathbf{q}'\alpha_1\lambda}^\dagger c_{\mathbf{k}\alpha_4\lambda} \right.$$

$$\left. - \gamma_{\mathbf{k},\mathbf{k}-\mathbf{q}'}^{\alpha_1\alpha_2} \sum_{\alpha_3} \gamma_{\mathbf{k}+\mathbf{q},\mathbf{k}}^{\alpha_3\alpha_1} c_{\mathbf{k}+\mathbf{q}\alpha_3\lambda}^\dagger c_{\mathbf{k}-\mathbf{q}'\alpha_2\lambda} \right). \qquad \text{(F.16)}$$

The last commutator involves the Coulomb interaction term, H_I, and reads

$$[H_I, n_\lambda(\mathbf{q})] = \frac{1}{2}\sum_{\mathbf{q}'} V_\lambda(\mathbf{q}')n_\lambda(-\mathbf{q}')$$

$$\left(\sum_{\mathbf{k},\alpha_1,\alpha_2,\alpha_4} \gamma_{\mathbf{k}+\mathbf{q}+\mathbf{q}',\mathbf{k}+\mathbf{q}}^{\alpha_1\alpha_2} \gamma_{\mathbf{k}+\mathbf{q},\mathbf{k}}^{\alpha_2\alpha_4} c_{\mathbf{k}+\mathbf{q}+\mathbf{q}'\alpha_1\lambda}^\dagger c_{\mathbf{k}\alpha_4\lambda} \right.$$

$$\left. - \sum_{\mathbf{k},\alpha_1,\alpha_2,\alpha_3} \gamma_{\mathbf{k},\mathbf{k}-\mathbf{q}'}^{\alpha_1\alpha_2} \gamma_{\mathbf{k}+\mathbf{q},\mathbf{k}}^{\alpha_3\alpha_1} c_{\mathbf{k}+\mathbf{q}\alpha_3\lambda}^\dagger c_{\mathbf{k}-\mathbf{q}'\alpha_2\lambda} \right)$$

$$+ \frac{1}{2}\sum_{\mathbf{q}'} V_\lambda(\mathbf{q}') \left(\sum_{\mathbf{k},\alpha_1,\alpha_2,\alpha_4} \gamma_{\mathbf{k}+\mathbf{q}-\mathbf{q}',\mathbf{k}+\mathbf{q}}^{\alpha_1\alpha_2} \gamma_{\mathbf{k}+\mathbf{q},\mathbf{k}}^{\alpha_2\alpha_4} c_{\mathbf{k}+\mathbf{q}-\mathbf{q}'\alpha_1\lambda}^\dagger c_{\mathbf{k}\alpha_4\lambda} \right.$$

$$\left. - \sum_{\mathbf{k},\alpha_1,\alpha_2,\alpha_3} \gamma_{\mathbf{k},\mathbf{k}+\mathbf{q}'}^{\alpha_1\alpha_2} \gamma_{\mathbf{k}+\mathbf{q},\mathbf{k}}^{\alpha_3\alpha_1} c_{\mathbf{k}+\mathbf{q}\alpha_3\lambda}^\dagger c_{\mathbf{k}+\mathbf{q}'\alpha_2\lambda} \right) n_\lambda(\mathbf{q}')$$

$$+ \sum_{\mathbf{q}'} V_{\bar\lambda\lambda}(q')n_{\bar\lambda}(-\mathbf{q}')$$

$$\left(\sum_{\mathbf{k},\alpha_1,\alpha_2,\alpha_4} \gamma_{\mathbf{k}+\mathbf{q}+\mathbf{q}',\mathbf{k}+\mathbf{q}}^{\alpha_1\alpha_2} \gamma_{\mathbf{k}+\mathbf{q},\mathbf{k}}^{\alpha_2\alpha_4} c_{\mathbf{k}+\mathbf{q}+\mathbf{q}'\alpha_1\lambda}^\dagger c_{\mathbf{k}\alpha_4\lambda} \right.$$

$$\left. - \sum_{\mathbf{k},\alpha_1,\alpha_2,\alpha_3} \gamma_{\mathbf{k},\mathbf{k}-\mathbf{q}'}^{\alpha_1\alpha_2} \gamma_{\mathbf{k}+\mathbf{q},\mathbf{k}}^{\alpha_3\alpha_1} c_{\mathbf{k}+\mathbf{q}\alpha_3\lambda}^\dagger c_{\mathbf{k}-\mathbf{q}'\alpha_2\lambda} \right), \qquad \text{(F.17)}$$

where $\bar\lambda \neq \lambda$. Next we make the average of the equation of motion and introduce the random phase approximation (RPA). This approximation replaces the average of terms with four operators, as for example,

$$\langle n_\lambda(-\mathbf{q}')c_{\mathbf{k}+\mathbf{q}+\mathbf{q}'\alpha_1\lambda}^\dagger c_{\mathbf{k}\alpha_4\lambda}\rangle \qquad \text{(F.18)}$$

by

$$\langle n_\lambda(\mathbf{q})\rangle f_{\mathbf{k}\alpha_1\lambda}\delta_{\alpha_1,\alpha_4}\delta_{\mathbf{q},-\mathbf{q}'}\,, \tag{F.19}$$

where $f_{\mathbf{k}\alpha_1\lambda}$ is the Fermi distribution function. Making similar averages in all the terms of the equation of motion we end up with

$$[\hbar\omega + E_{\mathbf{k}+\mathbf{q}\alpha} - E_{\mathbf{k}\alpha'}]\gamma^{\alpha\alpha'}_{\mathbf{k}+\mathbf{q},\mathbf{k}}\langle c^\dagger_{\mathbf{k}+\mathbf{q}\alpha\lambda}c_{\mathbf{k}\alpha'\lambda}\rangle =$$
$$[U(q) + V_\lambda(q)\langle n_\lambda(\mathbf{q})\rangle + V_{\bar{\lambda},\lambda}(q)\langle n_{\bar{\lambda}}(\mathbf{q})\rangle] \times$$
$$\gamma^{\alpha\alpha'}_{\mathbf{k}+\mathbf{q},\mathbf{k}}\gamma^{\alpha\alpha'}_{\mathbf{k},\mathbf{k}+\mathbf{q}}(f_{\mathbf{k}+\mathbf{q}\alpha\lambda} - f_{\mathbf{k}\alpha'\lambda})\,, \tag{F.20}$$

from where it follows that

$$\sum_{\mathbf{k}\alpha\alpha'}\gamma^{\alpha\alpha'}_{\mathbf{k}+\mathbf{q},\mathbf{k}}\langle c^\dagger_{\mathbf{k}+\mathbf{q}\alpha\lambda}c_{\mathbf{k}\alpha'\lambda}\rangle =$$
$$[U(q) + V_\lambda(q)\langle n_\lambda(\mathbf{q})\rangle + V_{\bar{\lambda},\lambda}(q)\langle n_{\bar{\lambda}}(\mathbf{q})\rangle]\,\chi_\lambda(\mathbf{q},\omega)\,, \tag{F.21}$$

where

$$\chi_\lambda(\mathbf{q},\omega) = \sum_{\mathbf{k}\alpha\alpha'}\gamma^{\alpha\alpha'}_{\mathbf{k}+\mathbf{q},\mathbf{k}}\gamma^{\alpha\alpha'}_{\mathbf{k},\mathbf{k}+\mathbf{q}}\frac{f_{\mathbf{k}+\mathbf{q}\alpha\lambda} - f_{\mathbf{k}+\mathbf{q}\alpha'\lambda}}{\hbar\omega + E_{\mathbf{k}+\mathbf{q}\alpha} - E_{\mathbf{k}\alpha'}}\,. \tag{F.22}$$

We now specialize Eq. (F.21) for $\lambda = 1, 2$:

$$\langle n_1(\mathbf{q})\rangle = [U(q) + V_1(q)\langle n_1(\mathbf{q})\rangle + V_{1,2}(q)\langle n_2(\mathbf{q})\rangle]\,\chi_1(\mathbf{q},\omega)\rangle\,, \tag{F.23}$$
$$\langle n_2(\mathbf{q})\rangle = [U(q) + V_2(q)\langle n_2(\mathbf{q})\rangle + V_{2,1}(q)\langle n_1(\mathbf{q})\rangle]\,\chi_2(\mathbf{q},\omega)\,, \tag{F.24}$$

from where it follows

$$\begin{bmatrix}\langle n_1(\mathbf{q})\rangle \\ \langle n_2(\mathbf{q})\rangle\end{bmatrix} = \frac{U(q)}{\varepsilon(\mathbf{q},\omega)}\begin{bmatrix}1 - V_2(q)\chi_2(\mathbf{q},\omega) & V_{12}(q)\chi_1(\mathbf{q},\omega) \\ V_{12}(q)\chi_2(\mathbf{q},\omega) & 1 - V_1(q)\chi_1(\mathbf{q},\omega)\end{bmatrix}\begin{bmatrix}\chi_1(\mathbf{q},\omega) \\ \chi_2(\mathbf{q},\omega)\end{bmatrix}\,, \tag{F.25}$$

where $\varepsilon(\mathbf{q},\omega)$ reads

$$\varepsilon(\mathbf{q},\omega) = [1 - V_1(q)\chi_1(\mathbf{q},\omega)][1 - V_2(q)\chi_2(\mathbf{q},\omega)] - V_{12}^2(q)\chi_1(\mathbf{q},\omega)\chi_2(\mathbf{q},\omega)\,. \tag{F.26}$$

Equation (F.25) can be rewritten as

$$\begin{bmatrix}\langle n_1(\mathbf{q})\rangle \\ \langle n_2(\mathbf{q})\rangle\end{bmatrix} = \chi^{\text{RPA}}\begin{bmatrix}U(q) \\ U(q)\end{bmatrix}\,, \tag{F.27}$$

where

$$\boxed{\chi^{\text{RPA}} = \frac{1}{\varepsilon(\mathbf{q},\omega)}\begin{bmatrix}[1 - V_2(q)\chi_2(\mathbf{q},\omega)]\chi_1(\mathbf{q},\omega) & V_{12}(q)\chi_1(\mathbf{q},\omega)\chi_2(\mathbf{q},\omega) \\ V_{12}(q)\chi_2(\mathbf{q},\omega)\chi_1(\mathbf{q},\omega) & [1 - V_1(q)\chi_1(\mathbf{q},\omega)]\chi_2(\mathbf{q},\omega)\end{bmatrix}}\,, \tag{F.28}$$

where χ^{RPA} is the susceptibility matrix of the double layer system.

On the other hand, if $V_{12} = 0$ the polarization and the density are related by

$$\begin{bmatrix} \langle n_1(\mathbf{q}) \rangle \\ \langle n_2(\mathbf{q}) \rangle \end{bmatrix} = \begin{bmatrix} \phi_1(\mathbf{q}, \omega)\chi_1(\mathbf{q}, \omega) \\ \phi_2(\mathbf{q}, \omega)\chi_2(\mathbf{q}, \omega) \end{bmatrix}, \tag{F.29}$$

where $\phi_j(\mathbf{q}, \omega)$ (with $j = 1, 2$) is the effective potential acting on the electrons. These are

$$\begin{bmatrix} \phi_1(\mathbf{q}, \omega) \\ \phi_2(\mathbf{q}, \omega) \end{bmatrix} = \frac{U(q)}{\varepsilon(\mathbf{q}, \omega)} \begin{bmatrix} 1 - V_2(q)\chi_2(\mathbf{q}, \omega) \\ 1 - V_1(q)\chi_1(\mathbf{q}, \omega) \end{bmatrix}. \tag{F.30}$$

Since the two layers are decoupled then the potential reads

$$\phi_1(\mathbf{q}, \omega) = \frac{U(q)}{1 - V_1(q)\chi_1(\mathbf{q}, \omega)}, \tag{F.31}$$

from where it follows the usual RPA dielectric susceptibility

$$\varepsilon^{\mathrm{RPA}}(\mathbf{q}, \omega) = 1 - V_1(q)\chi_1(\mathbf{q}, \omega), \tag{F.32}$$

given in Eq. (2.82), where $\chi_1(\mathbf{q}, \omega) = P(\mathbf{q}, \omega)$.

F.2 Diagrammatic Approach

The electronic Hamiltonian of the double layer system in the absence of an external potential can be schematically written as

$$H = H_0 + V_{11} + V_{22} + V_{12}, \tag{F.33}$$

where H_0 represents the independent electrons term, $V_{\lambda\lambda}$ (with $\lambda = 1, 2$) is the intra-layer Coulomb Hamiltonian and V_{12} is the inter-layer Coulomb term. Up to second order in perturbation theory the terms V_{11}^2, V_{22}^2, V_{12}^2, $V_{12}V_{11}$, and $V_{12}V_{22}$ are generated. On the-other-hand, since the averages are made using the non-interacting ground state the only finite polarization bubbles are those composed of Green's functions with the same λ−index at the two vertices. With this constraint, the effective intra-layer (say, for $\lambda = 1$) and inter-layer interactions can be represented in terms of diagrams as in Fig. F.1. Transforming the diagrammatic representation into equations we have

$$\begin{bmatrix} \phi_{11} & \phi_{12} \\ \phi_{12} & \phi_{22} \end{bmatrix} = \begin{bmatrix} V_{11} & V_{12} \\ V_{12} & V_{22} \end{bmatrix} + \begin{bmatrix} V_{11}\chi_1 & V_{12}\chi_2 \\ V_{12}\chi_1 & V_{22}\chi_2 \end{bmatrix} \begin{bmatrix} \phi_{11} & \phi_{12} \\ \phi_{12} & \phi_{22} \end{bmatrix}. \tag{F.34}$$

Solving the previous equation for the matrix

$$\begin{bmatrix} \phi_{11} & \phi_{12} \\ \phi_{12} & \phi_{22} \end{bmatrix}$$

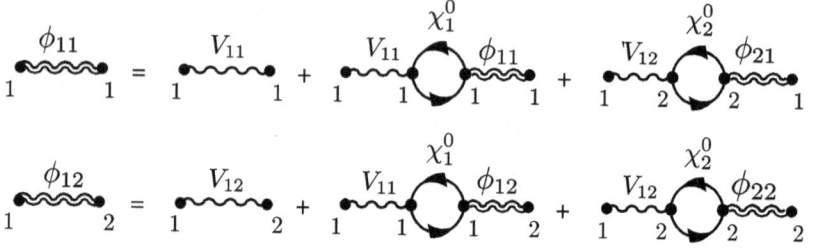

Figure F.1: RPA interaction in diagrammatic form. For simplicity of the drawing, only the intraband term ϕ_{11} is given (drawing the ϕ_{22} should be evident). The inter-layer effective interaction is represented as $\phi_{12} = \phi_{21}$, and therefore only one drawing is necessary.

we obtain

$$
\begin{bmatrix} \phi_{11} & \phi_{12} \\ \phi_{12} & \phi_{22} \end{bmatrix} = \frac{1}{\varepsilon(\mathbf{q},\omega)} \begin{bmatrix} V_1(1 - V_2\chi_2) + V_{12}^2\chi_2 + & V_{12} \\ V_{12} & V_2(1 - V_1\chi_1) + V_{12}^2\chi_1 \end{bmatrix},
$$

(F.35)

from where we can read the effective, $\phi_{\lambda\lambda'}$ ($\lambda, \lambda' = 1, 2$), intra- and inter-layer interactions. Finally, we note that the χ's are functions of $\mathbf{q}$ and ω, which we have omitted for simplicity of notation. As before, we recover Eq. (F.32) when $V_{12} = 0$. From the knowledge of χ^{RPA} obtained in the previous section we can also derive the result of Eq. (F.35). To that end we just have to note that in diagrammatic terms the effective interaction ϕ reads, schematically, $\phi = V + V\chi^{\mathrm{RPA}}V$, where V is given by

$$
V = \begin{bmatrix} V_{11} & V_{12} \\ V_{12} & V_{22} \end{bmatrix},
$$

(F.36)

and ϕ reads

$$
\phi = \begin{bmatrix} \phi_{11} & \phi_{12} \\ \phi_{12} & \phi_{22} \end{bmatrix}.
$$

(F.37)

The result obtained from this procedure agrees with that of Eq. (F.35). Thus we have found a connection between the RPA derivation of the susceptibility and the RPA effective interaction.

Appendix G

Effective Dielectric Constants for Coulomb-Coupled Double-Layer Graphene

In this Appendix we derive the intra- and inter-layer effective dielectric function ascribed to the Coulomb interaction amongst the same graphene sheet (intra-layer part) and also between distinct layers (inter-layer part). We shall accomplish this by solving the Poisson's equation in a double-layer system akin to the one depicted in Fig. 4.8.

Before moving on to the actual derivation, recall that for the single-layer case the Fourier transform of the Coulomb interaction, entering in the RPA dielectric function [cf. Eq. (2.82)], is just $\nu_q = e^2/(2q\epsilon_0\epsilon_r)$ where ϵ_r is the dielectric constant of the embedding medium (or, in the case of different cladding dielectrics: $\epsilon_r = \frac{\epsilon_1+\epsilon_2}{2}$). However, for the double-layer structure the situation is more complex since both graphene sheets interact to each other via the Coulomb interaction, and thus instead of ϵ_r we will have a ϵ_{intra} and ϵ_{inter}. Furthermore, these latter quantities are functions of the in-plane momentum (this will become more clear towards the end of the Appendix).

Let us then start our derivation by noting that the Coulomb potential created by a point charge at the origin is

$$V(\mathbf{r}) \equiv V(r) = \frac{Q}{4\pi\epsilon_0}\frac{1}{r} , \tag{G.1}$$

which is solution of the Poisson's equation $\nabla^2 V(\mathbf{r}) = -\frac{Q}{\epsilon_0}\delta(\mathbf{r})$. We are interested in the Fourier transform of the potential in the xy-plane only (which will then be graphene's plane), so that we write the potential in the real space as

$$V(\mathbf{r}) = \int \frac{d^2\mathbf{q}}{(2\pi)^2} V(\mathbf{q}, z)e^{i\mathbf{q}\cdot(x\hat{\mathbf{x}}+y\hat{\mathbf{y}})} , \tag{G.2}$$

where the two-dimensional vector $\mathbf{q} = (q_x, q_y)$ refers to the in-plane momentum, and $V(\mathbf{q}, z)$ is the Fourier transform of the potential in the x, y coordinates. By plugging in Eq. (G.2) into Poisson's equation, one obtains

$$\left(-\frac{\partial^2}{\partial z^2} + q^2\right) V(\mathbf{q}, z) = \frac{Q}{\epsilon_0}\delta(z) , \tag{G.3}$$

whose Fourier transformed solution is straightforwardly obtained as being

$$V(\mathbf{q}, \kappa) = \frac{Q}{\epsilon_0}\frac{1}{\kappa^2 + q^2} , \tag{G.4}$$

where we have performed the Fourier transform on z-component of the potential in solving for Eq. (G.4), which can be defined from

$$V(\mathbf{q}, z) = \int \frac{d\kappa}{2\pi} V(\mathbf{q}, \kappa) e^{i\kappa z} . \tag{G.5}$$

Combining Eqs. (G.4) and (G.5), we can compute the potential xy-transform:

$$V(\mathbf{q}, z) = \frac{Q}{2\pi\epsilon_0} \int_{-\infty}^{\infty} \frac{e^{i\kappa z}}{\kappa^2 + q^2} d\kappa = \frac{Q}{2\epsilon_0}\frac{e^{-q|z|}}{q} , \tag{G.6}$$

where in the last step we have performed the integral using the costumary techniques of contour integration in the complex plane[6], taking the contour in the upper-half plane (i.e. containing the $\kappa = iq$ pole) when $z > 0$, and taking the contour in the lower-half plane (i.e. around the $\kappa = -iq$ pole) for $z < 0$ —cf. figure G.1.

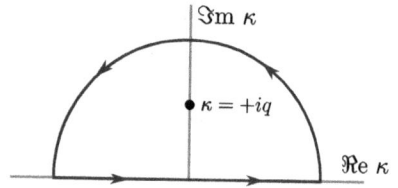

Figure G.1: Contour taken in the upper-half complex plane around the pole at $\kappa = iq$, which was used in the last step of Eq. (G.6), when assuming that $z > 0$. For negative values of z, the corresponding countour is taken in the lower-half plane around the $\kappa = -iq$ pole.

Having obtained $V(\mathbf{q}, z)$, our present goal is to arrive at an expression for the intra- and inter-layer (effective) dielectric constants, by solving the

---

[6]That is:

$$\int_C d\kappa \frac{e^{i\kappa z}}{(\kappa + iq)(\kappa - iq)} = \int_C d\kappa \frac{e^{i\kappa z}}{2iq}\left[\frac{1}{\kappa - iq} - \frac{1}{\kappa + iq}\right]$$

$$= \frac{\pi}{q}\begin{cases} \mathrm{Res}_{\kappa=iq} f(\kappa) & \text{for } z \geq 0 \\ \mathrm{Res}_{\kappa=-iq} f(\kappa) & \text{for } z < 0 \end{cases} = \pi\frac{e^{-q|z|}}{q} , \tag{G.7}$$

where the contour in the case of positive z is depicted in figure G.1.

electrostatic problem —also treated in [Katsnelson (2011)]— in a stratified media (along $\hat{z}$) containing three different insulators, arranged as:

$$\epsilon(z) = \begin{cases} \epsilon_1 & , z < 0 \\ \epsilon_2 & , 0 < z < d \\ \epsilon_3 & , z > d \end{cases} . \tag{G.8}$$

From Gauss's law, we have

$$\nabla \cdot \mathbf{D} = \rho_{\text{free}} \quad \Rightarrow \quad -\nabla \cdot [\epsilon(\mathbf{r})\nabla V(\mathbf{r})] = \frac{Q}{\epsilon_0}\delta(\mathbf{r}) . \tag{G.9}$$

Assuming that the point charge lies in the $z = 0$ plane and carrying out the xy-transform of the previous equation, we get the following differential equation:

$$-\frac{\partial}{\partial z}\left(\epsilon(z)\frac{\partial V(q, z)}{\partial z}\right) + q^2\epsilon(z)V(q, z) = \frac{Q}{\epsilon_0}\delta(z) , \tag{G.10}$$

whose first term is clearly discontinuous across an interface. This can be seen by integrating between $-\eta$ and η and sending $\eta \to 0$. Taking into account that the potential is continuous across the interface and that the dielectric function is constant (independent of z) on both sides of the interface we obtain

$$-\epsilon(\eta)V'(q, \eta) + \epsilon(-\eta)V'(q, -\eta) = \frac{Q}{\epsilon_0} . \tag{G.11}$$

Then, Eq. (G.11) allows to write two boundary conditions across $z = 0$ and $z = d$ (interfaces between adjacent dielectrics) in the form:

$$\begin{cases} -\epsilon_2 V_2'(q, z = 0) + \epsilon_1 V_1'(q, z = 0) = Q/\epsilon_0 \\ -\epsilon_3 V_3'(q, z = d) + \epsilon_2 V_2'(q, z = d) = 0 \end{cases} , \tag{G.12}$$

(recall that the point charge is only at $z = 0$) which together with continuity of the potential

$$\begin{cases} V_1(q, z = 0) = V_2(q, z = 0) \\ V_2(q, z = d) = V_3(q, z = d) \end{cases} , \tag{G.13}$$

where the V_j's are the potentials in the medium defined by the corresponding ϵ_j, allows the determination of the dielectric functions. Since we already had derived the form of $V(q, z)$ [cf. Eq. (G.7)], we can write the potential in the dielectrics as [satisfying the limit $\lim_{z\to\pm\infty} V_{\text{sys.}}(q, z) = 0$]

$$V_{\text{sys.}}(q, z) = \begin{cases} V_1(q, z) = Ae^{qz} & , z \leq 0 \\ V_2(q, z) = Be^{qz} + Ce^{-qz} & , 0 < z < d \\ V_3(q, z) = De^{-qz} & , z \geq d \end{cases} . \tag{G.14}$$

With this ansatz, the evaluation of the boundary conditions is straightforward, giving the following linear system of equations:

$$
\begin{pmatrix}
1 & -1 & -1 & 0 \\
0 & e^{qd} & e^{-qd} & -e^{-qd} \\
\epsilon_1 q & -\epsilon_2 q & \epsilon_2 q & 0 \\
0 & \epsilon_2 q e^{qd} & -\epsilon_2 q e^{-qd} & \epsilon_3 q e^{-qd}
\end{pmatrix}
\begin{pmatrix} A \\ B \\ C \\ D \end{pmatrix}
=
\begin{pmatrix} 0 \\ 0 \\ \frac{Q}{\epsilon_0} \\ 0 \end{pmatrix} . \tag{G.15}
$$

Notice that the "intra" (at $z = 0$) and "inter" (at $z = d$) potentials can be written as

$$
V_{\text{intra}}(q, 0) = A \equiv \frac{Q}{2q\epsilon_0} \frac{1}{\epsilon_{\text{intra}}}, \tag{G.16}
$$

$$
V_{\text{inter}}(q, d) = De^{-qd} \equiv \frac{Q}{2q\epsilon_0} \frac{1}{\epsilon_{\text{inter}}} e^{-qd}, \tag{G.17}
$$

where the functions $A \equiv A(q, d)$ and $D \equiv D(q, d)$ are obtained from the solution of the matricial equation (G.15), so that the sought intra- and inter-layer effective dielectric constants read[7]:

$$
\epsilon_{\text{intra}} = \frac{1}{2} \frac{(\epsilon_1 - \epsilon_2)(\epsilon_2 - \epsilon_3) + (\epsilon_1 + \epsilon_2)(\epsilon_2 + \epsilon_3)e^{2qd}}{\epsilon_2 - \epsilon_3 + (\epsilon_2 + \epsilon_3)e^{2qd}}, \tag{G.18}
$$

$$
\epsilon_{\text{inter}} = \frac{(\epsilon_1 - \epsilon_2)(\epsilon_2 - \epsilon_3) + (\epsilon_1 + \epsilon_2)(\epsilon_2 + \epsilon_3)e^{2qd}}{4\epsilon_2 e^{2qd}}. \tag{G.19}
$$

Note that ϵ_{inter} remains unchanged upon exchanging the outer insulators (i.e. $\epsilon_1 \leftrightarrow \epsilon_3$), while ϵ_{intra} does change. Finally, just for the sake of completeness, let us name graphene's sheet at $z = 0$ as "1", and the second sheet at $z = d$ as "2". Thus, in this language we can make the following identifications:

$$
\boxed{\epsilon_{11} \equiv \epsilon_{\text{intra}} \quad , \quad \epsilon_{12} = \epsilon_{21} \equiv \epsilon_{\text{inter}} \quad , \quad \epsilon_{22} \equiv \epsilon_{\text{intra}}\big|_{\epsilon_1 \leftrightarrow \epsilon_3}} . \tag{G.20}
$$

This notation with two sub-indexes demonstrated its usefulness in the study the random-phase approxmation for double-layer graphene (discussed in Appendix F for $\epsilon_1 = \epsilon_2 = \epsilon_3 = 1$). Also, this approach is essential in complex structures involving distinct dielectric media and several graphene sheets, as it is the case in few experimental studies, as for example in Coulomb Drag in double-layer graphene [Kim et al. (2011); Gorbachev et al. (2012)].

---

[7] Also, note that from Eqs. (G.18) and (G.19) one recovers the simple result, $\epsilon_{\text{intra}} = \epsilon_{\text{inter}} = \epsilon_r$, of the case in which $\epsilon_1 = \epsilon_2 = \epsilon_3 = \epsilon_r$.

Appendix H

Magneto-Optical Conductivity of Graphene: Boltzmann equation

Unstrained graphene can be treated as an uniform two-dimensional material whose conductivity is isotropic and described by a scalar quantity. However, if the system is subjected to an uniform static magnetic field, time-reversal symmetry is broken and graphene's conductivity becomes a 2-ranked tensor, and is termed *magneto-optical* conductivity. The aim of this Appendix is to determine the expressions for this tensor, within the semi-classical approach of Boltzmann's transport equation. The results derived below are valid for doped graphene only.

Within this semi-classical theory, the spinorial nature of the electronic wavefunctions is neglected and the spin and valley degeneracies are included as multiplicative factors. The transport Boltzmann's equation has the general form [Ashcroft and Mermin (1976); Ziman (1972)]

$$\nabla_r f_{\mathbf{k}} \cdot \mathbf{v}_{\mathbf{k}} + \nabla_{\mathbf{k}} f_{\mathbf{k}} \cdot \frac{\mathbf{F}}{\hbar} + \frac{\partial f_{\mathbf{k}}}{\partial t} = \left(\frac{\partial f_{\mathbf{k}}}{\partial t} \right)_{\text{coll.}} = -\frac{g_{\mathbf{k}}}{\tau} , \qquad \text{(H.1)}$$

where we have employed the relaxation-time approximation, in the limit that the scattering rate is momentum independent and $\mathbf{v}_{\mathbf{k}} = v_F (\cos \theta, \sin \theta)$. Here, $f_{\mathbf{k}}(\epsilon_{\mathbf{k}}) = f_{\mathbf{k}}^0(\epsilon_{\mathbf{k}}) + g_{\mathbf{k}}(\epsilon_{\mathbf{k}})$ is the distribution function of the charge-carriers, which is assumed to be the sum of the equilibrium distribution, $f_{\mathbf{k}}^0$ (Fermi-Dirac distribution function), with a *small* deviation function from equilibrium, $g_{\mathbf{k}} \equiv g(\mathbf{k}, \omega, B)$. We are interested in studying the case when a static magnetic field is applied perpendicularly to the graphene sheet (i.e. $\mathbf{B} = B\hat{\mathbf{z}}$, as in figure 4.20), while also being subjected to an alternating electric field with frequency ω, namely $\mathbf{E} = (E_{0,x}, E_{0,y}, 0)e^{-i\omega t}$. In such a context, the force experienced by the carriers is the Lorentz force, $\mathbf{F} = q_e [\mathbf{E} + (\mathbf{v}_{\mathbf{k}} \times \mathbf{B})]$, and Boltzmann's equation (H.1) now reads

$$\frac{q_e}{\hbar} \nabla_{\mathbf{k}} f_{\mathbf{k}}^0 \cdot \mathbf{E} + \frac{q_e}{\hbar} \nabla_{\mathbf{k}} g_{\mathbf{k}} \cdot (\mathbf{v}_{\mathbf{k}} \times \mathbf{B}) + \frac{\partial f_{\mathbf{k}}}{\partial t} = -\frac{g_{\mathbf{k}}}{\tau} , \qquad \text{(H.2)}$$

where q_e is the charge, we have assumed that the system is homogeneous ($\nabla_{\mathbf{r}} f_{\mathbf{k}} = 0$) and that the deviation from equilibrium is linear in the applied fields, so that we have only kept terms which are linear in the fields for consistency with the linear response regime. Noting that $\hbar^{-1} \nabla_{\mathbf{k}} f_{\mathbf{k}}^0 = \mathbf{v_k} \frac{\partial f_{\mathbf{k}}^0}{\partial \epsilon_{\mathbf{k}}}$ and using the cyclic properties of the tripe product figuring in the previous expression, we can rewrite it as

$$q_e \frac{\partial f_{\mathbf{k}}^0}{\partial \epsilon_{\mathbf{k}}} \mathbf{v_k} \cdot \mathbf{E} + \frac{q_e}{\hbar} \mathbf{v_k} \cdot (\mathbf{B} \times \nabla_{\mathbf{k}} g_{\mathbf{k}}) + \frac{\partial f_{\mathbf{k}}}{\partial t} = -\frac{g_{\mathbf{k}}}{\tau}. \qquad (H.3)$$

Assuming that the distribution away from equilibrium $g_{\mathbf{k}}$ has the explicit form,

$$g_{\mathbf{k}} = \mathbf{k} \cdot \mathbf{A} e^{-i\omega t}, \qquad (H.4)$$

where $\mathbf{A} = (A_x, A_y, 0)$ is a vector that needs to be determined. We are now in position to solve Eq. (H.3) exactly. It proves useful for what follows to realize that (using vectorial notation and Einstein's summation convention)

$$\mathbf{v_k} \cdot (\mathbf{B} \times \nabla_{\mathbf{k}} g_{\mathbf{k}}) = \varepsilon_{ijk} v^i B^j \frac{\partial}{\partial k^k} (k_l A_l) e^{-i\omega t}$$

$$= \varepsilon_{ijk} v^i B^j e^{-i\omega t} \left(\delta_l^k A^l + k^l \frac{\partial A^l}{\partial k^k} \right)$$

$$= \varepsilon_{ijk} v^i B^j e^{-i\omega t} \left(A^k + \hbar v^k k^l \frac{\partial A^l}{\partial \epsilon_{\mathbf{k}}} \right)$$

$$= \left[\varepsilon_{ijk} v^i B^j A^k + \hbar k^l \frac{\partial A^l}{\partial \epsilon_{\mathbf{k}}} \underbrace{\varepsilon_{ijk} v^i B^j v^k}_{=0} \right] e^{-i\omega t}$$

$$= \mathbf{v_k} \cdot (\mathbf{B} \times \mathbf{A}) e^{-i\omega t}, \qquad (H.5)$$

where ε_{ijk} is the Levi-Civita symbol. Therefore, we can write Boltzmann's equation in the suitable form:

$$\mathbf{v_k} \cdot \left[q_e \mathbf{E} \frac{\partial f_{\mathbf{k}}^0}{\partial \epsilon_{\mathbf{k}}} + \frac{q_e}{\hbar} (\mathbf{B} \times \mathbf{A}) e^{-i\omega t} + \mathbf{A} \frac{k}{v_F} (\tau^{-1} - i\omega) e^{-i\omega t} \right] = 0, \qquad (H.6)$$

which allow us to solve it for each component. Doing so, after some simple algebra one arrives at ($q_e \to -e$, with $e > 0$)

$$A_x = \frac{\tau v_F}{k} \frac{(1 - i\omega\tau)\Xi_x - \omega'\tau\Xi_y}{(1 - i\omega\tau)^2 + \omega'^2\tau^2}, \qquad (H.7)$$

$$A_y = \frac{\tau v_F}{k} \frac{(1 - i\omega\tau)\Xi_y + \omega'\tau\Xi_x}{(1 - i\omega\tau)^2 + \omega'^2\tau^2}, \qquad (H.8)$$

where $\omega'(k) = e v_F^2 B/(\hbar k)$ and $\Xi_\alpha = e E_{0,\alpha} \frac{\partial f_{\mathbf{k}}^0}{\partial \epsilon_{\mathbf{k}}}$.

Having obtained the components of the vector $\mathbf{A}$, the current follows from

$$\mathbf{J}(\omega) = g_s g_v \int d^2 k \, (-e) \mathbf{v_k} g_\mathbf{k} \;, \tag{H.9}$$

with $g_s = 2$ and $g_v = 2$ being the spin and valley degeneracies, respectively. Then the the x-component of the current reads

$$
\begin{aligned}
J_x(\omega) &= -\frac{e v_F}{\pi^2} \int_0^\infty dk \, k^2 \left(A_x \cdot \pi + A_y \cdot 0 \right) e^{-i\omega t} \\
&= \frac{e^2 v_F}{\pi \hbar} e^{-i\omega t} \int_0^{k_F} dk \, k \frac{(1 - i\omega\tau) E_{0,x} - \omega'(k)\tau E_{0,y}}{(1 - i\omega\tau)^2 + [\omega'(k)]^2 \tau^2} \delta(k_F - k) \\
&= \frac{e^2 v_F k_F}{\pi \hbar} \left[\frac{1 - i\omega\tau}{(1 - i\omega\tau)^2 + \omega_c^2 \tau^2} \mathbf{E} \cdot \hat{\mathbf{x}} - \frac{\omega_c \tau}{(1 - i\omega\tau)^2 + \omega_c^2 \tau^2} \mathbf{E} \cdot \hat{\mathbf{y}} \right] \;,
\end{aligned}
\tag{H.10}
$$

where in the above computation we have used the fact that, at $T = 0$ K, the derivative of the Fermi-Dirac distribution yields a delta function, i.e. $\frac{\partial f_k^0}{\partial \epsilon_k} = -\delta(k_F - k)/(\hbar v_F)$ and we have also introduced the parameter $\omega_c = e v_F^2 B / E_F$ which corresponds to graphene's *cyclotron frequency*. Moreover, Ohm's law allow us to write

$$J_x(\omega) = \sigma_{xx} E_x + \sigma_{xy} E_y \;, \tag{H.11}$$

from which, upon comparison of this latter expression with Eq. (H.10) one can make the following identifications for the components of the magneto-optical tensor:

$$\sigma_{xx}(\omega) = \sigma_0 \frac{4 E_F}{\pi} \frac{\Gamma - i\hbar\omega}{(\Gamma - i\hbar\omega)^2 + \Omega_c^2} \;, \tag{H.12}$$

and, for the Hall current,

$$\sigma_{xy}(\omega) = -\sigma_0 \frac{4 E_F}{\pi} \frac{\Omega_c}{(\Gamma - i\hbar\omega)^2 + \Omega_c^2} \;, \tag{H.13}$$

with $\Gamma = \hbar\gamma$ and $\Omega_c = \hbar\omega_c$. Using symmetry arguments, the remaining components of the conductivity tensor are found to be

$$\sigma_{yx}(\omega) = -\sigma_{xy}(\omega) \;, \tag{H.14}$$

$$\sigma_{xx}(\omega) = \sigma_{yy}(\omega) \;. \tag{H.15}$$

If the reader wants to convince herself/himself that the relations (H.14) and (H.15) are indeed true, she/he can work out the y-component of the current, J_y, following the above procedure, ending up with the components σ_{yx} and σ_{yy} directly.

Appendix I

Supplementary Material for Chapter 7

I.1 Derivation of the Expressions for the Transmittance and Reflectance

In Chapter 7, subsection 7.2.2, we claimed that the expressions for the transmittance and reflectance in terms of the electric-field amplitudes were given by equations (7.24) and (7.25), respectively. In what follows, we shall present a detailed derivation for these, as they constitute important physical quantities for the problem under consideration in the above-mentioned Chapter. With that end in mind, we introduce the Poynting vector (for non-magnetic media)

$$\mathbf{S} = \frac{1}{\mu_0} \mathbf{E} \times \mathbf{B} \, , \tag{I.1}$$

which describes the flux of electromagnetic energy per unit area. Assuming that both $\mathbf{E}$ and $\mathbf{B}$ can be written in terms of harmonic functions, $e^{-i\omega t}$, one may write the time-averaged Poynting vector as

$$\langle \mathbf{S} \rangle_t = \frac{1}{2\mu_0} \Re \left\{ \mathbf{E} \times \mathbf{B}^* \right\} \, , \tag{I.2}$$

which for a TM wave polarized in the xz-plane reads

$$\langle \mathbf{S} \rangle_t = \frac{1}{2\mu_0} \Re \left\{ E_x B_y^* \hat{\mathbf{z}} - E_z B_y^* \hat{\mathbf{x}} \right\} \, . \tag{I.3}$$

Thus, for the incident wave, we obtain

$$\langle S_z \rangle_t^{\text{inc}} \equiv S_z^{inc} = \frac{\omega \epsilon_2 \epsilon_0}{2k_z} |E_x^{inc}|^2 \, , \tag{I.4}$$

along the z-direction. Similarly, for the reflected and transmitted waves corresponding to n-th diffraction order, we have[8]

$$\langle S_{z,n} \rangle_t^{\text{ref}} \equiv S_{z,n}^{(2)} = -\frac{\omega \epsilon_2 \epsilon_0}{2|\kappa_{2,n}|} |E_{x,n}^{(2)}|^2 \, , \tag{I.5}$$

---

[8]Notice the minus sign in the expression for the reflected waves, indicating a power flow along the negative z-direction. Conversely, the transmitted waves carry energy in the positive direction along the z-axis.

and

$$\langle S_{z,n} \rangle_t^{\text{trans}} \equiv S_{z,n}^{(1)} = \frac{\omega \epsilon_1 \epsilon_0}{2|\kappa_{1,n}|} |E_{x,n}^{(1)}|^2 \,, \tag{I.6}$$

respectively. We note that in writing equations (I.5) and (I.6) we have assumed that $\kappa_{j,n}$ is purely imaginary, i.e. only propagating waves should be considered. For real $\kappa_{j,n}$, we get $S_{z,n}^{(1)} = S_{z,n}^{(2)} = 0$, meaning that that particular diffraction order carries no energy (along $\hat{\mathbf{z}}$) as expected for evanescent waves. Having obtained equations (I.4)-(I.6), we are now able to compute the reflectance

$$\mathcal{R}_n = \left| \frac{S_{z,n}^{(2)}}{S_z^{inc}} \right| = \frac{k_z}{|\kappa_{2,n}|} |r_n|^2 \,, \tag{I.7}$$

and transmittance

$$\mathcal{T}_n = \left| \frac{S_{z,n}^{(1)}}{S_z^{inc}} \right| = \frac{\epsilon_1}{\epsilon_2} \frac{k_z}{|\kappa_{1,n}|} |t_n|^2 \,, \tag{I.8}$$

for each n-th wave, where we have introduced the corresponding reflection and transmission amplitudes, given by

$$r_n = \frac{E_{x,n}^{(2)}}{E_x^{inc}} \,, \tag{I.9}$$

$$t_n = \frac{E_{x,n}^{(1)}}{E_x^{inc}} \,, \tag{I.10}$$

respectively. Therefore, we have demonstrated that equations (I.7) and (I.8) are equivalent to the relations (7.25) and (7.24) presented in subsection 7.2.2.

As previously mentioned, only diffraction orders corresponding to propagating modes should be included in the sums figuring in equations (7.25) and (7.24). In particular, it is worth pointing out that for the parameters chosen in this work, we have found that solely the specular ($n = 0$) reflectance and transmittance represent propagating modes; these read

$$\boxed{\mathcal{R}_0 = |r_0|^2} \,, \tag{I.11}$$

and

$$\boxed{\mathcal{T}_0 = \frac{\epsilon_1}{\epsilon_2} \frac{\cos\theta}{\sqrt{\epsilon_1/\epsilon_2 - \sin^2\theta}} |t_0|^2} \,, \tag{I.12}$$

with the latter being valid only for $\epsilon_1 > \epsilon_2$ or $\theta < \theta_c = \arcsin\left(\sqrt{\epsilon_1/\epsilon_2}\right)$.

I.2 A Closer Look on the Modes Contributing to Each One of the Resonances

We have seen in section 7.3 that the absorbance spectrum for a grating of graphene ribbons displays a prominent peak accompanied by a set of weaker, higher-order resonances. Here, we shall restrict ourselves to the case where $d_g = R/2$ (and $\theta = 0$), as this is the configuration considered in experiments so far [Ju *et al.* (2011); Yan *et al.* (2013); Strait *et al.* (2013)]. From the symmetry point of view, and using previous knowledge of the electromagnetic response (within the quasi-static limit) of a two-dimensional metallic stripe, we expect that only modes corresponding to $q \approx |nG| = n\pi/d_g$ with $n = 1, 3, 5, \ldots$ will couple significantly to GSPs, due to the symmetry of the incident field, i.e. $\mathbf{E}^{inc} = E_x^{inc}\hat{\mathbf{x}} = \text{const.}$ [Mikhailov and Savostianova (2005)]. In order to investigate how each one of the harmonics contribute to a particular resonance in this more complex geometry, we show in figure I.1 the amplitude squared for the modes corresponding to $n = \pm 1, \pm 2, \ldots, \pm 5$. First, we note that the spectral weights associated with different harmonics are located at the resonant frequencies which show up in the absorbance spectrum. More importantly, however, is that only certain modes contribute significantly to a specific peak. For instance, it is apparent from the figure that the major contributions to the first three resonances come from the (odd) harmonics $n = \pm 1, \pm 3, \pm 5$, as indicated in the figure by the arrows. In comparison, the even modes have much smaller spectral weights. Also, notice that there is no peak associated with $n = \pm 2$ between the ones corresponding (essentially) to $n = \pm 1$ and ± 3. Finally, we note that only modes with $|n| \geq 6$ will contribute to the last resonance visible in figure I.1 (top panel).

Thus, all of the above suggests that it is indeed reasonable to assign the wavevector $q \approx (2m-1)\pi/d_g$ with $m = 1, 3, 5, \ldots$ to the m-th GSP resonance.

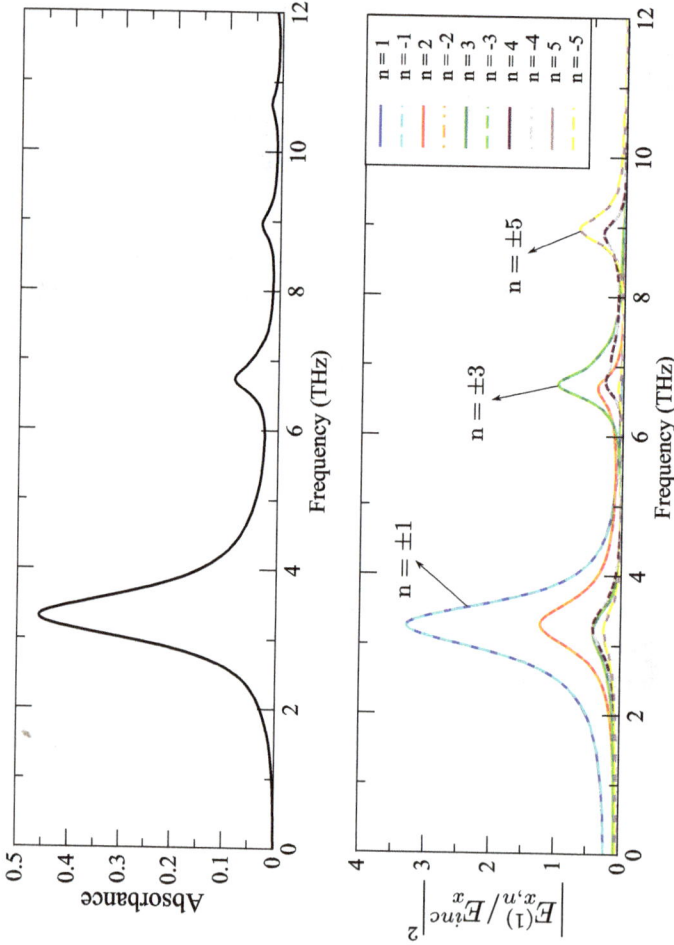

Figure I.1: Modulus squared of the amplitudes, $\left| E_{x,n}^{(1)} / E_x^{inc} \right|^2$, for modes with $n = \pm 1, \pm 2, \pm 3, \pm 4, \pm 5$. These illustrate the relative contribution of different harmonics to the observed resonances (bottom panel). The top panel shows the respective absorbance spectrum. We have used the following parameters: $R = d_g/2 = 8$ μm, $E_F = 0.45$ eV, $\Gamma = 2$ meV, $\epsilon_1 = 4$, $\epsilon_2 = 3$, $\theta = 0$.

Appendix J

The Method of Toigo

Toigo *et al.* provided a modification of Rayleigh method that leads to a simpler set of boundary conditions [Toigo *et al.* (1977)]. Here we adapt their method for dealing with graphene at an interface.

We can formally define a derivative along the normal to the surface by

$$\partial_n = \frac{1}{\sqrt{1+\dot{a}^2}}(-\dot{a}\hat{\mathbf{x}} + \hat{\mathbf{y}}) \cdot \nabla \,.\tag{J.1}$$

Following Chapter 9 notation, the boundary conditions are

$$B_z^+ - B_z^- = \frac{\mu\sigma}{\sqrt{1+\dot{a}^2}}(E_x + \dot{a}E_y)\,,\tag{J.2}$$

$$E_x^+ + \dot{a}E_y^+ = E_x^- + \dot{a}E_y^-\,,\tag{J.3}$$

where $a(x)$ is the curve defining the profile of the grating and $\dot{a} = \dot{a}(x) = da(x)/dx$. Maxwell equations for a TM wave read

$$E_x = \frac{i}{\omega\mu\epsilon}\partial_y B_z\,,\tag{J.4}$$

$$E_y = \frac{-i}{\omega\mu\epsilon}\partial_x B_z\,,\tag{J.5}$$

which allow to write the boundary conditions (J.2) and (J.3) as

$$B_z^+ - B_z^- = \frac{i\sigma}{\omega\epsilon_-}\partial_n B_z^- = \frac{i\sigma}{\omega\epsilon_-}\frac{1}{\sqrt{1+\dot{a}^2}}(\partial_y B_z^- - \dot{a}\partial_x B_z^-)\,,\tag{J.6}$$

$$\frac{1}{\epsilon_+}\partial_n B_z^+ = \frac{1}{\epsilon_-}\partial_n B_z^-$$

$$\Leftrightarrow \frac{1}{\epsilon_+}(\partial_y B_z^+ - \dot{a}\partial_x B_z^+) = \frac{1}{\epsilon_-}(\partial_y B_z^- - \dot{a}\partial_x B_z^-)\,,\tag{J.7}$$

where ϵ_+ (ϵ_-) is the dielectric permittivity above (below) the curve defined by $a(x)$. (The representation of the boundary conditions using ∂_n is useful for formal manipulations.) We note that the boundary conditions are

351

evaluated at $y = a(x)$. Since $a(x)$ is a periodic function we decompose it in Fourier series as

$$a(x) = \sum_n a_n e^{inGx},$$
(J.8)

where $G = 2\pi/D$ and D is the spatial period. The fields are expanded in a Fourier-Bloch series as

$$B^+ = B_i e^{i(xk\sin\theta - yk\cos\theta)} + \sum_n B_n e^{i(x\alpha_n + y\beta_n^+)},$$
(J.9)

$$B^- = \sum_n C_n e^{i(x\alpha_n - y\beta_n^-)},$$
(J.10)

where the first and second terms in Eq. (J.9) represent the incoming and reflected waves, respectively, and Eq. (J.10) represents the transmitted fields. In addition we have the definitions

$$k = k_+,$$
(J.11)

$$\alpha_n = k\sin\theta - nG,$$
(J.12)

$$\beta_n^\pm = \sqrt{k_\pm^2 - \alpha_n^2}, \quad k_\pm > \alpha_n,$$
(J.13)

$$\beta_n^\pm = i\sqrt{\alpha_n^2 - k_\pm^2}, \quad k_\pm < \alpha_n,$$
(J.14)

$$k_\pm = \omega/v_\pm,$$
(J.15)

with $v_\pm$ the velocity in upper (+) and lower (-) medium. The boundary condition (J.2) reads

$$B_i e^{i(xk\sin\theta - a(x)k\cos\theta)} + \sum_n B_n e^{i(x\alpha_n + a(x)\beta_n^+)} - \sum_n C_n e^{i(x\alpha_n - a(x)\beta_n^-)} =$$

$$= \frac{\sigma}{\omega\epsilon_-} \frac{1}{\sqrt{1 + \dot{a}^2}} \sum_n (\beta_n^+ + \dot{a}\alpha_n) C_n e^{i(x\alpha_n - a(x)\beta_n^-)}.$$
(J.16)

Multiplying Eq. (J.16) by $e^{-i\alpha_m x}$ and integrating over $\int_0^D dx/D$ we obtain

$$B_i M_m^i + \sum M_{m-n;n}^+ B_n - \sum M_{m-n;n}^- C_n =$$

$$= \frac{\sigma}{\omega\epsilon_-} \sum (\beta_n^- L_{m-n;n}^- + \alpha_n N_{m-n;n}^-) C_n,$$
(J.17)

where

$$M_m^i = \frac{1}{D} \int_0^D dx e^{imGx} e^{-ik\cos\theta a(x)} , \qquad (J.18)$$

$$M_{m-n;n}^{\pm} = \frac{1}{D} \int_0^D dx e^{i(m-n)Gx} e^{\pm i\beta_n^{\pm} a(x)} , \qquad (J.19)$$

$$L_{m-n;n}^- = \frac{1}{D} \int_0^D dx \frac{1}{\sqrt{1+\dot{a}^2}} e^{i(m-n)Gx} e^{-i\beta_n^- a(x)} \approx M_{m-n;n}^- , \quad (J.20)$$

$$N_{m-n;n}^- = \frac{1}{D} \int_0^D dx \frac{\dot{a}}{\sqrt{1+\dot{a}^2}} e^{i(m-n)Gx} e^{-i\beta_n^- a(x)}$$

$$\approx \frac{m-n}{\beta_n^-} M_{m-n;n}^- . \qquad (J.21)$$

After few manipulations we arrive at

$$\boxed{B_i M_m^i + \sum_n M_{m-n;n}^+ B_n = \sum_n \left(1 + \frac{\sigma}{\omega\epsilon_-} \frac{k_-^2 - \alpha_n\alpha_m}{\beta_n^-}\right) M_{m-n;n}^- C_n .}$$
$$(J.22)$$

The boundary condition (J.7) can be written as

$$\frac{1}{\epsilon_+} \left[(k\cos\theta + \dot{a}k\sin\theta) B_i e^{i(k\sin\theta x - ka(x)\cos\theta)} \right.$$

$$\left. - \sum_n (\beta_n^+ - \dot{a}\alpha_n) B_n e^{i(x\alpha_n + a(x)\beta_n^+)} \right] =$$

$$= \frac{1}{\epsilon_-} \sum_n (\beta_n^- + \dot{a}\alpha_n) C_n e^{i(\alpha_n x - a(x)\beta_n^-)} . \qquad (J.23)$$

Multiplying Eq. (J.16) by $e^{-i\alpha_m x}$ and integrating over $\int_0^D dx/D$, we obtain after few manipulations

$$\boxed{\frac{1}{\epsilon_+} \left[\frac{k_+^2 - k_+ \sin\theta\alpha_m}{k\cos\theta} M_m^i B_i - \sum_n \frac{k_+^2 - \alpha_n\alpha_m}{\beta_n^+} M_{m-n;n}^+ B_n \right] =}$$

$$\boxed{= \frac{1}{\epsilon_-} \frac{k_-^2 - \alpha_n\alpha_m}{\beta_n^-} M_{m-n;n}^- C_n .}$$
$$(J.24)$$

The solution of the boundary conditions (J.22) and (J.24) allow the calculation of the reflectance, transmittance, and absorbance spectra.

In the case where $a(x) = h\sin(Gx)$ we have

$$M_p^i = J_p(kh\cos\theta)\,, \tag{J.25}$$

$$M_{p;n}^+ = J_p(-\beta_n^+ h)\,, \tag{J.26}$$

$$M_{p;n}^- = J_p(\beta_n^- h)\,, \tag{J.27}$$

where $J_p(z)$ is the regular Bessel function of order p. The advantage of this method over that explained in Chapter 9 is that involves one less summation. The reflectance, transmittance, and absorbance are computed as in Chapter 9.

Appendix K

Boundary Condition in the Dipole Problem (*written with Jaime E. Santos*)

Let us consider a punctual dipole embedded, at a distance $z > 0$, in a dielectric medium of permittivity ϵ_1 above a doped graphene sheet ($E_F > 0$). The sheet rests on a dielectric of permittivity ϵ_2 (located at $z < 0$). We want to derive the explicit boundary conditions obeyed by the field (or the potential for that matter) at the graphene ($z = 0$). We define $\mathbf{s}$ an in-plane ($z = 0$) position vector. Since graphene is doped, the general boundary condition is [see Eq. (2.23)]:

$$\epsilon_1 E_z^1 - \epsilon_2 E_z^2 = \sigma_s(\mathbf{s}, \omega),$$

(K.1)

where the dependence of $\sigma_s(\mathbf{s}, \omega)$ in ω accounts for the presence of a time dependent electric field. Moreover the continuity equation must be obeyed, which is Fourier space reads

$$i\omega\sigma_s(\mathbf{s}, \omega) = \nabla_\mathbf{s} \cdot \mathbf{J}_s(\mathbf{s}, \omega).$$

(K.2)

Now, the current can be calculated from Boltzmann's equation (see Appendix C), and assuming an homogeneous temperature field, we have

$$-\frac{\partial g_\mathbf{k}}{\partial t} - \frac{\partial f^0(\epsilon_\mathbf{k})}{\partial \epsilon_\mathbf{k}} \mathbf{v}_\mathbf{k} \cdot [e\mathbf{E} - \nabla_\mathbf{s} E_F(\mathbf{s})] = \frac{g_\mathbf{k}}{\tau} + \mathbf{v}_\mathbf{k} \cdot \nabla_\mathbf{s} g_\mathbf{k},$$

(K.3)

where the notation $g_\mathbf{k}(\mathbf{s}) = g_\mathbf{k}$ was used.

Discarding the term $\nabla_\mathbf{s} g_\mathbf{k}$[9], and writing the gradient of the Fermi energy as

$$\nabla_\mathbf{s} E_F(\mathbf{s}) = \frac{\partial E_F^0}{\partial n_0}\nabla_\mathbf{s} n(\mathbf{s}),$$

(K.4)

---

[9]The dependence of $g_\mathbf{k}$ on $\mathbf{s}$ is only via the dependence of E_F on $\mathbf{s}$; this dependence is already included in the left-hand-side of the Boltzmann equation.

where E_F^0 is the Fermi energy of the homogeneous gas and n_0 is the corresponding electronic density, and going to frequency space, we obtain

$$i\omega g_{\mathbf{k}} - \frac{\partial f^0(\epsilon_{\mathbf{k}})}{\partial \epsilon_{\mathbf{k}}} \mathbf{v}_{\mathbf{k}} \cdot \left[e\mathbf{E} - \frac{\partial E_F^0}{\partial n_0} \nabla_{\mathbf{s}} n(\mathbf{s}) \right] = \frac{g_{\mathbf{k}}}{\tau}. \tag{K.5}$$

The previous equation can be solved for $g_{\mathbf{k}}$, giving

$$g_{\mathbf{k}} = -\frac{\partial f^0(\epsilon_{\mathbf{k}})}{\partial \epsilon_{\mathbf{k}}} \mathbf{v}_{\mathbf{k}} \cdot \left[e\mathbf{E} - \frac{\partial E_F^0}{\partial n_0} \nabla_{\mathbf{s}} n(\mathbf{s}) \right] \frac{\tau}{1 - i\omega\tau}. \tag{K.6}$$

Clearly, when $\nabla_{\mathbf{s}} n(\mathbf{s}) = 0$ we recover Drude's result. The current density follows from

$$J_s^j = g_s g_v \frac{e}{A} \sum_{\mathbf{k}} v_{\mathbf{k}}^j g_{\mathbf{k}}, \tag{K.7}$$

where g_s and g_v are the spin and valley degeneracies, respectively. To be specific, let us assume $\mathbf{E} = E_x \hat{\mathbf{x}}$ and compute J_s^x, then

$$
\begin{aligned}
J_s^x &= -g_s g_v \frac{e}{A} \sum_{\mathbf{k}} (v_{\mathbf{k}}^x)^2 \frac{\partial f^0(\epsilon_{\mathbf{k}})}{\partial \epsilon_{\mathbf{k}}} \left[eE_x - \frac{\partial E_F^0}{\partial n_0} \frac{\partial n(\mathbf{s})}{\partial x} \right] \frac{\tau}{1 - i\omega\tau} \\
&= g_s g_v \frac{e^2}{4\pi} \int_0^\infty k\, dk\, \delta(v_F \hbar k - E_F^0) \left[E_x - \frac{1}{e} \frac{\partial E_F^0}{\partial n_0} \frac{\partial n(\mathbf{s})}{\partial x} \right] \frac{\tau}{1 - i\omega\tau} \\
&= g_s g_v \frac{e^2}{2\pi v_F^2 \hbar^2} \frac{E_F^0}{1 - i\omega\tau} \left[E_x - \frac{1}{e} \frac{\partial E_F^0}{\partial n_0} \frac{\partial n(\mathbf{s})}{\partial x} \right] \\
&= \sigma_D(\omega) \left[E_x - \frac{1}{e} \frac{\partial E_F^0}{\partial n_0} \frac{\partial n(\mathbf{s})}{\partial x} \right],
\end{aligned}
\tag{K.8}
$$

where $\sigma_D(\omega)$ is the Drude conductivity of graphene. We can write the surface charge density as $\sigma_s(\mathbf{s}) = en(\mathbf{s})$ and using the previous expression for the current in the continuity equation, we obtain

$$i\omega \sigma_s(\mathbf{s}) = \sigma_D(\omega) \nabla_{\mathbf{s}} \cdot \left[\mathbf{E} - \frac{1}{e^2} \frac{\partial E_F^0}{\partial n_0} \nabla_{\mathbf{s}} \sigma_s(\mathbf{s}) \right]. \tag{K.9}$$

Clearly, the second term in square braces represents a current of charge due to a gradient in the charge density. Thus we identify the pre-factor with the diffusion coefficient:

$$D(\omega, E_F) = \frac{\sigma_D(\omega)}{e^2} \frac{\partial E_F^0}{\partial n_0}. \tag{K.10}$$

The continuity equation can be solved in Fourier space as

$$i\omega \sigma_s(\mathbf{q}, \omega) = i\mathbf{q} \cdot \mathbf{E}(\mathbf{q}) \sigma_D(\omega) + q^2 D(\omega, E_F) \sigma_s(\mathbf{q}, \omega), \tag{K.11}$$

or

$$\sigma_s(\mathbf{q}, \omega) = \sigma_D(\omega)\frac{i\mathbf{q} \cdot \mathbf{E}(\mathbf{q})}{i\omega - q^2 D(\omega, E_F)}. \qquad (K.12)$$

This result can now be used in the boundary conditions of the field:

$$\boxed{\epsilon_1 E_z^1(\mathbf{q}) - \epsilon_2 E_z^2(\mathbf{q}) = \sigma_D(\omega)\frac{i\mathbf{q} \cdot \mathbf{E}(\mathbf{q})}{i\omega - q^2 D(\omega, E_F)}.} \qquad (K.13)$$

If $q^2 D(\omega, E_F)/\omega \ll 1$ then the boundary condition is approximated to $\epsilon_1 E_z^1(\mathbf{q}) - \epsilon_2 E_z^2(\mathbf{q}) \approx \sigma_D(\omega)\mathbf{q} \cdot \mathbf{E}(\mathbf{q})/\omega$. This boundary condition is used in Sec. 10.1.

Appendix L

Fourier Transform of the Dipole Potential

In this Appendix we provide a derivation of the Fourier transform of a dipole potential in the in-plane coordinate (see [Swathi and Sebastian (2009)]). Referring to Fig. 10.1 the electrostatic potential of a dipole is given by

$$\varphi(z, \mathbf{s}) = \frac{1}{4\pi\epsilon_0} \frac{\boldsymbol{\mu} \cdot \mathbf{r}}{r^3} = \frac{1}{4\pi\epsilon_0} \frac{\boldsymbol{\mu} \cdot (\mathbf{s} - z\hat{u}_z)}{|\mathbf{s} - z\hat{u}_z|^3}, \tag{L.1}$$

where we have specialized to a point in the graphene plane. What we need to compute is the Fourier transform

$$\varphi(z, \mathbf{q}) = \frac{\boldsymbol{\mu}}{4\pi\epsilon_0} \cdot \int d^2s \, e^{-i\mathbf{q}\cdot\mathbf{s}} \frac{(\mathbf{s} - z\hat{u}_z)}{|\mathbf{s} - z\hat{u}_z|^3}. \tag{L.2}$$

Defining a vector $\mathbf{R} = (X, Y, z) = \mathbf{R}_\parallel + z\hat{z}$, we recall the identity

$$\nabla_\mathbf{R} \frac{1}{|\mathbf{R} - \mathbf{s}|} = \frac{\mathbf{s} - \mathbf{R}}{|\mathbf{R} - \mathbf{s}|^3}. \tag{L.3}$$

Thus

$$\frac{\mathbf{s} - z\hat{u}_z}{|z\hat{u}_z - \mathbf{s}|^3} = \left(\nabla_\mathbf{R} \frac{1}{|\mathbf{R} - \mathbf{s}|}\right)_{X=Y=0}. \tag{L.4}$$

Then, we can write

$$\int d^2s \, e^{-i\mathbf{q}\cdot\mathbf{s}} \frac{(\mathbf{s} - z\hat{u}_z)}{|z\hat{u}_z - \mathbf{s}|^3} = \left(\nabla_\mathbf{R} \int d^2s \, e^{-i\mathbf{q}\cdot\mathbf{s}} \frac{1}{|\mathbf{R} - \mathbf{s}|}\right)_{X=Y=0}. \tag{L.5}$$

The integral in braces has an analytical solution:

$$\int d^2s \, e^{-i\mathbf{q}\cdot\mathbf{s}} \frac{1}{|\mathbf{R} - \mathbf{s}|} = e^{-i\mathbf{q}\cdot\mathbf{R}_\parallel} \int d^2v \frac{e^{i\mathbf{q}\cdot\mathbf{v}}}{\sqrt{v^2 + z^2}}$$

$$= e^{-i\mathbf{q}\cdot\mathbf{R}_\parallel} \int_0^\infty \int_0^{2\pi} \frac{v\,dv\,d\theta}{\sqrt{v^2 + z^2}} e^{ivq\cos\theta}$$

$$= 2\pi e^{-i\mathbf{q}\cdot\mathbf{R}_\parallel} \int_0^\infty \frac{v\,dv}{\sqrt{v^2 + z^2}} J_0(qv)$$

$$= \frac{2\pi}{q} e^{-qz} e^{-i\mathbf{q}\cdot\mathbf{R}_\parallel}, \tag{L.6}$$

where we have written $\mathbf{R} = \mathbf{R}_\parallel + z\hat{u}_z$ and introduce the change of variables $\mathbf{s} = \mathbf{R}_\parallel - \mathbf{v}$.

We are now in position of computing the matrix element entering in Eq. (10.1). For a generic electrostatic potential of the form

$$\varphi(\mathbf{s}, z) = \frac{1}{4\pi\epsilon_0\epsilon(\omega)} \frac{\vec{\mu}\cdot(\mathbf{s} - z\hat{u}_z)}{|\mathbf{s} - z\hat{u}_z|^3}, \tag{L.7}$$

where $\epsilon(\omega)$ is some dielectric constant that can be frequency dependent and the vectors $\mathbf{s}$ and $z\hat{u}_z$ are those defined in Fig. 10.1, the matrix element reads $[V = -e\varphi(\mathbf{s}, z)]$

$$\langle \psi_{\mathbf{k}+\mathbf{q},+}|V|\psi_{\mathbf{k},+}\rangle$$
$$= -\frac{e}{2A}(1, e^{-i\theta_{\mathbf{k}+\mathbf{q}}}) \cdot (1, e^{-i\theta_\mathbf{q}})^T \int d^2s\, e^{-i\mathbf{s}\cdot(\mathbf{k}+\mathbf{q})}\varphi(\mathbf{s}, z)e^{-i\mathbf{s}\cdot\mathbf{k}}, \tag{L.8}$$

where $e > 0$ is the elementary charge. We see that what enters in the matrix element is the Fourier transform of the potential. Using the result (L.6), the matrix element can be written as

$$\boxed{\langle \psi_{\mathbf{k}+\mathbf{q},+}|V|\psi_{\mathbf{k},+}\rangle = \frac{e}{2A} \frac{1}{2\epsilon_0\epsilon(\omega)}(1 + e^{-i(\theta_{\mathbf{k}+\mathbf{q}}-\theta_\mathbf{k})})\vec{\mu}\cdot(i\mathbf{q}/q + \hat{u}_z)e^{-qz}}.$$

$$\tag{L.9}$$

The modulus square of the matrix element reads

$$|\langle \psi_{\mathbf{k}+\mathbf{q},+}|V|\psi_{\mathbf{k},+}\rangle|^2 = \frac{e^2}{8A^2\epsilon^2\epsilon(\omega)^2}[1 + \cos(\theta_{\mathbf{k}+\mathbf{q}} - \theta_\mathbf{k})]e^{-2qz}|\vec{\mu}\cdot(i\mathbf{q}/q + \hat{u}_z)|^2,$$

$$\tag{L.10}$$

where we can still write

$$|\vec{\mu}\cdot(i\mathbf{q}/q + \hat{u}_z)|^2 = \mu_z^2 + \mu_\parallel^2\cos^2\theta, \tag{L.11}$$

with $\mu_\parallel^2 = \mu_x^2 + \mu_y^2$.

Appendix M

Non-Radiative Decaying Rate: Electrostatic Calculation

M.1 Doped Graphene

In this appendix we give the details of the calculations of the non-radiative decay rate, taking into account that we have a frequency and momentum dependent dielectric constant, $\epsilon(\mathbf{q}, \omega)$ [see Eq. (10.19)]. The decay rate (10.1) is given by [see Eq. L.9 for the matrix element of the dipole potential]

$$\frac{1}{\tau} = g_s g_v \frac{2\pi}{\hbar} \frac{e^2}{2\epsilon_0^2} \frac{1}{(2\pi)^4} \int d^2\mathbf{q} \frac{(\mu_z^2 + \mu_\parallel^2 \cos^2\theta)e^{-2qz}}{|\epsilon(\mathbf{q}, \omega)|^2} \int d^2\mathbf{k}[1 + \cos(\theta_{\mathbf{k+q}} - \theta_{\mathbf{k}})]$$
$$f(v_F \hbar k)[1 - f(v_F \hbar |\mathbf{k} + \mathbf{q}|)]\delta(v_F \hbar |\mathbf{k} + \mathbf{q}|) - v_F \hbar q - \hbar \omega), \qquad \text{(M.1)}$$

where $f(x)$ is the Fermi distribution function, θ is the angle between $\mathbf{q}$ and $\vec{\mu}_\parallel = (\mu_x, \mu_y)$, and $\epsilon(\mathbf{q}, \omega)$ reads

$$\epsilon(\mathbf{q}, \omega) = \epsilon_2 + \epsilon_1 + \lambda/(\epsilon_0 q). \qquad \text{(M.2)}$$

The Fermi functions guarantee that the state $\mathbf{k}$ is occupied and the state $\mathbf{k} + \mathbf{q}$ is empty. With the help of Fig. M.1 we can write

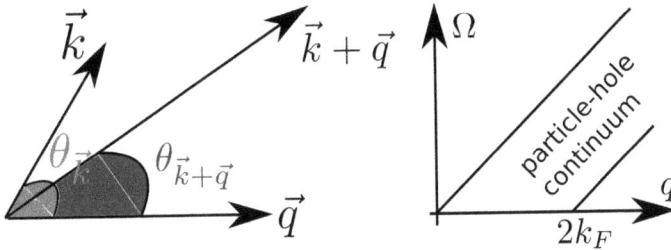

Figure M.1: Left: Orientation of the vectors $\mathbf{q}$, $\mathbf{k}$, and $\mathbf{k} + \mathbf{q}$ for computing $\cos(\theta_{\mathbf{k+q}} - \theta_{\mathbf{k}})$. Right: plane (q, Ω) with the domain available for particle-hole excitations in the calculation of the decay rate $1/\tau$ ($\Omega = \omega/v_F$).

$$\cos(\theta_{\mathbf{k+q}} - \theta_{\mathbf{k}}) = \frac{\mathbf{k} \cdot (\mathbf{k+q})}{k|\mathbf{k+q}|}, \tag{M.3}$$

and for simplicity of notation we define $\theta_{\mathbf{k}} = \psi$. Then

$$\cos(\theta_{\mathbf{k+q}} - \theta_{\mathbf{k}}) = \frac{k + q\cos\psi}{\sqrt{k^2 + q^2 + 2kq\cos\psi}}. \tag{M.4}$$

We are faced with the calculation of the integral

$$\begin{aligned}
I(\omega, q) &= \int d^2\mathbf{k}[1 + \cos(\theta_{\mathbf{k+q}} - \theta_{\mathbf{k}})]f(v_F\hbar k)[1 - f(v_F\hbar|\mathbf{k+q}|)] \\
&\quad \delta(v_F\hbar|\mathbf{k+q}| - v_F\hbar k - \hbar\omega) \\
&= \frac{1}{v_F\hbar} \int_0^{k_F} kdk \int_0^{2\pi} d\psi[1 + \cos(\theta_{\mathbf{k+q}} - \theta_{\mathbf{k}})] \\
&\quad [1 - f(v_F\hbar q + v_F\hbar\Omega)]\delta(|\mathbf{k+q}| - k - \Omega), \tag{M.5}
\end{aligned}$$

where $\Omega = \omega/v_F$ and k_F is the Fermi momentum. The function $[1 - f(v_F\hbar k + v_F\hbar\Omega)]$ imposes the constraint $k > k_F - \Omega$. If $\Omega > k_F$ the lower limit of the integral in k is zero, if $\Omega < k_F$ the lower limit of the integral is $k_F - \Omega$. Then we can write

$$\begin{aligned}
I(\omega, q) &= \frac{1}{v_F\hbar} \int_{\max(k_F-\Omega,0)}^{k_F} kdk \int_0^{2\pi} d\psi[1 + \cos(\theta_{\mathbf{k+q}} - \theta_{\mathbf{k}})]\delta[g(\psi)] \\
&= \frac{1}{v_F\hbar} \int_{\max(k_F-\Omega,0)}^{k_F} kdk \int_0^{2\pi} d\psi\left(1 + \frac{k + q\cos\psi}{k + \Omega}\right)\delta[g(\psi)], \tag{M.6}
\end{aligned}$$

with $g(\psi) = \sqrt{k^2 + q^2 + 2kq\cos\psi} - k - \Omega$. Simplifying the $\delta-$function, we write (after some algebra)

$$\left(1 + \frac{k + q\cos\psi}{k + \Omega}\right)\delta[g(\psi)] = \frac{1}{k}\frac{\sqrt{(2k + \Omega)^2 - q^2}}{\sqrt{q^2 - \Omega^2}}[\delta(\psi - \psi_0) + \delta(\psi + \psi_0)], \tag{M.7}$$

where ψ_0 follows from $g(\psi_0) = 0$. The integral in ψ is now elementary, and we obtain

$$I(\omega, q) = \frac{2}{v_F\hbar} \int_{\max(k_F-\Omega,0)}^{k_F} dk \frac{\sqrt{(2k + \Omega)^2 - q^2}}{\sqrt{q^2 - \Omega^2}}\Theta(q - \Omega), \tag{M.8}$$

where $\Theta(x)$ is the step function. A further constraint is imposed by the square root in the numerator: $k > (q - \Omega)/2$ and since $0 < k < k_F$ we determine the domain of q available for excitation of particle-hole pairs as

$$\Omega < q < 2k_F + \Omega, \tag{M.9}$$

as shown in the right panel of Fig. M.1. Then, the integral $I(\omega, q)$ reads

$$I(\omega, q) = \frac{2}{v_F \hbar} \frac{1}{\sqrt{q^2 - \Omega^2}} \int_{\max[k_F - \Omega, (q - \Omega)/2]}^{k_F} dk \sqrt{(2k + \Omega)^2 - q^2} \,. \qquad (\text{M.10})$$

We now make the transformation of variable $2k + \Omega = x$ followed by the transformation $x = q \cosh \phi$. The integral becomes elementary and we obtain

$$I(\omega, q) = \frac{1}{v_F \hbar} \frac{q^2}{\sqrt{q^2 - \Omega^2}}$$
$$\frac{1}{2} \left[\phi_- - \phi_+ + \frac{1}{2} \sinh(2\phi_+) - \frac{1}{2} \sinh(2\phi_-) \right]$$
$$\equiv \frac{1}{v_F \hbar} \frac{q^2}{\sqrt{q^2 - \Omega^2}} G(q) \,, \qquad (\text{M.11})$$

where

$$\phi_+ = \text{arccosh} \frac{2k_F + \Omega}{q} \,, \qquad (\text{M.12})$$

$$\phi_- = \text{arccosh} \frac{\max(2k_F - \Omega, q)}{q} \,. \qquad (\text{M.13})$$

Finally, the decay rate reads

$$\frac{1}{\tau} = 2 \frac{2\pi}{\hbar} \frac{e^2}{\epsilon_0^2} \frac{1}{(2\pi)^4} \int d\mathbf{q} \frac{(\mu_z^2 + \mu_\parallel^2 \cos^2 \theta)}{|\epsilon(\mathbf{q}, \omega)|^2} \frac{1}{v_F \hbar} \frac{q^2 e^{-2qz}}{\sqrt{q^2 - \Omega^2}} G(q) \,. \qquad (\text{M.14})$$

Integrating over the angle θ we obtain

$$\boxed{ \frac{1}{\tau} = 2 \frac{2\pi}{\hbar} \frac{e^2}{\epsilon_0^2} \frac{2\pi\mu_z^2 + \pi\mu_\parallel^2}{(2\pi)^4 \hbar v_F} \int_\Omega^{2k_F + \Omega} \frac{dq}{|\epsilon(\mathbf{q}, \omega)|^2} \frac{q^3 e^{-2qz}}{\sqrt{q^2 - \Omega^2}} G(q) } \,. \qquad (\text{M.15})$$

The integral in q cannot be done analytically. We note that the integral vanishes for $k_F = 0$, which is a consequence of the fact that only intraband transitions were considered.

M.2 A Detour: Neutral Graphene

We now take a detour of our main line —plasmons in graphene— and show, following [Swathi and Sebastian (2009)], that neutral graphene also plays a role in the non-radiative decay rate of a quantum emitter, although it does not support plasmons. This role happens because graphene has no energy-gap and, therefore, particle-hole excitations from the valence band to the conduction band can take place at arbitrary low energies. As in the previous section the quantity to be computed is the non-radiative decay rate, which for this case takes the form

$$\frac{1}{\tau} = g_s g_v \frac{2\pi}{\hbar} \sum_{\mathbf{k}} \sum_{\mathbf{q}} |\langle \psi_{\mathbf{k}+\mathbf{q},+} | V | \psi_{\mathbf{k},-} \rangle|^2 \delta(E_{\mathbf{k}+\mathbf{q},+} - E_{\mathbf{k},-} - \hbar\omega). \quad \text{(M.16)}$$

Note the differences between Eq. (10.1) and Eq. (M.16). We consider the case of graphene in vacuum. Following the same steps as in the previous section, we are faced with the calculation of an integral of the form [see Eq. (M.5)]

$$I(\omega, q) = \frac{1}{v_F \hbar} \int_0^\infty k \, dk \int_0^{2\pi} d\psi [1 - \cos(\theta_{\mathbf{k}+\mathbf{q}} - \theta_{\mathbf{k}})] \delta(|\mathbf{k} + \mathbf{q}| + k - \Omega). \tag{M.17}$$

Following the same procedure as in the previous section, the integrand in Eq. (M.17) can be transformed into to

$$[1 - \cos(\theta_{\mathbf{k}+\mathbf{q}} - \theta_{\mathbf{k}})] \delta[g(\psi)] = \frac{1}{k} \frac{\sqrt{q^2 - (\Omega - 2k)^2}}{\sqrt{\Omega^2 - q^2}} [\delta(\psi - \psi_0) + \delta(\psi + \psi_0)],$$

$$\tag{M.18}$$

where $g(\psi) = |\mathbf{k} + \mathbf{q}| + k - \Omega$ and ψ_0 follows from $g(\psi_0) = 0$. The following conditions have to be enforced: $\Omega > q$ and $q^2 > (\Omega - 2k)^2$, from which follows that $(\Omega - q)/2 < k < (\Omega + q)/2$. The integral in ψ becomes elementary, and the integral over k transforms into

$$I(\omega, q) = \frac{2}{v_F \hbar} \frac{1}{\sqrt{\Omega^2 - q^2}} \int_{(\Omega - q)/2}^{(\Omega + q)/2} dk \sqrt{q^2 - (\Omega - 2k)^2}. \tag{M.19}$$

Introducing the change of variables $\Omega - 2k = x$, followed by $x = q \sin \theta$, we obtain

$$I(\omega, q) = \frac{\pi}{2v_F \hbar} \frac{q^2}{\sqrt{\Omega^2 - q^2}}. \tag{M.20}$$

Thus, the expression for the decaying rate becomes

$$\frac{1}{\tau} = \frac{\pi^2 e^2}{2\epsilon_0^2 \hbar (2\pi)^3 v_F \hbar} (\mu_z^2 + \mu_\parallel^2/2) \int_0^\Omega dq \frac{q^3 e^{-2qz}}{\sqrt{\Omega^2 - q^2}}. \tag{M.21}$$

Let us now compute the integral over q in an approximate way. We introduce the change a variables $q = x\Omega$. This leads to

$$\int_0^\Omega dq \frac{q^3 e^{-2qz}}{\sqrt{\Omega^2 - q^2}} = \Omega^3 \int_0^1 dx \frac{x^3 e^{-2x\Omega z}}{\sqrt{1 - x^2}}. \tag{M.22}$$

In the limit $\Omega z \gg 1$, the integral over x is dominated by values of $x \ll 1$. Then, we approximate the square root by one, following

$$\Omega^3 \int_0^1 dx \frac{x^3 e^{-2x\Omega z}}{\sqrt{1-x^2}} \approx \Omega^3 \int_0^\infty dx x^3 e^{-2x\Omega z} = \frac{3}{8\Omega} \frac{1}{z^4}. \tag{M.23}$$

Finally, the decaying rate reads

$$\frac{1}{\tau} = \frac{\pi^2 e^2}{2\epsilon_0^2 \hbar (2\pi)^3 \hbar} (\mu_z^2 + \mu_\parallel^2/2) \frac{3}{8\omega} \frac{1}{z^4} = \alpha \frac{3\tilde{\mu}^2}{64\pi\epsilon_0 \hbar} \frac{\lambda}{z^4}, \tag{M.24}$$

where λ is the wavelength of the radiation. Recalling the form of the decaying rate of a dipole in vacuum, Eq. (10.27), we obtain

$$\boxed{\frac{\tau_{\text{vac}}}{\tau} = \alpha \frac{3^2}{2^9 \pi^3} \frac{\tilde{\mu}^2}{\mu^2} \left(\frac{\lambda}{z}\right)^4}, \tag{M.25}$$

where $\tilde{\mu}^2 = \mu_z^2 + \mu_\parallel^2/2$. In conclusion, the non-radiative decaying rate decreases with a z^{-4} dependence in the distance of the quantum emitter to graphene [Swathi and Sebastian (2009)].

Appendix N

Reflection Amplitude of an Electromagnetic Wave due to a Graphene Interface

In this Appendix we derive the transmittance and reflectance amplitudes for TM ($p-$polarized) and TE ($s-$polarized) waves, when electromagnetic radiation impinges on graphene resting on a dielectric. First we consider the simple case of free standing graphene. Later we generalize to graphene clad by two different dielectrics.

N.1 The Simple Case of Free Standing Graphene

In this section we solve a very simple problem as a warm up exercise to the next section. The geometry of the problem is given in Fig. N.1. Let us first work out the scattering amplitudes for the $p-$polarized wave. We define the electric field to the left of graphene as $\mathbf{E}_< = \mathbf{E}_i + \mathbf{E}_t$ and to the

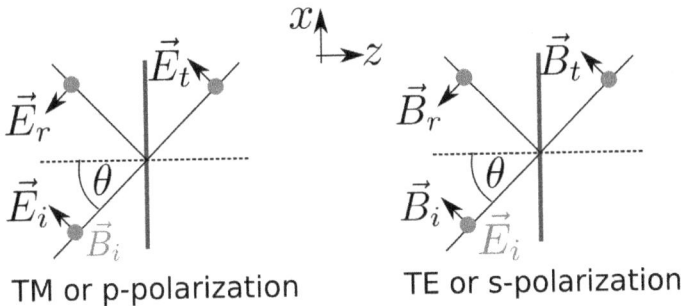

TM or p-polarization   TE or s-polarization

Figure N.1: Left: Geometry of the scattering problem for a $p-$polarized wave (the magnetic field is perpendicular to the plane of the drawing). Right: Geometry of the scattering problem for a $s-$polarized wave (the electric field is perpendicular to the plane of the drawing).

right as $\mathbf{E}_> = \mathbf{E}_t$. In terms of coordinates we have

$$\mathbf{E}_i = E_i(-\sin\theta, \cos\theta)e^{i(\mathbf{k}_i \cdot \mathbf{r} - \omega t)}, \tag{N.1}$$

$$\mathbf{E}_r = E_r(-\sin\theta, -\cos\theta)e^{i(\mathbf{k}_r \cdot \mathbf{r} - \omega t)}, \tag{N.2}$$

$$\mathbf{E}_t = E_t(-\sin\theta, \cos\theta)e^{i(\mathbf{k}_t \cdot \mathbf{r} - \omega t)}, \tag{N.3}$$

where the wavevectors are defined as

$$\mathbf{k}_i = k(\cos\theta, \sin\theta), \tag{N.4}$$

$$\mathbf{k}_r = k(-\cos\theta, \sin\theta), \tag{N.5}$$

$$\mathbf{k}_f = k(\cos\theta, \sin\theta), \tag{N.6}$$

where k is the wave number of the impinging radiation. The boundary conditions imply that the component of the field parallel to the interface must be continuous. This condition gives

$$E_{>,\|} = E_{<,\|} \Leftrightarrow E_i - E_t = E_r. \tag{N.7}$$

The other boundary is the discontinuity at the interface of the parallel component of the magnetic field, that is, $B_{<,\|} - B_{>,\|} = \omega\mu_0\sigma_L(\omega, q)E_{>,\|}$. The magnetic field is determined by Maxwell's equation. Once this is known, the boundary condition reduces to

$$k(E_i + E_r - E_t) = \omega\mu_0\sigma_L(\omega, q)E_t \cos\theta. \tag{N.8}$$

Combining the two boundary conditions we obtain

$$t_p = \frac{E_t}{E_i} = \frac{2}{2 + \pi\alpha f_L(\omega, q)\cos\theta}, \tag{N.9}$$

$$r_p = 1 - t_p = \frac{\pi\alpha f_L(\omega, q)\cos\theta}{2 + \pi\alpha f_L(\omega, q)\cos\theta}, \tag{N.10}$$

where we have written $\sigma_L(\omega, q) = \sigma_0 f_L(\omega, q)$, with $\sigma_0 = \pi e^2/(2h)$ and α the fine structure constant. For the s-polarized geometry, the boundary conditions are $E_{i,y} + E_{r,y} = E_{t,y}$, where the subscript y denotes the component of the electric field perpendicular to the plane of the drawing in Fig. N.1. The boundary condition for the magnetic field reads

$$k\cos\theta(E_{i,y} - E_{r,y} - E_{t,y}) = \omega\mu_0\sigma_T(\omega, q)E_{t,y}. \tag{N.11}$$

Solving the boundary conditions, we obtain

$$t_s = \frac{E_{t,y}}{E_{i,y}} = \frac{2\cos\theta}{2\cos\theta + \pi\alpha f_T(\omega, q)}, \tag{N.12}$$

$$r_s = t_s - 1 = -\frac{\pi\alpha f_T(\omega, q)}{2\cos\theta + \pi\alpha f_T(\omega, q)}, \tag{N.13}$$

where $\sigma_T(\omega,q) = \sigma_0 f_T(\omega,q)$. We note that $\cos\theta = k_{z,i}/k = \sqrt{k^2 - k_x^2}/k$. Thus, if the plane of incidence is rotate around the z−axis, we have $\cos\theta = \sqrt{k^2 - k_x^2 - k_y^2}/k \equiv \sqrt{k^2 - k_\rho^2}/k \equiv \sqrt{1 - s^2}$. For the Drude model, and neglecting the momentum dependence of $f_{L/T}(\omega,q)$, r_p and r_s have the explicit forms

$$r_p = \left| \frac{2E_F\alpha\cos\theta}{2E_F\alpha\cos\theta + \gamma - i\omega} \right|, \tag{N.14}$$

and

$$r_s = -\left| \frac{2E_F\alpha}{2E_F\alpha + \cos\theta(\gamma - i\omega)} \right|. \tag{N.15}$$

For neutral graphene we take $f_{L/T}(\omega) = 1$.

N.2 Reflection and Transmission Amplitudes: Fresnel Coefficients in the General Case

We start by computing the Fresnel coefficients for an interface containing a graphene sheet, as represented in Fig. N.2. We consider a planar interface

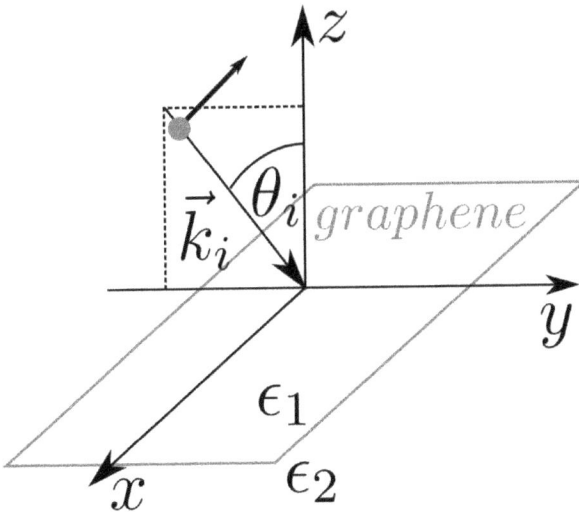

Figure N.2: Geometry of an incoming field, with general polarization, at a graphene interface, clad between two different dielectric media of relative permittivity ϵ_1 and ϵ_2. The electric field has $s-$ and $p-$components. The magnetic field is not shown.

at $z = 0$ between two media. Medium 1, $z > 0$, has a dielectric constant ϵ_1, while medium 2 has dielectric constant ϵ_2. We assume that there is a source of electromagnetic field in region 1. In each region the electric field can be written in general as

$$\mathbf{E}(\mathbf{r}, t) = \int \frac{d\omega}{2\pi} \int \frac{d^2 k}{(2\pi)^2} \mathbf{E}(\omega, \mathbf{k}, z) e^{i(\mathbf{k}\cdot\mathbf{r}_{\parallel} - \omega t)}, \tag{N.16}$$

where $|\mathbf{k}| = k$, with $\mathbf{k}$ an in-plane two-dimensional vector. In a medium with relative dielectric constant ϵ_n and relative permeability μ_n, the wave equation of the electric field $\mathbf{E}$ in the absence of source currents reads

$$\nabla \times \nabla \times \mathbf{E} + \mu_n \mu_0 \epsilon_n \epsilon_0 \frac{\partial^2 \mathbf{E}}{\partial t^2} = 0 \tag{N.17}$$

Using the fact that $\nabla \times \nabla \times = \nabla \cdot \nabla \cdot - \nabla^2$ and the fact that if there are no free charges, we have $\nabla \cdot \mathbf{D} = \epsilon_n \epsilon_0 \nabla \cdot \mathbf{E} = 0$, we obtain

$$- \nabla^2 \mathbf{E}(\mathbf{r}) + \mu_n \mu_0 \epsilon_n \epsilon_0 \frac{\partial^2 \mathbf{E}}{\partial t^2} = 0, \tag{N.18}$$

which in terms of the Fourier components becomes

$$\left(-\frac{\partial^2}{\partial z^2} + k^2\right) \mathbf{E}(\omega, k, z) - \frac{\omega^2}{c_n^2} \mathbf{E}(\omega, k, z) = 0. \tag{N.19}$$

The previous equation implies that $\mathbf{E}(\omega, \mathbf{k}, z) = \mathbf{E}(\omega, k, z)$. The general solution of the previous differential equation is given by

$$\mathbf{E}(\omega, k, z) = \mathbf{E}^+(\omega, k) e^{i k_{z,n} z} + \mathbf{E}^-(\omega, k) e^{-i k_{z,n} z},$$

where

$$k_{z,n} = \text{sgn}(\omega)\sqrt{k_n^2 - k^2 + i\text{sgn}(\omega)}, \tag{N.20}$$

and we have introduced the notation $k_n = \omega/c_n$, and $c_n = 1/\sqrt{\epsilon_n \mu_n \epsilon_0 \mu_0}$. For $k_n^2 > k^2$, the solutions correspond to propagating waves, while for $k_n^2 < k^2$ the solutions correspond to evanescent waves with

$$k_{z,n} = i\sqrt{k^2 - k_n^2}. \tag{N.21}$$

The solution $\mathbf{E}^+(\omega, k)$ corresponds to a wave propagating in the $+z$ direction, while $\mathbf{E}^-(\omega, k)$ propagates in the $-z$ direction. The condition $\nabla \cdot \mathbf{E} = 0$ still has to be enforced. This is automatically satisfied if we write the field in terms of the polarization vectors:

$$\mathbf{E}^\pm(\omega, k) = E^\pm_{s,n}(\omega, k) \mathbf{e}^\pm_{s,n} + E^\pm_{p,n}(\omega, k) \mathbf{e}^\pm_{p,n}, \tag{N.22}$$

where we have introduced the polarization vectors (see Fig. P.1)

$$\mathbf{e}_{s,n}^{\pm} = \mathbf{e}_k \times \mathbf{e}_z, \tag{N.23}$$

$$\mathbf{e}_{p,n}^{\pm} = \frac{k}{k_n}\mathbf{e}_z \mp \frac{k_{z,n}}{k_n}\mathbf{e}_k, \tag{N.24}$$

where $\mathbf{e}_k = (k_x, k_y, 0)/k$ and $\mathbf{e}_z$ is the unit vector along the z direction. These two polarization vectors correspond to $s-$ (transverse electric, TE) and $p-$ (transverse magnetic, TM) polarizations of the electric field. For concreteness, in a coordinate system such that $(k_x, k_y) = (0, k_\parallel)$, the polarization vectors are given by

$$\mathbf{e}_{s,n}^{\pm} = (1, 0, 0), \tag{N.25}$$

$$\mathbf{e}_{p,n}^{\pm} = \left(0, \mp\frac{k_{z,n}}{k_n}, \frac{k}{k_n}\right). \tag{N.26}$$

The magnetic field strength follows from $\mathbf{H} = (i\mu\omega)^{-1}\nabla \times \mathbf{E}$ and is written as

$$\mathbf{H}^{\pm}(\omega, k) = \sqrt{\frac{\epsilon_0\epsilon_n}{\mu_0\mu_n}}\left(-E_{s,n}^{\pm}(\omega, k)\mathbf{e}_{p,n}^{\pm} + \sqrt{\frac{\epsilon_0\epsilon_n}{\mu_0\mu_n}}E_{p,n}^{\pm}(\omega, k)\mathbf{e}_{s,n}^{\pm}\right). \tag{N.27}$$

From this we see the justification of the names TE and TM. For the s/TE polarization, the electric field is transverse to the incidence plane (which is spanned by the vectors $\mathbf{e}_k$ and $\mathbf{e}_z$), while for the p/TM polarizations it is $\mathbf{H}$ that is perpendicular to the plane of incidence. For the situation where we have isotropy and no Hall conductivity in the interface, both polarizations are decoupled from each other. Thus, in the following, we analyze then separately.

Assuming that the 2D material, located at the interface, is isotropic, we want to study the situation where we have an incident electric field from $+z$, $\mathbf{E}_1^-$, a reflected wave $\mathbf{E}_1^+$, and a transmitted wave $\mathbf{E}_2^-$. We write the TE and TM components of the electric field electric as

$$\mathbf{E}_{1/2}^{\pm(TE)} = E_{1/2}^{\pm(TE)}(1, 0, 0), \tag{N.28}$$

$$\mathbf{E}_{1/2}^{\pm(TM)} = E_{1/2}^{\pm(TM)}(0, \mp k_{z,1/2}, k)/k_{1/2}, \tag{N.29}$$

where we are assuming that $k_x = 0$, for simplicity. The choice of signs in the polarization vectors obeys the condition that $\nabla \cdot \mathbf{E} = 0$ (for later use, it is good to have this in mind). Furthermore, recall that the in-plane momentum $\mathbf{k}$ reads $\mathbf{k} = (0, k_y) = (0, k)$, since $k_x = 0$. With this convention, the magnetic field strength components read respectively

$$\mathbf{H}_{1/2}^{\pm(TE)} = \sqrt{\frac{\epsilon_0\epsilon_1}{\mu_0\mu_1}}E_{1/2}^{\pm(TE)}(0, \pm k_{z,1/2}, -k)/k_{1/2}, \tag{N.30}$$

$$\mathbf{H}_{1/2}^{\pm(TM)} = \sqrt{\frac{\epsilon_0\epsilon_1}{\mu_0\mu_1}}E_{1/2}^{\pm(TE)}(1, 0, 0). \tag{N.31}$$

Let us analyse each component separately.

N.2.1   *TE (transverse electric) / s−components*

With the results (N.28) and (N.30) in mind, let us discuss the scattering of a transverse electric wave by a graphene sheet embeded in two different dielectrics. We assume the electric field is coming from $+z$. Therefore, we write

$$\mathbf{E}_1^{-(TE)} = E_i^{(TE)} (1,0,0), \tag{N.32}$$

$$\mathbf{E}_1^{+(TE)} = r_s E_i^{(TE)} (1,0,0), \tag{N.33}$$

$$\mathbf{E}_2^{-(TE)} = t_s E_i^{(TE)} (1,0,0), \tag{N.34}$$

$$\mathbf{E}_2^{+(TE)} = (0,0,0), \tag{N.35}$$

where r_s and t_s are the reflection and transmission amplitudes, respectively. The boundary condition (see Chap. 2)

$$\hat{\mathbf{n}} \times (\mathbf{E}_1 - \mathbf{E}_2) = 0, \tag{N.36}$$

implies that

$$E_s^i + E_s^r = E_s^t, \tag{N.37}$$

where the notation has been simplified, and E_s^i, E_s^r, and E_s^t are the incident, reflected, and transmitted amplitudes of the electric field, respectively, for and $s-$polarized electric field. The boundary condition (see Chap. 2)

$$\hat{\mathbf{n}} \times (\mathbf{H}_1 - \mathbf{H}_2) = \mathbf{J}_{f,2D}, \tag{N.38}$$

implies

$$\frac{k_{z1}}{\mu_1} \left(E_s^i - E_s^r \right) - \frac{k_{z2}}{\mu_2} E_s^t = \mu_0 \omega J_{f,2D}^x. \tag{N.39}$$

We introduce the conductivity tensor, such that $\mathbf{J}_{f,2D} = \sigma_T \cdot \mathbf{E}_{2D}$. Therefore we have two boundary conditions

$$E_s^i + E_s^r = E_s^t, \tag{N.40}$$

and

$$\frac{k_{z1}}{\mu_1} \left(E_s^i - E_s^r \right) - \frac{k_{z2}}{\mu_2} E_s^t = \mu_0 \omega \sigma_T E_s^t. \tag{N.41}$$

Solving for E_s^r and E_s^t we obtain:

$$r_s = \frac{E_s^r}{E_s^i} = \frac{\mu_2 k_{z1} - \mu_1 k_{z2} - \mu_1 \mu_2 \mu_0 \omega \sigma_T}{\mu_2 k_{z1} + \mu_1 k_{z2} + \mu_1 \mu_2 \mu_0 \omega \sigma_T}, \tag{N.42}$$

$$t_s = \frac{E_s^t}{E_s^i} = \frac{2 \mu_2 k_{z1}}{\mu_2 k_{z1} + \mu_1 k_{z2} + \mu_1 \mu_2 \mu_0 \omega \sigma_T}, \tag{N.43}$$

where σ_T is the transverse conductivity (see Appendix C). It is easy to see that

$$t_s = 1 + r_s. \tag{N.44}$$

We can show that these last three results reduce to the case discussed in Sec. N.1 when $\epsilon_1 = \epsilon_2 = \epsilon_0$.

N.2.2 TM (transverse magnetic) / p−components

With the results (N.29) and (N.31) in mind, let us discuss the scattering of a transverse magnetic wave by a graphene sheet embeded in two different dielectrics. The magnetic field strength is transverse and is given by

$$\mathbf{H}_p^i = \sqrt{\frac{\epsilon_1 \epsilon_0}{\mu_1 \mu_0}} \left(E_p^i, 0, 0 \right) , \tag{N.45}$$

$$\mathbf{H}_p^r = \sqrt{\frac{\epsilon_1 \epsilon_0}{\mu_1 \mu_0}} \left(E_p^r, 0, 0 \right) , \tag{N.46}$$

$$\mathbf{H}_p^t = \sqrt{\frac{\epsilon_2 \epsilon_0}{\mu_2 \mu_0}} \left(E_p^t, 0, 0 \right) . \tag{N.47}$$

The boundary condition

$$\hat{n} \times (\mathbf{E}_1 - \mathbf{E}_2) = 0 , \tag{N.48}$$

tells us that

$$E_{py}^i + E_{py}^r = E_{py}^t \Leftrightarrow \frac{k_{z1}}{k_1} E_p^i - \frac{k_{z1}}{k_1} E_p^r = \frac{k_{z2}}{k_2} E_p^t$$

$$\Leftrightarrow \frac{ck_{z1}}{\sqrt{\epsilon_1 \mu_1} \omega} E_p^i - \frac{ck_{z1}}{\sqrt{\epsilon_1 \mu_1} \omega} E_p^r = \frac{ck_{z2}}{\sqrt{\epsilon_2 \mu_2} \omega} E_p^t , \tag{N.49}$$

while the boundary condition

$$\hat{n} \times (\mathbf{H}_1 - \mathbf{H}_2) = \mathbf{J}_{f,2D}, \tag{N.50}$$

tells us that

$$\sqrt{\frac{\epsilon_1 \epsilon_0}{\mu_1 \mu_0}} E_p^i + \sqrt{\frac{\epsilon_1 \epsilon_0}{\mu_1 \mu_0}} E_p^r - \sqrt{\frac{\epsilon_2 \epsilon_0}{\mu_2 \mu_0}} E_p^t = J_{f,2D}^y = \sigma_L \frac{k_{z2}}{k_2} E_p^t = \sigma_L \frac{ck_{z2}}{\sqrt{\epsilon_2 \mu_2} \omega} E_p^t , \tag{N.51}$$

where σ_L is the longitudinal conductivity (see Appendix C). Therefore, solving the two following conditions

$$\frac{k_{z1}}{\sqrt{\epsilon_1 \mu_1}} E_p^i - \frac{k_{z1}}{\sqrt{\epsilon_1 \mu_1}} E_p^r = \frac{k_{z2}}{\sqrt{\epsilon_2 \mu_2}} E_p^t , \tag{N.52}$$

and

$$\sqrt{\frac{\epsilon_1}{\mu_1}}E_p^i + \sqrt{\frac{\epsilon_1}{\mu_1}}E_p^r - \sqrt{\frac{\epsilon_2}{\mu_2}}E_p^t = \frac{\sigma_L}{\omega\epsilon_0}\frac{k_{z2}}{\sqrt{\epsilon_2\mu_2}}E_p^t \qquad (N.53)$$

for E_p^r and E_p^t we obtain:

$$\begin{aligned}
r_p = \frac{E_p^r}{E_p^i} &= \frac{k_{z1}\epsilon_2 - k_{z2}\epsilon_1 + k_{z1}k_{z2}\sigma_L/(\epsilon_0\omega)}{k_{z1}\epsilon_2 + k_{z2}\epsilon_1 + k_{z1}k_{z2}\sigma_L/(\epsilon_0\omega)} \\
&= \frac{\epsilon_2/k_{z2} - \epsilon_1/k_{z1} + \sigma_L/(\epsilon_0\omega)}{\epsilon_2/k_{z2} + \epsilon_1/k_{z1} + \sigma_L/(\epsilon_0\omega)}, \\
t_p = \frac{E_p^t}{E_p^i} &= \frac{2k_{z1}\epsilon_2}{k_{z1}\epsilon_2 + k_{z2}\epsilon_1 + k_{z1}k_{z2}\sigma_L/(\epsilon_0\omega)}\sqrt{\frac{\epsilon_1\mu_2}{\mu_1\epsilon_2}} \\
&= \frac{2\epsilon_2/k_{z2}}{\epsilon_2/k_{z2} + \epsilon_1/k_{z1} + \sigma_L/(\epsilon_0\omega)}\sqrt{\frac{\epsilon_1\mu_2}{\mu_1\epsilon_2}}.
\end{aligned}$$

$$(N.54)$$
$$(N.55)$$

The coefficients obey the following relation:

$$t_p = \frac{k_{z1}}{k_{z2}}\sqrt{\frac{\epsilon_2\mu_2}{\epsilon_1\mu_1}}\left(1 - r_p\right). \qquad (N.56)$$

We can show that these last three results reduce to the case discussed in Sec. N.1 when $\epsilon_1 = \epsilon_2 = \epsilon_0$.

It is also possible to generalize the above expressions for the reflection and transmission amplitudes to the case where a magnetic field perpendicular to the graphene plane is present [Kort-Kamp *et al.* (2015)]. This is important in the discussion of the effect of a magnetic field in the Purcell factor.

Appendix O

Green's Functions of a Rope with Two Different Mass Densities Subject to a Point-Like Excitation
(*written with Bruno Amorim*)

It is perhaps surprising to find a problem of mechanics in a book about plasmonics. However, this appendix in meant as a prelude to the next one. Here we discuss a problem that is quite similar to the calculation of the Green's functions of a radiating dipole in the presence of an interface, but without the complexity of matrix Green's functions. The study of this appendix may facilitate the reading of the following one.

O.1 The Homogeneous Rope

Let us consider a uniform rope under stress, which we describe by tbe Lagrangian density

$$\mathcal{L} = \frac{1}{2}\rho\dot{\phi}^2 - \frac{1}{2}Y\left(\partial_x\phi\right)^2, \tag{O.1}$$

where $\phi(x,t)$ is the height profile of the rope, ρ is the mass density, Y is the tension of the rope, $\dot{\phi} = \partial\phi/\partial t$ and $\partial_x\phi = \partial\phi/\partial x$. The waves propagate with a velocity given by $c = \sqrt{Y/\rho}$. The equation of motion for the height field is given by

$$\rho\ddot{\phi} = Y\partial_x^2\phi. \tag{O.2}$$

We now want to consider the response of the rope to an external perturbation.

O.1.1 *Response to a localized force*

Let us subject the rope to an external force at $x = 0$ and $t = 0$, $F(x,t) = f\delta(x)\delta(t)$, such that the Lagrangian density reads

$$\mathcal{L} = \frac{1}{2}\rho\dot{\phi}^2 - \frac{1}{2}Y\left(\partial_x\phi\right)^2 - \phi F, \tag{O.3}$$

and the equation of motion become

$$\rho\ddot{\phi} = Y\partial_x^2\phi + f\delta(x)\delta(t). \tag{O.4}$$

We want to obtain ϕ as a function of f. Performing a Fourier transform of the equation of motion and using the definition

$$\phi(x,t) = \int \frac{d\omega dk}{(2\pi)^2}\phi(k,\omega)e^{ikx}e^{-i\omega t}, \tag{O.5}$$

we obtain

$$\phi(k,\omega) = \frac{1}{-\rho\omega^2 + Yk^2}f, \tag{O.6}$$

which implies

$$\phi(x,t) = f\int \frac{d\omega dk}{(2\pi)^2}\frac{e^{ikx}e^{-i\omega t}}{-\rho\omega^2 + Yk^2}. \tag{O.7}$$

This allows to write

$$\phi(x,t) = \int_{-\infty}^{t} dt'dx'G(x-x';t-t')F(x',t'), \tag{O.8}$$

where we have defined the Green's function for the rope by

$$G(x;t) = \int \frac{d\omega dk}{(2\pi)^2}\frac{e^{ikx}e^{-i\omega t}}{-\rho\omega^2 + Yk^2}. \tag{O.9}$$

To obtain a retarded response we perform the usual replacement $\omega \to \omega + i0^+$. It is useful for later use to invert the Fourier transform in space, thus obtaining

$$G(x;\omega) = \int \frac{dk}{2\pi}\frac{e^{ikx}}{-\rho\omega^2 + Yk^2} = \frac{1}{Y}\int \frac{dk}{2\pi}\frac{e^{ikx}}{(k-\omega/c-i0^+)(k+\omega/c+i0^+)}. \tag{O.10}$$

For $x > 0$ we can close the contour in the upper half plane, collecting the contribution from the pole at $k = \omega/c + i0^+$ and obtaining (for $x > 0$)

$$\boxed{G(x;\omega) = i\frac{e^{i\omega x/c}}{2Y\omega/c}}. \tag{O.11}$$

For $x < 0$ we can close the contour in the lower half plane, collecting the contribution from the pole at $k = -\omega/c - i0^+$ and obtaining

$$G(x;\omega) = \int \frac{dk}{2\pi}\frac{e^{ikx}}{-\rho\omega^2 + Yk^2} = i\frac{e^{-i\omega x/c}}{2Y\omega/c}. \ x < 0. \tag{O.12}$$

Writing both cases in a unified way we have

$$G(x; \omega) = i\frac{e^{i\omega|x|/c}}{2Y\omega/c}. \tag{O.13}$$

Finally, the height field is given by

$$\phi(x, \omega) = G(x; \omega)f. \tag{O.14}$$

Notice that integrating the equation of motion Eq. (O.4), around $x = 0$ we would obtain the condition

$$Y\partial_x\phi(0^+) - Y\partial_x\phi(0^-) = -f. \tag{O.15}$$

Therefore the Green's function should obey such discontinuity condition:

$$Y\partial_x G(x = 0^+) - Y\partial_x G(x = 0^-) = -1. \tag{O.16}$$

Indeed, we can check that Eq. (O.13) satisfies the previous condition

$$\lim_{x\to 0^+} \frac{\partial}{\partial x} G(x; \omega) = -\frac{1}{2Y} \tag{O.17}$$

$$\lim_{x\to 0^-} \frac{\partial}{\partial x} G(x; \omega) = \frac{1}{2Y}. \tag{O.18}$$

O.2   The Inhomogeneous Rope

Let us suppose now that we have an inhomogeneous rope such that for $x > 0$ it has a mass density ρ_1 and tension Y_1 and for $x < 0$ it has a mass density ρ_2 and tension Y_2. The Lagrangian density describing this situation is given by

$$\mathcal{L} = \frac{1}{2}\rho_1\dot{\phi}^2 - \frac{1}{2}Y_1\left(\partial_x\phi\right)^2, \qquad x > 0, \tag{O.19}$$

$$\mathcal{L} = \frac{1}{2}\rho_2\dot{\phi}^2 - \frac{1}{2}Y_2\left(\partial_x\phi\right)^2, \qquad x < 0. \tag{O.20}$$

Having two different tension values in the two different parts of the rope is perhaps a bit artificial.

O.2.1   *Reflection and transmition coefficients*

We now want to discuss the scattering of an incoming wave at the interface between the two different pieces of a rope (in terms of the electric field problem of the next appendix, this corresponds to the scattering at the interface between two different dielectrics). Let us assume we have an incoming wave from $x \to \infty$

$$\phi_i e^{-i\omega t} e^{-i\omega x/c_1}, \tag{O.21}$$

where $c_i = \sqrt{Y_i/\rho_i}$, with $i = 1, 2$, refers to the speed of the wave in the medium i. The solution everywhere is of the form

$$\phi(x > 0, t) = \phi_i e^{-i\omega t} e^{-i\omega x/c_1} + r\phi_i e^{-i\omega t} e^{+i\omega x/c_1}, \qquad (\text{O.22})$$
$$\phi(x < 0, t) = t\phi_i e^{-i\omega t} e^{-i\omega x/c_2}, \qquad (\text{O.23})$$

where r and t are the reflection and transmission coefficients, which need to be determined. At $x = 0$ we have the following boundary conditions apply:

$$\phi(0^+) = \phi(0^-), \qquad (\text{O.24})$$
$$Y_1 \partial_x \phi(0^+) = Y_2 \partial_x \phi(0^-), \qquad (\text{O.25})$$

which translate into

$$1 + r = t, \qquad (\text{O.26})$$
$$-Y_1 k_1 + Y_1 k_1 r = -Y_2 k_2 t, \qquad (\text{O.27})$$

and where we have defined $k_i = \sqrt{\rho_i/Y_i}\omega$ for $i = 1, 2$. Solving this set of equations we obtain

$$\boxed{\begin{aligned} r &= \frac{Y_1 k_1 - Y_2 k_2}{Y_1 k_1 + Y_2 k_2}, \\ t &= \frac{2Y_1 k_1}{Y_1 k_1 + Y_2 k_2}. \end{aligned}} \qquad \begin{aligned} (\text{O.28}) \\ (\text{O.29}) \end{aligned}$$

These coefficients will emerge again in what follows.

O.2.2 *Response to an external force in the inhomogeneous regime*

Low let us apply an external force at $x = x_0 > 0$ of the form $F = f\delta(x - x_0)$ (in the electromagnetic problem of the following appendix, a radiating dipole plays the role of the force in this mechanical problem). Then, the equations of motion become

$$\rho_1 \ddot{\phi} = Y_1 \partial_x^2 \phi + f\delta(x - x_0), \; x > 0, \qquad (\text{O.30})$$
$$\rho_2 \ddot{\phi} = Y_2 \partial_x^2 \phi, \; x < 0. \qquad (\text{O.31})$$

Integrating the first equations from $x_0 - \epsilon$ to $x_0 + \epsilon$ an taking the limit $\epsilon \to 0$ we obtain the boundary condition at $x = x_0$:

$$Y_1 \partial_x \phi(x_0^+) - Y_1 \partial_x \phi(x_0^-) + f = 0 \qquad (\text{O.32})$$

Therefore we now have four boundary conditions (two at $x = 0$ and other two at $x = x_0$)

$$\phi(0^+) - \phi(0^-) = 0 \,, \tag{O.33}$$

$$Y_1 \partial_x \phi(0^+) - Y_2 \partial_x \phi(0^-) = 0 \,, \tag{O.34}$$

$$\phi(x_0^+) - \phi(x_0^-) = 0 \,, \tag{O.35}$$

$$Y_1 \partial_x \phi(x_0^+) - Y_1 \partial_x \phi(x_0^-) = -f \,. \tag{O.36}$$

We write the solution in the whole space as

$$\phi(x_0 < x, t) = \phi_I^> e^{-i\omega t} e^{ik_1 x} \,, \tag{O.37}$$

$$\phi(0 < x < x_0, t) = \phi_{II}^> e^{-i\omega t} e^{ik_1 x} + \phi_{II}^< e^{-i\omega t} e^{-ik_1 x} \,, \tag{O.38}$$

$$\phi(x < 0, t) = \phi_{III}^< e^{-i\omega t} e^{-ik_2 x} \,, \tag{O.39}$$

where $k_{1/2} = \omega/c_{1/2}$, and $\phi_{I,II}^>$ represents the amplitude of the waves that are propagating towards $x \to \infty$ and $\phi_{II,III}^<$ the amplitudes of the waves that are propagating towards $x \to -\infty$. With this choice for the fields, the boundary conditions become

$$\phi_{II}^> + \phi_{II}^< - \phi_{III}^< = 0 \,, \tag{O.40}$$

$$Y_1 \left[k_1 \phi_{II}^> - k_1 \phi_{II}^< \right] - Y_2 \left[-k_2 \phi_{III}^< \right] = 0 \,, \tag{O.41}$$

$$\phi_I^> - \left[\phi_{II}^> + \phi_{II}^< e^{-i2k_1 x_0} \right] = 0 \,, \tag{O.42}$$

$$\phi_I^> - \left[\phi_{II}^> - \phi_{II}^< e^{-i2k_1 x_0} \right] = -e^{-ik_1 x_0} f/(iY_1 k_1) \,. \tag{O.43}$$

It is easy to solve this system of equations obtaining

$$\phi_I^> = \frac{if}{2Y_1 k_1} e^{-ik_1 x_0} + \frac{if}{2Y_1 k_1} e^{ik_1 x_0} \frac{Y_1 k_1 - Y_2 k_2}{Y_1 k_1 + Y_2 k_2} \,, \tag{O.44}$$

$$\phi_{II}^> = \frac{if}{2Y_1 k_1} e^{ik_1 x_0} \frac{Y_1 k_1 - Y_2 k_2}{Y_1 k_1 + Y_2 k_2} \,, \tag{O.45}$$

$$\phi_{II}^< = \frac{if}{2Y_1 k_1} e^{ik_1 x_0} \,, \tag{O.46}$$

$$\phi_{III}^< = \frac{if}{2Y_1 k_1} e^{ik_1 x_0} \frac{2Y_1 k_1}{Y_1 k_1 + Y_2 k_2} \,. \tag{O.47}$$

In the above expressions we recognize the presence of the reflection and transmission coefficients that we have obtained in the previous section. The full solution to ϕ is therefore given by

$$\phi^I(x_0 < x) = \frac{if}{2Y_1 k_1} e^{-ik_1 x_0} e^{ik_1 x} + r \frac{if}{2Y_1 k_1} e^{ik_1 x_0} e^{ik_1 x} \,, \tag{O.48}$$

$$\phi^{II}(0 < x < x_0) = \frac{if}{2Y_1 k_1} e^{ik_1 x_0} e^{-ik_1 x} + r \frac{if}{2Y_1 k_1} e^{ik_1 x_0} e^{ik_1 x} \,, \tag{O.49}$$

$$\phi^{III}(x < 0) = t \frac{if}{2Y_1 k_1} e^{ik_1 x_0} e^{-ik_2 x} \,, \tag{O.50}$$

The solution can be recast in the appealing form [cf. P.70-P.72]

$$\phi^I(x_0 < x) = \left[G^{i>}(x, x_0; \omega) + G^r(x, x_0; \omega)\right] f \tag{O.51}$$
$$\phi^{II}(0 < x < x_0) = \left[G^{i<}(x, x_0; \omega) + G^r(x, x_0; \omega)\right] f \tag{O.52}$$
$$\phi^{III}(x < 0) = G^t(x, x_0; \omega) f \tag{O.53}$$

where we have introduced the Green's functions

$$G^{i>}(x, x_0; \omega) = \frac{i}{2Y_1 k_1} e^{-ik_1 x_0} e^{ik_1 x}, \tag{O.54}$$

$$G^{i<}(x, x_0; \omega) = \frac{i}{2Y_1 k_1} e^{ik_1 x_0} e^{-ik_1 x}, \tag{O.55}$$

$$G^r(x, x_0; \omega) = r \frac{i}{2Y_1 k_1} e^{ik_1 x_0} e^{ik_1 x}, \tag{O.56}$$

$$G^t(x, x_0; \omega) = t \frac{i}{2Y_1 k_1} e^{ik_1 x_0} e^{-ik_2 x}. \tag{O.57}$$

We recognize that the Green's functions $G^{i>}(x, x_0; \omega)$ and $G^{i<}(x, x_0; \omega)$ are actually just the Green's function of a homogeneous rope characterized by ρ_1 and Y_1:

$$G^{i>}(x, x_0; \omega) = G^{i<}(x, x_0; \omega) = i \frac{e^{ik_i|x-x_0|}}{2Y_1 k_1}. \tag{O.58}$$

Since the Green's functions represent the response to an external stimulus, Eqs. (O.51)-(O.53) have the following physical interpretation: the external force f at $x = x_0$ emits a primary wave in both directions along the x axis with amplitude $G^{i \gtrless}(x, x_0; \omega) f$. The wave that goes towards $x \to -\infty$ is reflected and transmitted at the interface $x = 0$. The amplitude of the reflected wave is obtained by multiplying the primary wave by the reflection coefficient r. The reflected wave seems to be emitted from the position $x = -x_0$. The amplitude of the transmitted wave is obtained by multiplying the amplitude of the primary wave by the transmission coefficient t. The transmitted wave seems to be emitted from position $x = k_1 x_0 / k_2$. The last result is a consequence of the change of the material properties of the rope as we pass through $x = 0$. These results can be applied to other kinds of waves, provided we know the form of the Green's functions and the form of the reflection and transmission coefficients, as we shall see in the following appendix. For future use, we note that the physical quantities introduce in this appendix relate to those of the next one by the following mapping:

$$\phi \rightarrow \mathbf{E}, \tag{O.59}$$

$$Y \rightarrow \frac{1}{\mu\mu_0}, \tag{O.60}$$

$$\rho \rightarrow \epsilon\epsilon 0, \tag{O.61}$$

$$f \rightarrow i\omega \mathbf{j}_f(\omega), \tag{O.62}$$

$$G(x, x_0; \omega) \rightarrow \mu\mu_0 \mathbf{G}(r, r_0; \omega), \tag{O.63}$$

where $\mathbf{E}$ is the electric field, $\mu\mu_0$ and $\epsilon\epsilon 0$ are the magnetic permeability and the dielectric permittivity of the medium, respectively, ω is the frequency of the emitting dipole, $\mathbf{j}_f(\omega)$ is the current associated with the emitting dipole, and $\mathbf{G}(r, r_0; \omega)$ is the matrix Green's function. The reason why we have a matrix Green's function for the radiating dipole problem is due to the vectorial nature of the electric field.

Appendix P

Derivation of the Transition Rate of a Quantum Emitter Near an Interface
(written with Bruno Amorim)

Here consider the problem of deriving Eq. (10.38). The final expression needs the general results of Appendix N, as they define the reflectance amplitudes for the $s-$ and $p-$polarizations. We will see that the problem of determining the transition rate of a dipole can be cast in terms of matrix Green's functions (also termed dyadic Green's functions). The reader may want to read first Appendix O, which is meant as prelude to this one.

Using first order perturbation theory, we can prove[Novotny and Hecht (2014)] that the lifetime of a two-level system is given by

$$\frac{1}{\tau} = \frac{2\omega^2 \mu \mu_0}{\hbar} \mu_{ge}^* \cdot \Im m\mathbf{G}\left(\mathbf{r}_0, \mathbf{r}_0; \omega\right) \cdot \mu_{ge}, \tag{P.1}$$

where μ is the relative permeability, μ_0 is the vacuum permeability, $\mathbf{G}\left(\mathbf{r}_0, \mathbf{r}_0; \omega\right)$ is the electromagnetic Green's function at the position of the dipole, $\mu_{ge}^T = \langle g| \hat{\mathbf{d}} |e\rangle \equiv (\mu_x, \mu_y, \mu_z)$ is the dipole operator matrix element (and T stands for the transpose operation), $\omega = \epsilon_e - \epsilon_g$, is the frequency of the emitting dipole, ϵ_g and ϵ_e are the energies of the ground and excited states, respectively, and $\mathbf{r}_0$ is the position of the dipole. The goal is to compute $1/\tau$ for a dipole near a flat interface covered with graphene.

This is a rather long Appendix, but its study will pay off, since the reader will be, in the end, equipped with all the tools needed to tackle more complex problems. A note on notation is in order: $d^2\mathbf{k}$ stands for a 2D in-plane integration, whereas d^3k stands for a 3D integration. We also use $n = \sqrt{\epsilon}$ for the refractive index of the medium where the dipole is embedded. When used as a subindex, n refers to the medium with relative dielectric constant ϵ_n.

P.1 Polarization Vectors and Green's Functions

The aim of this section is to define the polarization vectors, depicted in Fig. P.1, to write the electric and magnetic fields in terms of these vectors, and, finally, to obtain the free space Green's function and the Green's functions in the presense of an interface in terms of the polarization vectors.

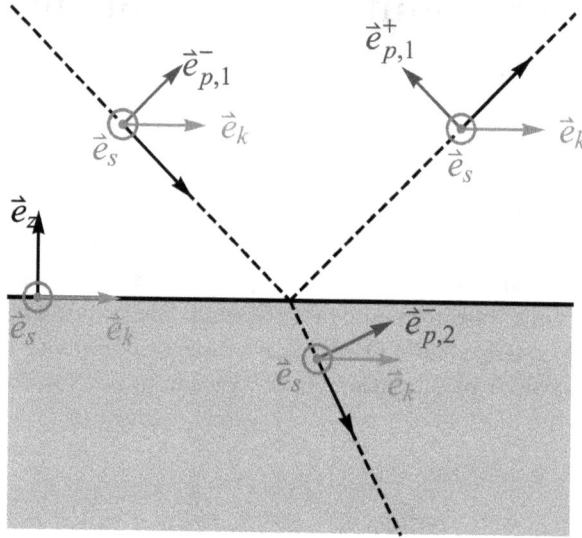

Figure P.1: Definition of the polarization vectors in the interface problem. These are: $\mathbf{e}_{s,n}^{\pm} = (1,0,0)$ and $\mathbf{e}_{p,n}^{\pm} = (0, \mp k_{z,n}/k_n, k/k_n)$. Note that k is the absolute value of the in-plane momentum and should not be confused with $k_n = \omega/v_n$, with v_n the velocity in medium n.

P.1.1 Definition of s− and p−polarization vectors and angular spectrum representation

For the sake of completeness, here we repeat the beginning of Sec. N.2. Any electric field can be written in general form

$$\mathbf{E}(\mathbf{r},t) = \int \frac{d\omega}{2\pi} \int \frac{d^2\mathbf{k}}{(2\pi)^2} \mathbf{E}(\omega, \mathbf{k}, z)e^{i(\mathbf{k}\cdot\mathbf{r}_\parallel - \omega t)}, \qquad (P.2)$$

where $|\mathbf{k}| = k$, with $\mathbf{k}$ an in-plane two-dimensional vector. In a medium with relative dielectric constant ϵ_n and relative permeability μ_n, the wave equation of the electric field $\mathbf{E}$ in the absence of source currents reads

$$\nabla \times \nabla \times \mathbf{E} + \mu_n \mu_0 \epsilon_n \epsilon_0 \frac{\partial^2 \mathbf{E}}{\partial t^2} = 0 \qquad \text{(P.3)}$$

Using the fact that $\nabla \times \nabla \times = \nabla \cdot \nabla \cdot - \nabla^2$ and the fact that if there are no free charges, we have $\nabla \cdot \mathbf{D} = \epsilon_n \epsilon_0 \nabla \cdot \mathbf{E} = 0$, we obtain

$$-\nabla^2 \mathbf{E}(\mathbf{r}) + \mu_n \mu_0 \epsilon_n \epsilon_0 \frac{\partial^2 \mathbf{E}}{\partial t^2} = 0, \qquad \text{(P.4)}$$

which in terms of the Fourier components becomes

$$\left(-\frac{\partial^2}{\partial z^2} + k^2\right) \mathbf{E}(\omega, k, z) - \frac{\omega^2}{c_n^2} \mathbf{E}(\omega, k, z) = 0. \qquad \text{(P.5)}$$

The previous equation implies that $\mathbf{E}(\omega, \mathbf{k}, z) = \mathbf{E}(\omega, k, z)$. The general solution of the previous differential equation is given by

$$\mathbf{E}(\omega, k, z) = \mathbf{E}^+(\omega, k) e^{i k_{z,n} z} + \mathbf{E}^-(\omega, k) e^{-i k_{z,n} z},$$

where

$$k_{z,n} = \text{sgn}(\omega) \sqrt{k_n^2 - k^2 + i\,\text{sgn}(\omega)}, \qquad \text{(P.6)}$$

and we have introduced the notation $k_n = \omega/c_n$, and $c_n = 1/\sqrt{\epsilon_n \mu_n \epsilon_0 \mu_0}$. For $k_n^2 > k^2$, the solutions correspond to propagating waves, while for $k_n^2 < k^2$ the solutions correspond to evanescent waves with

$$k_{z,n} = i\sqrt{k^2 - k_n^2}. \qquad \text{(P.7)}$$

The solution $\mathbf{E}^+(\omega, k)$ corresponds to a wave propagating in the $+z$ direction, while $\mathbf{E}^-(\omega, k)$ propagates in the $-z$ direction. The condition $\nabla \cdot \mathbf{E} = 0$ still has to be enforced. This is automatically satisfied if we write the field in terms of the polarization vectors:

$$\boxed{\mathbf{E}^\pm(\omega, k) = E_{s,n}^\pm(\omega, k) \mathbf{e}_{s,n}^\pm + E_{p,n}^\pm(\omega, k) \mathbf{e}_{p,n}^\pm}, \qquad \text{(P.8)}$$

where we have introduced the polarization vectors (see Fig. P.1)

$$\boxed{\mathbf{e}_{s,n}^\pm = \mathbf{e}_k \times \mathbf{e}_z,} \qquad \text{(P.9)}$$

$$\boxed{\mathbf{e}_{p,n}^\pm = \frac{k}{k_n} \mathbf{e}_z \mp \frac{k_{z,n}}{k_n} \mathbf{e}_k,} \qquad \text{(P.10)}$$

where $\mathbf{e}_k = (k_x, k_y, 0)/k$ and $\mathbf{e}_z$ is the unit vector along the z direction. These two polarization vectors correspond to $s-$ (transverse electric, TE) and $p-$ (transverse magnetic, TM) polarizations of the electric field. For concreteness, in a coordinate system such that $(k_x, k_y) = (0, k_\parallel)$, the polarization vectors are given by

$$\mathbf{e}^\pm_{s,n} = (1, 0, 0),$$
(P.11)

$$\mathbf{e}^\pm_{p,n} = \left(0, \mp\frac{k_{z,n}}{k_n}, \frac{k}{k_n}\right).$$
(P.12)

The magnetic field strength follows from $\mathbf{H} = (i\mu\omega)^{-1}\nabla \times \mathbf{E}$ and is written as

$$\mathbf{H}^\pm(\omega, k) = \sqrt{\frac{\epsilon_0\epsilon_n}{\mu_0\mu_n}}\left(-E^\pm_{s,n}(\omega, k)\mathbf{e}^\pm_{p,n} + \sqrt{\frac{\epsilon_0\epsilon_n}{\mu_0\mu_n}}E^\pm_{p,n}(\omega, k)\mathbf{e}^\pm_{s,n}\right).$$
(P.13)

From this we see the justification of the names TE and TM. For the s/TE polarization, the electric field is transverse to the incidence plane (which is spanned by the vectors $\mathbf{e}_k$ and $\mathbf{e}_z$), while for the p/TM polarizations it is $\mathbf{H}$ that is perpendicular to the plane of incidence. For the situation where we have isotropy and no Hall conductivity in the interface, both polarizations are decoupled from each other. Thus, in the following, we analyze then separately.

P.1.2   *The free space Green's function*

In the presence of a free source current $\mathbf{J}_f$ in a medium with dielectric constant ϵ_n and permeability μ_n, the wave equation of the electric field $\mathbf{E}$ reads

$$\nabla \times \nabla \times \mathbf{E} + \mu_n\mu_0\epsilon_n\epsilon_0\frac{\partial^2\mathbf{E}}{\partial t^2} = -\mu_n\mu_0\frac{\partial\mathbf{J}_f}{\partial t}.$$
(P.14)

The electric field can be written in terms of the source current as

$$\mathbf{E}(t, r) = \int dt' \int d^3r'\mathbf{G}(t, \mathbf{r}; t', \mathbf{r}') \cdot \left(-\mu_n\mu_0\frac{\partial\mathbf{J}_f}{\partial t}(t', \mathbf{r}')\right),$$
(P.15)

where $\mathbf{G}(\mathbf{r}, t; \mathbf{r}', t')$ is the electric field dyadic Green's function, defined by the following equation

$$\nabla \cdot \nabla \cdot \mathbf{G} - \nabla^2\mathbf{G} + \frac{1}{c^2}\frac{\partial^2\mathbf{G}}{\partial t^2} = \mathrm{Id}\delta(\mathbf{r})\delta(t),$$
(P.16)

where Id is the 3×3 identity matrix. If we have translational invariance, $\mathbf{G}(t, \mathbf{r}; t', \mathbf{r}') = \mathbf{G}(t - t', \mathbf{r} - \mathbf{r}')$ and performing a Fourier transform in space and time, we obtain

$$\mathbf{E}(\omega, k) = i\mu_n\mu_0\omega\mathbf{G}(\omega, k) \cdot \mathbf{J}_f(\omega, k),$$
(P.17)

and

$$\left[(k^2 - k_n^2)\,\delta_{ij} - k_i k_j\right] G_{jk}(\omega, k) = \delta_{ik}. \qquad \text{(P.18)}$$

Isotropy imposes that G_{ij} must be of the form $G_{ij} = a\delta_{ij} + bk_i k_j$ [10]. Using this form, the previous equation can be easily inverted and we obtain

$$G_{ij}(\omega, k) = \left[\delta_{ij} - \frac{k_i k_j}{k_n^2}\right]\frac{1}{k^2 - k_n^2}. \qquad \text{(P.19)}$$

Note that $G_{ij}(\omega, k)$ has longitudinal a component that will disappear when we perform below the Fourier transform to real-space in the k_z coordinate. At will soon be clear the final result is purely transverse. Performing and inverse Fourier transform in the z coordinate we obtain

$$\begin{aligned}
G_{ij}(\omega, k_\parallel, z) &= \int \frac{dk_z}{2\pi} G_{ij}(\omega, k) e^{ik_z z} \\
&= \int \frac{dk_z}{2\pi}\left[\delta_{ij} - \frac{k_i k_j}{k_n^2}\right]\frac{e^{ik_z z}}{k_z^2 + k_\parallel^2 - k_n^2}. \\
&= \int \frac{dk_z}{2\pi}\left[\delta_{ij} - \frac{k_i k_j}{k_n^2}\right]\frac{e^{iq_z z}}{(k_z - k_{z,n})(k_z + k_{z,n})} \qquad \text{(P.20)}
\end{aligned}$$

The definition of $k_{z,n} = \sqrt{k_n^2 - k_\parallel^2}$ is such that $\Im m k_{z,n} > 0$. Therefore, for $z > 0$, the integration can be deformed into the upper complex half-space and collecting the pole at $k_{z,n}$. The obtained result is

$$G_{ij}(\omega, k_\parallel, z) = i\left[\delta_{ij} - \frac{c^2}{\omega^2}k_i k_j\right]\Bigg|_{k_z \to k_{z,n}}\frac{e^{ik_{z,n}z}}{2k_{z,n}}. \qquad \text{(P.21)}$$

In a coordinate system such that $(k_x, k_y) = (0, k_\parallel)$ we obtain

$$\begin{aligned}
\left[\delta_{ij} - \frac{k_i k_j}{k_n^2}\right]\Bigg|_{k_z \to k_{z,n}} &= \begin{bmatrix} 1 & & \\ & 1 - \frac{k_\parallel^2}{k_n^2} & -\frac{k_\parallel k_{z,n}}{k_n^2} \\ & -\frac{k_\parallel k_{z,n}}{k_n^2} & 1 - \frac{k_{z,n}^2}{k_n^2} \end{bmatrix} = \begin{bmatrix} 1 & & \\ & \frac{k_{z,n}^2}{k_n^2} & -\frac{k_\parallel k_{z,n}}{k_n^2} \\ & -\frac{k_\parallel k_{z,n}}{k_n^2} & \frac{k_\parallel^2}{k_n^2} \end{bmatrix} \\
&= \begin{bmatrix} 1 \\ 0 \\ 0 \end{bmatrix}\cdot\begin{bmatrix} 1 & 0 & 0 \end{bmatrix} + \begin{bmatrix} 0 \\ -\frac{k_{z,n}}{k_n} \\ \frac{k_\parallel}{k_n} \end{bmatrix}\cdot\begin{bmatrix} 0 & -\frac{k_{z,n}}{k_n} & \frac{k_\parallel}{k_n} \end{bmatrix}. \text{(P.22)}
\end{aligned}$$

Comparing with Eqs. (P.11) and (P.12), we recognize that

$$\begin{bmatrix} 1 & & \\ & \frac{q_z^2}{k_n^2} & -\frac{k_\parallel k_{z,n}}{k_n^2} \\ & -\frac{k_\parallel k_{z,n}}{k_n^2} & \frac{k_\parallel^2}{k_n^2} \end{bmatrix} = \mathbf{e}_{s,n}^+ \otimes \mathbf{e}_{s,n}^+ + \mathbf{e}_{p,n}^+ \otimes \mathbf{e}_{p,n}^+. \qquad \text{(P.23)}$$

---

[10]This is because $\mathcal{R} \cdot \mathbf{G}(\mathbf{k}) \cdot \mathcal{R}^T = \mathbf{G}(\mathcal{R} \cdot \mathbf{k})$, where $\mathcal{R}$ is a rotation matrix, if $\mathbf{G}(\mathbf{k})$ has the given form.

For $z < 0$, the integral in Eq. (P.20) can be performed by deforming the contour into the lower half-plane and collecting the contribution at $-k_{z,n}$. Therefore, the Green's function can be written as

$$\mathbf{G}^0(\omega, k_\parallel) = \frac{i}{2k_{z,n}} e^{\pm i k_{z,n} z} \left(\mathbf{e}_{s,n}^\pm \otimes \mathbf{e}_{s,n}^\pm + \mathbf{e}_{p,n}^\pm \otimes \mathbf{e}_{p,n}^\pm \right), \qquad \text{(P.24)}$$

where $\otimes$ represents a dyadic or tensor product and the $+$ sign is valid for $z > 0$ and the $-$ sign for $z < 0$. Note that in the derivation we have assumed that the emitting dipole is located at $z = 0$.

It is also instructive to express the local electric field in terms of the real-space Green's functions. Let us now see the connection between the electric field of a dipole $\mathbf{d}$ with the free space Green's function. If we have a free dipole described by the polarization

$$\mathbf{P}_f(r, t) = \mathbf{d}(t) \delta(r - r_0), \qquad \text{(P.25)}$$

then we have a free current[11]

$$\mathbf{J}_f(r, t) = \frac{\partial \mathbf{d}(t)}{\partial t} \delta(r - r_0), \qquad \text{(P.26)}$$

and the electric field generated by the dipole is, from Eq. (P.15), given by

$$\mathbf{E}(\mathbf{r}, t) = -\mu\mu_0 \int dt' \mathbf{G}(\mathbf{r}, \mathbf{r}_0; t, t') \cdot \frac{\partial^2 \mathbf{d}(t')}{\partial t'^2}. \qquad \text{(P.27)}$$

or, after making a Fourier transform in time, by

$$\mathbf{E}(\mathbf{r}, \omega) = \mu\mu_0 \omega^2 \mathbf{G}(\mathbf{r}, \mathbf{r}_0; \omega) \cdot \mathbf{d}(\omega). \qquad \text{(P.28)}$$

This result will be used ahead.

P.1.3   *Green's functions for problems with flat interfaces I*

Eq. (P.24) is very useful as allows one to naturally interpret results in terms of emitted waves with $s-$ and $p-$polarization. For example, suppose we wave a dipole, $\mathbf{d}$, located at $z = z_0$. The electric field emitted by the dipole is given by

$$\begin{aligned}
\mathbf{E}^0(\omega, k) &= \mu_n \mu_0 \omega^2 \mathbf{G}^0(\omega, k) \cdot \mathbf{d}(\omega, k) \\
&= \frac{i}{2k_{z,n}} e^{\pm i k_{z,n}(z - z_0)} \mathbf{e}_{s,n}^\pm \left[\mathbf{e}_{s,n}^\pm \cdot \mathbf{d}(\omega, k) \right] \\
&\quad + \frac{i}{2k_{z,n}} e^{\pm i k_{z,n}(z - z_0)} \mathbf{e}_{p,n}^\pm \left[\mathbf{e}_{p,n}^\pm \cdot \mathbf{d}(\omega, k) \right],
\end{aligned} \qquad \text{(P.29)}$$

and we have an emitted electric field in the $\pm z$ direction and s-polarization with amplitude

$$E_{s,n}^{0,\pm}(\omega, k) = \frac{i}{2k_{z,n}} \mathbf{e}_{s,n}^\pm \cdot \mathbf{d}(\omega, k), \qquad \text{(P.30)}$$

---

[11]In frequency space the current density reads $\mathbf{J}_f(r, t) = -i\omega \mathbf{d}\delta(r - r_0) \equiv \mathbf{j}\delta(r - r_0)$.

and an electric field with p-polarization with amplitude

$$E_{p,n}^{0,\pm}(\omega,k) = \frac{i}{2k_{z,n}} \mathbf{e}_{p,n}^{\pm} \cdot \mathbf{d}(\omega,k). \tag{P.31}$$

If we have an interface located at $z=0$, the field emitted in the $-z$ direction will be reflected at the interface. At the interface the emitted electric field is given by

$$E_{s,n}^{0,-}(\omega,k)e^{ik_{z,1}z_0}\mathbf{e}_{s,1}^{-} + E_{p,n}^{0,-}(\omega,k)e^{ik_{z,1}z_0}\mathbf{e}_{p,1}^{-}. \tag{P.32}$$

Since we already have the field decomposed in $s-$ and $p-$polarizations, the reflected wave can be written in terms of Fresnel coefficients. The reflected field at the interface is simply given by

$$r_{2\leftarrow1}^{(s)}E_{s,n}^{0,-}(\omega,k)e^{ik_{z,1}z_0}\mathbf{e}_{s,1}^{-} + r_{2\leftarrow1}^{(p)}E_{p,n}^{0,-}(\omega,k)e^{ik_{z,1}z_0}\mathbf{e}_{p,1}^{-}. \tag{P.33}$$

Therefore the total electric field for $z_0 > z > 0$ is given by

$$
\begin{aligned}
\mathbf{E}(z,z_0,k,\omega) &= E_{s,n}^{0,-}(\omega,k)e^{ik_{z,1}(z_0-z)}\mathbf{e}_{s,1}^{-} + E_{p,n}^{0,-}(\omega,k)e^{ik_{z,1}(z_0-z)}\mathbf{e}_{p,1}^{-} \\
&+ r_{2\leftarrow1}^{(s)}E_{s,n}^{0,-}(\omega,k)e^{ik_{z,1}(z+z_0)}\mathbf{e}_{s,1}^{-} \\
&+ r_{2\leftarrow1}^{(p)}E_{p,n}^{0,-}(\omega,k)e^{ik_{z,1}(z+z_0)}\mathbf{e}_{p,1}^{-} \\
&= i\frac{\mu_1\mu_0\omega^2}{2k_{z,1}}\left\{ e^{-ik_{z,1}(z-z_0)}\mathbf{e}_{s,1}^{-}\otimes\mathbf{e}_{s,1}^{-} \right. \\
&\left. + e^{-ik_{z,1}(z-z_0)}\mathbf{e}_{p,1}^{-}\otimes\mathbf{e}_{p,1}^{-} \right\}\cdot\mathbf{d} \\
&+ i\frac{\mu_1\mu_0\omega_0^2}{2k_{z,1}}\left\{ r_{2\leftarrow1}^{(s)}e^{ik_{z,1}(z_0+z)}\mathbf{e}_{s,1}^{+}\otimes\mathbf{e}_{s,1}^{-} \right. \\
&\left. + r_{2\leftarrow1}^{(p)}e^{ik_{z,1}(z_0+z)}\mathbf{e}_{p,1}^{+}\otimes\mathbf{e}_{p,1}^{-} \right\}\cdot\mathbf{d},
\end{aligned}
\tag{P.34}
$$

where in the second equality we have used Eqs. (P.30) and (P.31). For $z > z_0$, the total electric field is given by

$$
\begin{aligned}
\mathbf{E}(z,z_0,k,\omega) &= i\frac{\mu_1\mu_0\omega_0^2}{2k_{z,1}}\left\{ e^{ik_{z,1}(z-z_0)}\mathbf{e}_{s,1}^{+}\otimes\mathbf{e}_{s,1}^{+} + e^{ik_{z,1}(z-z_0)}\mathbf{e}_{p,1}^{+}\otimes\mathbf{e}_{p,1}^{+} \right\}\cdot\mathbf{d} \\
&+ i\frac{\mu_1\mu_0\omega_0^2}{2k_{z,1}}\left\{ r_{2\leftarrow1}^{(s)}e^{ik_{z,1}(z_0+z)}\mathbf{e}_{s,1}^{+}\otimes\mathbf{e}_{s,1}^{-} \right. \\
&\left. + r_{2\leftarrow1}^{(p)}e^{ik_{z,1}(z_0+z)}\mathbf{e}_{p,1}^{+}\otimes\mathbf{e}_{p,1}^{-} \right\}\cdot\mathbf{d}.
\end{aligned}
\tag{P.35}
$$

For $z < 0$, we have a transmitted electric field that is given by

$$
\begin{aligned}
\mathbf{E}(z,z_0,k,\omega) &= t_{(12)}^{(s)}E_0^{-(s)}e^{ik_{z,1}z_0}e^{-ik_{z,2}z}\mathbf{e}_{s,2}^{-} + E_0^{-(p)}e^{ik_{z,1}z_0}e^{-ik_{z,2}z}\mathbf{e}_{p,2}^{-} \\
&= i\frac{\mu_1\mu_0\omega_0^2}{2k_{z,1}}\left\{ t_{2\leftarrow1}^{(s)}e^{ik_{z,1}z_0}e^{-ik_{z,2}z}\mathbf{e}_{s,2}^{-}\otimes\mathbf{e}_{s,1}^{-} \right. \\
&\left. + t_{2\leftarrow1}^{(p)}e^{ik_{z,1}z_0}e^{-ik_{z,2}z}\mathbf{e}_{p,2}^{-}\otimes\mathbf{e}_{p,1}^{-} \right\}\cdot\mathbf{d},
\end{aligned}
\tag{P.36}
$$

where in the second equality we have used Eqs. (P.30) and (P.31). Note that for the transmitted field both the polarization vectors of the media 1 and 2 appear. The previous equations must be read from right to left. Thus, in the case of Eq. (P.36) the dipole vector contracts, for example, with the polarization vector $\mathbf{e}_{p,1}^{-}$ to give the amplitude of the incident field at the interface, via Eq. (P.31). The final result comes proportional to the polarization vector $\mathbf{e}_{p,2}^{-}$ which gives the polarization of the transmitted field in medium 2. In summary, we can write

$$\mathbf{E}(z, z_0, k, \omega) = \mu_1 \mu_0 \omega^2 \mathbf{G}(z, z_0, k, \omega_0) \cdot \mathbf{d}, \qquad (P.37)$$

where the Green's function is written for $z > 0$ as

$$\mathbf{G}(z, z_0, k, \omega_0) = \mathbf{G}^0(z, z_0, k, \omega_0) + \mathbf{G}_r(z, z_0, k, \omega_0), \qquad (P.38)$$

with

$$\begin{aligned}
\mathbf{G}_r(z, z_0, k, \omega_0) = \frac{i}{2k_{z,1}} \Big\{ &r_{2\leftarrow1}^{(s)} e^{ik_{z,1}z_0} e^{ik_{z,1}z} \mathbf{e}_{s,1}^{+} \otimes \mathbf{e}_{s,1}^{-} \\
&+ r_{2\leftarrow1}^{(p)} e^{ik_{z,1}z_0} e^{ik_{z,1}z} \mathbf{e}_{p,1}^{+} \otimes \mathbf{e}_{p,1}^{-} \Big\}.
\end{aligned} \qquad (P.39)$$

and for $z < 0$

$$\begin{aligned}
\mathbf{G}(z, z_0, k, \omega_0) &= \mathbf{G}_t(z, z_0, k, \omega_0) \\
&= \frac{i}{2k_{z,1}} \Big\{ t_{2\leftarrow1}^{(s)} e^{ik_{z,1}z_0} e^{-ik_{z,2}z} \mathbf{e}_{s,2}^{-} \otimes \mathbf{e}_{s,1}^{-} \\
&\quad + t_{2\leftarrow1}^{(p)} e^{ik_{z,1}z_0} e^{-ik_{z,2}z} \mathbf{e}_{p,2}^{-} \otimes \mathbf{e}_{p,1}^{-} \Big\}.
\end{aligned} \qquad (P.40)$$

This concludes the relation between the matrix Green's function and the polarization vectors, with the former being written in terms of the latter.

We end this section by noting that in the Weyl gauge the electromagnetic field scalar potential is set to zero, $\phi = 0$, and $\mathbf{E} = -\partial \mathbf{A}/\partial t$. Therefore, we have that

$$\mathbf{E} = i\omega \mathbf{A}, \qquad (P.41)$$

and Eq. (P.17) becomes

$$\mathbf{A}(\omega, k) = \mu_n \mu_0 \mathbf{G}(\omega, k) \cdot \mathbf{J}_f(\omega, k), \qquad (P.42)$$

therefore $\mathbf{G}$ coincides with the Green's function for the vector potential $\mathbf{A}$ in the Weyl gauge.

P.2 Dipole Decaying Rate

To compute Eq. (P.1) we need the Green's function in real space at the position of dipole. For a dipole close to an interface, at position $\mathbf{r}_0 = (0, 0, z_0)$, we have

$$\mathbf{G}(\mathbf{r}_0, \mathbf{r}_0; \omega) = \mathbf{G}^0(\mathbf{r}_0, \mathbf{r}_0; \omega) + \int \frac{d^2\mathbf{k}}{(2\pi)^2} \mathbf{G}_s^r(z_0, z_0; \omega; \mathbf{k})$$

$$+ \int \frac{d^2\mathbf{k}}{(2\pi)^2} \mathbf{G}_p^r(z_0, z_0; \omega; \mathbf{k}), \quad \text{(P.43)}$$

where $\mathbf{G}^0(\mathbf{r}_0, \mathbf{r}_0; \omega)$ is the free Green's function in real space, $\mathbf{G}_s^r(z_0, z_0; \omega; \mathbf{k})$ is given by Eq. (P.84), and $\mathbf{G}_p^r(z_0, z_0; \omega; \mathbf{k})$ is given by Eq. (P.92). Once $\mathbf{G}(\mathbf{r}_0, \mathbf{r}_0; \omega)$ is determined we known how to compute Eq. (P.1). For having an explicit matrix form for the Green's functions we need to write the tensor product of the polarization vectors explicitly. This is done in Sec. P.3. Once is possession of these results, we progress by performing the angular integration implicit in $d^2\mathbf{k}$, obtaining

$$\int \frac{d^2\mathbf{k}}{(2\pi)^2} \mathbf{G}_s^r(z_0, z_0; \omega; \mathbf{k}) = \frac{r_s}{4\pi} \int dk_\rho k_\rho \frac{ie^{i2z_0 k_{z1}}}{2k_{z1}} \begin{bmatrix} 1 & 0 & 0 \\ 0 & 1 & 0 \\ 0 & 0 & 0 \end{bmatrix}, \quad \text{(P.44)}$$

and

$$\int \frac{d^2\mathbf{k}}{(2\pi)^2} \mathbf{G}_p^r(z_0, z_0; \omega; \mathbf{k}) = \frac{r_p}{4\pi} \int dk_\rho k_\rho \frac{ie^{i2k_{z1} z_0}}{2k_{z1}} \begin{bmatrix} f_\rho & 0 & 0 \\ 0 & f_\rho & 0 \\ 0 & 0 & 2f_{z1} \end{bmatrix}, \quad \text{(P.45)}$$

where

$$f_\rho = -1 + \frac{k_\rho^2}{(n_1\omega/c)^2} = 1 + \frac{k_\rho^2}{k_1^2}, \quad \text{(P.46)}$$

$$f_{z1} = 1 - \frac{k_{z1}^2}{(n_1\omega/c)^2} = 1 - \frac{k_{z1}^2}{k_1^2}, \quad \text{(P.47)}$$

where $n_1 = \sqrt{\epsilon_1}$ is the refractive index of medium 1. Using the these results for determining $\mathbf{G}(\mathbf{r}_0, \mathbf{r}_0; \omega)$ and replacing it in Eq. P.1 we obtain

$$\frac{\gamma}{\gamma_{01}} = 1 + \frac{3c}{2\sqrt{\mu_1\epsilon_1}\omega} \frac{\mu_x^2 + \mu_y^2}{\mu^2} \Im \int_0^\infty dk_\rho k_\rho \left[r_s + r_p \left(-1 + \frac{k_\rho^2}{(n_1\omega/c)^2} \right) \right] \frac{ie^{i2k_{z1} z_0}}{2k_{z1}}$$

$$+ \frac{3c}{2\sqrt{\mu_1\epsilon_1}\omega} \frac{2\mu_z^2}{\mu^2} \Im \int_0^\infty dk_\rho k_\rho r_p \left(1 - \frac{k_{z1}^2}{(n_1\omega/c)^2} \right) \frac{ie^{i2k_{z1} z_0}}{2k_{z1}}, \quad \text{(P.48)}$$

where $\gamma = 1/\tau$ and γ_{01} is the dipole decaying rate if the whole space was embedded with the dielectric medium 1, which reads

$$\gamma_{01} = \frac{2\omega^2\mu_1\mu_0}{\hbar} \frac{\sqrt{\mu_1\epsilon_1}\omega}{6\pi c} \mu^2 = \mu_1\mu_0 \frac{\omega^3\mu^2}{3\pi\hbar} \frac{\sqrt{\mu_1\epsilon_1}}{c}. \quad \text{(P.49)}$$

Now we define $k_1 = \sqrt{\mu_1 \epsilon_1}\omega/c$, and perform a change of coordinates $k_\rho = k_1 s$, and $k_{z1} = k_1\sqrt{1 - s^2 + i0^+}$, thus obtaining

$$
\begin{aligned}
\frac{\gamma}{\gamma_{01}} = 1 &+ \frac{3}{4}\frac{\mu_x^2 + \mu_y^2}{\mu^2}\Im m \int_0^\infty ds \frac{s}{\sqrt{1 - s^2}} r_p \left(-1 + s^2\right) i e^{i2k_1 z_0 \sqrt{1 - s^2}} \\
&+ \frac{3}{4}\frac{\mu_x^2 + \mu_y^2}{\mu^2}\Im m \int_0^\infty ds \frac{s}{\sqrt{1 - s^2}} r_s i e^{i2k_1 z_0 \sqrt{1 - s^2}} \\
&+ \frac{3}{2}\frac{\mu_z^2}{\mu^2}\Im m \int_0^\infty ds \frac{s^3}{\sqrt{1 - s^2}} r_p e^{i2k_{z1} z_0 \sqrt{1 - s^2}} ,
\end{aligned}
$$

(P.50)

a result that agrees with Eq. (10.38).

P.3 Explicit Form of the Tensor Product of the Polarization Vectors

For converting the Green's functions, written in terms of a tensor product of the polarization vectors, in a matrix form, we need to make the tensor product explicit. This is done below. We recall that the polarization vectors are given by Eqs. (P.9) and (P.10). Then we have

$$
\mathbf{e}_{s,n}^\pm \otimes \mathbf{e}_{s,n}^\pm = \frac{1}{k^2}\begin{bmatrix} k_y^2 & -k_x k_y & 0 \\ -k_x k_y & k_x^2 & 0 \\ 0 & 0 & 0 \end{bmatrix} ,
$$

(P.51)

$$
\mathbf{e}_{p,1}^+ \otimes \mathbf{e}_{p,1}^+ = \frac{1}{k_1^2}\begin{bmatrix} k_x^2 k_{z1}^2/k^2 & k_x k_y k_{z1}^2/k^2 & -k_x k_{z1} \\ k_x k_y k_{z1}^2/k^2 & k_y^2 k_{z1}^2/k^2 & -k_y k_{z1} \\ -k_x k_{z1} & -k_y k_{z1} & k^2 \end{bmatrix} ,
$$

(P.52)

$$
\mathbf{e}_{p,1}^- \otimes \mathbf{e}_{p,1}^- = \frac{1}{k_1^2}\begin{bmatrix} k_x^2 k_{z1}^2/k^2 & k_x k_y k_{z1}^2/k^2 & k_x k_{z1} \\ k_x k_y k_{z1}^2/k^2 & k_y^2 k_{z1}^2/k^2 & k_y k_{z1} \\ k_x k_{z1} & k_y k_{z1} & k^2 \end{bmatrix} ,
$$

(P.53)

$$
\mathbf{e}_{p,1}^+ \otimes \mathbf{e}_{p,1}^- = \frac{1}{k_1^2}\begin{bmatrix} -k_x^2 k_{z1}^2/k^2 & -k_x k_y k_{z1}^2/k^2 & -k_x k_{z1} \\ -k_x k_y k_{z1}^2/k^2 & -k_y^2 k_{z1}^2/k^2 & -k_y k_{z1} \\ k_x k_{z1} & k_y k_{z1} & k^2 \end{bmatrix} ,
$$

(P.54)

and

$$
\mathbf{e}_{p,2}^- \otimes \mathbf{e}_{p,1}^- = \frac{1}{k_1 k_2}\begin{bmatrix} k_x^2 k_{z1} k_{z2}/k^2 & k_x k_y k_{z1} k_{z2}/k^2 & k_x k_{z2} \\ k_x k_y k_{z1} k_{z2}/k^2 & k_y^2 k_{z1} k_{z2}/k^2 & k_y k_{z2} \\ k_x k_{z1} & k_y k_{z1} & k^2 \end{bmatrix} .
$$

(P.55)

The expressions given above enter the Green's functions of the flat interface problem. Other combinations of the tensor product of the polarization vectors can be computed in similar form.

P.4 The Free Green's Function in Real Space

Using Eq. (P.19), the Green's function in real space is obtained from (note that in this section $\mathbf{k}$ is a 3D vector)

$$G_{ij}(\mathbf{r}, t) = \int \frac{d\omega d^3 k}{(2\pi)^4} \left[\delta_{ij} - \frac{c^2}{n^2 \omega^2} k_i k_j \right] \frac{1}{k^2 - n^2 \omega^2 / c^2} e^{i\mathbf{k}\cdot\mathbf{r}} e^{-i\omega t} \quad (P.56)$$

$$= \int \frac{d\omega d^3 k}{(2\pi)^4} \left[\delta_{ij} + \frac{c^2}{n^2 \omega^2} \partial_i \partial_j \right] \frac{1}{k^2 - n^2 \omega^2 / c^2} e^{i\mathbf{k}\cdot\mathbf{r}} e^{-i\omega t} \quad (P.57)$$

where $n = \sqrt{\epsilon}$ is the refractive index of the medium where the dipole is embedded. Making a Fourier transform back to real space, we obtain

$$\int \frac{d^3 k}{(2\pi)^3} \frac{1}{k^2 - n^2 \omega^2 / c^2} e^{i\mathbf{k}\cdot\mathbf{r}}$$

$$= \frac{1}{4\pi^2} \int_0^\pi d\theta \sin\theta \int_0^\infty dk \frac{k^2 e^{ikr\cos\theta}}{(k - n\omega/c - i0^+)(k + n\omega/c + i0^+)}$$

$$= \frac{1}{2\pi r} \int_{-\infty}^\infty \frac{dk}{2\pi i} \frac{k e^{ikr}}{(k - n\omega/c - i0^+)(k + n\omega/c + i0^+)}$$

$$= \frac{e^{in\omega r/c}}{4\pi r} . \quad (P.58)$$

Where the prescription $\omega \to \omega + i0^+$ is imposed because we are interested in a retarded response function. Therefore, we obtain

$$G_{ij}(\mathbf{r}, \omega) = G_{ij}^{NF}(\mathbf{r}, \omega) + G_{ij}^{IF}(\mathbf{r}, \omega) + G_{ij}^{FF}(\mathbf{r}, \omega) \quad (P.59)$$

with

$$G_{ij}^{NF}(\mathbf{r}, \omega) = -\frac{e^{in\omega r/c}}{4\pi r^3} \frac{c^2}{n^2 \omega^2} \left[\delta_{ij} - 3\frac{x_i x_j}{r^2} \right], \quad (P.60)$$

$$G_{ij}^{IF}(\mathbf{r}, \omega) = i\frac{e^{in\omega r/c}}{4\pi r^2} \frac{c}{n\omega} \left[\delta_{ij} - 3\frac{x_i x_j}{r^2} \right], \quad (P.61)$$

$$G_{ij}^{FF}(\mathbf{r}, \omega) = \frac{e^{in\omega r/c}}{4\pi r} \left[\delta_{ij} - \frac{x_i x_j}{r^2} \right]. \quad (P.62)$$

Note that r should be interpreted as $|\mathbf{r} - \mathbf{r}_0|$. From these, the contribution to the total decaying rate coming from $\mathbf{G}^0(\mathbf{r}_0, \mathbf{r}_0; \omega)$, and leading to the free space decaying rate, can be computed. Some care must be exercised since there are cancelling divergences when $\mathbf{r} \to \mathbf{r}_0$. With this section, we have all we need to compute the Purcell factor.

P.5   Green's Function for Problems With Flat Interfaces II

For the sake of completeness, here we re-derive the same results obtained above using a more traditional approach [Novotny and Hecht (2014)], where everything is expressed in terms matrix Green's functions right from the start, with no reference to polarization vectors (see also [Johansson (2011)]).

P.5.1   *The free space Green's function revisited*

Combing Eqs. (P.23) and (P.24) it is possible to write the free space Green's function as a 3×3 matrix. It will also be useful to split the free Green's function in $s-$ and $p-$components. Let us choose $\mathbf{k} = (k_x, k_y) = (0, k_\rho)$, we then obtain the matrix Green's function (or dyadic Green's function)

$$\mathbf{G}(\omega, k_y = k_\rho, z) = \frac{i}{2k_z} \begin{bmatrix} 1 & 0 & 0 \\ 0 & 1 - \frac{k_\rho^2}{n^2\omega^2/c^2} & \mp\frac{k_\rho k_z}{n^2\omega^2/c^2} \\ 0 & \mp\frac{k_\rho k_z}{n^2\omega^2/c^2} & 1 - \frac{k_z^2}{n^2\omega^2/c^2} \end{bmatrix} e^{ik_z|z|}, \qquad (\text{P.63})$$

where the $-$ sign is for $z > 0$, while the $+$ sign is for $z < 0$, where $n = \sqrt{\epsilon}$ is the refractive index of the medium where the dipole is embedded. We now split the Green's function in $s-$ and $p-$components, such that

$$G_{ij}(\omega, k_y = k_\rho, z) = G_{ij}^s(\omega, k_y = k_\rho, z) + G_{ij}^p(\omega, k_y = k_\rho, z), \qquad (\text{P.64})$$

where (specializing in the $z > 0$ case)

$$\mathbf{G}^s(\omega, k_y = k_\rho, z) = \frac{ie^{ik_z|z|}}{2k_z} \begin{bmatrix} 1 & 0 & 0 \\ 0 & 0 & 0 \\ 0 & 0 & 0 \end{bmatrix} \qquad (\text{P.65})$$

$$\mathbf{G}^p(\omega, k_y = k_\rho, z) = \frac{ie^{ik_z|z|}}{2k_z} \begin{bmatrix} 0 & 0 & 0 \\ 0 & 1 - \frac{k_\rho^2}{n^2\omega^2} & -\frac{k_\rho k_z}{n^2\omega^2} \\ 0 & -\frac{k_\rho k_z}{n^2\omega^2} & 1 - \frac{k_z^2}{n^2\omega^2} \end{bmatrix}. \qquad (\text{P.66})$$

We can recover the the Green's function for general (k_x, k_y) by applying the rotation matrix

$$\mathbf{R} = \begin{bmatrix} \sin\theta & \cos\theta & 0 \\ -\cos\theta & \sin\theta & 0 \\ 0 & 0 & 1 \end{bmatrix}, \qquad (\text{P.67})$$

such that $\mathbf{R} \cdot (0, k_\rho) = k_\rho(\cos\theta, \sin\theta)$. We have (omitting the exponential pre-factors in the rotation of the matrices, but reintroducing them in the final expression):

$$\mathbf{G}^s(\omega, \mathbf{k}, z) = \mathbf{R} \cdot \begin{bmatrix} 1 & 0 & 0 \\ 0 & 0 & 0 \\ 0 & 0 & 0 \end{bmatrix} \cdot \mathbf{R}^t = \frac{ie^{ik_z|z|}}{2k_z k_\rho^2} \begin{bmatrix} k_y^2 & -k_x k_y & 0 \\ -k_x k_y & k_x^2 & 0 \\ 0 & 0 & 0 \end{bmatrix}, \qquad (\text{P.68})$$

and

$$\mathbf{G}^p(\omega, \mathbf{k}, z) = \mathbf{R} \cdot \begin{bmatrix} 0 & 0 & 0 \\ 0 & 1 - \frac{k_\rho^2}{n^2\omega^2/c^2} & -\frac{k_\rho k_z}{n^2\omega^2/c^2} \\ 0 & -\frac{k_\rho k_z}{n^2\omega^2/c^2} & 1 - \frac{k_z^2}{n^2\omega^2/c^2} \end{bmatrix} \cdot \mathbf{R}^t$$

$$= \frac{ie^{ik_z|z|}}{2k_z} \begin{bmatrix} \frac{k_x^2}{k_\rho^2}\left(1 - \frac{k_\rho^2}{(n\omega/c)^2}\right) & \frac{k_x k_y}{k_\rho^2}\left(1 - \frac{k_\rho^2}{(n\omega/c)^2}\right) & -\frac{k_x k_z}{(n\omega/c)^2} \\ \frac{k_x k_y}{k_\rho^2}\left(1 - \frac{k_\rho^2}{(n\omega/c)^2}\right) & \frac{k_y^2}{k_\rho^2}\left(1 - \frac{k_\rho^2}{(n\omega/c)^2}\right) & -\frac{k_y k_z}{(n\omega/c)^2} \\ -\frac{k_x k_z}{(n\omega/c)^2} & -\frac{k_y k_z}{(n\omega/c)^2} & 1 - \frac{k_z^2}{(n\omega/c)^2} \end{bmatrix} \quad \text{(P.69)}$$

P.5.2   Electric field from a dipole in terms of Green's functions

We want to determine the form of the electromagnetic field generated by a dipole (at position $z = z_0$) close to an interface (at position $z = 0$). Following Eq. (P.28), we write the electric field as [cf. Eqs. (O.51)- (O.53)]

$$\mathbf{E}^I(z_0 < z) = (i\omega\mu_1\mu_0)\left[\mathbf{G}^{i>}(z, z_0; \omega; \mathbf{k})\right.$$
$$\left. + \mathbf{G}^r(z, z_0; \omega; \mathbf{k})\right] \cdot \mathbf{j} \quad \text{(P.70)}$$
$$\mathbf{E}^{II}(0 < z < z_0) = (i\omega\mu_1\mu_0)\left[\mathbf{G}^{i<}(z, z_0; \omega; \mathbf{k})\right.$$
$$\left. + \mathbf{G}^r(z, z_0; \omega; \mathbf{k})\right] \cdot \mathbf{j} \quad \text{(P.71)}$$
$$\mathbf{E}^{III}(z < 0) = (i\omega\mu_1\mu_0)\left[\mathbf{G}^t(z, z_0; \omega; \mathbf{k})\right] \cdot \mathbf{j} \quad \text{(P.72)}$$

where $\mathbf{j} = -i\omega\mathbf{d}$ is the current density in frequency space generated by the dipole (recall Sec. P.1.2). The reason for the preceding expressions for the fields can be intuitively obtained from Fig. P.2. For example, $\mathbf{E}^I(z_0 < z)$

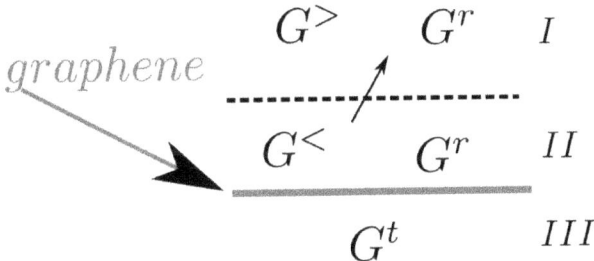

Figure P.2: Three regions of the scattering problem (I, II, and III) and corresponding Green's functions: G_t and G_r are the transmitted and reflected Green's functions, respectively, $G^<$ is the contribution that propagates downwards from the dipole (represented by an arrow), and $G^>$ is the contribution that propagates upwards, from the same source. The dipole is located at $z = z_0$.

refer to the field above the dipole. Then, it is composed from the contribution of the Green's function that propagates in the forward direction, $\mathbf{G}^{i>}(z, z_0; \omega; \mathbf{k})$, plus the Green's functions that has been reflected at the interface, $\mathbf{G}^r(z, z_0; \omega; \mathbf{k})$. The same reasoning applies to the other spatial regions.

We recall that the Green's function is the response to a point-like source. As such, it can be loosely understood as a field. This allows to write the Green's functions entering in the three previous equations as follows:

P.5.3 Green's functions: $s-component$

For the $s-$component, the three previous equations translate into:

$$\mathbf{E}_s^I(z_0 < z) = (i\omega\mu_1\mu_0)\left[\mathbf{G}_s^{i>}(z, z_0; \omega; \mathbf{k}) + \mathbf{G}_s^r(z, z_0; \omega; \mathbf{k})\right] \cdot \mathbf{j}, \quad \text{(P.73)}$$

$$\mathbf{E}_s^{II}(0 < z < z_0) = (i\omega\mu_1\mu_0)\left[\mathbf{G}_s^{i<}(z, z_0; \omega; \mathbf{k}) + \mathbf{G}_s^r(z, z_0; \omega; \mathbf{k})\right] \cdot \mathbf{j}, \quad \text{(P.74)}$$

$$\mathbf{E}_s^{III}(z < 0) = (i\omega\mu_1\mu_0)\left[\mathbf{G}_s^t(z, z_0; \omega; \mathbf{k})\right] \cdot \mathbf{j}. \quad \text{(P.75)}$$

The $s-$component of the free Green's function is given by [remember that $\mathbf{k} = (0, k_y) = (0, k_\rho)$]

$$\mathbf{G}_s^0(z; \omega; k_\rho) = \frac{ie^{ik_{z1}|z|}}{2k_{z1}}\begin{bmatrix} 1 & 0 & 0 \\ 0 & 0 & 0 \\ 0 & 0 & 0 \end{bmatrix}. \quad \text{(P.76)}$$

From Eq. (P.76) we can easily write the different Green's functions appearing in Eqs. (P.73), (P.74), and (P.75) as

$$\mathbf{G}_s^{i>}(z, z_0; \omega; k_\rho) = \frac{i}{2k_{z1}}\begin{bmatrix} 1 & 0 & 0 \\ 0 & 0 & 0 \\ 0 & 0 & 0 \end{bmatrix} e^{ik_{z1}(z-z_0)}, \quad \text{(P.77)}$$

$$\mathbf{G}_s^{i<}(z, z_0; \omega; k_\rho) = \frac{i}{2k_{z1}}\begin{bmatrix} 1 & 0 & 0 \\ 0 & 0 & 0 \\ 0 & 0 & 0 \end{bmatrix} e^{-ik_{z1}(z-z_0)}, \quad \text{(P.78)}$$

$$\mathbf{G}_s^r(z, z_0; \omega; k_\rho) = r_s\frac{i}{2k_{z1}}\begin{bmatrix} 1 & 0 & 0 \\ 0 & 0 & 0 \\ 0 & 0 & 0 \end{bmatrix} e^{ik_{z1}(z+z_0)}, \quad \text{(P.79)}$$

$$\mathbf{G}_s^t(z, z_0; \omega; k_\rho) = t_p\frac{i}{2k_{z1}}\begin{bmatrix} 1 & 0 & 0 \\ 0 & 0 & 0 \\ 0 & 0 & 0 \end{bmatrix} e^{-ik_{z2}z}e^{ik_{z1}z_0}. \quad \text{(P.80)}$$

The origin of the phase in the exponentials of the Green's functions can be understood as follows. Take, for example, the Green's function (P.77): since $z > z_0$, the propagation from z_0 to z picks up a phase $k_{z1}(z - z_0)$. By the same reasoning, the phase pick up by the Green's function (P.80) is $k_{z1}z_0 + k_{z2}|z| = k_{z1}z_0 - k_{z2}z$, since $z < 0$. The simplicity of the matrix structure is a consequence of the polarization of the field, which is transverse, being in the graphene plane. To obtain the Green's function for general $\mathbf{k} = (k_x, k_y)$ we just need to apply the rotation matrix

$$\mathbf{R} = \begin{bmatrix} \sin\theta & \cos\theta & 0 \\ -\cos\theta & \sin\theta & 0 \\ 0 & 0 & 1 \end{bmatrix}, \tag{P.81}$$

and we obtain:

$$\mathbf{G}_s^{i>}(z, z_0; \omega; \mathbf{k}) = \frac{i}{2k_{z1}k_\rho^2} \begin{bmatrix} k_y^2 & -k_x k_y & 0 \\ -k_x k_y & k_x^2 & 0 \\ 0 & 0 & 0 \end{bmatrix} e^{ik_{z1}(z-z_0)}, \tag{P.82}$$

$$\mathbf{G}_s^{i<}(z, z_0; \omega; \mathbf{k}) = \frac{i}{2k_{z1}k_\rho^2} \begin{bmatrix} k_y^2 & -k_x k_y & 0 \\ -k_x k_y & k_x^2 & 0 \\ 0 & 0 & 0 \end{bmatrix} e^{-ik_{z1}(z-z_0)}, \tag{P.83}$$

$$\mathbf{G}_s^r(z, z_0; \omega; \mathbf{k}) = r_s \frac{i}{2k_{z1}k_\rho^2} \begin{bmatrix} k_y^2 & -k_x k_y & 0 \\ -k_x k_y & k_x^2 & 0 \\ 0 & 0 & 0 \end{bmatrix} e^{ik_{z1}(z+z_0)}, \tag{P.84}$$

$$\mathbf{G}_s^t(z, z_0; \omega; \mathbf{k}) = t_s \frac{i}{2k_{z1}k_\rho^2} \begin{bmatrix} k_y^2 & -k_x k_y & 0 \\ -k_x k_y & k_x^2 & 0 \\ 0 & 0 & 0 \end{bmatrix} e^{-ik_{z2}z} e^{ik_{z1}z_0}, \tag{P.85}$$

where $k_\rho = \sqrt{k_x^2 + k_y^2}$. Notice how the boundary conditions at $z = z_0$ and $z = 0$ are satisfied by the Green's functions. At $z = z_0$, $\mathbf{G}^r(z, z_0; \omega; \mathbf{k})$ and its derivatives are continuous, while the Green's functions $\mathbf{G}^{i \lessgtr}(z, z_0; \omega; \mathbf{k})$ satisfies the boundary conditions imposed by the current at $z = z_0$. At $z = 0$, we have the Fresnel coefficients r_s and t_s that enforce the boundary conditions due to the change of medium.

P.5.4   *Green's functions: p−component*

As for the $p-$component, we can write the electric field in the whole space as

$$\mathbf{E}_p^I(z_0 < z) = (i\omega\mu_1\mu_0)\left[\mathbf{G}_p^{i>}(z, z_0; \omega; \mathbf{k}) + \mathbf{G}_p^r(z, z_0; \omega; \mathbf{k})\right]\cdot\mathbf{j} \quad \text{(P.86)}$$
$$\mathbf{E}_p^{II}(0 < z < z_0) = (i\omega\mu_1\mu_0)\left[\mathbf{G}_p^{i<}(z, z_0; \omega; \mathbf{k}) + \mathbf{G}_p^r(z, z_0; \omega; \mathbf{k})\right]\cdot\mathbf{j} \quad \text{(P.87)}$$
$$\mathbf{E}_p^{III}(z < 0) = (i\omega\mu_1\mu_0)\left[\mathbf{G}_p^t(z, z_0; \omega; \mathbf{k})\right]\cdot\mathbf{j}. \quad \text{(P.88)}$$

The $p-$component of the free space Green's function is given by $[\mathbf{k} = (0, k_\rho)]$

$$\mathbf{G}_p^0(z, z_0; \omega; k_\rho) = \frac{ie^{ik_z|z-z_0|}}{2k_z k_1^2}\begin{bmatrix}0 & 0 & 0 \\ 0 & k_{z1}^2 & -\operatorname{sgn}(z-z_0)k_\rho k_{z1} \\ 0 & -\operatorname{sgn}(z-z_0)k_\rho k_{z1} & k_\rho^2\end{bmatrix}.$$
$$\text{(P.89)}$$

Contrary to the $s-$polarization, it is not immediately obvious how to write from $\mathbf{G}_p^0$ the different Green's functions. This happens because the polarization vectors are not contained in the graphene plane. For $\mathbf{k} = (0, k_y) = (0, k_\rho)$ the results are (See Sec. P.5.5)

$$\mathbf{G}_p^{i>}(z, z_0; \omega; k_\rho) = \frac{i}{2k_{z1}k_1^2}\begin{bmatrix}0 & 0 & 0 \\ 0 & k_{z1}^2 & -k_\rho k_{z1} \\ 0 & -k_\rho k_{z1} & k_\rho^2\end{bmatrix}e^{ik_{z1}(z-z_0)}, \quad \text{(P.90)}$$

$$\mathbf{G}_p^{i<}(z, z_0; \omega; k_\rho) = \frac{i}{2k_{z1}k_1^2}\begin{bmatrix}0 & 0 & 0 \\ 0 & k_{z1}^2 & k_\rho k_{z1} \\ 0 & k_\rho k_{z1} & k_\rho^2\end{bmatrix}e^{-ik_{z1}(z-z_0)}, \quad \text{(P.91)}$$

$$\mathbf{G}_p^r(z, z_0; \omega; k_\rho) = r_p\frac{i}{2k_{z1}k_1^2}\begin{bmatrix}0 & 0 & 0 \\ 0 & -k_{z1}^2 & -k_\rho k_{z1} \\ 0 & k_\rho k_{z1} & k_\rho^2\end{bmatrix}e^{ik_{z1}(z+z_0)}, \quad \text{(P.92)}$$

$$\mathbf{G}_p^t(z, z_0; \omega; k_\rho) = t_p\frac{i}{2k_{z1}k_1 k_2}\begin{bmatrix}0 & 0 & 0 \\ 0 & k_{z1}k_{z2} & k_\rho k_{z2} \\ 0 & k_\rho k_{z1} & k_\rho^2\end{bmatrix}e^{-ik_{z2}z+ik_{z1}z_0}. \quad \text{(P.93)}$$

To obtain the Green's function for general $\mathbf{k} = (k_x, k_y)$ we just need to apply the rotation matrix Eq. (P.81). We obtain (note that upon rotation the length of k_ρ does not change)

$$\mathbf{G}_p^{i>}(z, z_0; \omega; \mathbf{k}) = \frac{i}{2k_{z1}k_1^2}\begin{bmatrix}k_x^2 k_{z1}^2/k_\rho^2 & k_x k_y k_{z1}^2/k_\rho^2 & -k_x k_{z1} \\ k_x k_y k_{z1}^2/k_\rho^2 & k_y^2 k_{z1}^2/k_\rho^2 & -k_y k_{z1} \\ -k_x k_{z1} & -k_y k_{z1} & k_\rho^2\end{bmatrix}e^{ik_{z1}(z-z_0)},$$
$$\text{(P.94)}$$

$$\mathbf{G}_p^{i<}(z, z_0; \omega; \mathbf{k}) = \frac{i}{2k_{z1}k_1^2} \begin{bmatrix} k_x^2 k_{z1}^2/k_\rho^2 & k_x k_y k_{z1}^2/k_\rho^2 & k_x k_{z1} \\ k_x k_y k_{z1}^2/k_\rho^2 & k_y^2 k_{z1}^2/k_\rho^2 & k_y k_{z1} \\ k_x k_{z1} & k_y k_{z1} & k_\rho^2 \end{bmatrix} e^{-ik_{z1}(z-z_0)},$$

$$(\text{P.95})$$

$$\mathbf{G}_p^{r}(z, z_0; \omega; \mathbf{k}) = r_p \frac{i}{2k_{z1}k_1^2} \begin{bmatrix} -k_x^2 k_{z1}^2/k_\rho^2 & -k_x k_y k_{z1}^2/k_\rho^2 & -k_x k_{z1} \\ -k_x k_y k_{z1}^2/k_\rho^2 & -k_y^2 k_{z1}^2/k_\rho^2 & -k_y k_{z1} \\ k_x k_{z1} & k_y k_{z1} & k_\rho^2 \end{bmatrix}$$

$$\times e^{ik_{z1}(z+z_0)},$$

$$(\text{P.96})$$

$$\mathbf{G}_p^{t}(z, z_0; \omega; \mathbf{k}) = t_p \frac{i}{2k_{z1}k_1 k_2} \begin{bmatrix} k_x^2 k_{z1} k_{z2}/k_\rho^2 & k_x k_y k_{z1} k_{z2}/k_\rho^2 & k_x k_{z2} \\ k_x k_y k_{z1} k_{z2}/k_\rho^2 & k_y^2 k_{z1} k_{z2}/k_\rho^2 & k_y k_{z2} \\ k_x k_{z1} & k_y k_{z1} & k_\rho^2 \end{bmatrix}$$

$$\times e^{-ik_{z2}z+ik_{z1}z_0}.$$

$$(\text{P.97})$$

P.5.5   *Derivation of the matrix structure of the Green's functions for the p-polarization component in the interface problem*

For convenience we rewrite here the p-component of the free Green's function:

$$\mathbf{G}_p^0(z; z_0; \omega; k_y = k_\rho)$$

$$= \frac{ie^{ik_{z1}|z-z_0|}}{2k_{z1}k_1^2} \begin{bmatrix} 0 & 0 & 0 \\ 0 & k_{z1}^2 & -\operatorname{sgn}(z-z_0)k_\rho k_{z1} \\ 0 & -\operatorname{sgn}(z-z_0)k_\rho k_{z1} & k_\rho^2 \end{bmatrix}. \quad (\text{P.98})$$

If follows that the Green's functions that emanate from z_0 are simply given by

$$\mathbf{G}_p^{i>}(z; z_0; \omega; k_y = k_\rho) = \frac{ie^{ik_{z1}(z-z_0)}}{2k_{z1}k_1^2} \begin{bmatrix} 0 & 0 & 0 \\ 0 & k_{z1}^2 & -k_\rho k_{z1} \\ 0 & -k_\rho k_{z1} & k_\rho^2 \end{bmatrix}, \quad (\text{P.99})$$

and

$$\mathbf{G}_p^{i<}(z; z_0; \omega; k_y = k_\rho) = \frac{ie^{ik_{z1}(z_0-z)}}{2k_{z1}k_1^2} \begin{bmatrix} 0 & 0 & 0 \\ 0 & k_{z1}^2 & k_\rho k_{z1} \\ 0 & k_\rho k_{z1} & k_\rho^2 \end{bmatrix}. \quad (\text{P.100})$$

It is the latter Green's function that gives rise to the reflected and transmitted Green's functions. To derive the reflected and transmitted Green's functions one route is to start by writing the electric field in the three regions (see Fig. N.2) as

$$\mathbf{E}_p^I(z_0 < z) = E_p^{i>} e^{ik_{z1}z} \begin{bmatrix} 0 \\ -k_{z1}/k_1 \\ k_\rho/k_1 \end{bmatrix}$$

$$+ r_p E_p^{i<} e^{ik_{z1}z} \begin{bmatrix} 0 \\ -k_{z1}/k_1 \\ k_\rho/k_1 \end{bmatrix} \qquad \text{(P.101)}$$

$$\mathbf{E}_p^{II}(0 < z < z_0) = E_p^{i<} e^{-ik_{z1}z} \begin{bmatrix} 0 \\ k_{z1}/k_1 \\ k_\rho/k_1 \end{bmatrix}$$

$$+ r_p E_p^{i<} e^{ik_{z1}z} \begin{bmatrix} 0 \\ -k_{z1}/k_1 \\ k_\rho/k_1 \end{bmatrix} \qquad \text{(P.102)}$$

$$\mathbf{E}_p^{III}(z < 0) = t_p E_s^{i<} e^{-ik_{z2}z} \begin{bmatrix} 0 \\ k_{z2}/k_2 \\ k_\rho/k_2 \end{bmatrix} \qquad \text{(P.103)}$$

where $E_p^{i<}$ is the amplitude of the primary field at $z = z_0$. We can connect the Green's function $\mathbf{G}_p^{i<}(z; z_0; \omega; k_y)$ with

$$E_p^{i<} e^{-ik_{z1}z} \begin{bmatrix} 0 \\ k_{z1}/k_1 \\ k_\rho/k_1 \end{bmatrix} \qquad \text{(P.104)}$$

using Eq. (P.28). Clearly we must have

$$E_p^{i<} \begin{bmatrix} 0 \\ k_{z1}/k_1 \\ k_\rho/k_1 \end{bmatrix} = \mu\mu_0\omega^2 e^{ik_{z1}z_0} \mathbf{G}_p^{i<}(z_0; z_0; \omega; k_y) \cdot \mathbf{d}. \qquad \text{(P.105)}$$

This allow us to write the reflected Green's function as follows

$$r_p E_p^{i<} e^{ik_{z1}z} \begin{bmatrix} 1 \\ -k_{z1}/k_1 \\ k_\rho/k_1 \end{bmatrix} = r_p E_p^{i<} e^{ik_{z1}z} \begin{bmatrix} 0 & 0 & 0 \\ 0 & -1 & 0 \\ 0 & 0 & 1 \end{bmatrix} \begin{bmatrix} 0 \\ k_{z1}/k_1 \\ k_\rho/k_1 \end{bmatrix} =$$

$$\mu\mu_0\omega^2 r_p e^{ik_{z1}(z+z_0)} \begin{bmatrix} 0 & 0 & 0 \\ 0 & -1 & 0 \\ 0 & 0 & 1 \end{bmatrix} \mathbf{G}_p^{i<}(z_0; z_0; \omega; k_y) \cdot \mathbf{d}, \qquad \text{(P.106)}$$

which implies that

$$\mathbf{G}_p^r(z; z_0; \omega; k_y) = r_p e^{ik_{z1}(z+z_0)} \begin{bmatrix} 0 & 0 & 0 \\ 0 & -1 & 0 \\ 0 & 0 & 1 \end{bmatrix} \mathbf{G}_p^{i<}(z_0; z_0; \omega; k_y), \qquad \text{(P.107)}$$

which upon matrix multiplication yields Eq. (P.92) The transmitted Green's function is built is a similar manner. First we write

$$
t_p E_s^{i<} e^{-ik_{z2}z} \begin{bmatrix} 0 \\ k_{z2}/k_2 \\ k_\rho/k_2 \end{bmatrix} = t_p E_s^{i<} e^{-ik_{z2}z} \frac{k_1}{k_2} \begin{bmatrix} 0 & 0 & 0 \\ 0 & k_{z2}/k_{z1} & 0 \\ 0 & 0 & 1 \end{bmatrix} \begin{bmatrix} 0 \\ k_{z1}/k_1 \\ k_\rho/k_1 \end{bmatrix}
$$

$$
= \mu\mu_0\omega^2 e^{ik_{z1}z_0} t_p e^{-ik_{z2}z} \frac{k_1}{k_2} \begin{bmatrix} 0 & 0 & 0 \\ 0 & k_{z2}/k_{z1} & 0 \\ 0 & 0 & 1 \end{bmatrix} \mathbf{G}_p^{i<}(z_0; z_0; \omega; k_y) \cdot \mathbf{d} . \quad (\text{P.108})
$$

From the latter result it follows

$$
\mathbf{G}_p^t(z; z_0; \omega; k_y) = e^{ik_{z1}z_0} t_p e^{-ik_{z2}z} \frac{k_1}{k_2} \begin{bmatrix} 0 & 0 & 0 \\ 0 & k_{z2}/k_{z1} & 0 \\ 0 & 0 & 1 \end{bmatrix} \mathbf{G}_p^{i<}(z_0; z_0; \omega; k_y) ,
$$

$$
(\text{P.109})
$$

which upon matrix multiplication yields Eq. (P.93). The same technique could have been used to construct the s−polarized Green's function, although in this case this is not strictly necessary.

Bibliography

Abedinpour, S. H., Vignale, G., Principi, A., Polini, M., Tse, W.-K., and MacDonald, A. H. (2011). Drude weight, plasmon dispersion, and ac conductivity
in doped graphene sheets, *Phys. Rev. B* **84**, p. 045429.

Abergel, D., Apalkov, V., Berashevich, J., Ziegler, K., and Chakrabort, T. (2010).
Properties of graphene: a theoretical perspective, *Advances in Physics* **59**,
pp. 261–482.

Ablowitz, M. J. and Fokas, A. S. (1997). *Complex Variables* (Cambridge University Press, New York).

Abramowitz, M. and Stegun, I. A. (1965). *Handbook of Mathematical Functions:
with Formulas, Graphs, and Mathematical Tables* (Dover, New York).

Aćimović, S. S., Ortega, M. A., Sanz, V., Berthelot, J., Garcia-Cordero, J. L.,
Renger, J., Maerkl, S. J., Kreuzer, M. P., and Quidant, R. (2014). Lspr
chip for parallel, rapid, and sensitive detection of cancer markers in serum,
Nano Lett. **14**, pp. 2636–2641.

Alonso-González, P., Nikitin, A. Y., Golmar, F., Centeno, A., Pesquera, A., Vélez,
S., Chen, J., Navickaite, G., Koppens, F., Zurutuza, A., Casanova, F.,
Hueso, L. E., and Hillenbrand, R. (2014). Controlling graphene plasmons
with resonant metal antennas and spatial conductivity patterns, *Science*
344.

Amorim, B., Schiefele, J., Sols, F., and Guinea, F. (2012). Coulomb drag
in graphene/boron nitride heterostructures: Effect of virtual phonon exchange, *Phys. Rev. B* **86**, p. 125448.

Anker, J. N., Hall, W. P., Lyandres, O., Shah, N. C., Zhao, J., and Duyn, R.
P. V. (2008). Biosensing with plasmonic nanosensors, *Nature Materials* **7**,
pp. 442 – 453.

Archambault, A., Teperik, T. V., Marquier, F., and Greffet, J. J. (2009). Surface
plasmon fourier optics, *Phys. Rev. B* **79**, p. 195414.

Arfken, G. B., Weber, H. J., and Harris, F. E. (2012). *Mathematical Methods for
Physicists*, 7th edn. (Academic Press).

Armstrong, S. (2012). Plasmonics: Diffraction-free surface waves, *Nature Photonics* **6**, p. 720.

Ashcroft, N. W. and Mermin, N. D. (1976). *Solid State Physics*, 1st edn. (Saunders

College, New York).

Atwater, H. A. (2007). The promise of plasmonics, *Scientific American* **4**, pp. 1–5.

Aversa, C. and Sipe, J. E. (1995). Nonlinear optical susceptibilities of semiconductors: Results with a length-gauge analysis, *Phys. Rev. B* **52**, pp. 14636–14645.

Balandin, A. A., Ghosh, S., Bao, W., Calizo, I., Teweldebrhan, D., Miao, F., and Lau, C. N. (2008). Superior thermal conductivity of single-layer graphene, *Nano Lett.* **8**, pp. 902–907.

Bao, Q., Zhang, H., Wang, B., Ni, Z., Lim, C. H. Y. X., Wang, Y., Tang, D. Y., and Loh, K. P. (2011). Broadband graphene polarizer, *Nature Photonics* **5**, pp. 411–415.

Barnes, W. L. (2006). Surface plasmon-polariton length scales: a route to subwavelength optics, *J. Opt. A: Pure Appl. Opt.* **8**, p. S87.

Barnes, W. L., Dereux, A., and Ebbesen, T. W. (2003). Surface plasmon subwavelength optics, *Nature* **424**, pp. 824–830.

Baskin, I., Ashkinadze, B. M., Cohen, E., and Pfeiffer, L. N. (2011). Imaging of magnetoplasmons excited in a two-dimensional electron gas, *Phys. Rev. B* **84**, p. 041305.

Bass, F. G. and Yakovenko, V. M. (1965). Theory of radiation from a charge passing through an electrically inhomogeneous medium, *Sov. Phys. Usp.* **8**, p. 420.

Beletskii, N. N. and Bludov, Y. V. (2005). Edge magnetoplasmons in a finite array of two-dimensional electron gas strips, *Telecommunications and Radio Engineering* **63**, pp. 405–417.

Berini, P. and Leon, I. D. (2012). Surface plasmon-polariton amplifiers and lasers, *Nature Photonics* **6**, pp. 16–24.

Biehs, S.-A. and Agarwal, G. S. (2013). Large enhancement of förster resonance energy transfer on graphene platforms, *ppl. Phys. Lett.* **103**, p. 243112.

Billings, L. (2013). Exotic optics: Metamaterial world, *Nature* **500**, pp. 138–140.

Blount, E. I. (1962). Formalisms of band theory, in F. Seitz and D. Turnbull (eds.), *Solid state physics*, Vol. 13 (Academic Press), pp. 839–839.

Bludov, Y. V., Ferreira, A., Peres, N. M. R., and Vasilevskiy, M. I. (2013). A primer on surface plasmon-polaritons in graphene, *Int. J. Mod. Phys. B* **27**, p. 1341001.

Bludov, Y. V., Peres, N. M. R., and Vasilevskiy, M. I. (2012a). Graphene-based polaritonic crystal, *Phys. Rev. B* **85**, p. 245409.

Bludov, Y. V., Smirnova, D. A., Kivshar, Y. S., Peres, N. M. R., and Vasilevskiy, M. I. (2014). Nonlinear te-polarized surface polaritons on graphene, *Phys. Rev. B* **89**, p. 035406.

Bludov, Y. V., Vasilevskiy, M. I., and Peres, N. M. R. (2010). Mechanism for graphene-based optoelectronic switches by tuning surface plasmon-polaritons in monolayer graphene, *EPL* **92**, p. 68001.

Bludov, Y. V., Vasilevskiy, M. I., and Peres, N. M. R. (2012b). Tunable graphene-based polarizer, *Journal of Appl. Phys.* **112**, 8, p. 084320.

Bolotin, K., Sikes, K., Jiang, Z., Klima, M., Fudenberg, G., Hone, J., Kim, P., and Stormer, H. (2008). Ultrahigh electron mobility in suspended graphene, *Solid State Communications* **146**, pp. 51–355.

Boriskina, S. V., Ghasemi, H., and Chen, G. (2013). Plasmonic materials for energy: From physics to applications, *Materials Today* **16**, pp. 375–386.

Bracewell, R. N. (1999). *The Fourier Transform & Its Applications*, 3rd edn. (McGraw-Hill, New York).

Brar, V. W., Jang, M. S., Sherrott, M., Kim, S., Lopez, J. J., Kim, L. B., Choi, M., and Atwater, H. (2014). Hybrid surface-phonon-plasmon polariton modes in graphene/monolayer h-bn heterostructures, *Nano Lett.* **14**, pp. 3876–3880.

Brar, V. W., Jang, M. S., Sherrott, M., Lopez, J. J., and Atwater, H. A. (2013). Highly confined tunable mid-infrared plasmonics in graphene nanoresonators, *Nano Lett.* **13**, pp. 2541–2547.

Brolo, A. G. (2012). Plasmonics for future biosensors, *Nature Photonics* **6**, pp. 709–713.

Cai, X., Sushkov, A. B., Jadidi, M. M., Nyakiti, L. O., Myers-Ward, R. L., Gaskill, D. K., Murphy, T. E., Fuhrer, M. S., and Drew, H. D. (2015). Plasmon-enhanced terahertz photodetection in graphene, *Nano Lett.* **15**, pp. 4295–4302.

Cai, Y.-J., Li, M., Ren, X.-F., Zou, C.-L., Xiong, X., Lei, H.-L., Liu, B.-H., Guo, G.-P., and Guo, G.-C. (2014). High-visibility on-chip quantum interference of single surface plasmons, *Phys. Rev. Appl.* **2**, p. 014004.

Caldwell, J. D., Kretinin, A. V., Chen, Y., Giannini, V., Fogler, M. M., Francescato, Y., Ellis, C. T., Tischler, J. G., Woods, C. R., Giles, A. J., Hong, M., Watanabe, K., Taniguchi, T., Maier, S. A., and Novoselov, K. S. (2014). Sub-diffractional volume-confined polaritons in the natural hyperbolic material hexagonal boron nitride, *Nature Communications* **5**, pp. 1–9.

Carbotte, J. P., LeBlanc, J. P. F., and Nicol, E. J. (2012). Emergence of plasmaronic structure in the near-field optical response of graphene, *Phys. Rev. B* **85**, p. 201411(R).

Castro Neto, A. H., Guinea, F., Peres, N. M. R., Novoselov, K. S., and Geim, A. K. (2009). The electronic properties of graphene, *Rev. Mod. Phys.* **81**.

Castro Neto, A. H. and Novoselov, K. (2011). New directions in science and technology: two-dimensional crystals, *Reports on Progress in Physics* **74**, p. 082501.

Chaikin, P. M. and Lubensky, T. C. (2000). *Principles of Condensed Matter Physics* (Cambridge University Press, Cambridge).

Chang, D. E., Sørensen, A. S., Hemmer, P. R., and Lukin, M. D. (2006). Quantum optics with surface plasmons, *Phys. Rev. Lett.* **97**, p. 053002.

Chaves, F. A., Jiménez, D., Cummings, A. W., and Roche, S. (2014). Physical model of the contact resistivity of metal-graphene junctions, *Journal of Appl. Phys.* **115**, p. 164513.

Chen, J., Badioli, M., Alonso-González, P., Thongrattanasiri, S., Huth, F., Osmond, J., Spasenović, M., Centeno, A., Pesquera, A., Godignon, P., Elorza, A. Z., Camara, N., de Abajo, F. J. G., Hillenbrand, R., and Koppens, F. H. L. (2012). Optical nano-imaging of gate-tunable graphene plasmons, *Nature* **487**, pp. 77–81.

Chen, P.-Y. and Alù, A. (2011). Atomically thin surface cloak using graphene monolayers, *ACS Nano* **5**, pp. 5855 – 5863.

Chen, Z., Berciaud, S., Nuckolls, C., Heinz, T. F., and Brus, L. E. (2010). Energy transfer from individual semiconductor nanocrystals to graphene, *ACS Nano* **4**, pp. 2964–2968.

Cheng, H., Chen, S., Yu, P., Duan, X., Xie, B., and Tian, J. (2013). Dynamically tunable plasmonically induced transparency in periodically patterned graphene nanostrips, *Appl. Phys. Lett.* **103**, p. 203112.

Cheng, J., Wang, W. L., Mosallaei, H., and Kaxiras, E. (2014). Surface plasmon engineering in graphene functionalized with organic molecules: A multiscale theoretical investigation, *Nano Lett.* **14**, pp. 50 – 56.

Chiu, K. W. and Quinn, J. J. (1974). Plasma oscillations of a two-dimensional electron gas in a strong magnetic field, *Phys. Rev. B* **9**, pp. 4724–4732.

Christensen, J., Manjavacas, A., Thongrattanasiri, S., Koppens, F. H. L., and de Abajo, F. J. G. (2012). Graphene plasmon waveguiding and hybridization in individual and paired nanoribbons, *ACS Nano* **6**, 1, pp. 431–440.

Christensen, T., Jauho, A.-P., Wubs, M., and Mortensen, N. A. (2015a). Localized plasmons in graphene-coated nanospheres, *Phys. Rev. B* **91**, p. 125414.

Christensen, T., Wang, W., Jauho, A.-P., Wubs, M., and Mortensen, N. A. (2014). Classical and quantum plasmonics in graphene nanodisks: the role of edge states, *Phys. Rev. B* **90**, p. 241414(R).

Christensen, T., Yan, W., Jauho, A.-P., Wubs, M., and Mortensen, N. A. (2015b). Kerr nonlinearity and plasmonic bistability in graphene nanoribbons, *Phys. Rev. B* **92**, 121407(R).

Constant, T. J., Hornett, S. M., Chang, D. E., and Hendry, E. (2015). All-optical generation of surface plasmons in graphene, *Nat. Phys.* **11**, p. 1.

Cox, J. D. and de Abajo, F. J. G. (2014). Electrically tunable nonlinear plasmonics in graphene nanoislands, *Nature Communications* **5**, 5725.

Cox, J. D. and de Abajo, F. J. G. (2015). Plasmon-enhanced nonlinear wave mixing in nanostructured graphene, *ACS Photonics* **2**, pp. 306–312.

Cox, J. D., Singh, M. R., Gumbs, G., Anton, M. A., and Carreno, F. (2012). Dipole-dipole interaction between a quantum dot and a graphene nanodisk, *Phys. Rev. B* **86**, p. 125452.

Crassee, I., Orlita, M., Potemski, M., Walter, A. L., Ostler, M., Seyller, T., Gaponenko, I., Chen, J., and Kuzmenko, A. B. (2012). Intrinsic terahertz plasmons and magnetoplasmons in large scale monolayer graphene, *Nano Lett.* **12**, pp. 2470–2474.

Crozier, K. B., Zhu, W., Wang, D., Lin, S., Best, M. D., and Camden, J. P. (2014). Plasmonics for surface enhanced raman scattering: Nanoantennas for single molecules, *IEEE Journal of Selected Topics in Quantum Electronics* **20**.

Dai, S., Fei, Z., Ma, Q., Rodin, A. S., Wagner, M., McLeod, A. S., Liu, M. K., Gannett, W., Regan, W., Watanabe, K., Taniguchi, T., Thiemens, M., Dominguez, G., Castro Neto, A. H., Zettl, A., Keilmann, F., Jarillo-Herrero, P., Fogler, M. M., and Basov, D. N. (2014). Tunable phonon polaritons in atomically thin van der waals crystals of boron nitride, *Science* **343**, pp. 1125–1129.

Dai, S., Ma, Q., Zhu, S.-E., Liu, M. K., Andersen, T., Fei, Z., Goldflam, M., Wagner, M., Watanabe, K., Taniguchi, T., Thiemens, M., Keilmann, F., Janssen, G. C. A. M., Jarillo-Herrero, P., Fogler, M. M., and Basov, D. N. (2015). Graphene on hexagonal boron nitride as an agile hyperbolic metamaterial, *Nature Nanotechnology* **10**, pp. 682–686.

Danos, M. (1955). Čerenkov radiation from extended electron beams, *J. Appl. Phys.* **26**, p. 2.

Das, A. K. (1974). The linear response of a one-dimensional electron gas, *Solid State Communications* **15**, pp. 475–477.

Das, A. K. (1975). The relaxation-time approximation in the rpa dielectric formulation, *J. Phys. F: Met. Phys.* **5**, p. 2035.

de Abajo, F. J. G. (2014). Graphene plasmonics: Challenges and opportunities, *ACS Photonics* **1**, 135-152.

Di Martino, G., Sonnefraud, Y., Tame, M. S., Kéna-Cohen, S., Dieleman, F., Ozdemir, S. K., Kim, M. S., and Maier, S. A. (2014). Observation of quantum interference in the plasmonic hong-ou-mandel effect, *Phys. Rev. Appl.* **1**, p. 034004.

Dionne, J. A. and Atwater, H. A. (2012). Plasmonics: Metal-worthy methods and materials in nanophotonics, *MRS Bulletin* **37**, pp. 717–724.

Dionne, J. A., Sweatlock, L. A., Atwater, H. A., and Polman, A. (2005). Planar metal plasmon waveguides: frequency-dependent dispersion, propagation, localization, and loss beyond the free electron model, *Phys. Rev. B* **72**, p. 075405.

Dressel, M. and Grüner, G. (2002). *Electrodynamics of Solids: Optical Properties of Electrons in Matter*, 1st edn. (Cambridge University Press, Cambridge).

Drude, P. (1909). Zur elektronentheorie der metalle, *Annalen der Physik* **306**, pp. 566–613.

Duffy, D. G. (2015). *Green's Functions with Applications*, 2nd edn. (CRC Press, Boca Raton).

Ebbesen, T. W., Genet, C., and Bozhevolnyi, S. I. (2008). Surface plasmon circuitry, *Nature* **61**, pp. 44–50.

Ebbesen, T. W., Lezec, H. J., Ghaemi, H. F., Thio, T., and Wolff, P. A. (1998). Extraordinary optical transmission through sub-wavelength hole arrays, *Nature* **391**, pp. 667–669.

Editorial (2012). Surface plasmon resurrection, *Nature Photonics* **6**, p. 707.

Ehrenreich, H. and Cohen, M. H. (1959). Self-consistent field approach to the many-electron problem, *Phys. Rev.* **115**, p. 786.

Enoch, S. and Bonod, N. (2012). *Plasmonics: From Basics to Advanced Topics* (Springer Series in Optical Sciences, New York).

Fakonas, J. S., Lee, H., Kelaita, Y. A., and Atwater, H. A. (2014). Two-plasmon quantum interference, *Nature Photonics* **8**, pp. 317–320.

Falkovsky, L. A. (2008). Optical properties of graphene, *Journal of Physics: Conference Series* **129**, p. 012004.

Falkovsky, L. A. and Varlamov, A. A. (2007). Space-time dispersion of graphene conductivity, *Eur. Phys. J. B* **56**, pp. 281–284.

Fallahi, A. and Perruisseau-Carrier, J. (2012). Design of tunable biperiodic graphene metasurfaces, *Phys. Rev. B* **86**, p. 195408.

Fang, Z., Liu, Z., Wang, Y., Ajayan, P. M., Nordlander, P., and Halas, N. J. (2012a). Graphene-antenna sandwich photodetector, *Nano Lett.* **12**, pp. 3808–3813.

Fang, Z., Thongrattanasiri, S., Schlather, A., Liu, Z., Ma, L., Wang, Y., Ajayan, P. M., Nordlander, P., Halas, N. J., and de Abajo, F. J. G. (2013). Gated tunability and hybridization of localized plasmons in nanostructured graphene, *ACS Nano* **7**, pp. 2388–2395.

Fang, Z., Wang, Y., Liu, Z., Schlather, A., Ajayan, P. M., Koppens, F. H. L., Nordlander, P., and Halas, N. J. (2012b). Plasmon-induced doping of graphene, *ACS Nano* **6**, pp. 10222–10228.

Fang, Z., Wang, Y., Schlather, A. E., Liu, Z., Ajayan, P. M., de Abajo, F. J. G., Nordlander, P., Zhu, X., and Halas, N. J. (2014). Active tunable absorption enhancement with graphene nanodisk arrays, *Nano Lett.* **14**, pp. 299–304.

Fano, U. (1941). The theory of anomalous diffraction gratings and of quasi-stationary waves on metallic surfaces (sommerfeld's waves), *J. Opt. Soc. Am.* **31**, pp. 213–222.

Farhat, M., Guenneau, S., and Bağcı, H. (2011). Exciting graphene surface plasmon polaritons through light and sound interplay, *Phys. Rev. Lett.* **111**, p. 237404.

Farias, G. A. and Maradudin, A. A. (1983). Surface plasmons on a randomly rough surface, *Phys. Rev. B* **28**, pp. 5675–5687.

Fedorych, O. M., Moreau, S., Potemski, M., Studenikin, S. A., Saku, T., and Hirayama, Y. (2009). Microwave magnetoplasmon absorption by a 2deg stripe, *Int. J. Mod. Phys. B* **23**, pp. 2698–2702.

Fei, Z., Andreev, G. O., Bao, W., Zhang, L. M., McLeod, A. S., Wang, C., Stewart, M. K., Zhao, Z., Dominguez, G., Thiemens, M., Fogler, M. M., Tauber, M. J., Castro Neto, A. H., Lau, C. N., Keilmann, F., and Basov, D. N. (2011). Infrared nanoscopy of dirac plasmons at the graphene–sio2 interface, *Nano Lett.* **11**, pp. 4701–4705.

Fei, Z., Iwinski, E. G., Ni, G. X., Zhang, L. M., Bao, W., Rodin, A. S., Lee, Y., Wagner, M., Liu, M. K., Dai, S., Goldflam, M. D., Thiemens, M., Keilmann, F., Lau, C. N., Castro Neto, A. H., Fogler, M. M., and Basov, D. N. (2015). Tunneling plasmonics in bilayer graphene, *Nano Lett.* **15**, pp. 4973–4978.

Fei, Z., Rodin, A. S., Andreev, G. O., Bao, W., McLeod, A. S., Wagner, M., Zhang, L. M., Zhao, Z., Thiemens, M., Dominguez, G., Fogler, M. M., Castro Neto, A. H., Lau, C. N., Keilmann, F., and Basov, D. N. (2012). Gate-tuning of graphene plasmons revealed by infrared nano-imaging, *Nature* **487**, pp. 82–85.

Fei, Z., Rodin, A. S., Gannett, W., Dai, S., Regan, W., Wagner, M., Liu, M. K., McLeod, A. S., Dominguez, G., Thiemens, M., Castro Neto, A. H., Keilmann, F., Zettl, A., Hillenbrand, R., Fogler, M. M., and Basov, D. N. (2013). Electronic and plasmonic phenomena at graphene grain boundaries, *Nature Nanotechnology* **8**, pp. 821 –825.

Ferreira, A. and Peres, N. M. R. (2012). Complete light absorption in graphene-metamaterial corrugated structures, *Phys. Rev. B* **86**, p. 205401.

Ferreira, A., Peres, N. M. R., and Castro Neto , A. H. (2012). Confined magneto-optical waves in graphene, *Phys. Rev. B* **85**, p. 205426.

Ferreira, A., Viana-Gomes, J., Bludov, Y. V., Pereira, V. M., Peres, N. M. R., and Castro Neto, A. H. (2011). Faraday effect in graphene enclosed in an optical cavity and the equation of motion method for the study of magneto-optical transport in solids, *Phys. Rev. B* **84**, p. 235410.

Fetter, A. L. (1986). Magnetoplasmons in a two-dimensional electron fluid: Disk geometry, *Phys. Rev. B* **33**, pp. 5221–5227.

Fetter, A. L. and Walecka, J. D. (2003). *Quantum Theory of Many-Particle Systems* (Dover, New York).

Fischetti, M. V., Neumayer, D. A., and Cartier, E. A. (2001). Effective electron mobility in si inversion layers in metal–oxide–semiconductor systems with a high-κ insulator: The role of remote phonon scattering, *Journal of Appl. Phys.* **90**, 9, pp. 4587–4608.

Flensberg, K., Hu, B. Y.-K., Jauho, A.-P., and Kinaret, J. M. (1995). Linear-response theory of coulomb drag in coupled electron systems, *Phys. Rev. B* **52**, pp. 14761–14774.

Fox, M. (2010). *Optical Properties of Solids* (Oxford University Press, New York).

Fratini, S. and Guinea, F. (2008). Substrate-limited electron dynamics in graphene, *Phys. Rev. B* **77**, p. 195415.

Freitag, M., Low, T., Zhu, W., Yan, H., Xia, F., and Avouris, P. (2013). Photocurrent in graphene harnessed by tunable intrinsic plasmons, *Nature Communications* **4**, 1951.

Fuchs, R. and Kliewer, K. L. (1965). Optical modes of vibration in an ionic crystal slab, *Phys. Rev.* **140**, pp. A2076–A2088.

Gao, W., Shi, G., Jin, Z., Shu, J., Zhang, Q., Vajtai, R., Ajayan, P. M., Kono, J., and Xu, Q. (2013). Excitation and active control of propagating surface plasmon polaritons in graphene, *Nano Lett.* **13**, pp. 3698–3702.

Garcia, M. A. (2011). Surface plasmons in metallic nanoparticles: fundamentals and applications, *J. Phys. D: Appl. Phys.* **44**, p. 283001.

Garcia de Abajo, F. J. and Manjavacas, A. (2015). Plasmonics in atomically thin materials, *Faraday Discuss.* **178**, pp. 87–107.

Gaudreau, L., Tielrooij, K. J., Prawiroatmodjo, G. E. D. K., Osmond, J., de Abajo, F. J. G., and Koppens, F. H. L. (2013). Universal distance-scaling of nonradiative energy transfer to graphene, *Nano Lett.* **13**, pp. 2030–2035.

Geim, A. K. (2009). Graphene: Status and prospects, *Science* **324**, pp. 1530–1534.

Geim, A. K. (2012). Graphene prehistory, *Physica Scripta* **2012**, p. 014003.

Geim, A. K. and Grigorieva, I. V. (2013). Van der waals heterostructures, *Nature* **499**, pp. 419 – 425.

Gilbertson, A. M., Francescato, Y., Roschuk, T., Shautsova, V., Chen, Y., Sidiropoulos, T. P. H., Hong, M., Giannini, V., Maier, S. A., Cohen, L. F., and Oulton, R. F. (2015). Plasmon-induced optical anisotropy in hybrid graphene–metal nanoparticle systems, *Nano Lett.* **15**, pp. 3458–3464.

Gómez-Santos, G. and Stauber, T. (2011). Fluorescence quenching in graphene: a fundamental ruler and evidence for transverse plasmons, *Phys. Rev. B*

84, p. 165438.

Goerbig, M. O. (2011). Electronic properties of graphene in a strong magnetic field, *Rev. Mod. Phys.* **83**, pp. 1193–1243.

Goldflam, M. D., Ni, G.-X., Post, K. W., Fei, Z., Yeo, Y., Tan, J. Y., Rodin, A. S., Chapler, B. C., Ozyilmaz, B., Castro Neto, A. H., Fogler, M. M., and Basov, D. N. (2015). Tuning and persistent switching of graphene plasmons on a ferroelectric substrate, *Nano Lett.* **15**, pp. 4859–4864.

Gong, S., Hu, M., Zhong, R., Chen, X., Zhang, P., Zhao, T., and Liu, S. (2014). Electron beam excitation of surface plasmon polaritons, *Opt. Express* **22**, pp. 19252–19261.

Gorbach, A. V. (2013). Nonlinear graphene plasmonics: Amplitude equation for surface plasmons, *Phys. Rev. A* **87**, p. 013830.

Gorbachev, R. V., Geim, A. K., Katsnelson, M. I., Novoselov, K. S., Tudorovskiy, T., Grigorieva, I. V., MacDonald, A. H., Watanabe, K., Taniguchi, T., and Ponomarenko, L. A. (2012). Strong coulomb drag and broken symmetry in double-layer graphene, *Nature Phys.* **8**, pp. 896–901.

Green, M. A. and Pillai, S. (2012). Harnessing plasmonics for solar cells, *Nature Photonics* **6**, pp. 130–132.

Grigorenko, A. N., Polini, M., and Novoselov, K. S. (2012). Graphene plasmonics, *Nature Photonics* **6**, pp. 749–758.

Gu, B., Kwong, N. H., and Binder, R. (2013). Relation between the interband dipole and momentum matrix elements in semiconductors, *Phys. Rev. B* **87**, p. 125301.

Gullans, M., Chang, D. E., Koppens, F. H. L., de Abajo, F. J. G., and Lukin, M. D. (2013). Single-photon nonlinear optics with graphene plasmons, *Phys. Rev. Lett.* **111**, p. 247401.

Gupta, B. D., Srivastava, S. K., and Verma, R. (2015). *Fiber Optic Sensors Based on Plasmonics* (World Scientific, Singapore).

Gusynin, V. P., Sharapov, S. G., and Carbotte, J. P. (2009). On the universal ac optical background in graphene, *New J. Phys.* **11**, p. 095013.

Haes, A. J., Haynes, C. L., McFarland, A. D., Schatz, G. C., Duyne, R. P. V., and Zou, S. (2005). Plasmonic materials for surface-enhanced sensing and spectroscopy, *MRS Bulletin* **30**, pp. 368–375.

Hall, E. H. (1879). On a new action of the magnet on electric currents, *American Journal of Mathematics* **2**, pp. 287–292.

Hanson, G. W., Hassani Gangaraj, S. A., Lee, C., Angelakis, D. G., and Tame, M. (2015). Quantum plasmonic excitation in graphene and loss-insensitive propagation, *Phys. Rev. A* **92**, p. 013828.

Heeg, S., Fernandez-Garcia, R., Oikonomou, A., Schedin, F., Narula, R., Maier, S. A., Vijayaraghavan, A., and Reich, S. (2013). Polarized plasmonic enhancement by au nanostructures probed through raman scattering of suspended graphene, *Nano Lett.* **13**, pp. 301–308.

Heeres, R. W., Kouwenhoven, L. P., and Zwiller, V. (2013). Quantum interference in plasmonic circuits, *Nature Nanotechnology* **8**, pp. 719–722.

Hessel, A. and Oliner, A. A. (1965). A new theory of wood's anomalies on optical gratings, *Appl. Optics* **4**, pp. 1275–1297.

Hoggard, A., Wang, L.-Y., Ma, L., Fang, Y., You, G., Olson, J., Liu, Z., Chang, W.-S., Ajayan, P. M., and Link, S. (2013). Using the plasmon linewidth to calculate the time and efficiency of electron transfer between gold nanorods and graphene, *ACS Nano* **7**, pp. 11209–11217.

Hopfield, J. J. (1958). Theory of the contribution of excitons to the complex dielectric constant of crystals, *Phys. Rev.* **112**, p. 1555.

Huidobro, P. A., Nesterov, M. L., Martín-Moreno, L., and García-Vidal, F. J. (2010). Transformation optics for plasmonics, *Nano Lett.* **10**, pp. 1985–1990.

Huidobro, P. A., Nikitin, A. Y., González-Ballestero, C., Martín-Moreno, L., and García-Vidal, F. J. (2012). Superradiance mediated by graphene surface plasmons, *Phys. Rev. B* **85**, p. 155438.

Hwang, E. H. and Sarma, S. D. (2007). Dielectric function, screening, and plasmons in two-dimensional graphene, *Phys. Rev. B* **75**, p. 205418.

Hwang, E. H., Sensarma, R., and Sarma, S. D. (2010). Plasmon-phonon coupling in graphene, *Phys. Rev. B* **82**, p. 195406.

Ishikawa, A. and Tanaka, T. (2013). Plasmon hybridization in graphene metamaterials, *Appl. Phys. Lett.* **102**, p. 253110.

Jablan, M., Buljan, H., and Soljačić, M. (2009). Plasmonics in graphene at infrared frequencies, *Phys. Rev. B* **80**, p. 245435.

Jablan, M., Soljačić, M., and Buljan, H. (2013). Plasmons in graphene: Fundamental properties and potential applications, *Proc. IEEE* **101**, pp. 1689 – 1704.

Jackson, J. D. (1998). *Classical Electrodynamics*, 3rd edn. (Wiley, New York).

Jacob, Z. (2012). Quantum plasmonics, *MRS Bulletin* **37**, pp. 761–767.

Jacob, Z. (2014). Hyperbolic phonon-polaritons, *Nature Materials* **13**, pp. 1081–1083.

Jacob, Z. and Shalaev, V. M. (2011). Plasmonics goes quantum, *Science* **28**, pp. 463–464.

Jauho, A. (1998). Quantum transport theory, in E. Schöll (ed.), *Theory of Transport Properties of Semiconductor Nanostructures, Electronic Materials Series*, Vol. 4 (Springer), pp. 127–171.

Jishi, R. A. (2014). *Feynman Diagram Techniques in Condensed Matter Physics* (Cambridge University Press, Cambridge).

Johansson, P. (2011). Electromagnetic green's function for layered systems: Applications to nanohole interactions in thin metal films, *Phys. Rev. B* **83**, p. 195408.

Johnson, P. B. and Christy, R. W. (1972). Optical constants of the noble metals, *Phys. Rev. B* **6**, pp. 4370–4379.

Ju, L., Geng, B., Horng, J., Girit, C., Martin, M., Hao, Z., Bechtel, H. A., Liang, X., Zettl, A., Shen, Y. R., and Wang, F. (2011). Graphene plasmonics for tunable terahertz metamaterials, *Nature Nanotechnology* **6**, pp. 630–634.

Juan, M. L., Righini, M., and Quidant, R. (2011). Plasmon nano-optical tweezers, *Nature Photonics* **5**, pp. 349–356.

Kabashin, A. V., Evans, P., Pastkovsky, S., Hendren, W., Wurtz, G. A., Atkinson, R., Pollard, R., Podolskiy, V. A., and Zayats, A. V. (2009). Plasmonic nanorod metamaterials for biosensing, *Nature Materials* **8**, pp. 867 – 871.

Kaipa, C. S. R., Yakovlev, A. B., Hanson, G. W., Padooru, Y. R., Medina, F., and Mesa, F. (2012). Enhanced transmission with a graphene-dielectric microstructure at low-terahertz frequencies, *Phys. Rev. B* **85**, p. 245407.

Kamat, P. V. (2010). Graphene-based nanoarchitectures. anchoring semiconductor and metal nanoparticles on a two-dimensional carbon support, *J. Phys. Chem. Lett.* **1**, pp. 520 – 527.

Kaminer, I., Katan, Y. T., Buljan, H., Shen, Y., Ilic, O., López, J. J., Wong, L. J., Joannopoulos, J. D., and Soljačić, M. (2015). Quantum Čerenkov effect from hot carriers in graphene: An efficient plasmonic source, *arXiv:1510.00883 [physics.optics]* , 1.

Kang, N. (2015). *Green's Function for Poisson Equation in Layered Nano-Structures including Graphene*, Master's thesis, University of Waterloo.

Kasry, A., Ardakani, A. A., Tulevski, G. S., Menges, B., Copel, M., and Vyklicky, L. (2012). Highly efficient fluorescence quenching with graphene, *J. Phys. Chem. C* **116**, pp. 2858–2862.

Katsnelson, M. I. (2011). Coulomb drag in graphene single layers separated by a thin spacer, *Phys. Rev. B* **84**, p. 041407(R).

Kauranen, M. and Zayats, A. V. (2012). Nonlinear plasmonics, *Nature Photonics* **6**, pp. 737–748.

Khavasi, A. (2015). Design of ultra-broadband graphene absorber using circuit theory, *J. Opt. Soc. Am. B* **32**, pp. 1941–1946.

Khavasi, A. and Rejaei, B. (2014). Analytical modeling of graphene ribbons as optical circuit elements, *IEEE J. Quantum Electron.* **50**, pp. 397–403.

Khurgin, J. B. (2015). How to deal with the loss in plasmonics and metamaterials, *Nature Nanotechnology* **10**, pp. 2–6.

Khurgin, J. B. and Boltasseva, A. (2012). Reflecting upon the losses in plasmonics and metamaterials, *MRS Bulletin* **37**, pp. 768–779.

Kim, J. T., Yu, Y.-J., Choi, H., and Choi, C.-G. (2014). Graphene-based plasmonic photodetector for photonic integrated circuits, *Opt. Express* **22**, pp. 803–808.

Kim, S., Jo, I., Nah, J., Yao, Z., Banerjee, S. K., and Tutuc, E. (2011). Coulomb drag of massless fermions in graphene, *Phys. Rev. B* **83**, p. 161401.

Kitamura, R., Pilon, L., and Jonasz, M. (2007). Optical constants of silica glass from extreme ultraviolet to far infrared at near room temperature, *Appl. Opt.* **46**, pp. 8118–8133.

Klimov, V. (2014). *Nanoplasmonics* (Pan Stanford).

Kneipp, K. (2007). Surface-enhanced raman scattering, *Phys. Today* **60**, pp. 40–46.

Koch, R. J., Seyller, T., and Schaefer, J. A. (2010). Strong phonon-plasmon coupled modes in the graphene/silicon carbide heterosystem, *Phys. Rev. B* **82**, p. 201413.

Konstantatos, G., Badioli, M., Gaudreau, L., Osmond, J., Bernechea, M., de Arquer, F. P. G., Gatti, F., and Koppens, F. H. L. (2012). Hybrid graphene-quantum dot phototransistors with ultrahigh gain, *Nature Nanotechnology* **7**, pp. 363 – 368.

Koppens, F. H. L., Chang, D. E., and de Abajo, F. J. G. (2011). Graphene plasmonics: A platform for strong light-matter interactions, *Nano Lett.* **11**, pp. 3370 – 3377.

Koppens, F. H. L., Mueller, T., Avouris, P., Ferrari, A. C., Vitiello, M. S., and Polini, M. (2014). Photodetectors based on graphene, other two-dimensional materials and hybrid systems, *Nature Nanotechnology* **9**, pp. 780–793.

Kort-Kamp, W. J. M., Amorim, B., Bastos, G., Pinheiro, F. A., Rosa, F. S. S., Peres, N. M. R., and Farina, C. (2015). Active magneto-optical control of spontaneous emission in graphene, *Phys. Rev. B* **92**, p. 205415.

Kotov, V. N., Uchoa, B., Pereira, V. M., Guinea, F., and Castro Neto, A. H. (2012). Electron-electron interactions in graphene: Current status and perspectives, *Rev. Mod. Phys.* **84**, pp. 1067–1125.

Kragler, R. and Thomas, H. (1980). Dielectric function in the relaxation-time approximation generalized to electronic multiple-band systems, *Z. Physik B - Condensed Matter* **39**, pp. 99–107.

Kravets, V. G., Jalil, R., Kim, Y.-J., Ansell, D., Aznakayeva, D. E., Thackray, B., Britnell, L., Belle, B. D., Withers, F., Radko, I. P., Han, Z., Bozhevolnyi, S. I., Novoselov, K. S., Geim, A. K., and Grigorenko, A. N. (2014). Graphene-protected copper and silver plasmonics, *Scientific Reports* **4**, p. 5517.

Kravets, V. G., Schedin, F., Jalil, R., Britnell, L., Novoselov, K. S., and Grigorenko, A. N. (2012). Surface hydrogenation and optics of a graphene sheet transferred onto a plasmonic nanoarray, *The Journal of Physical Chemistry C* **116**, pp. 3882–3887.

Kretinin, A. V., Cao, Y., Tu, J. S., Yu, G. L., Jalil, R., Novoselov, K. S., Haigh, S. J., Gholinia, A., Mishchenko, A., Lozada, M., Georgiou, T., Woods, C. R., Withers, F., Blake, P., Eda, G., Wirsig, A., Hucho, C., Watanabe, K., Taniguchi, T., Geim, A. K., and Gorbachev, R. V. (2014). Electronic properties of graphene encapsulated with different two-dimensional atomic crystals, *Nano Lett.* **14**, pp. 3270 – 3276.

Kretschmann, E. and Raether, H. (1968). Radiative decay of nonradiative surface plasmon excited by light, *Z. Naturf. A* **23**, pp. 2135–2136.

Kumar, A., Low, T., Fung, K. H., Avouris, P., and Fang, N. X. (2015). Tunable light-matter interaction and the role of hyperbolicity in graphene-hbn system, *Nano Lett.* **15**, pp. 3172–3180.

Lee, C., Kim, J. Y., Bae, S., Kim, K. S., Hong, B. H., and Choi, E. J. (2011). Optical response of large scale single layer graphene, *Appl. Phys. Lett.* **98**, p. 071905.

Lee, C., Wei, X., Kysar, J. W., and Hone, J. (2008). Measurement of the elastic properties and intrinsic strength of monolayer graphene, *Science* **321**, 5887, pp. 385–388.

Lee, S. H., Choi, M., Kim, T.-T., Lee, S., Liu, M., Yin, X., Choi, H. K., Lee, S. S., Choi, C.-G., Choi, S.-Y., Zhang, X., and Min, B. (2012). Switching terahertz waves with gate-controlled active graphene metamaterials, *Nature Materials* **11**, pp. 936 – 941.

Lekner, J. (1991). Reflection and refraction by uniaxial crystals, *J. Phys.: Condens. Matter* **3**, pp. 6121–6133.

Li, E.-P. and Chu, H.-S. (2014). *Plasmonic Nanoelectronics and Sensing* (Cambridge).

Li, Z. Q., Henriksen, E. A., Jiang, Z., Hao, Z., Martin, M. C., Kim, P., Stormer, H. L., and Basov, D. N. (2008). Dirac charge dynamics in graphene by infrared spectroscopy, *Nature Phys.* **4**, pp. 532–535.

Liang, Z., Sun, J., Jiang, Y., Jiang, L., and Chen, X. (2014). Plasmonic enhanced optoelectronic devices, *Plasmonics* **9**, pp. 859–866.

Lim, D.-K., Barhoumi, A., Wylie, R. G., Reznor, G., Langer, R. S., and Kohane, D. S. (2013). Enhanced photothermal effect of plasmonic nanoparticles coated with reduced graphene oxide, *Nano Lett.* **13**, pp. 4075–4079.

Lin, I.-T. and Liu, J.-M. (2014). Optimization of double-layer graphene plasmonic waveguides, *Appl. Phys. Lett.* **105**, p. 061116.

Lindhard, J. (1954). On the properties of a gas of charged particles, *Mat.-Fys. Medd. K. Dan. Vidensk. Selsk.* **28**, pp. 1 – 57.

Ling, X., Xie, L., Fang, Y., Xu, H., Zhang, H., Kong, J., Dresselhaus, M. S., Zhang, J., and Liu, Z. (2010). Can graphene be used as a substrate for raman enhancement? *Nano Lett.* **10**, pp. 553–561.

Liu, C.-Y., Liang, K.-C., Chen, W., hao Tu, C., Liu, C.-P., and Tzeng, Y. (2011a). Plasmonic coupling of silver nanoparticles covered by hydrogen-terminated graphene for surface-enhanced raman spectroscopy, *Opt. Express* **19**.

Liu, F., Qian, C., and Chong, Y. D. (2015a). Directional excitation of graphene surface plasmons, *Opt. Express* **23**, pp. 2383–2391.

Liu, H., Sun, S., Wu, L., and Bai, P. (2014a). Optical near-field enhancement with graphene bowtie antennas, *Plasmonics* **9**, pp. 845–850.

Liu, P. Q., Valmorra, F., Maissen, C., and Faist, J. (2015b). Electrically tunable graphene anti-dot array terahertz plasmonic crystals exhibiting multi-band resonances, *Optica* **2**, pp. 135 – 140.

Liu, S., Hu, M., Zhang, Y., Li, Y., and Zhong, R. (2009). Electromagnetic diffraction radiation of a subwavelength-hole array excited by an electron beam, *Phys. Rev. E* **80**, p. 036602.

Liu, S., Zhang, C., Hu, M., Chen, X., Zhang, P., Gong, S., Zhao, T., and Zhong, R. (2014b). Coherent and tunable terahertz radiation from graphene surface plasmon polaritons excited by an electron beam, *Applied Physics Letters* **104**, 201104.

Liu, Y., Cheng, R., Liao, L., Zhou, H., Bai, J., Liu, G., Liu, L., Huang, Y., and Duan, X. (2011b). Plasmon resonance enhanced multicolour photodetection by graphene, *Nature Communications* **2**, 579.

Liu, Y. and Willis, R. F. (2010). Plasmon-phonon strongly coupled mode in epitaxial graphene, *Phys. Rev. B* **81**, p. 081406.

Liu, Y., Willis, R. F., Emtsev, K. V., and Seyller, T. (2008). Plasmon dispersion and damping in electrically isolated two-dimensional charge sheets, *Phys. Rev. B* **78**.

Liua, Y. and Zhang, X. (2011). Metamaterials: a new frontier of science and technology, *Chem. Soc. Rev.* **40**, pp. 2494–2507.

Llatser, I., Kremers, C., Cabellos-Aparicio, A., Jornet, J. M., Alarcón, E., and Chigrin, D. N. (2012). Graphene-based nano-patch antenna for terahertz radiation, *Photonics and Nanostructures - Fundamentals and Applications* **10**, pp. 353–358.

Low, T. and Avouris, P. (2014). Graphene plasmonics for terahertz to mid-infrared applications, *ACS Nano* **8**, pp. 1086–1101.

Low, T., Guinea, F., Yan, H., Xia, F., and Avouris, P. (2014a). Novel mid-infrared plasmonic properties of bilayer graphene, *Phys. Rev. Lett.* **112**, p. 116801.

Low, T., Roldán, R., Wang, H., Xia, F., Avouris, P., Moreno, L. M., and Guinea, F. (2014b). Plasmons and screening in monolayer and multilayer black phosphorus, *Phys. Rev. Lett.* **113**, p. 106802.

Lu, J.-K. (1993). *Boundary Value Problems for Analytic Functions* (World Scientific, Singapore).

Luo, X., Qiu, T., Lu, W., and Ni, Z. (2013). Plasmons in graphene: Recent progress and applications, *Materials Science and Engineering R* **74**, pp. 351 – 376.

Luxmoore, I. J., Gan, C. H., Liu, P. Q., Valmorra, F., Li, P., Faist, J., and Nash, G. R. (2014). Strong coupling in the far-infrared between graphene plasmons and the surface optical phonons of silicon dioxide, *ACS Photonics* **1**, p. 1151.

Ma, R.-M., Oulton, R. F., Sorger, V. J., and Zhang, X. (2013). Plasmon lasers: coherent light source at molecular scales, *Laser Photon. Rev.* **7**, pp. 1–21.

MacDonald, K. F., Sámson, Z. L., Stockman, M. I., and Zheludev, N. I. (2009). Ultrafast active plasmonics, *Nature Photonics* **3**, pp. 55–58.

Mahan, G. D. (2000). *Many-Particle Physics*, 3rd edn. (Springer, New York).

Maier, S. A. (2007). *Plasmonics: Fundamentals and Applications* (Springer, New York).

Manjavacas, A., Marchesin, F., Thongrattanasiri, S., Koval, P., Nordlander, P., Sánchez-Portal, D., and de Abajo, F. J. G. (2013). Tunable molecular plasmons in polycyclic aromatic hydrocarbons, *ACS Nano* **7**, pp. 3635–3643.

Manjavacas, A., Nordlander, P., and de Abajo, F. J. G. (2012a). Plasmon blockade in nanostructured graphene, *ACS Nano* **6**, pp. 1724 – 1731.

Manjavacas, A., Thongrattanasiri, S., Chang, D. E., and de Abajo, F. J. G. (2012b). Temporal quantum control with graphene, *New J. Phys.* **14**, p. 123020.

Manjavacas, A., Thongrattanasiri, S., and de Abajo, F. (2015). Plasmons driven by single electrons in graphene nanoislands, *Nanophotonics* **2**, pp. 139–151.

Manjavacas, A., Thongrattanasiri, S., Greffet, J.-J., and de Abajo, F. J. G. (2014). Graphene optical-to-thermal converter, *Appl. Phys. Lett.* **105**, p. 211102.

Manzoni, M. T., Silveiro, I., de Abajo, F. J. G., and Chang, D. E. (201). Second-order quantum nonlinear optical processes in graphene nanostructures, *New J. Phys.* **17**, p. 083031.

Maradudin, A. A. (1986). Electromagnetic surface excitations on rough surfaces, in R. F. Wallis, G. I. Stegeman, and T. Tamir (eds.), *Electromagnetic Surface Excitations, Springer Series on Wave Phenomena*, Vol. 3 (Springer Berlin Heidelberg), pp. 57–131.

Maradudin, A. A. (2007a). The design of randomly rough surfaces that scatter waves in a specified manner, in A. A. Maradudin (ed.), *Light Scattering and Nanoscale Surface Roughness*, Nanostructure Science and Technology (Springer), pp. 467–488.

Maradudin, A. A. (ed.) (2007b). *Light Scattering and Nanoscale Surface Roughness* (Springer).

Maradudin, A. A. (2014). Surface electromagnetic waves on structured perfectly conducting surfaces, in A. A. Maradudin, W. L. Barnes, and J. R. Sambles (eds.), *Modern Plasmonics* (Elsevier), pp. 223–251.

Maradudin, A. A., Barnes, W. L., and Sambles, J. R. (eds.) (2014). *Modern Plasmonics* (Elsevier).

Markos, P. and Soukoulis, C. M. (2008). *Wave Propagation: From Electrons to Photonic Crystals and Left-Handed Materials* (Princeton University Press, New Jersey).

Martín-Moreno, L., García-Vidal, F. J., Lezec, H. J., Pellerin, K. M., Thio, T., Pendry, J. B., and Ebbesen, T. W. (2001). Theory of extraordinary optical transmission through subwavelength hole arrays, *Phys. Rev. Lett.* **86**, pp. 1114–1117.

Maxwell, J. C. (1865). A dynamical theory of the electromagnetic field, *Philosophical Transactions of the Royal Society of London* **155**, pp. 459–512.

Menabde, S., Mason, D., Kornev, E., Lee, C., and Park, N. (2016). Direct optical probing of transverse electric mode in graphene, *Scientific Reports* **6**, p. 21523.

Mermin, N. D. (1970). Lindhard dielectric function in the relaxation-time approximation, *Phys. Rev. B* **1**, pp. 2362–2363.

Mihaila, B. (2011). Lindhard function of a d-dimensional fermi gas, *arXiv:1111.5337* , pp. 1–8.

Mikhailov, S. A. and Savostianova, N. A. (2005). Microwave response of a two-dimensional electron stripe, *Phys. Rev. B* **71**, p. 035320.

Mikhailov, S. A. and Ziegler, K. (2007). New electromagnetic mode in graphene, *Phys. Rev. Lett.* **99**, p. 016803.

Mishchenko, A., Tu, J. S., Cao, Y., Gorbachev, R. V., Wallbank, J. R., Greenaway, M., Morozov, V. E., Morozov, S. V., Zhu, M. J., Wong, S. L., Withers, F., Woods, C. R., Kim, Y.-J., Watanabe, K., Taniguchi, T., Vdovin, E. E., Makarovsky, O., Fromhold, T. M., Falko, V. I., Geim, A. K., Eaves, L., and Novoselov, K. S. (2014). Twist-controlled resonant tunnelling in graphene-boron nitride-graphene heterostructures, *Nature Nanotechnology* **9**, pp. 808–813.

Mitrofanov, O., Yu, W., Thompson, R. J., Jiang, Y., Brener, I., Pan, W., Berger, C., de Heer, W. A., and Jiang, Z. (2013). Probing terahertz surface plasmon waves in graphene structures, *Appl. Phys. Lett.* **103**, 11, p. 111105.

Mittleman, D. M. (2013). Frontiers in terahertz sources and plasmonics, *Nature Photonics* **7**, pp. 666–669.

Mönch, W. (2010). *Semiconductor Surfaces and Interfaces*, 3rd edn. (Springer).

Mooradian, A. and Wright, G. B. (1966). Observation of the interaction of plasmons with longitudinal optical phonons in gaas, *Phys. Rev. Lett.* **16**, pp.

999–1001.

Moriya, R., Yamaguchi, T., Inoue, Y., Sata, Y., Morikawa, S., Masubuchi, S., and Machida, T. (2015). Influence of the density of states of graphene on the transport properties of graphene/mos2/metal vertical field-effect transistors, *Appl. Phys. Lett.* **106**, p. 223103.

Nakajima, S. (1955). Perturbation theory in statistical mechanics, *Advances in Physics* **4**, pp. 363–380.

Nene, P., Strait, J. H., Chan, W.-M., Manolatou, C., Kevek, J. W., Tiwari, S., McEuen, P. L., and Rana, F. (2014). Coupling of plasmon modes in graphene microstructures, *Appl. Phys. Lett.* **105**, p. 143108.

Nikitin, A. Y., Guinea, F., Garcia-Vidal, F. J., and Martin-Moreno, L. (2012a). Surface plasmon enhanced absorption and suppressed transmission in periodic arrays of graphene ribbons, *Phys. Rev. B* **85**, p. 081405(R).

Nikitin, A. Y., Guinea, F., and Martin-Moreno, L. (2012b). Resonant plasmonic effects in periodic graphene antidot arrays, *Applied Physics Letters* **101**, p. 151119.

Nikitin, A. Y., Low, T., and Martin-Moreno, L. (2014). Anomalous reflection phase of graphene plasmons and its influence on resonators, *Phys. Rev. B* **90**, p. 041407.

Niu, J., Shin, Y. J., Son, J., Lee, Y., Ahn, J.-H., and Yang, H. (2012). Shifting of surface plasmon resonance due to electromagnetic coupling between graphene and au nanoparticles, *Opt. Express* **20**, pp. 19690 – 19696.

Novoselov, K. S. and Castro Neto, A. H. (2012). Two-dimensional crystals-based heterostructures: materials with tailored properties, *Physica Scripta* **2012**, p. 014006.

Novoselov, K. S., Geim, A. K., Morozov, S. V., Jiang, D., Zhang, Y., Dubonos, S. V., Grigorieva, I. V., and Firsov, A. A. (2004). Electric field effect in atomically thin carbon films, *Science* **306**, pp. 666–669.

Novotny, L. and Hecht, B. (2014). *Principles of Nano-Optics*, 2nd edn. (Cambridge University Press, Cambridge).

Ochiai, T. (2015). Spatially periodic modulation of optical conductivity in doped graphene by two-dimensional diffraction grating, *J. Opt. Soc. Am. B* **32**, pp. 701–707.

Otto, A. (1968). Excitation of nonradiative surface plasma waves in silver by the method of frustrated total reflection, *Z. Phys.* **216**, pp. 398–410.

Ozbay, E. (2006). Plasmonics: Merging photonics and electronics at nanoscale dimensions, *Science* **311**, pp. 189–193.

Padooru, Y. R., Yakovlev, A. B., Kaipa, C. S. R., Hanson, G. W., Medina, F., and Mesa, F. (2013). Dual capacitive-inductive nature of periodic graphene patches: Transmission characteristics at low-terahertz frequencies, *Phys. Rev. B* **87**, p. 115401.

Palik, E. D. (1997). *Handbook of Optical Constants of Solids*, 1st edn. (Academic Press).

Pelton, M., Aizpurua, J., and Bryant, G. (2008). Metal-nanoparticle plasmonics, *Laser & Photon. Rev.* **2**, pp. 136–159.

Pelton, M. and Bryant, G. W. (2013). *Introduction to Metal-Nanoparticle Plas-*

monics, 1st edn. (John Wiley & Sons, Inc, New Jersey).

Peres, N. M. R. (2010). Colloquium: The transport properties of graphene: An introduction, *Rev. Mod. Phys.* **82**.

Peres, N. M. R., Bludov, Y. V., Ferreira, A., and Vasilevskiy, M. I. (2013). Exact solution for square-wave grating covered with graphene: surface plasmon-polaritons in the terahertz range, *J. Phys.: Condens. Matter* **25**, p. 125303.

Peres, N. M. R., Bludov, Y. V., Santos, J. E., Jauho, A.-P., and Vasilevskiy, M. I. (2014). Optical bistability of graphene in the terahertz range, *Phys. Rev. B* **90**, p. 125425.

Peres, N. M. R., dos Santos, J. M. B. L., and Stauber, T. (2007). Phenomenological study of the electronic transport coefficients of graphene, *Phys. Rev. B* **76**, p. 073412.

Peres, N. M. R., Ferreira, A., Bludov, Y. V., and Vasilevskiy, M. I. (2012). Light scattering by a medium with a spatially modulated optical conductivity: the case of graphene, *J. Phys.: Condens. Matter* **24**, p. 245303.

Pettit, R. B., Silcox, J., and Vincent, R. (1975). Measurement of surface-plasmon dispersion in oxidized aluminum films, *Phys. Rev. B* **11**, pp. 3116–3123.

Polanco, J., Fitzgerald, R. M., and Maradudin, A. A. (2013). Scattering of surface plasmon polaritons by one-dimensional surface defects, *Phys. Rev. B* **87**, p. 155417.

Politano, A. and Chiarello, G. (2014). Plasmon modes in graphene: status and prospect, *Nanoscale* **6**, pp. 10927–10940.

Poole Jr. , C. P. (2004). *Encyclopedic Dictionary of Condensed Matter Physics, Vol. 1*, 1st edn. (Academic Press, New York).

Powell, C. J. and Swan, J. B. (1959). Origin of the characteristic electron energy losses in aluminum, *Phys. Rev.* **115**, p. 869.

Pratesi, R., Toraldo di Francia, G., Ronchi, L., and Scheggi, A. M. (1962). Radiation from a charged particle in uniform straight motion through a particular stratified medium, *Il Nuovo Cimento* **XXV**, p. 756.

Principi, A., Vignale, G., Carrega, M., and Polini, M. (2013). Intrinsic lifetime of dirac plasmons in graphene, *Phys. Rev. B* **88**, p. 195405.

Prodan, E., Radloff, C., Halas, N. J., and Nordlander, P. (2003). A hybridization model for the plasmon response of complex nanostructures, *Science* **302**, pp. 419–422.

Purcell, E. M. (1995). Spontaneous emission probabilities at radio frequencies, in E. Burstein and C. Weisbuch (eds.), *Confined Electrons and Photons, NATO ASI Series*, Vol. 340 (Springer), pp. 839–839.

Quidant, R. (2012). Plasmonic tweezers—the strength of surface plasmons, *MRS Bulletin* **37**, pp. 739–744.

Raether, H. (1988). *Surface Plasmons on Smooth and Rough Surfaces and on Gratings*, 1st edn. (Springer Tracts in Modern Physics, New York).

Rahman, T. S. and Maradudin, A. A. (1980). Surface-plasmon dispersion relation in the presence of surface roughness, *Phys. Rev. B* **21**, pp. 2137–2143.

Rayleigh, L. (1907). On the dynamical theory of gratings, *Proc. R. Soc. Lond. A* **79**, pp. 399–416.

Reece, P. J. (2008). Plasmonics: Finer optical tweezers, *Nature Photonics* **2**, pp.

333–334.

Rejaei, B. and Khavasi, A. (2015). Scattering of surface plasmons on graphene by a discontinuity in surface conductivity, *J. of Opt.* **17**, p. 075002.

Ritchie, R. H. (1957). Plasma losses by fast electrons in thin films, *Phys. Rev.* **106**, p. 874.

Ritchie, R. H., Arakawa, E. T., Cowan, J. J., and Hamm, R. N. (1968). Surface-plasmon resonance effect in grating diffraction, *Phys. Rev. Lett.* **21**, pp. 1530–1533.

Rodrigo, D., Limaj, O., DavideJanner, Etezadi, D., de Abajo, F. J. G., Pruneri, V., and Altug, H. (2015). Mid-infrared plasmonic biosensing with graphene, *Science* **349**, 6244, pp. 165–168.

Rothwell, E. J. and Cloud, M. J. (2009). *Electromagnetics*, 2nd edn. (CRC Press).

Rouhi, N., Capdevila, S., Jain, D., Zand, K., Wang, Y., Brown, E., Jofre, L., and Burke, P. (2012). Terahertz graphene optics, *Nano Research* **5**, pp. 667–678.

Saleh, B. E. A. and Teich, M. C. (2012). *Fundamentals of Photonics*, 2nd edn. (Wiley).

Sambles, J. R., Bradberya, G. W., and Yang, F. (1991). Optical excitation of surface plasmons: An introduction, *Contemporary Physics* **32**, pp. 173–183.

Sánchez-Gil, J. A. and Maradudin, A. A. (1999). Near-field and far-field scattering of surface plasmon polaritons by one-dimensional surface defects, *Phys. Rev. B* **60**, pp. 8359–8367.

Santos, J. E., Vasilevskiy, M. I., Peres, N. M. R., Smirnov, G., and Bludov, Y. V. (2014). Renormalization of nanoparticle polarizability in the vicinity of a graphene-covered interface, *Phys. Rev. B* **90**, p. 235420.

Sarid, D. and Challener, W. (2010). *Modern Introduction to Surface Plasmons: Theory, Mathematica Modeling, and Applications* (Cambridge University Press, Cambridge).

Satou, A. and Mikhailov, S. A. (2007). Excitation of two-dimensional plasmon polaritons by an incident electromagnetic wave at a contact, *Phys. Rev. B* **75**, p. 045328.

Schieber, D. and Schächter, L. (2001). Wake field of an electron bunch moving parallel to a dielectric cylinder, *Phys. Rev. E* **64**, p. 056503.

Schiefele, J., Pedrós, J., Sols, F., Calle, F., and Guinea, F. (2013). Coupling light into graphene plasmons through surface acoustic waves, *Phys. Rev. Lett.* **111**, p. 237405.

Schiefele, J., Sols, F., and Guinea, F. (2012). Temperature dependence of the conductivity of graphene on boron nitride, *Phys. Rev. B* **85**, p. 195420.

Schlücker, S. (2014). Surface-enhanced raman spectroscopy: Concepts and chemical applications, *Angew. Chem. Int. Ed.* **53**, pp. 4756 – 4795.

Sensale-Rodriguez, B., Yan, R., Zhu, M., Jena, D., Liu, L., and Grace Xing, H. (2012). Efficient terahertz electro-absorption modulation employing graphene plasmonic structures, *Appl. Phys. Lett.* **101**, 26, p. 261115.

Sharma, B., Frontiera, R. R., Henry, A.-I., Ringe, E., and Duyne, R. P. V. (2012). Sers: Materials, applications, and the future, *Materials Today* **15**, pp. 16 – 25.

Sheng, P., Stepleman, R. S., and Sanda, P. N. (1982). Exact eigenfunctions for square-wave gratings: Application to diffraction and surface-plasmon calculations, *Phys. Rev. B* **26**, pp. 2907–2916.

Shung, K. W. K. (1986). Dielectric function and plasmon structure of stage-1 intercalated graphite, *Phys. Rev. B* **34**, p. 979.

Shylau, A. A., Badalyan, S. M., Peeters, F. M., and Jauho, A. P. (2015). Electron polarization function and plasmons in metallic armchair graphene nanoribbons, *Phys. Rev. B* **91**, p. 205444.

Silveiro, I. and de Abajo, F. J. G. (2014). Plasmons in inhomogeneously doped neutral and charged graphene nanodisks, *Appl. Phys. Lett.* **104**, p. 131103.

Silveiro, I., Manjavacas, A., Thongrattanasiri, S., and Abajo, F. J. G. (2013). Plasmonic energy transfer in periodically doped graphene, *New Journal of Physics* **15**, p. 033042.

Slipchenko, T. M., Nesterov, M. L., Martin-Moreno, L., and Nikitin, A. Y. (2013a). Analytical solution for the diffraction of an electromagnetic wave by a graphene grating, *J. Opt.* **15**, p. 114008.

Slipchenko, T. M., Nesterov, M. L., Martin-Moreno, L., and Nikitin, A. Y. (2013b). Analytical solution for the diffraction of an electromagnetic wave by a graphene grating, *J. Opt. 15 114008* **15**, p. 114008.

Sommerfeld, A. (1899). Über die fortpflanzung elektrodynamischer wellen an längs eines drahtes, *Ann. der Physik und Chemie* **67**, pp. 233–290.

Song, B., Li, D., Qi, W., Elstner, M., Fan, C., and Fang, H. (2010). Graphene on au(111): A highly conductive material with excellent adsorption properties for high-resolution bio/nanodetection and identification, *Chem. Phys. Chem.* **11**, pp. 585–589.

Sorger, C., Preu, S., Schmidt, J., Winnerl, S., Bludov, Y. V., Peres, N. M. R., Vasilevskiy, M. I., and Weber, H. B. (2015). Terahertz response of patterned epitaxial graphene, *New J. of Phys.* **17**, p. 053045.

Sorger, V. J., Oulton, R. F., Ma, R.-M., and Zhang, X. (2012). Toward integrated plasmonic circuits, *MRS Bulletin* **37**, pp. 728–738.

Stauber, T. (2014). Plasmonics in dirac systems: from graphene to topological insulators, *J. Phys.: Condens. Matter* **26**, p. 123201.

Stauber, T., Peres, N. M. R., and Geim, A. K. (2008). Optical conductivity of graphene in the visible region of the spectrum, *Phys. Rev. B* **78**, p. 085432.

Stauber, T., Peres, N. M. R., and Guinea, F. (2007). Electronic transport in graphene: A semi-classical approach including midgap states, *Phys. Rev. B* **76**, p. 205423.

Stern, F. (1967). Polarizability of a two-dimensional electron gas, *Phys. Rev. Lett.* **18**, p. 546.

Stiles, P. L., Dieringer, J. A., Shah, N. C., and Duyne, R. P. V. (2008). Surface-enhanced raman spectroscopy, *Annu. Rev. Anal. Chem.* **1**, pp. 601 – 626.

Stockman, M. I. (2011). Nanoplasmonics: The physics behind the applications, *Phys. Today* **64**, pp. 39–44.

Strait, J. H., Nene, P., Chan, W.-M., Manolatou, C., Tiwari, S., Rana, F., Kevek, J. W., and McEuen, P. L. (2013). Confined plasmons in graphene microstructures: Experiments and theory, *Phys. Rev. B* **87**, p. 241410.

Sule, N., Willis, K. J., Hagness, S. C., and Knezevic, I. (2014). Terahertz-frequency electronic transport in graphene, *Phys. Rev. B* **90**, p. 045431.

Sun, S., Gao, L., Liu, Y., and Sun, J. (2011a). Assembly of cdse nanoparticles on graphene for low-temperature fabrication of quantum dot sensitized solar cell, *Appl. Phys. Lett.* **98**.

Sun, Z., Gutierrez-Rubio, A., Basov, D. N., and Fogler, M. M. (2015). Hamiltonian optics of hyperbolic polaritons in nanogranules, *Nano Lett.* **15**, pp. 4455–4460.

Sun, Z., Pint, C. L., Marcano, D. C., Zhang, C., Yao, J., Ruan, G., Yan, Z., Zhu, Y., Hauge, R. H., and Tour, J. M. (2011b). Towards hybrid superlattices in graphene, *Nature Communications* **2**, p. 1.

Swathi, R. S. and Sebastian, K. L. (2009). Long range resonance energy transfer from a dye molecule to graphene has distance[4] dependence, *The J. of Chem. Phys.* **130**, p. 086101.

Tame, M. S., McEnery, K. R., Ozdemir, S. K., Lee, J., Maier, S. A., and Kim, M. S. (2013). Quantum plasmonics, *Nature Phys.* **9**, pp. 329–340.

Tao, Z., Ren-Bin, Z., Min, H., Xiao-Xing, C., Ping, Z., Sen, G., and Sheng-Gang, L. (2015). Tunable terahertz radiation from arbitrary profile dielectric grating coated with graphene excited by an electron beam, *Chin. Phys. B* **24**, p. 094102.

Tassin, P., Koschny, T., and Soukoulis, C. M. (2013). Graphene for terahertz applications, *Science* **341**, pp. 620–621.

Temnov, V. V. (2012). Ultrafast acousto-magneto-plasmonics, *Nature Photonics* **6**, pp. 728–736.

Teng, Y. Y. and Stern, E. A. (1967). Plasma radiation from metal grating surfaces, *Phys. Rev. Lett.* **19**, p. 511.

Thongrattanasiri, S. and de Abajo, F. J. G. (2013). Optical field enhancement by strong plasmon interaction in graphene nanostructures, *Phys. Rev. Lett.* **110**, p. 187401.

Thongrattanasiri, S., Koppens, F. H. L., and de Abajo, F. J. G. (2012a). Complete optical absorption in periodically patterned graphene, *Phys. Rev. Lett.* **108**, p. 047401.

Thongrattanasiri, S., Manjavacas, A., and de Abajo, F. J. G. (2012b). Quantum finite-size effects in graphene plasmons, *ACS Nano* **6**, pp. 1766–1775.

Tielrooij, K. J., Orona, L., Ferrier, A., Badioli, M., Navickaite, G., Coop, S., Nanot, S., Kalinic, B., Cesca, T., Gaudreau, L., Ma, Q., Centeno, A., Pesquera, A., Zurutuza, A., de Riedmatten, H., Goldner, P., de Abajo, F. J. G., Jarillo-Herrero, P., and Koppens, F. H. L. (2015). Electrical control of optical emitter relaxation pathways enabled by graphene, *Nature Phys.* **11**, pp. 281 – 287.

Toigo, F., Marvin, A., Celli, V., and Hill, N. R. (1977). Optical properties of rough surfaces: General theory and the small roughness limit, *Phys. Rev. B* **15**, pp. 5618–5626.

Tokunaga, E., Ivanov, A. L., Nair, S. V., and Masumoto, Y. (2001). Hopfield coefficients measured by inverse polariton series, *Phys. Rev. B* **63**, p. 233203.

Tomadin, A., Guinea, F., and Polini, M. (2014). Generation and morphing of plasmons in graphene superlattices, *Phys. Rev. B* **89**, p. 224512.

Toropov, A. A. and Shubina, T. V. (2015). *Plasmonic Effects in Metal-Semiconductor Nanostructures* (Oxford).

Torre, I., Tomadin, A., Krahne, R., Pellegrini, V., and Polini, M. (2015). Electrical plasmon detection in graphene waveguides, *Phys. Rev. B* **91**, p. 081402(R).

Turbadar, T. (1959). Complete absorption of light by thin metal films, *Proceedings of the Physical Society* **73**, p. 40.

Tymchenko, M., Nikitin, A. Y., and Martín-Moreno, L. (2013). Faraday rotation due to excitation of magnetoplasmons in graphene microribbons, *ACS Nano* **7**, 11, pp. 9780–9787.

Urich, A., Pospischil, A., Furchi, M. M., Dietze, D., Unterrainer, K., and Mueller, T. (2012). Silver nanoisland enhanced raman interaction in graphene, *Appl. Phys. Lett.* **101**, p. 153113.

Vakil, A. and Engheta, N. (2011). Transformation optics using graphene, *Science* **332**, pp. 1291 – 1294.

van Hulst, N. F. (2012). Nanophotonics: Plasmon quantum limit exposed, *Nature Nanotechnology* **7**, pp. 775 – 777.

Vasić, B., Isić, G., and Gajić, R. (2013). Localized surface plasmon resonances in graphene ribbon arrays for sensing of dielectric environment at infrared frequencies, *J. Appl. Phys.* **113**, p. 013110.

Velizhanin, K. A. (2015). Geometric universality of plasmon modes in graphene nanoribbon arrays, *Phys. Rev. B* **91**, p. 125429.

Velizhanin, K. A. and Efimov, A. (2011). Probing plasmons in graphene by resonance energy transfer, *Phys. Rev. B* **84**, p. 085401.

Vesseur, E. J. R., Aizpurua, J., Coenen, T., Reyes-Coronado, A., Batson, P. E., and Polman, A. (2012). Plasmonic excitation and manipulation with an electron beam, *MRS Bulletin* **37**, pp. 752–760.

Vladimirov, V. S. (1971). *Equations of Mathematical Physics* (Dekker, New York).

Vo-Dinh, T., Fales, A. M., Griffin, G. D., Khoury, C. G., Liu, Y., Ngo, H., Norton, S. J., Register, J. K., Wang, H.-N., and Yuan, H. (2013). Plasmonic nanoprobes: from chemical sensing to medical diagnostics and therapy, *Nanoscale* **5**, pp. 10127–10140.

Voronovich, A. G. (2007). Rayleigh hypothesis, in A. A. Maradudin (ed.), *Light Scattering and Nanoscale Surface Roughness*, Nanostructure Science and Technology (Springer), pp. 93–105.

Wachsmuth, P., Hambach, R., Benner, G., and Kaiser, U. (2014). Plasmon bands in multilayer graphene, *Phys. Rev. B* **90**, p. 235434.

Wang, P., Liang, O., Zhang, W., Schroeder, T., and Xie, Y.-H. (2013). Ultra-sensitive graphene-plasmonic hybrid platform for label-free detection, *Advanced Materials* **25**, pp. 4918 – 4924.

Wang, P., Zhang, W., Liang, O., Pantoja, M., Katzer, J., Schroeder, T., and Xie, Y.-H. (2012a). Giant optical response from graphene–plasmonic system, *ACS Nano* **6**, pp. 6244–6249.

Wang, W. (2012). Plasmons and optical excitations in graphene rings, *J. of Phys.: Cond. Matt.* **24**, p. 402202.

Wang, W., Apell, P., and Kinaret, J. (2011). Edge plasmons in graphene nanostructures, *Phys. Rev. B* **84**, p. 085423.

Wang, W., Apell, S. P., and Kinaret, J. M. (2012b). Edge magnetoplasmons and the optical excitations in graphene disks, *Phys. Rev. B* **86**, p. 125450.

Wang, W., Christensen, T., Jauho, A.-P., Thygesen, K. S., Wubs, M., and Mortensen, N. A. (2015). Plasmonic eigenmodes in individual and bow-tie graphene nanotriangles, *Scientific Reports* **5**, 9535.

Wang, W. and Kinaret, J. M. (2013). Plasmons in graphene nanoribbons: Interband transitions and nonlocal effects, *Phys. Rev. B* **87**, p. 195424.

Wang, W., Kinaret, J. M., and Apell, S. P. (2012c). Excitation of edge magnetoplasmons in semi-infinite graphene sheets: Temperature effects, *Phys. Rev. B* **85**, p. 235444.

Wang, W., Li, B.-H., Mortensen, N. A., and Christensen, J. (2016). Localized surface plasmons in vibrating graphene nanodisks, *Nanoscale* **8**, p. 3809.

Weber, M. L. and Willets, K. A. (2012). Nanoscale studies of plasmonic hot spots using super-resolution optical imaging, *MRS Bulletin* **37**, pp. 745–751.

Weinberg, S. (1996). *The Quantum Theory of Fields, Vols. 1, 2 and 3* (Cambridge University Press, New York).

Wijaya, E., Lenaerts, C., Maricot, S., Hastanin, J., Habraken, S., Vilcot, J.-P., Boukherroub, R., and Szunerits, S. (2011). Surface plasmon resonance-based biosensors: From the development of different {SPR} structures to novel surface functionalization strategies, *Current Opinion in Solid State and Materials Science* **15**, pp. 208 – 224.

Willets, K. A. and Duyne, R. P. V. (2007). Localized surface plasmon resonance spectroscopy and sensing, *Annu. Rev. Phys. Chem.* **58**, pp. 267–297.

Withers, F., Bointon, T. H., Dubois, M., Russo, S., and Craciun, M. F. (2011). Nanopatterning of fluorinated graphene by electron beam irradiation, *Nano Lett.* **11**, pp. 3912 – 3916.

Woessner, A., Lundeberg, M. B., Gao, Y., Principi, A., Alonso-González, P., Carrega, M., Watanabe, K., Taniguchi, T., Vignale, G., Polini, M., Hone, J., Hillenbrand, R., and Koppens, F. H. L. (2015). Highly confined low-loss plasmons in graphene-boron nitride heterostructures, *Nature Materials* **14**, pp. 421–425.

Wood, R. W. (1902). On a remarkable case of uneven distribution of light in a diffraction grating spectrum, *Philosophical Magazine* **4**, pp. 396–402.

Wooten, F. (1972). *Optical Properties of Solids* (Academic Press, New York).

Wu, J., Zhou, C., Yu, J., Cao, H., Li, S., and Jia, W. (2014a). Design of infrared surface plasmon resonance sensors based on graphene ribbon arrays, *Optics and Laser Technology* **59**, pp. 99 – 103.

Wu, J.-W., Hawrylak, P., Eliasson, G., Quinn, J., and Fetter, A. (1986a). Magnetoplasma surface waves on the lateral surface of a semiconductor superlattice, *Solid State Communications* **58**, 11, pp. 795–798.

Wu, J.-W., Hawrylak, P., Eliasson, G., and Quinn, J. J. (1986b). Theory of the lateral surface magnetoplasmon in a semiconductor superlattice, *Phys. Rev. B* **33**, pp. 7091–7098.

Wu, J. W., Hawrylak, P., and Quinn, J. J. (1985). Charge-density excitation on a lateral surface of a semiconductor superlattice and edge plasmons of a two-dimensional electron gas, *Phys. Rev. Lett.* **55**, pp. 879–882.

Wu, J.-Y., Gumbsand, G., and Lin, M.-F. (2014b). Combined effect of stacking and magnetic field on plasmon excitations in bilayer graphene, *Phys. Rev. B* **89**, p. 165407.

Wu, L., Chu, H. S., Koh, W. S., and Li, E. P. (2010). Highly sensitive graphene biosensors based on surface plasmon resonance, *Opt. Express* **18**, pp. 14395 – 14400.

Wunsch, B., Stauber, T., Sols, F., and Guinea, F. (2006). Dynamical polarization of graphene at finite doping, *New J. Phys.* **8**, p. 318.

Xiao, S., Zhu, X., Li, B.-H., and Mortensen, N. A. (2016). Graphene-plasmon polaritons: From fundamental properties to potential applications, *Front. Phys.* **11**, p. 117801.

Xu, G., Liu, J., Wang, Q., Hui, R., Chen, Z., Maroni, V. A., and Wu, J. (2012a). Graphene: Plasmonic graphene transparent conductors, *Adv. Mater.* **24**, pp. 1521 – 4095.

Xu, W., Ling, X., Xiao, J., Dresselhaus, M. S., Kong, J., Xu, H., Liu, Z., and Zhang, J. (2012b). Surface enhanced raman spectroscopy on a flat graphene surface, *Proceedings of the National Academy of Sciences* **109**, 24, pp. 9281–9286.

Yamaguchi, T., Moriya, R., Inoue, Y., Morikawa, S., Masubuchi, S., Watanabe, K., Taniguchi, T., and Machida, T. (2014). Tunneling transport in a few monolayer-thick ws2/graphene heterojunction, *Appl. Phys. Lett.* **105**, p. 223109.

Yamamoto, K., Tani, M., and Hangyo, M. (2007). Terahertz time-domain spectroscopy of imidazolium ionic liquids, *J. Phys. Chem. B* **111**, pp. 4854–4859.

Yan, H., Li, X., Chandra, B., Tulevski, G., Wu, Y., Freitag, M., Zhu, W., Avouris, P., and Xia, F. (2012a). Tunable infrared plasmonic devices using graphene/insulator stacks, *Nature Nanotechnology* **7**, pp. 330–334.

Yan, H., Li, Z., Li, X., Zhu, W., Avouris, P., and Xia, F. (2012b). Infrared spectroscopy of tunable dirac terahertz magneto-plasmons in graphene, *Nano Lett.* **12**, pp. 3766–3771.

Yan, H., Low, T., Zhu, W., Wu, Y., Freitag, M., Li, X., Guinea, F., Avouris, P., and Xia, F. (2013). Damping pathways of mid-infrared plasmons in graphene nanostructures, *Nat. Photonics* **7**, pp. 394–399.

Yan, H., Xia, F., Li, Z., and Avouris, P. (2012c). Plasmonics of coupled graphene micro-structures, *New J. Phys.* **14**, p. 125001.

Yao, Y., Kats, M. A., Genevet, P., Yu, N., Song, Y., Kong, J., and Capasso, F. (2013). Broad electrical tuning of graphene-loaded plasmonic antennas, *Nano Lett.* **13**, pp. 1257–1264.

Yeung, K. Y. M., Chee, J., Song, Y., Kong, J., and Ham, D. (2015). Symmetry engineering of graphene plasmonic crystals, *Nano Lett.* **15**, pp. 5001–5009.

Yeung, K. Y. M., Chee, J., Yoon, H., Song, Y., Kong, J., and Ham, D. (2014). Far-infrared graphene plasmonic crystals for plasmonic band engineering, *Nano Lett.* **14**, pp. 2479–2484.

Yin, P. T., Kim, T.-H., Choi, J.-W., and Lee, K.-B. (2013). Prospects for graphene-nanoparticle-based hybrid sensors, *Phys. Chem. Chem. Phys.* **15**, pp. 12785 – 12799.

Yusoff, A. R. B. M. (2014). *Graphene Optoelectronics: Synthesis, Characterization, Properties, and Applications* (Wiley).

Zayats, A. V., Smolyaninov, I. I., and Maradudin, A. A. (2005). Nano-optics of surface plasmon polaritons, *Physics Reports* **408**, pp. 131–314.

Zeng, S., Hu, S., Xia, J., Anderson, T., Dinh, X.-Q., Meng, X.-M., Coquet, P., and Yong, K.-T. (2015). Graphene-mos2 hybrid nanostructures enhanced surface plasmon resonance biosensors, *Sensors and Actuators B: Chemical* **207**, pp. 801–810.

Zenneck, J. (1907). Uber die fortpflanztmg ebener elektro-magnetischer wellen langs einer ebenen leiterflache und ihre beziehung zur drahtlosen telegraphie, *Annalen der Physik* **23**, pp. 846–866.

Zettili, N. (2009). *Quantum Mechanics*, 2nd edn. (Wiley, West Sussex).

Zhan, T., Han, D., Hu, X., Liu, X., Chui, S.-T., and Zi, J. (2014). Tunable terahertz radiation from graphene induced by moving electrons, *Phys. Rev. B* **89**, p. 245434.

Zhan, T., Shi, X., Dai, Y., Liu, X., and Zi, J. (2013). Transfer matrix method for optics in graphene layers, *J. Phys.: Condens. Matter* **25**, p. 215301.

Zhang, J., Sun, Y., Xu, B., Zhang, H., Gao, Y., Zhang, H., and Song, D. (2013). A novel surface plasmon resonance biosensor based on graphene oxide decorated with gold nanorod-antibody conjugates for determination of transferrin, *Biosensors and Bioelectronics* **45**, pp. 230 – 236.

Zhang, J., Xiao, S., Wubs, M., and Mortensen, N. A. (2011). Surface plasmon wave adapter designed with transformation optics, *ACS Nano* **5**, pp. 4359–4364.

Zhang, Q., Li, X., Hossain, M. M., Xue, Y., Zhang, J., Song, J., Liu, J., Turner, M. D., Fan, S., Bao, Q., and Gu, M. (2014). Graphene surface plasmons at the near-infrared optical regime, *Scientific Reports* **4**, p. 6559.

Zhao, Y., Chen, G., Du, Y., Xu, J., Wu, S., Qu, Y., and Zhu, Y. (2014). Plasmonic-enhanced raman scattering of graphene on growth substrates and its application in sers, *Nanoscale* **6**, pp. 13754 – 13760.

Zhao, Y., Hu, X., Chen, G., Zhang, X., Tan, Z., Chen, J., Ruoff, R. S., Zhu, Y., and Lu, Y. (2013). Infrared biosensors based on graphene plasmonics: modeling, *Phys. Chem. Chem. Phys.* **15**, pp. 17118 – 17125.

Zheludev, N. I. and Kivshar, Y. S. (2012). From metamaterials to metadevices, *Nature Materials* **11**, pp. 917 – 924.

Zheng, Y. B., Kiraly, B., Weiss, P. S., and Huang, T. J. (2012). Molecular plasmonics for biology and nanomedicine, *Nanomedicine* **7**, pp. 751 – 770.

Zhu, A. Y. and Cubukcu, E. (2015). Graphene nanophotonic sensors, *2D Materials* **2**, p. 032005.

Zhu, J., Liu, Q. H., and Lin, T. (2013a). Manipulating light absorption of graphene using plasmonic nanoparticles, *Nanoscale* **5**, pp. 7785 – 7789.

Zhu, J. J., Badalyan, S. M., and Peeters, F. M. (2013b). Plasmonic excitations in coulomb-coupled n-layer graphene structures, *Phys. Rev. B* **8**, p. 085401.

Zhu, X., Shi, L., Schmidt, M. S., Boisen, A., Hansen, O., Zi, J., Xiao, S., and Mortensen, N. A. (2013c). Enhanced light–matter interactions in graphene-covered gold nanovoid arrays, *Nano Lett.* **13**, pp. 4690–4696.

Zhu, X., Wang, W., Yan, W., Larsen, M. B., Bøggild, P., Pedersen, T. G., Xiao, S., Zi, J., and Mortensen, N. A. (2014). Plasmon-phonon coupling in large-area graphene dot and antidot arrays fabricated by nanosphere lithography, *Nano Lett.* **14**, pp. 2907 – 2913.

Zhu, X., Yan, W., Uhd Jepsen, P., Hansen, O., Asger Mortensen, N., and Xiao, S. (2013d). Experimental observation of plasmons in a graphene monolayer resting on a two-dimensional subwavelength silicon grating, *Appl. Phys. Lett.* **102**, p. 131101.

Ziman, J. M. (1972). *Principles of the theory of solids*, 2nd edn. (Cambridge University Press, Cambridge).

Index

www.ingramcontent.com/pod-product-compliance
Lightning Source LLC
Chambersburg PA
CBHW052010230326
41598CB00078B/2309